Shop Manual for

Automotive Engine Performance

Shop Manual for Automotive Engine Performance

Don Knowles

Knowles Automotive Training
Moose Jaw, Saskatchewan
CANADA

Jack Erjavec

Series Advisor

Columbus State Community College
Columbus, Ohio

Delmar Publishers
I(T)P An International Thomson Publishing Company

Albany • Bonn • Boston • Cincinnati • Detroit • London • Madrid • Melbourne
Mexico City • New York • Pacific Grove • Paris • San Francisco • Singapore • Tokyo
Toronto • Washington

NOTICE TO THE READER

COVER PHOTO: Courtesy of Chevrolet Motor Division, General Motors Corporation
PHOTO SEQUENCES: Photography by Rod Dixon Associates

Portions of materials contained herein have been reprinted with permission of General Motors Corporation, Service Technology Group.

DELMAR STAFF

Senior Administrative Editor: Vernon Anthony
Developmental Editor: Catherine Eads
Project Editor: Eleanor Isenhart
Production Coordinator: Karen Smith
Art/Design Coordinator: Heather Brown

Printed in the United States of America

For information, contact:

Delmar Publishers
3 Columbia Circle, Box 15015
Albany, New York 12212-5015

International Thomson Editors
Campos Eliseos 385, Piso 7
Col Polanco
11560 Mexico DF Mexico

International Thomson Publishing Europe
Berkshire House 168-173
High Holborn
London, WC1V7AA
England

International Thomson Publishing GmbH
Königswinterer Strasse 418
53227 Bonn
Germany

Thomas Nelson Australia
102 Dodds Street
South Melbourne, 3205
Victoria, Australia

International Thomson Publishing Asia
221 Henderson Road
#05-10 Henderson Building
Singapore 0315

Nelson Canada
1120 Birchmont Road
Scarborough, Ontario
Canada M1K 5G4

International Thomson Publishing Japan
Hirakawacho Kyowa Building, 3F
2-2-1 Hirakawacho
Chiyoda-ku, Tokyo 102
Japan

2 3 4 5 6 7 8 9 10 XXX 00 99 98 97 96 95

Library of Congress Cataloging-in-Publication Data

Knowles, Don.
Automotive engine performance / Don Knowles.
p. cm. --(Today's technician)
Includes index.
Contents: v. 1. Classroom manual -- v. 2. Shop manual.
ISBN 0-8273-6186-6(set)
1. Automobiles -- Motors -- Modification. 2. Automobiles --Performance. I. Title. II. Series.
TL210.K56 1995
629.25'04 -- dc20 94-31203
CIP

CONTENTS

Photo Sequences

PREFACE

Unlike yesterday's mechanic, the technician of today and for the future must know the underlying theory of all automotive systems and be able to service and maintain those systems. Today's technician must also know how these individual systems interact with each other. Standards and expectations have been set for today's technician, and these must be met in order to keep the world's automobiles running efficiently and safely.

The *Today's Technician* series, by Delmar Publishers, features textbooks that cover all mechanical and electrical systems of automobiles and light trucks. Principal titles correspond with the eight major areas of ASE (National Institute for Automotive Service Excellence) certification. Additional titles include remedial skills and theories common to all of the certification areas and advanced or specialized subject areas that reflect the latest technological trends.

Each title is divided into two manuals: a Classroom Manual and a Shop Manual. Dividing the material into two manuals provides the reader with the information needed to begin a successful career as an automotive technician without interrupting the learning process by mixing cognitive and performance-based learning objectives.

Each Classroom Manual contains the principles of operation for each system and subsystem. It also discusses the design variations used by different manufacturers. The Classroom Manual is organized to build upon basic facts and theories. The primary objective of this manual is to allow the reader to gain an understanding of how each system and subsystem operates. This understanding is necessary to diagnose the complex automobile systems.

The understanding acquired by using the Classroom Manual is required for competence in the skill areas covered in the Shop Manual. All of the high priority skills, as identified by ASE, are explained in the Shop Manual. The Shop Manual also includes step-by-step instructions for diagnostic and repair procedures. Photo Sequences are used to illustrate many of the common service procedures. Other common procedures are listed and are accompanied with fine-line drawings and photographs that allow the reader to visualize and conceptualize the finest details of the procedure. The Shop Manual also contains the reasons for performing the procedures, as well as when that particular service is appropriate.

The two manuals are designed to be used together and are arranged in corresponding chapters. Not only are the chapters in the manuals linked together, the contents of the chapters are also linked. Both manuals contain clear and thoughtfully selected illustrations. Many of the illustrations are original drawings or photos prepared for inclusion in this series. This means that the art is a vital part of each manual.

The page layout is designed to include information that would otherwise break up the flow of information presented to the reader. The main body of the text includes all of the "need-to-know" information and illustrations. In the side margins are many of the special features of the series. Items such as definitions of new terms, common trade jargon, tools lists, and cross-references are placed in the margin, out of the normal flow of information so as not to interrupt the thought process of the reader. Each manual in this series is organized in a like manner and contains the same features.

Jack Erjavec, Series Advisor

Classroom Manual

To stress the importance of safe work habits, the Classroom Manual dedicates one full chapter to safety. Included in this chapter are common safety practices, safety equipment, and safe handling of hazardous materials and wastes. This includes information on MSDS sheets and OSHA regulations. Other features of this manual include:

Cognitive Objectives

These objectives define the contents of the chapter and define what the student should have learned upon completion of the chapter.
Each topic is divided into small units to promote easier understanding and learning.

Marginal Notes

Page numbers for cross-referencing appear in the margin. Some of the common terms used for components, and other bits of information, also appear in the margin. This provides an understanding of the language of the trade and helps when conversing with an experienced technician.

Cautions and Warnings

Throughout the text, cautions are given to alert the reader to potentially hazardous materials or unsafe conditions. Warnings are also given to advise the student of things that can go wrong if instructions are not followed or if a nonacceptable part or tool is used.

References to the Shop Manual

Reference to the appropriate page in the Shop Manual is given whenever necessary. Although the chapters of the two manuals are synchronized, material covered in other chapters of the Shop Manual may be fundamental to the topic discussed in the Classroom Manual.

A Bit of History

This feature gives the student a sense of the evolution of the automobile. This feature not only contains nice-to-know information, but also should spark some interest in the subject matter.

Summaries

Each chapter concludes with summary statements that contain the important topics of the chapter. These are designed to help the reader review the contents.

Terms to Know

A list of new terms appears next to the Summary. Definitions for these terms can be found in the Glossary at the end of the manual.

Review Questions

Short answer essay, fill-in-the-blank, and multiple-choice type questions follow each chapter. These questions are designed to accurately assess the student's competence in the stated objectives at the beginning of the chapter.

Shop Manual

To stress the importance of safe work habits, the Shop Manual also dedicates one full chapter to safety. Other important features of this manual include:

Performance Objectives

These objectives define the contents of the chapter and define what the student should have learned upon completion of the chapter. These objectives also correspond with the list of required tasks for ASE certification. *Each ASE task is addressed.*

Although this textbook is not designed to simply prepare someone for the certification exams, it is organized around the ASE task list. These tasks are defined generically when the procedure is commonly followed and specifically when the procedure is unique for specific vehicle models. Imported and domestic model automobiles and light trucks are included in the procedures.

Photo Sequences

Many procedures are illustrated in detailed photo sequences. These detailed photographs show the students what to expect when they perform particular procedures. They also can provide a student a familiarity with a system or type of equipment, which the school may not have.

Marginal Notes

New terms are pulled out and defined. Common trade jargon also appears in the margin and gives some of the common terms used for components. This allows the reader to speak and understand the language of the trade, especially when conversing with an experienced technician.

Cautions and Warnings

Throughout the text, cautions are given to alert the reader to potentially hazardous materials or unsafe conditions. Warnings are also given to advise the student of things that can go wrong if instructions are not followed or if a nonacceptable part or tool is used.

References to the Classroom Manual

Reference to the appropriate page in the Classroom Manual is given whenever necessary. Although the chapters of the two manuals are synchronized, material covered in other chapters of the Classroom Manual may be fundamental to the topic discussed in the Shop Manual.

Customer Care

This feature highlights those little things a technician can do or say to enhance customer relations.

Tools Lists

Each chapter begins with a list of the Basic Tools needed to perform the tasks included in the chapter. Whenever a Special Tool is required to complete a task, it is listed in the margin next to the procedure.

Service Tips

Whenever a special procedure is appropriate, it is described in the text. These tips are generally those things commonly done by experienced technicians.

Case Studies

Case Studies concentrate on the ability to properly diagnose the systems. Each chapter ends with a case study in which a vehicle has a problem, and the logic used by a technician to solve the problem is explained.

Terms to Know

Terms in this list can be found in the Glossary at the end of the manual.

Diagnostic Chart

Chapters include detailed diagnostic charts linked with the appropriate ASE task. These charts list common problems and most probable causes. They also list a page reference in the Classroom Manual for better understanding of the system's operation and a page reference in the Shop Manual for details on the procedure necessary for correcting the problem.

ASE Style Review Questions

Each chapter contains ASE style review questions that reflect the performance objectives listed at the beginning of the chapter. These questions can be used to review the chapter as well as to prepare for the ASE certification exam.

Instructor's Guide

The Instructor's Guide is provided free of charge as part of the *Today's Technician Series* of automotive technology textbooks. It contains Lecture Outlines, Answers to Review Questions, Pretest and Test Bank including ASE style questions.

Classroom Manager

The complete ancillary package is designed to aid the instructor with classroom preparation and provide tools to measure student performance. For an affordable price, this comprehensive package contains:

Instructor's Guide
200 Transparency Masters
Answers to Review Questions
Lecture Outlines and Lecture Notes
Printed and Computerized Test Bank
Laboratory Worksheets and Practicals

Reviewers

Special thanks to the following instructors for reviewing this material:

Robert D. Brunken
Portland Community College

Tom Fitch
Monroe Community College

Earl J. Friedell, Jr.
DeKalb Technical Institute

Anthony Hoffman
Arizona Western College

Norris Martin
Texas State Technical College-Waco

John Thorp
Illinois Central College

Shop Practices

Upon completion and review of this chapter, you should be able to:

- ❑ Describe three requirements for shop layout, and explain why these requirements are important.
- ❑ Observe all shop rules when working in the shop.
- ❑ Operate vehicles in the shop according to shop driving rules.
- ❑ Observe all shop housekeeping rules.
- ❑ Follow the necessary procedures to maintain satisfactory shop air quality.
- ❑ Fulfill employee obligations when working in the shop.
- ❑ Accept job responsibilities for each job completed in the shop.
- ❑ Describe the ASE technician testing and certification process, including the eight areas of certification.

Shop Layout

There are many different types of shops in the automotive service industry including:

New car dealer
Independent repair shop
Specialty shop
Service station
Fleet shop

CAUTION: Always know the location of all safety equipment in the shop and be familiar with the operation of this equipment.

The shop layout in any shop is important to maintaining shop efficiency and safety. Shop layout includes bays for various types of repairs, space for equipment storage, and office locations. Safety equipment such as fire extinguishers, first aid kits, and eye wash fountains must be in easily accessible locations. The location of each piece of safety equipment must be clearly marked. Areas such as the parts department and the parts cleaning area must be located so they are easily accessible from all areas of the shop. The service manager's office should also be centrally located. All shop personnel should familiarize themselves with the shop layout, especially the location of safety equipment. If you know the exact fire extinguisher locations, you may get an extinguisher into operation a few seconds faster. These few seconds could make the difference between a fire that is quickly extinguished and one that gets out of control, causing extensive damage and personal injury! Most shops have specific bays for certain types of work, such as electrical repair, wheel alignment and tires, and machining (Figure 1-1).

The tools and equipment required for a certain type of work are stored in that specific bay. For example, the equipment for electrical and electronic service work is stored in the bay allotted to that type of repair. When certain bays are allotted to specific types of repair work, unnecessary equipment movement is eliminated. Each technician has his or her own tools on a portable roll cabinet that is moved to the vehicle being repaired. Special tools are provided by the shop, and these tools may be located on tool boards attached to the wall. Other shops may have a tool room where special tools are located. Adequate work bench space must be provided in those bays where bench work is required.

Shop Rules

▲ **WARNING:** Shop rules, vehicle operation in the shop, and shop housekeeping are serious business. Each year a significant number of technicians are injured and vehicles are damaged by disregarding shop rules, careless vehicle operation, and sloppy housekeeping.

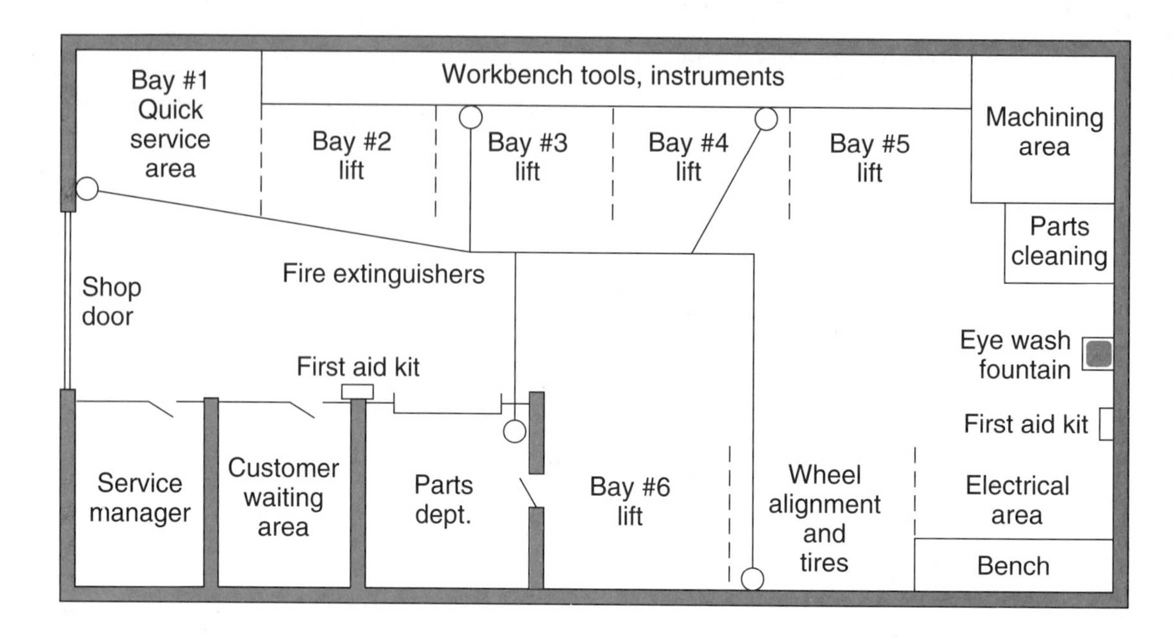

Figure 1-1 Typical shop layout

The application of some basic shop rules helps to prevent serious, expensive accidents. Failure to comply with shop rules may cause personal injury or expensive damage to vehicles and shop facilities. It is the responsibility of the employer and all shop employees to make sure that shop rules are understood and followed until these rules become automatic habits. These basic shop rules should be followed:

1. Always wear safety glasses and other protective equipment that is required by a service procedure. For example, a special parts washer must be used to avoid breathing asbestos dust into the lungs. *Asbestos dust is a known cause of lung cancer.* This dust is encountered in manual transmission clutch facings and brake linings.
2. Tie long hair securely behind the head, and do not wear loose or torn clothing.
3. Do not wear rings, watches, or loose hanging jewelry. If jewelry such as a ring, metal watch band, or chain makes contact between an electrical terminal and ground, the jewelry becomes extremely hot, which results in severe burns.
4. Set the parking brake when working on a vehicle. If the vehicle has an automatic transmission, place the gear selector in park unless a service procedure requires another selector position. When the vehicle is equipped with a manual transmission, position the gear selector in neutral with the engine running, or reverse with the engine stopped.
5. Always connect a shop exhaust hose to the vehicle tail pipe, and be sure the shop exhaust fan is running. If it is absolutely necessary to operate a vehicle without a shop exhaust pipe connected to the tail pipe, open the large shop door to provide adequate ventilation. Carbon monoxide in the vehicle exhaust may cause severe headaches and other medical problems. High concentrations of carbon monoxide may result in death!
6. Keep hands, clothing, and wrenches away from rotating parts such as cooling fans. Remember that electric-drive fans may start turning at any time, even with the ignition off.
7. Always leave the ignition switch off unless a service procedure requires another switch position.
8. Do not smoke in the shop. If the shop has designated smoking areas, smoke only in these areas.
9. Store oily rags and other discarded combustibles in regulation, covered metal garbage containers.
10. Always use a wrench or socket that fits properly on the bolt. Do not substitute metric for English wrenches or vice versa.

Carbon monoxide is a poisonous gas. When breathed into the lungs, it may cause headaches, nausea, ringing in the ears, tiredness, and heart flutter. In strong concentrations, it causes death.

11. Keep tools in good condition. For example, do not use a punch or chisel with a mushroomed end. When struck with a hammer, a piece of the mushroomed metal could break off, resulting in severe eye or other injury.
12. Do not leave power tools running and unattended.
13. Serious burns may be prevented by avoiding contact with hot metal components such as exhaust manifolds, other exhaust system components, radiators, and some air conditioning hoses.
14. When lubricant such as engine oil is drained, always use caution because the oil could be hot enough to cause burns.
15. Prior to getting under a vehicle, be sure the vehicle is placed securely on safety stands.
16. Operate all shop equipment, including lifts, according to the equipment manufacturer's recommended procedure. Do not operate equipment unless you are familiar with the correct operating procedure.
17. Do not run or engage in horseplay in the shop.
18. Obey all state and federal fire, safety, and environmental regulations.
19. Do not stand in front of, or behind, vehicles.
20. Always place fender covers and a seat cover on a customer's vehicle before working on the car.
21. Inform the shop foreman of any safety dangers and suggestions for safety improvement.

Vehicle Operation

When driving a customer's vehicle, certain precautions must be observed to prevent accidents and maintain good customer relations.

1. Prior to driving a vehicle, always make sure the brakes are operating and fasten the safety belt.
2. Check to be sure there is no person or object under the car before you start the engine.
3. If the vehicle is parked on a lift, be sure the lift is fully down and the lift arms or components are not contacting the vehicle chassis.
4. Check to see if there are any objects directly in front of, or behind, the vehicle before driving away.
5. Always drive slowly in the shop and watch carefully for personnel and other moving vehicles.
6. Make sure the shop door is up high enough so there is plenty of clearance between the top of the vehicle and the door.
7. Watch the shop door to be certain that it is not coming down as you attempt to drive under the door.
8. If a road test is necessary, obey all traffic laws, and never drive in a reckless manner.
9. Do not squeal tires when accelerating or turning corners.

If the customer observes that service personnel take good care of his or her car by driving carefully and installing fender and seat covers, the service department image is greatly enhanced in the customer's eyes. These procedures impress upon the customer that shop personnel respect his or her car. Conversely, if grease spots are found on upholstery or fenders after service work is completed, the customer will probably think the shop is very careless, not only in car care, but also in service work quality.

Housekeeping

CUSTOMER CARE: When a customer sees that you are concerned about his or her vehicle, and that you operate a shop with excellent housekeeping habits, the customer will be impressed and will likely keep returning for service.

Excellent housekeeping involves general shop cleanliness, proper shop safety equipment in good working condition, and the proper maintenance of all shop equipment and tools.

Careful housekeeping habits prevent accidents and increase worker efficiency. Good housekeeping also helps to impress upon the customer that quality work is a priority in this shop. Follow these housekeeping rules:

1. Keep aisles and walkways clear of tools, equipment, and other items.
2. Be sure all sewer covers are securely in place.
3. Keep floor surfaces free of oil, grease, water, and loose material.
4. Place proper garbage containers in convenient locations, and empty the containers regularly.
5. Keep access to fire extinguishers unobstructed at all times, and check fire extinguishers for proper charge at regular intervals.

SERVICE TIP: When you are finished with a tool, never set it on the customer's car. After using a tool, the best place for it is in your tool box or on the work bench. Many tools have been lost by leaving them on customers' vehicles.

6. Keep tools clean and in good condition.
7. When not in use, store tools in their proper location.
8. Place oily rags and other combustibles in proper covered garbage containers.
9. Install guards on rotating components on equipment and machinery, and be sure that all shop equipment has regular service and adjustment schedules.
10. Maintain benches and seats in a clean condition.
11. Keep parts and materials in their proper location.
12. When not in use, creepers must not be left on the shop floor. Store creepers in a specific location.
13. Keep the shop well lighted, and be sure all lights are in working order.
14. Replace frayed electric cords on lights or equipment.
15. Clean walls and windows regularly.
16. Keep stairs clean, well lighted, and free of loose material.

If these housekeeping rules are followed, the shop will be a safer place to work, and customers will be impressed with the appearance of the premises.

Air Quality

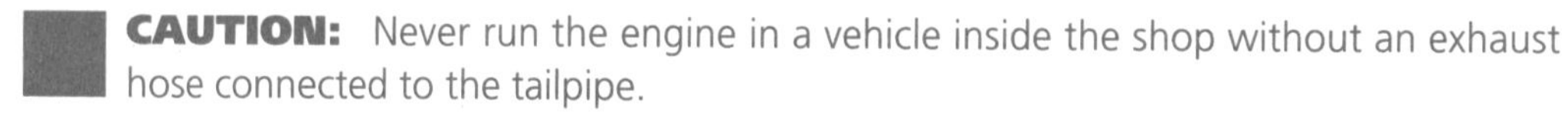

CAUTION: Never run the engine in a vehicle inside the shop without an exhaust hose connected to the tailpipe.

Vehicle exhaust contains small amounts of carbon monoxide, which is a poisonous gas. Strong concentrations of carbon monoxide may be fatal for human beings. All shop personnel are responsible for air quality in the shop.

Shop management is responsible for an adequate exhaust system to remove exhaust fumes from the maximum number of vehicles that may be running in the shop at the same time.

Technicians should never run a vehicle in the shop unless a shop exhaust hose is installed on the tailpipe of the vehicle. The exhaust fan must be switched on to remove exhaust fumes.

If shop heaters or furnaces have restricted chimneys, they release carbon monoxide emissions into the shop air. Therefore, chimneys should be checked periodically for restriction and proper ventilation.

Monitors are available to measure the level of carbon monoxide in the shop. Some of these monitors read the amount of carbon monoxide present in the shop air, and other monitors provide an audible alarm if the concentration of carbon monoxide exceeds the danger level.

Diesel exhaust contains some carbon monoxide, but particulates are also present in the exhaust from these engines. Particulates are basically small carbon particles, which can be harmful to the lungs.

The sulfuric acid solution in car batteries is a very corrosive, poisonous liquid. If a battery is charged with a fast charger at a high rate for a period of time, the battery becomes hot, and the sulfuric acid solution begins to boil. Under this condition, the battery may emit a strong sulfuric acid smell, and these fumes may be harmful to the lungs. If this condition occurs in the shop, the battery charger should be turned off, or the charger rate should be reduced considerably.

WARNING: When an automotive battery is charged, hydrogen gas and oxygen gas escape from the battery. If these gases are combined, they form water, but hydrogen gas by itself is very explosive. While a battery is charged, sparks, flames, and other sources of ignition must not be allowed near the battery.

WARNING: Breathing asbestos dust must be avoided because this dust is a known contributor to lung cancer.

Some automotive clutch facings and brake linings contain asbestos. Never use an air hose to blow dirt from these components, because this action disperses asbestos dust into the shop where it may be inhaled by technicians and other people in the shop. Parts washers approved by Occupational Safety and Health Act (OSHA) must be used to clean the dust from these components (Figure 1-2). Brake washer concentrate is mixed with water in these parts washers (Figure 1-3). A catch basin with a removable liner is placed under the parts to be washed, and the washer sprays the cleaning solution on the parts. After the washing operation is completed, the liner containing the cleaning solution and asbestos dust is removed, sealed, and labeled for proper handling and disposal according to waste disposal laws.

Even though technicians take every precaution to maintain air quality in the shop, some undesirable gases may still get in the air. For example, exhaust manifolds may get oil on them during an engine overhaul. When the engine is started and these manifolds get hot, the oil burns off the manifolds and pollutes the shop air with oil smoke. Adequate shop ventilation must be provided to take care of this type of air contamination.

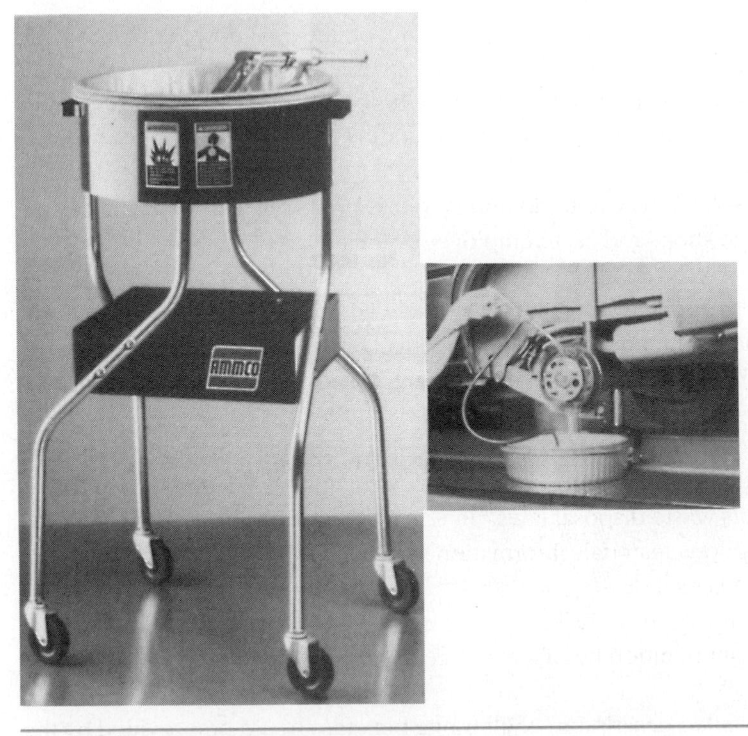

Figure 1-2 Brake assembly washer for asbestos dust (Courtesy of Hennessy Industries Inc.)

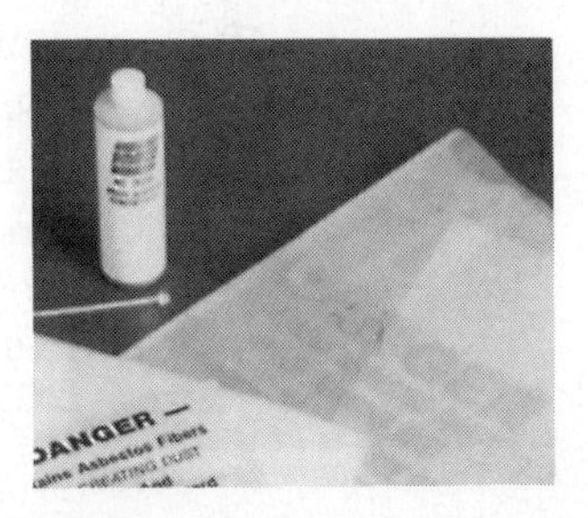

Figure 1-3 Brake washer concentrate (Courtesy of Hennessy Industries Inc.)

Employer and Employee Obligations

When you begin employment, you enter into a business agreement with your employer. A business agreement involves an exchange of goods or services that have value. Although the automotive technician may not have a written agreement with his or her employer, the technician exchanges time, skills, and effort for money paid by the employer. Both the employee and the employer have obligations. The automotive technician's obligations include the following:

1. Productivity—As an automotive technician you have a responsibility to your employer to make the best possible use of time on the job. Each job should be done in a reasonable length of time. Employees are paid for their skills, effort, and time.
2. Quality—Each repair job should be a quality job! Work should never be done in a careless manner. Nothing improves customer relations like quality workmanship.
3. Teamwork—The shop staff are a team, and everyone, including technicians and management personnel, are team members. You should cooperate with, and care about, other team members. Each member of the team should strive for harmonious relations with fellow workers. Cooperative teamwork helps to improve shop efficiency, productivity, and customer relations. Customers may be "turned off" by bickering between shop personnel.
4. Honesty—Employers and customers expect, and deserve, honesty from automotive technicians. Honesty creates a feeling of trust between technicians, employers, and customers.
5. Loyalty—As an employee, you are obliged to act in the best interests of your employer, both on and off the job.
6. Attitude—Employees should maintain a positive attitude at all times. As in other professions, automotive technicians have days when it may be difficult to maintain a positive attitude. For example, there will be days when the technical problems on a certain vehicle are difficult to solve. However, developing a negative attitude certainly will not help the situation! A positive attitude has a positive effect on the job situation as well as on the customer and employer.
7. Responsibility—You are responsible for your conduct on the job and your work-related obligations. These obligations include always maintaining good workmanship and customer relations. Attention to details such as always placing fender and seat covers on customer vehicles prior to driving or working on the vehicle greatly improve customer relations.
8. Following directions—All of us like to do things "our way." Such action may not be in the best interests of the shop, and as an employee you have an obligation to follow the supervisor's directions.
9. Punctuality and regular attendance—Employees have an obligation to be on time for work and to be regular in attendance on the job. It is very difficult for a business to operate successfully if they cannot count on their employees to be on the job at the appointed time.
10. Regulations—Automotive technicians should be familiar with all state and federal regulations pertaining to their job situation, such as the Occupational Safety and Health Act (OSHA) and hazardous waste disposal laws. In Canada, employees should be familiar with work place hazardous materials information systems (WHMIS).

Employer to employee obligations include:

1. Wages—The employer has a responsibility to inform the employee regarding the exact amount of financial remuneration he or she will receive and when that wage will be paid.
2. Fringe benefits—A detailed description of all fringe benefits should be provided by the employer. These benefits may include holiday pay, sickness and accident insurance, and pension plans.

3. Working conditions—A clean, safe work place must be provided by the employer. The shop must have adequate safety equipment and first aid supplies. Employers must be certain that all shop personnel maintain the shop area and equipment to provide adequate safety and a healthy work place atmosphere.
4. Employee instruction—Employers must provide employees with clear job descriptions, and be sure that each worker is aware of his or her obligations.
5. Employee supervision—Employers should inform their workers regarding the responsibilities of their immediate supervisors and other management personnel.
6. Employee training—Employers must make sure that employees are familiar with the safe operation of all the equipment that they are required to use in their job situation. Since automotive technology is changing rapidly, employers should provide regular update training for their technicians. Under the right-to-know laws, employers are required to inform all employees about hazardous materials in the shop. Employees should be familiar with work place hazardous materials information systems, which detail the labeling and handling of hazardous waste, and the health problems if exposed to hazardous waste.

Job Responsibilities

An automotive technician has specific responsibilities regarding each job performed on a customer's vehicle. These job responsibilities include:

1. Do every job to the best of your ability. There is no place in the automotive service industry for careless workmanship! Automotive technicians and students must realize they have a very responsible job. During many repair jobs, you, as a student or technician working on a customer's vehicle, actually have the customer's life and the safety of his or her vehicle in your hands. For example, if you are doing a brake job and leave the wheel nuts loose on one wheel, that wheel may fall off the vehicle at high speed. This could result in serious personal injury for the customer and others, plus extensive vehicle damage. If this type of disaster occurs, the individual who worked on the vehicle and the shop may be involved in a very expensive legal action. As a student or technician working on customer vehicles, you are responsible for the safety of every vehicle that you work on! Even when careless work does not create a safety hazard, it leads to dissatisfied customers who often take their business to another shop, and nobody benefits when that happens.
2. Treat customers fairly and honestly on every repair job. Do not install parts that are unnecessary to complete the repair job.
3. Use published specifications; do not guess at adjustments.
4. Follow the service procedures in the service manual provided by the vehicle manufacturer or an independent manual publisher.
5. When the repair job is completed, always be sure the customer's complaint has been corrected.
6. Do not be too concerned with work speed when you begin working as an automotive technician. Speed comes with experience.

National Institute for Automotive Service Excellence (ASE) Certification

ASE has provided voluntary testing and certification of automotive technicians on a national basis for many years. The image of the automotive service industry has been enhanced by the ASE certification program. More than 265,000 ASE certified automotive technicians now work in a wide variety of automotive service shops. ASE provides certification in these eight areas of automotive repair:

Figure 1-4 ASE certification shoulder patches worn by automotive technicians and master technicians (Courtesy of National Institute for Automotive Service Excellence [ASE])

1. Engine repair
2. Automatic transmissions/transaxles
3. Manual drivetrain and axles
4. Suspension and steering
5. Brakes
6. Electrical systems
7. Heating and air conditioning
8. Engine performance

A technician may take the ASE test and become certified in any, or all, of the eight areas. When a technician passes an ASE test in one of the eight areas, an Automotive Technician's shoulder patch is issued by ASE. If a technician passes all eight tests, he or she receives a Master Technician's shoulder patch (Figure 1-4). Retesting at five-year intervals is required to remain certified.

The certification test in each of the eight areas contains 40 to 80 multiple choice questions. The test questions are written by a panel of automotive service experts from various areas of automotive service such as automotive instructors, service managers, automotive manufacturer's representatives, test equipment representatives, and certified technicians. The test questions are pretested and checked for quality by a national sample of technicians. Most questions have the Technician A and Technician B format similar to the questions at the end of each chapter in this book. ASE regulations demand that each technician must have two years of working experience in the automotive service industry prior to taking a certification test or tests. However, relevant formal training may be substituted for one of the years of working experience. Contact ASE for details regarding this substitution. The contents of the Engine Performance test are listed in Table 1-1.

Classroom Manual
Chapter 1, page 1

Shops that employ ASE certified technicians display an official ASE blue seal of excellence. This blue seal increases the customer's awareness of the shop's commitment to quality service and the competency of certified technicians.

Shop Projects to Enhance the Theories in Chapter 1 of the Classroom Manual

1. Strike the tread of an inflated tire with a rubber hammer. What reaction did you get? Why did this reaction occur?
2. Place some cold water in a graduated heat-resistant container supported on a metal stand, and record the exact quantity of water. Heat the water with a propane torch until it boils, and record the exact quantity of water. What happened to the water when heated?
3. Obtain an empty, metal one-gallon container with a tight-fitting top. Place about one pint of water in the container, and heat the container until the water is boiling. Then install the top securely, and allow the container to cool. Cooling time may be reduced by placing ice on the container. What happened to the container? Why did this happen?
4. Obtain a vacuum hand pump and a vacuum advance from a distributor. Gradually operate the pump and observe the vacuum diaphragm. How much vacuum is required to

Table 1-1 ENGINE PERFORMANCE TEST SUMMARY

Content Area	Questions in Test	Percentage of Test
A. General Engine Diagnosis	19	24.0%
B. Ignition System Diagnosis and Repair	16	20.0%
C. Fuel, Air Induction, and Exhaust Systems Diagnosis and Repair	21	26.0%
D. Emissions Control Systems Diagnosis and Repair	19	24.0%
1. Positive Crankcase Ventilation (1)		
2. Spark Timing Controls (3)		
3. Idle Speed Controls (3)		
4. Exhaust Gas Recirculation (4)		
5. Exhaust Gas Treatment (2)		
6. Inlet Air Temperature Controls (2)		
7. Intake Manifold Temperature Controls (2)		
8. Fuel Vapor Controls (2)		
E. Engine Related Service	2	2.5%
F. Engine Electrical Systems Diagnosis and Repair	3	3.5%
1. Battery (1)		
2. Starting System (1)		
3. Charging System (1)		
Total	**80**	**100.0%**

(Courtesy of National Institute for Automotive Service Excellence [ASE])

begin moving the vacuum diaphragm? How much vacuum is required to move this diaphragm to the fully advanced position? What two forces move the diaphragm?

Guidelines for Following Proper Shop Practices

1. Technicians must be familiar with shop layout, especially the location of safety equipment. This knowledge provides a safer, more efficient shop.
2. Shop rules must be observed by everyone in the shop to provide adequate shop safety, personal health protection, and vehicle protection.
3. The application of driving rules in the shop increases safety, protects customer vehicles and shop property, and improves the shop image in the eyes of the customer.
4. When good housekeeping habits are developed, shop safety is improved, worker efficiency is increased, and customers are impressed.
5. If some basic rules are followed to maintain shop air quality, the personal health of shop employees is improved.
6. When employers and employees accept and fulfill their obligations, personal relationships and general attitudes are greatly improved, shop productivity is increased, and customer relations are improved.
7. If a technician accepts certain job responsibilities, job quality improves and customer satisfaction increases.
8. ASE technician certification improves the quality of automotive repair and improves the image of the profession.

CASE STUDY

A technician was removing and replacing the alternator on a General Motors car. After installing the replacement alternator and connecting the alternator battery wire, the technician proceeded to install the alternator belt. The rubber boot was still removed from the alternator battery terminal. While installing this belt, the technician's wrist watch expansion bracelet made electrical contact from the alternator battery terminal to ground on the alternator housing. Even though the alternator battery wire was protected with a fuse link, which melted, a high current flowed through the wrist watch bracelet. This heated the bracelet to a very high temperature and severely burned the technician's arm. This technician forgot two safety rules:

1. Never wear jewelry such as watches and rings while working in an automotive shop.
2. Before performing electrical work on a vehicle, disconnect the negative battery cable. If the vehicle is equipped with an air bag, wait one minute after this cable is disconnected.

CASE STUDY

A technician was removing and replacing the starting motor with the solenoid mounted on top of the motor. The technician began removing the battery cable from the solenoid, and the ring on one of his fingers made contact between the end of the wrench and the engine block. The current flow through the ring was so high that the positive battery terminal melted out of the battery. The technician's finger was so badly burned that a surgeon had to cut the ring from his finger and repair the finger. This technician forgot to disconnect the negative battery cable before working on the vehicle!

CASE STUDY

A technician had just replaced the engine in a Ford product, and she was performing final adjustments such as timing and air-fuel mixture. In this shop, the cars were parked in the work bays at an angle on both sides of the shop. With the engine running at fast idle, the automatic transmission suddenly slipped into reverse. The car went backwards across the shop and collided with a car in one of the electrical repair bays. Both vehicles were damaged to a considerable extent. Fortunately, no personnel were injured. This technician forgot to apply the parking brake while working on the vehicle!

Terms to Know

Shop layout
Carbon monoxide
Asbestos parts washer
Diesel particulates
Sulfuric acid
National Institute for Automotive Service Excellence (ASE)
ASE technician certification
ASE blue seal of excellence

ASE Style Review Questions

1. While discussing shop layout and safety:
   Technician A says the location of each piece of safety equipment should be clearly marked.
   Technician B says some shops have a special tool room where special tools are located.
   Who is correct?
   A. A only **C.** Both A and B
   B. B only **D.** Neither A nor B

2. While discussing shop rules:
   Technician A says breathing carbon monoxide may cause arthritis.
   Technician B says breathing carbon monoxide may cause headaches.
   Who is correct?
   A. A only **C.** Both A and B
   B. B only **D.** Neither A nor B

3. While discussing shop rules:
   Technician A says breathing asbestos dust may cause heart defects.
   Technician B says oily rags should be stored in uncovered garbage containers.
   Who is correct?
   A. A only **C.** Both A and B
   B. B only **D.** Neither A nor B

4. While discussing shop rules:
   Technician A says English tools may be substituted for metric tools.
   Technician B says that foot injuries may be caused by loose sewer covers.
   Who is correct?
   A. A only **C.** Both A and B
   B. B only **D.** Neither A nor B

5. While discussing air quality:
   Technician A says a restricted chimney on a shop furnace may cause carbon monoxide gas in the shop.
   Technician B says monitors are available to measure the level of carbon monoxide in the shop air.
   Who is correct?
   A. A only **C.** Both A and B
   B. B only **D.** Neither A nor B

6. While discussing air quality:
   Technician A says a battery gives off hydrogen gas during the charging process.
   Technician B says a battery gives off oxygen gas during the charging process.
   Who is correct?
   A. A only **C.** Both A and B
   B. B only **D.** Neither A nor B

7. While discussing air quality:
   Technician A says diesel exhaust contains particulates.
   Technician B says particulate emissions contain oxides of nitrogen.
   Who is correct?
   A. A only **C.** Both A and B
   B. B only **D.** Neither A nor B

8. While discussing employer and employee responsibilities:
   Technician A says employers are required to inform their employees about hazardous materials in the shop.
   Technician B says that employers have no obligation to inform their employees about hazardous materials in the shop.
   Who is correct?
   A. A only **C.** Both A and B
   B. B only **D.** Neither A nor B

9. While discussing ASE certification:
   Technician A says a technician must pass four of the eight ASE certification tests to receive a master technician's shoulder patch.
   Technician B says a technician must pass all eight ASE certification tests to receive a master technician's shoulder patch.
   Who is correct?
   A. A only **C.** Both A and B
   B. B only **D.** Neither A nor B

10. While discussing ASE certification:
    Technician A says retesting is required at five-yea intervals to remain certified.
    Technician B says a technician must have four years of automotive repair experience before writing an ASE certification test.
    Who is correct?
    A. A only **C.** Both A and B
    B. B only **D.** Neither A nor B

CHAPTER 2

Tools and Safety Practices

Upon completion and review of this chapter, you should be able to:

- ❑ Perform automotive measurements using United States customary (USC) and international system (SI) systems of weights and measures.
- ❑ Observe all personal safety precautions while working in the automotive shop.
- ❑ Demonstrate proper lifting procedures and precautions.
- ❑ Demonstrate proper vehicle lift operating and safety procedures.
- ❑ Observe all safety precautions when hydraulic tools are used in the automotive shop.
- ❑ Follow the recommended procedure while operating hydraulic tools such as presses, floor jacks, and vehicle lifts to perform automotive service tasks.
- ❑ Follow safety precautions regarding the use of power tools.
- ❑ Demonstrate proper safety precautions during the use of compressed air equipment.
- ❑ Follow safety precautions while using cleaning equipment in the automotive shop.

Measuring Systems

The United States customary (USC) system of weights and measures is commonly referred to as the English system.

The international system (SI) of weights and measures is called the metric system.

Two systems of weights and measures are commonly used in the United States. One system of weights and measures is the United States customary system (USC). This system is commonly referred to as the English system. Well-known measurements for length in the USC system are the inch, foot, yard, and mile. In this system, the quart and gallon are common measurements for volume, and ounce, pound, and ton are measurements for weight. A second system of weights and measures is called the international system (SI). This system is often referred to as the metric system.

In the USC system, the basic linear measurement is the yard, whereas the corresponding linear measurement in the metric system is the meter. Each unit of measurement in the metric system is related to the other metric units by a factor of 10. Thus, every metric unit can be multiplied or divided by 10 to obtain larger units (multiples) or smaller units (submultiples). For example, the meter may be divided by 10 to obtain centimeters (1/100 meter) or millimeters (1/1,000 meter).

The United States government passed the Metric Conversion Act in 1975 in an attempt to move American industry and the general public to accept and adopt the metric system. The automotive industry has adopted the metric system and in recent years most bolts, nuts, and fittings on vehicles have been changed to metric. During the early 1980s, some vehicles had a mix of USC and metric bolts. Import vehicles have used the metric system for many years. Although the automotive industry has changed to the metric system, the general public in the United States has been slow to convert from the USC system to the metric system. One of the factors involved in this change is cost. What would it cost to change every highway distance and speed sign in the United States to read kilometers? Of course, the answer to that question is probably hundreds of millions or billions of dollars.

Service technicians must be able to work with both the USC and the metric system. One meter (m) in the metric system is equal to 39.37 inches (in) in the USC system. Some common equivalents between the metric and USC systems are these:

1 meter (m) = 39.37 inches (in)
1 centimeter (cm) = 0.3937 inch
1 millimeter (mm) = 0.03937 inch
1 inch = 2.54 cm
1 inch = 25.4 mm

SERVICE TIP: A metric conversion calculator provides fast, accurate metric-to-USC conversions or vice versa.

In the USC system, phrases such as 1/8 of an inch are used for measurements. The metric system uses a set of prefixes. For example, in the word *kilometer*, the prefix kilo indicates 1,000, and this prefix indicates there are 1,000 meters in a kilometer. Common prefixes in the metric system follow:

NAME	SYMBOL	MEANING
mega	M	one million
kilo	k	one thousand
hecto	h	one hundred
deca	da	ten
deci	d	one tenth of
centi	c	one hundredth of
milli	m	one thousandth of
micro	µ	one millionth of

Measurement of Mass

In the metric system, mass is measured in grams, kilograms, or tonnes. 1,000 grams (g) = 1 kilogram (kg). In the USC system, mass is measured in ounces, pounds, or tons. When converting pounds to kilograms, 1 pound = 0.453 kilogram.

Measurement of Length

In the metric system, length is measured in millimeters, centimeters, meters, or kilometers. 10 millimeters (mm) = 1 centimeter (cm). In the USC system, length is measured in inches, feet, yards, or miles. When distance conversions are made between the two systems, some of the conversion factors are these:

1 inch = 25.4 millimeters
1 foot = 30.48 centimeters
1 yard = 0.91 meter
1 mile = 1.60 kilometers

Measurement of Volume

In the metric system, volume is measured in milliliters, cubic centimeters, and liters. 1 cubic centimeter = 1 milliliter. If a cube has a length, depth, and height of 10 centimeters (cm), the volume of the cube is 10 cm x 10 cm x 10 cm = 1,000 cm^3 = 1 liter. When volume conversions are made between the two systems, 1 cubic inch = 16.38 cubic centimeters. If an engine has a displacement of 350 cubic inches, 350 x 16.38 = 5,733 cubic centimeters, and 5,733 ÷ 1,000 = 5.7 liters.

Personal Safety

Personal safety is the responsibility of each technician in the shop. Always follow these safety practices:

1. Always use the correct tool for the job. If the wrong tool is used, it may slip and cause hand injury.
2. Follow the car manufacturer's recommended service procedures.
3. Always wear eye protection such as safety glasses or a face shield (Figure 2-1).
4. Wear protective gloves when cleaning parts in hot or cold tanks and when handling hot parts such as exhaust manifolds.
5. Do not smoke when working on a vehicle. A spark from a cigarette or lighter may ignite flammable materials in the work area.
6. When working on a running engine, keep hands and tools away from rotating parts. Remember that electric-drive fans may start turning at any time.
7. Do not wear loose clothing, and keep long hair tied behind your head. Loose clothing or long hair is easily entangled in rotating parts.
8. Wear safety shoes or boots.

Figure 2-1 Safety glasses (Courtesy of Siebe North, Inc.)

9. Do not wear watches, jewelry, or rings when working on a vehicle. Severe burns occur when jewelry makes contact between an electric terminal and ground.
10. Always place a shop exhaust hose on the vehicle tail pipe if the engine is running, and be sure the exhaust fan is running. Carbon monoxide in the vehicle exhaust is harmful or fatal to the human body.
11. Be sure that the shop has adequate ventilation.
12. Make sure the work area has adequate lighting.
13. Use trouble lights with steel or plastic cages around the bulb. If an unprotected bulb breaks, it may ignite flammable materials in the area.
14. When servicing a vehicle, always apply the parking brake, and place the transmission in park with an automatic transmission, or neutral with a manual transmission.
15. Avoid working on a vehicle parked on an incline.
16. Never work under a vehicle unless the vehicle chassis is supported securely on safety stands.
17. When one end of a vehicle is raised, place wheel chocks on both sides of the wheels remaining on the floor.
18. Be sure that you know the location of shop first aid kits and eye wash fountains.
19. Familiarize yourself with the location of all shop fire extinguishers.
20. Do not use any type of open flame heater to heat the work area.
21. Collect oil, fuel, brake fluid, and other liquids in the proper safety containers.
22. Use only approved cleaning fluids and equipment. Do not use gasoline to clean parts.
23. Obey all state and federal safety, fire, and hazardous material regulations.
24. Always operate equipment according to the equipment manufacturer's recommended procedure.
25. Do not operate equipment unless you are familiar with the correct operating procedure.
26. Do not leave running equipment unattended.
27. Do not use electrical equipment, including trouble lights, with frayed cords.
28. Be sure the safety shields are in place on rotating equipment.
29. Before operating electric equipment, be sure the power cord has a ground connection.
30. When working in an area where extreme noise levels are encountered, wear ear plugs or covers.

Classroom Manual
Chapter 2, page 22

Lifting and Carrying

Many automotive service jobs require heavy lifting. You should know your maximum weight lifting ability, and do not attempt to lift more than this weight. If a heavy part exceeds your weight lifting ability, have a co-worker help with the lifting job. Follow these steps when lifting or carrying an object:

1. If the object is to be carried, be sure your path is free from loose parts or tools.
2. Position your feet close to the object and position your back reasonably straight for proper balance.
3. Keep your back and elbows as straight as possible. Continue to bend your knees until your hands reach the best lifting location on the object to be lifted.
4. Be certain the container is in good condition. If a container falls apart during the lifting operation, parts may drop out of the container and result in foot injury or part damage.
5. Maintain a firm grip on the object, and do not attempt to change your grip while lifting is in progress.
6. Straighten your legs to lift the object, and keep the object close to your body. Use leg muscles rather than back muscles.
7. If you have to change direction of travel, turn your whole body instead of twisting it.

8. Do not bend forward to place an object on a work bench or table. Position the object on the front surface of the work bench and slide it back. Do not pinch your fingers under the object while setting it on the front of the bench.
9. If the object must be placed on the floor or a low surface, bend your legs to lower the object. Do not bend your back forward, because this movement strains back muscles.
10. When a heavy object must be placed on the floor, locate suitable blocks under the object to prevent jamming your fingers under the object.

Hand Tool Safety

Many shop accidents are caused by improper use and care of hand tools. These hand tool safety steps must be followed:

1. Maintain tools in good condition and keep them clean. Worn tools may slip and result in hand injury. If a hammer is used with a loose head, the head may fly off and cause personal injury or vehicle damage. Your hand may slip off a greasy tool, and this action may cause some part of your body to hit the vehicle. For example, your head may hit the vehicle hood.
2. Using the wrong tool for the job may cause damage to the tool, fastener, or your hand, if the tool slips. If you use a screwdriver for a chisel or pry bar, the blade may shatter, causing serious personal injury.
3. Use sharp, pointed tools with caution. Always check your pockets before sitting on the vehicle seat. A screwdriver, punch, or chisel in the back pocket may put an expensive tear in the upholstery. Do not lean over fenders with sharp tools in your pockets.
4. Tool tips that are intended to be sharp should be kept in a sharp condition. Sharp tools, such as chisels, will do the job faster with less effort.

Power Tool Safety

Power tools use electricity, shop air, or hydraulic pressure as a power source. Careless operation of power tools may cause personal injury and vehicle damage. Follow these steps for safe power tool operation:

1. Do not operate power tools with frayed power cords.
2. Be sure the power tool cord has a proper ground connection.
3. Do not stand on a wet floor while operating an electric power tool.
4. Always unplug an electric power tool before servicing the tool.
5. Do not leave a power tool running and unattended.
6. When using a power tool on small parts, do not hold the part in your hand. The part must be secured in a bench vise or with locking pliers.
7. Do not use a power tool on a job where the maximum capacity of the tool is exceeded.
8. Be sure that all power tools are in good condition, and always operate these tools according to the tool manufacturer's recommended procedure.
9. Make sure all protective shields and guards are in position.
10. Maintain proper body balance while using a power tool.
11. Always wear safety glasses or a face shield.
12. Wear ear protection.
13. Follow the equipment manufacturer's recommended maintenance schedule for all shop equipment.
14. Never operate a power tool unless you are familiar with the tool manufacturer's recommended operating procedure. Serious accidents occur from improper operating procedures.

15. Always make sure that the wheels on the electric grinder are securely attached and in good condition.
16. Keep fingers and clothing away from grinding and buffing wheels. When grinding or buffing a small part, it should be held with a pair of locking pliers.
17. Always make sure the sanding or buffing disc is securely attached to the sander pad.
18. Special heavy-duty sockets must be used on impact wrenches. If ordinary sockets are used on an impact wrench, they may break and cause serious personal injury.
19. Never operate an air chisel unless the tool is securely connected to the chisel with the proper retaining device.
20. Never direct a blast of air from an air gun against any part of your body. If air penetrates the skin and enters the blood stream, it may cause very serious health problems and even death.

Compressed Air Equipment Safety

The shop air supply contains high-pressure air in the shop compressor and air lines. Serious injury or property damage may result from careless operation of compressed air equipment. When these steps are followed, safety is improved regarding compressed air equipment:

1. Safety glasses or a face shield should be worn for all shop tasks, including those tasks involving the use of compressed air equipment.
2. Wear ear protection when using compressed air equipment.
3. Always maintain air hoses and fittings in good condition. If an end suddenly blows off an air hose, the hose will whip around, and this may cause personal injury.
4. Do not direct compressed air against the skin. This air may penetrate the skin, especially through small cuts or scratches. If air penetrates the skin and enters the blood stream, it can be fatal or cause serious health complications. Use only air gun nozzles approved by Occupational Safety and Health Act (OSHA).
5. Do not use an air gun to blow off clothing or hair.
6. Do not clean the work bench or floor with compressed air. This action may blow very small parts against your skin or into your eye. Small parts blown by compressed air may cause vehicle damage. For example, if the car in the next stall has the air cleaner removed, a small part may go into the carburetor or throttle body. When the engine is started, this part will likely be pulled into the cylinder by engine vacuum, and the part will penetrate through the top of a piston.
7. Never spin bearings with compressed air, because the bearing will rotate at extremely high speed. Under this condition, the bearing may be damaged or it may disintegrate causing personal injury.
8. All pneumatic tools must be operated according to the tool manufacturer's recommended operating procedure.
9. Follow the equipment manufacturer's recommended maintenance schedule for all compressed air equipment.

Basic Engine Diagnostic Tools

Cooling System Pressure Tester

A cooling system pressure tester contains a hand pump and a pressure gauge. A hose is connected from the hand pump to a special adapter that fits on the radiator filler neck (Figure 2-2). This tester is used to pressurize the cooling system and check for coolant leaks. Additional adapters are available to connect the tester to the radiator cap. With the tester connected to the radiator cap, the pressure relief action of the cap may be checked.

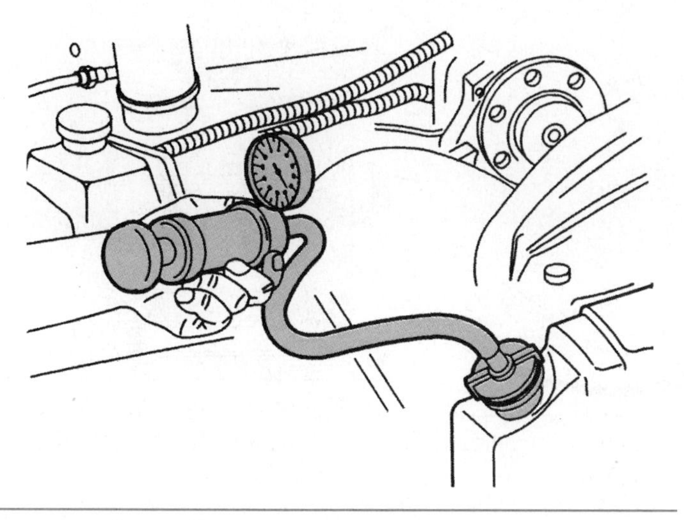

Figure 2-2 Cooling system pressure tester (Courtesy of Oldsmobile Division, General Motors Corporation)

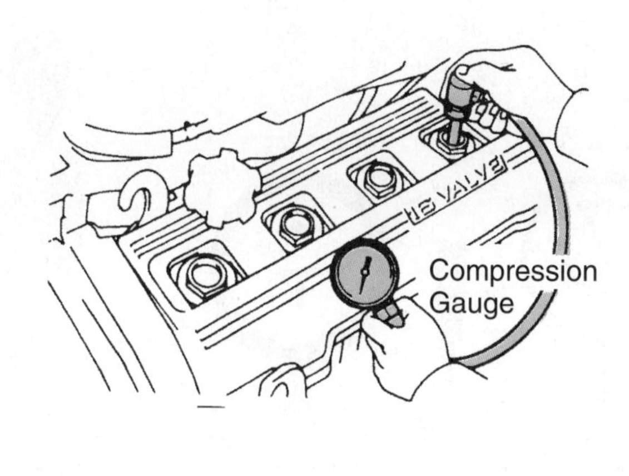

Figure 2-3 Cylinder compression tester (Courtesy of Toyota Motor Corporation)

Compression Tester

A compression tester is used to test cylinder compression, which is an indication of piston ring, valve, and combustion chamber condition (Figure 2-3). Various adapters are supplied with the compression gauge to fit different spark plug openings.

Stethoscope

A stethoscope is used to locate the source of engine and other noises. The stethoscope pickup is placed on the suspected component, and the stethoscope receptacles are placed in the technician's ears (Figure 2-4).

Oil Pressure Gauge

The oil pressure gauge may be connected to the engine to check the oil pressure (Figure 2-5). Various fittings are usually supplied with the oil pressure gauge to fit different openings in the lubrication system.

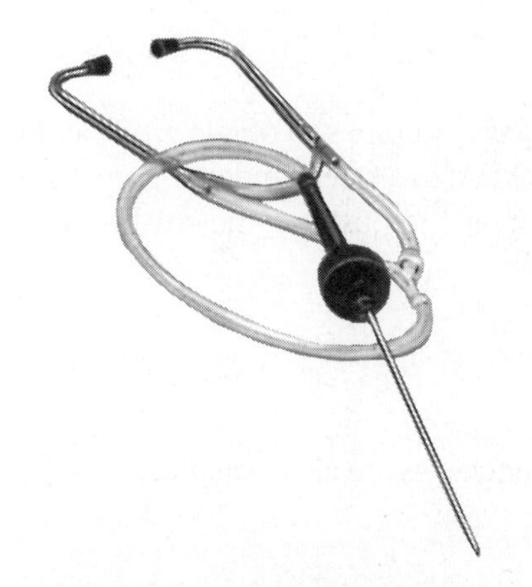

Figure 2-4 Stethoscope (Courtesy of Snap-on Tools)

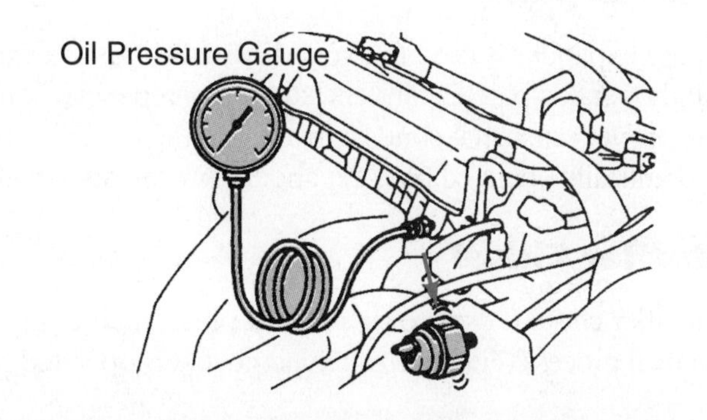

Figure 2-5 Oil pressure gauge connected to test engine oil pressure (Courtesy of Toyota Motor Corporation)

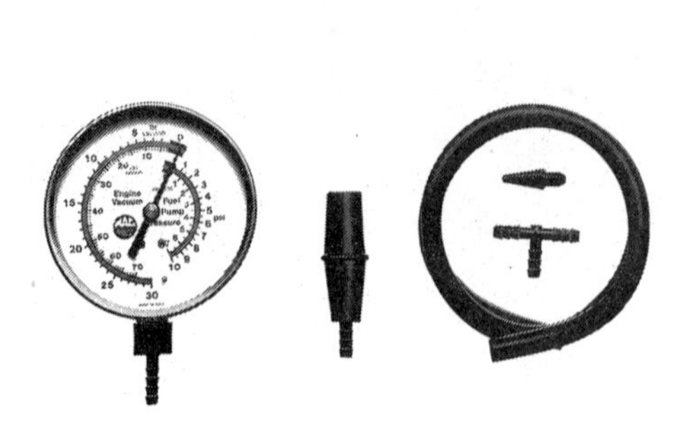

Figure 2-6 Vacuum pressure gauge (Courtesy of Mac Tools, Inc.)

Figure 2-7 Coolant hydrometer (Courtesy of Mac Tools, Inc.)

Figure 2-8 Thermostat tester (Courtesy of Toyota Motor Corporation)

Vacuum Pressure Gauge

A vacuum pressure gauge may be used to check intake manifold vacuum or low pressure such as turbocharger boost pressure or mechanical fuel pump pressure (Figure 2-6). The vacuum pressure gauge is usually supplied with a hose and various fittings and grommets.

Coolant Hydrometer

A coolant hydrometer is used to check the amount of antifreeze in the coolant. This tester contains a pickup hose, coolant reservoir, and squeeze bulb. The pickup hose is placed in the radiator coolant. When the squeeze bulb is squeezed and released, coolant is drawn into the reservoir. As coolant enters the reservoir, a pivoted float moves upward with the coolant level. A pointer on the float indicates the freezing point of the coolant on a scale located on the reservoir housing (Figure 2-7).

Thermostat Tester

The thermostat tester is a heat-resistant container with a thermometer. The thermostat to be tested is placed with some water and the thermometer in the container (Figure 2-8). When the container and the water are heated, the thermostat should open at the rated temperature.

Belt Tension Gauge

A belt tension gauge is used to measure drive belt tension. The belt tension gauge is installed over the belt, and the gauge indicates the amount of belt tension (Figure 2-9).

Pipe Expander

A pipe expander is used to expand partially crushed exhaust system pipes to obtain a proper fit. Hand-operated pipe expanders have a tapered metal cone attached to a steel handle (Figure 2-10). The proper size metal cone is driven into the partially crushed pipe to push out the crushed area. Hydraulically operated pipe expander tools are also available.

Muffler Chisel

A muffler chisel is used to slit exhaust pipes, muffler pipes, or catalytic converter pipes during the removal process (Figure 2-11). Electric- or air-operated muffler chisels are also available.

Feeler Gauge

A feeler gauge is a thin strip of metal with a precision thickness. Feeler gauges with different thicknesses usually are sold in sets. These gauges are pivoted on one end and mounted inside a metal

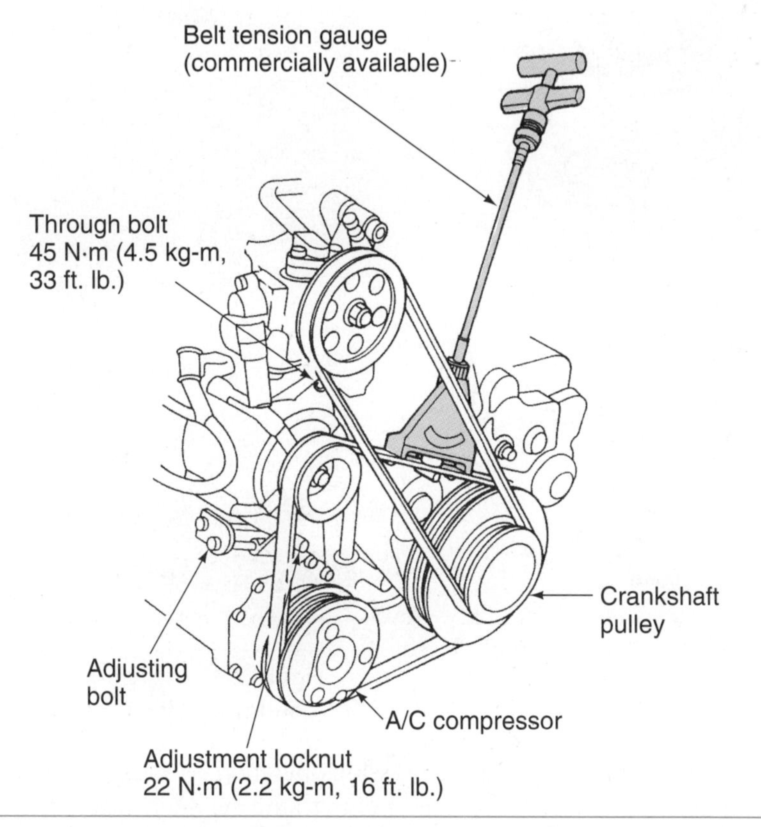

Figure 2-9 Belt tension gauge (Courtesy of American Honda Motor Co., Inc.)

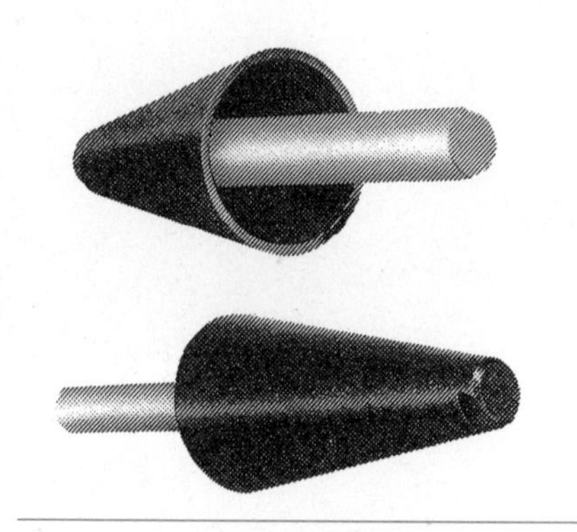

Figure 2-10 Pipe expander (Courtesy of Mac Tools, Inc.)

Figure 2-11 Muffler chisel (Courtesy of Mac Tools, Inc.)

holder. When a specific feeler gauge is required, that feeler gauge is pivoted out of the set (Figure 2-12). Feeler gauges are used to make precision measurements of small gaps. Individual feeler gauges are used for certain service operations such as measuring piston clearance in a cylinder. Feeler gauge thickness in the English system is in thousandths of an inch, whereas metric feeler gauge thickness is in millimeters (mm). Round wire feeler gauges are recommended for measuring spark plug gaps (Figure 2-13).

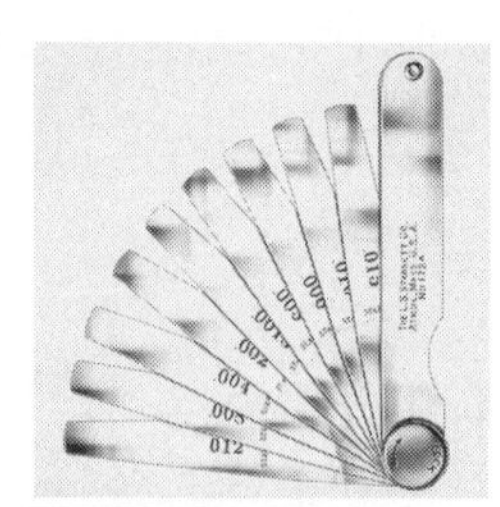

Figure 2-12 Feeler gauge set (Courtesy of L.S. Starrett Co.)

Hydraulic Pressing and Lifting Equipment

Hydraulic Press

WARNING: When operating a hydraulic press, always be sure that the components being pressed are supported properly on the press bed with steel supports.

CAUTION: When using a hydraulic press, never operate the pump handle until the pressure gauge exceeds the maximum pressure rating of the press. If this pressure is exceeded, some part of the press may suddenly break and cause severe personal injury.

When two components have a tight precision fit between them, a hydraulic press is used to separate these components or press them together. The hydraulic press rests on the shop floor, and an adjustable steel beam bed is retained to the lower press frame with heavy steel pins. A hydraulic cylinder and ram is mounted on the top part of the press with the ram facing downward toward the press bed (Figure 2-14). The component being pressed is placed on the press bed with appropriate steel supports. A hand-operated hydraulic pump is mounted on the side of the press. When the handle is pumped, hydraulic fluid is forced into the cylinder, and the ram is extended against

Figure 2-13 Round wire feeler gauges (Courtesy of Mac Tools, Inc.)

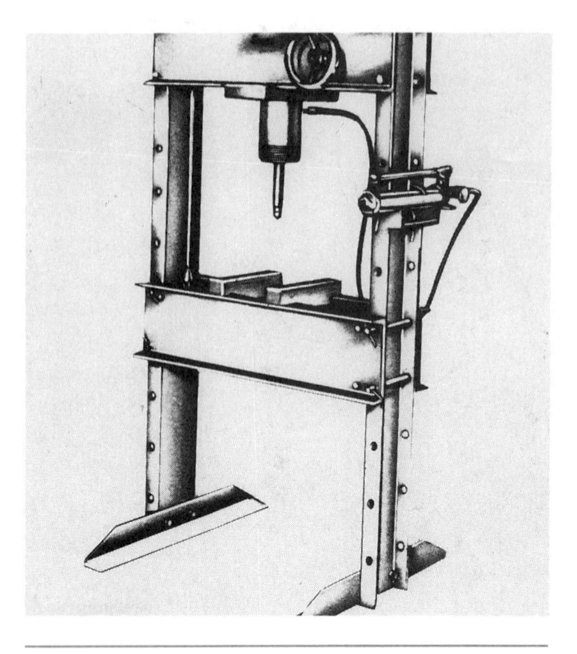

Figure 2-14 Hydraulic press (Courtesy of Snap-on Tools)

Figure 2-15 Hydraulic floor jack (Courtesy of Mac Tools, Inc.)

the component on the press bed to complete the pressing operation. A pressure gauge on the press indicates the pressure applied from the hand pump to the cylinder. The press frame is designed for a certain maximum pressure, and this pressure must not be exceeded during hand pump operation.

Floor Jack

WARNING: The maximum lifting capacity of the floor jack is usually written on a jack decal. Never lift a vehicle that exceeds the jack lifting capacity. This action may cause the jack to break or collapse, resulting in vehicle damage or personal injury.

A floor jack is a portable unit mounted on wheels. The lifting pad on the jack is placed under the chassis of the vehicle, and the jack handle is operated with a pumping action (Figure 2-15). This jack handle operation forces fluid into a hydraulic cylinder in the jack, and this cylinder extends to force the jack lift pad upward and lift the vehicle. Always be sure that the lift pad is positioned securely under one of the car manufacturer's recommended lift points. To release the hydraulic pressure and lower the vehicle, the handle or release lever must be turned slowly. Do not leave the jack handle where someone can trip over it.

Lift

WARNING: Always be sure that the lift arms are securely positioned under the car manufacturer's recommended lift points before raising a vehicle. These lift points are shown in the service manual.

WARNING: The maximum capacity of the vehicle lift is placed on an identification plate. Never lift a vehicle that is heavier than the maximum capacity of the lift.

A lift may be referred to as a hoist.

A lift is used to raise a vehicle so the technician can work under the vehicle. The lift arms must be placed under the car manufacturer's recommended lift points prior to raising a vehicle. Twin posts are used on some lifts, whereas other lifts have a single post (Figure 2-16). Some lifts have an electric motor that drives a hydraulic pump to create fluid pressure and force the lift upward. Other lifts use air pressure from the shop air supply to force the lift upward. If shop air pressure is used for this purpose, the air pressure is applied to fluid in the lift cylinder. A control lever or switch is

Figure 2-16 Lifts are used to raise a vehicle. (Courtesy of Bear Automotive Service Equipment Company)

placed near the lift. The control lever supplies shop air pressure to the lift cylinder, and the switch turns on the lift pump motor. Always be sure that the safety lock is engaged after the lift is raised. When the safety lock is released, a release lever is operated slowly to lower the vehicle.

Lift Safety

WARNING: Do not raise a vehicle on a lift if the vehicle weight exceeds the maximum capacity of the lift.

WARNING: When a vehicle is raised on a lift, the vehicle must be raised high enough to allow engagement of the lift locking mechanism.

Special precautions and procedures must be followed when a vehicle is raised on a lift. Follow these steps for lift safety:

1. Always be sure the lift is completely lowered before driving a vehicle on or off the lift.
2. Do not hit or run over lift arms and adaptors when driving a vehicle on or off the lift. Have a co-worker guide you when driving a vehicle onto the lift. Do not stand in front of a lift with the car coming towards you.
3. Be sure the lift pads on the lift are contacting the car manufacturer's recommended lift points shown in the service manual. If the proper lift points are not used, components under the vehicle such as brake lines or body parts may be damaged. Failure to use the recommended lift points may cause the vehicle to slip off the lift, resulting in severe vehicle damage and personal injury.
4. Before a vehicle is raised or lowered, close the doors, hood, and trunk lid.
5. When a vehicle is lifted a short distance off the floor, stop the lift and check the contact between the hoist lift pads and the vehicle to be sure the lift pads are still on the recommended lift points.
6. When a vehicle is raised on a lift, be sure the safety mechanism is in place to prevent the lift from dropping accidentally.
7. Prior to lowering a vehicle on a lift, always make sure there are no objects, tools, or people under the vehicle.
8. Do not rock a vehicle on a lift during a service job.

9. When a vehicle is raised on a lift, removal of some heavy components may cause vehicle imbalance on the lift. Since front wheel drive cars have the engine and transaxle located at the front of the vehicle, these cars have most of their weight on the front end. Removing a heavy rear end component on these cars may cause the back end of the car to rise off the lift. If this action is allowed to happen, the vehicle could fall off the lift!
10. Do not raise a vehicle on a lift with people in the vehicle.
11. When raising pickup trucks and vans on a lift, remember these vehicles are higher than a passenger car. Be sure there is adequate clearance between the top of the vehicle and the shop ceiling or components under the ceiling.
12. *Do not raise a four wheel drive vehicle with a frame contact lift, because this may damage axle joints.*
13. Do not operate a front wheel drive vehicle that is raised on a frame contact lift. This action may damage the front drive axles.

Engine Lift

CAUTION: An engine lift has a maximum lifting capacity that is usually indicated on a decal. Never lift anything heavier than the maximum capacity of the lift. This action may result in lift damage or personal injury.

An engine lift may be called a crane or a cherry picker.

An engine lift is used to remove and replace automotive engines. A long pivoted arm is mounted on the top of the engine lift (Figure 2-17). When the lift handle is pumped, hydraulic fluid is forced into a cylinder under the lift arm. This action extends the cylinder arm and forces the arm upward to lift the engine. A lifting chain is attached to the lift arm, and this chain is bolted securely to the engine. Always be sure that these retaining bolts are strong enough to support the engine weight.

Hydraulic Jack and Jack Stand Safety

WARNING: Always make sure the jack stand weight capacity rating exceeds the vehicle weight that is lowered onto the stands.

WARNING: Never lift a vehicle with a floor jack if the weight of the vehicle exceeds the rated capacity of the jack.

Accidents involving the use of floor jacks and jack stands may be avoided if these safety precautions are followed:

1. Never work under a vehicle unless jack stands are placed securely under the vehicle chassis and the vehicle is resting on these stands (Figure 2-18).
2. Prior to lifting a vehicle with a floor jack, be sure that the jack lift pad is positioned securely under a recommended lift point on the vehicle. Lifting the front end of a vehi-

Figure 2-17 Engine lift (Courtesy of Mac Tools, Inc.)

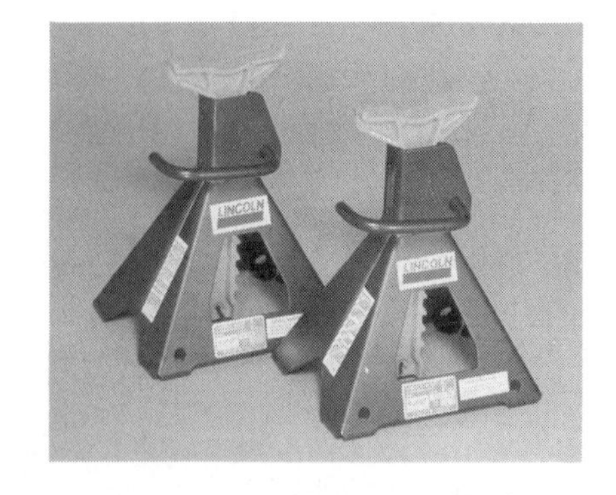

Figure 2-18 Jack stands (Courtesy of Mac Tools, Inc.)

cle with the jack placed under a radiator support may cause severe damage to the radiator and support.

3. Position the jack stands under a strong chassis member such as the frame or axle housing. The jack stands must contact the vehicle manufacturer's recommended lift points.
4. Since the floor jack is on wheels, the vehicle and jack tend to move as the vehicle is lowered from a floor jack onto jack stands. Always be sure the jack stands remain under the chassis member during this operation, and be sure the jack stands do not tip. All the jack stand legs must remain in contact with the shop floor.
5. When the vehicle is lowered from the floor jack onto jack stands, remove the floor jack from under the vehicle. Never leave a jack handle sticking out from under a vehicle. Someone may trip over the handle and be injured.

Cleaning Equipment Safety and Environmental Considerations

All technicians are required to clean parts during their normal work routines. Face shields and protective gloves must be worn while operating cleaning equipment. In most states, environmental regulations require that the runoff from steam cleaning must be contained in the steam cleaning system. This runoff cannot be dumped into the sewer system. Since it is expensive to contain this runoff in the steam cleaner system, the popularity of steam cleaning has decreased. The solution in hot and cold cleaning tanks may be caustic, and contact between this solution and skin or eyes must be avoided. Parts cleaning often creates a slippery floor, and care must be taken when walking in the parts cleaning area. The floor in this area should be cleaned frequently. When the caustic cleaning solution in hot or cold cleaning tanks is replaced, environmental regulations require that the old solution be handled as hazardous waste. Use caution when placing aluminum or aluminum alloy parts in a cleaning solution. Some cleaning solutions will damage these components. Always follow the cleaning equipment manufacturer's recommendations.

Classroom Manual
Chapter 2, page 31

Parts Washers with Electro-Mechanical Agitation

Some parts washers provide electro-mechanical agitation of the parts to provide improved cleaning action (Figure 2-19). These parts washers may be heated with gas or electricity, and various water-based hot tank cleaning solutions are available, depending on the type of metals being cleaned. For example, Kleer-Flo Greasoff® number 1 powdered detergent is available for cleaning iron and steel. Nonheated electro-mechanical parts washers are also available, and these washers use cold cleaning solutions such as Kleer-Flo Degreasol® formulas.

Many cleaning solutions, such as Kleer-Flo Degreasol® 99R, contain no ingredients listed as hazardous by the Environmental Protection Agency's RCRA Act. This cleaning solution is a blend of sulphur-free hydrocarbons, wetting agents, and detergents. Degreasol® 99R does not contain aromatic or chlorinated solvents, and it conforms to California's Rule 66 for clean air. Always use the cleaning solution recommended by the equipment manufacturer.

Cold Parts Washer with Agitated Immersion Tank

Some parts washers have an agitated immersion chamber under the shelves, which provides thorough parts cleaning. Folding workshelves provide a large upper cleaning area with a constant flow of solution from the dispensing hose (Figure 2-20). This cold parts washer operates on Degreasol® 99R cleaning solution.

Aqueous Parts Cleaning Tank

The aqueous parts cleaning tank uses a water-based, environmentally friendly cleaning solution such as Greasoff® 2, rather than traditional solvents. The immersion tank is heated and agitated for

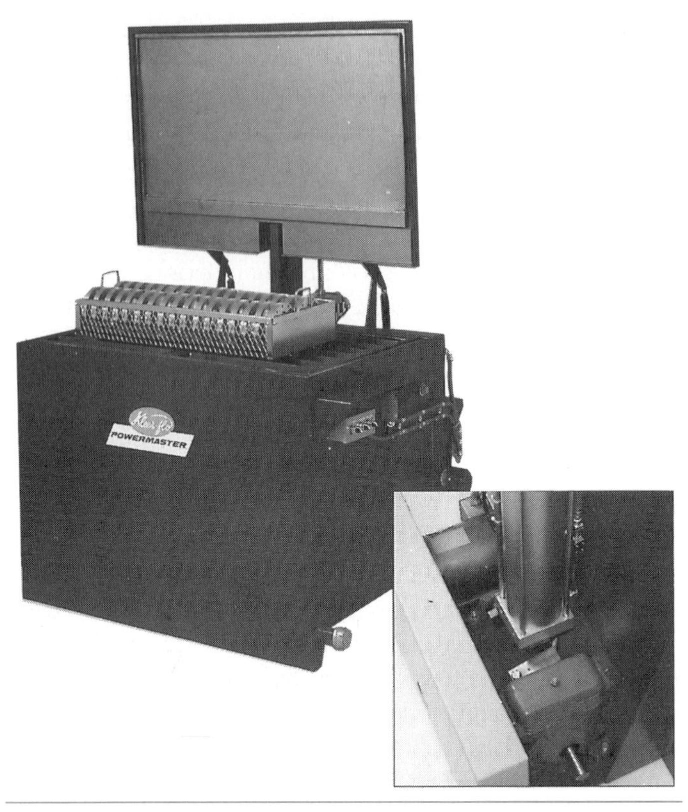

Figure 2-19 Parts washer with electro-mechanical agitator (Courtesy of Kleer-Flo Company)

Figure 2-20 Cold parts washer with agitated immersion tank (Courtesy of Kleer-Flo Company)

Figure 2-21 Aqueous parts cleaning tank (Courtesy of Kleer-Flo Company)

effective parts cleaning (Figure 2-21). A sparger bar pumps a constant flow of cleaning solution across the surface to push floating oils away, and an integral skimmer removes these oils. This action prevents floating surface oils from redepositing on cleaned parts.

Safety Training Exercises

After the equipment operation and safety practices are explained by your instructor, complete these safety training exercises to demonstrate your understanding of equipment operation and safety procedures:

1. Find the lift points on a specific vehicle in the vehicle manufacturer's service manual. Under the supervision of your instructor, position this vehicle properly on a lift, then use the proper lift operating procedures to raise and lower the vehicle on the lift.
2. Find the lift points on a specific vehicle in the vehicle manufacturer's service manual. Under the supervision of your instructor, raise the front and rear suspension of this vehicle with a floor jack, and lower the vehicle onto jack stands. Raise the vehicle with a floor jack, remove the jack stands, and lower the vehicle.
3. Find the lift points on a specific vehicle in the vehicle manufacturer's service manual. Under the supervision of your instructor, position this vehicle properly on a lift, and raise the vehicle. Follow proper service and safety procedures to remove and replace all four wheels. Remove two wheels with an electric impact wrench, and the other two wheels with an air impact wrench.
4. Draw a layout diagram of your automotive shop, or shops, indicating the major service areas and service equipment, and clearly identify the location of all safety equipment.

CUSTOMER CARE: Some automotive service centers have a policy of performing some minor service as an indication of their appreciation to the customer. This service may include cleaning all the windows and/or vacuuming the floors before the car is returned to the customer.

Although this service involves more labor costs for the shop, it may actually improve profits over a period of time. When customers find their windows cleaned and/or the floors vacuumed, it impresses them with the quality of work you do and the fact that you care about their vehicle. They will likely return for service, and tell their friends about the quality of service your shop performs.

Terms to Know

United States customary (USC)
International system (SI)
Compression gauge
Cooling system pressure tester
Stethoscope
Oil pressure gauge
Vacuum pressure gauge
Coolant hydrometer
Thermostat tester
Belt tension gauge
Pipe expander
Muffler chisel
Feeler gauge
Blowgun
Hand press
Hydraulic press
Floor jack
Jack stand
Engine lift
Lift

ASE Style Review Questions

1. While discussing systems of weights and measures:
   Technician A says the international system (SI) is called the metric system.
   Technician B says every unit in the metric system can be divided or multiplied by 10.
   Who is correct?
   A. A only **C.** Both A and B
   B. B only **D.** Neither A nor B

2. While discussing measurements in the metric system:
   Technician A says one decimeter is equal to 1/10 of a meter.
   Technician B says one decimeter is equal to 1/100 of a meter.
   Who is correct?
   A. A only **C.** Both A and B
   B. B only **D.** Neither A nor B

3. While discussing personal safety:
   Technician A says rings and jewelry may be worn in the automotive shop.
   Technician B says some electric-drive cooling fans may start turning at any time.
   Who is correct?
   A. A only **C.** Both A and B
   B. B only **D.** Neither A nor B

4. While discussing the proper way to lift heavy objects:
   Technician A says you should bend your back to pick up a heavy object.
   Technician B says you should bend your knees to pick up a heavy object.
   Who is correct?
   A. A only **C.** Both A and B
   B. B only **D.** Neither A nor B

5. While discussing hand tool safety:
   Technician A says a screwdriver may be used as a chisel.
   Technician B says a screwdriver may be used as a pry bar.
   Who is correct?
   A. A only **C.** Both A and B
   B. B only **D.** Neither A nor B

6. While discussing power tool safety:
   Technician A says an electric power tool cord does not require a ground.
   Technician B says frayed electric cords should be replaced.
   Who is correct?
   A. A only **C.** Both A and B
   B. B only **D.** Neither A nor B

7. While discussing shop cleaning equipment safety:
   Technician A says some hot tanks contain caustic solutions.
   Technician B says some metals such as aluminum may dissolve in hot tanks.
   Who is correct?
   A. A only **C.** Both A and B
   B. B only **D.** Neither A nor B

8. While discussing hydraulic equipment:
   Technician A says jack stands have a maximum weight capacity.
   Technician B says prior to raising a vehicle on a lift, the doors, trunk, and hood should be closed.
   Who is correct?
   A. A only **C.** Both A and B
   B. B only **D.** Neither A nor B

9. While discussing vehicle lifts:
   Technician A says a four wheel drive vehicle may be lifted with a frame contact lift.
   Technician B says a vehicle should not be raised on a lift with people in the vehicle.
   Who is correct?
   A. A only **C.** Both A and B
   B. B only **D.** Neither A nor B

10. While discussing hydraulic equipment:
    Technician A says jack stands have a maximum weight capacity.
    Technician B says the floor jack release lever must be turned slowly to lower the jack.
    Who is correct?
    A. A only **C.** Both A and B
    B. B only **D.** Neither A nor B

General Engine Condition Diagnosis

Upon completion and review of this chapter, you should be able to:

- ❑ List the steps in a general diagnostic procedure that may be used in any diagnostic situation.
- ❑ Diagnose fuel leaks, and determine needed repairs.
- ❑ Diagnose engine oil leaks, and determine needed repairs.
- ❑ Diagnose engine coolant leaks, and determine necessary repairs.
- ❑ Pressure test the cooling system.
- ❑ Diagnose engine exhaust odor, color, and noise, and determine needed repairs.
- ❑ Diagnose engine defects from intake manifold vacuum readings.
- ❑ Perform an engine power balance test, and determine the needed repairs.
- ❑ Perform an engine compression test, and determine the necessary repairs.
- ❑ Perform a cylinder leakage test, and determine the needed repairs.
- ❑ Diagnose engine noises and vibration problems, and determine the necessary repairs.
- ❑ Test engine oil pressure, and determine the cause of low oil pressure.
- ❑ Diagnose oil leaks and oil consumption.

General Diagnostic Procedure

Basic Tools

Basic technician's tool set

Service manual

Jumper wire

As automobiles become increasingly complex, the automotive technician's diagnostic capabilities become more important. Following is a general diagnostic procedure that technicians can apply to any diagnostic situation:

1. Always be sure the customer's complaint about his or her vehicle is accurately identified. Listen to the customer's description of the vehicle operation, and question the customer to determine the exact problems. If necessary, road test the vehicle under the conditions when the problem occurs.
2. Think of the possible causes of the problem from your experience with similar defects on other vehicles.
3. Perform the appropriate diagnostic tests to determine the exact cause of the problem. Always begin the diagnostic procedure with the diagnostic tests that can be performed quickly. Advise the customer regarding the cost of repairs.
4. Repair the cause of the problem, and perform any other necessary maintenance to prevent other problems from occurring.
5. Be sure the original problem is corrected and the vehicle operation is normal. Road test the vehicle if necessary.

Engine Fuel Leak Diagnosis

CAUTION: Never attempt to repair a metal fuel tank until it is drained, removed, and steamed out for at least 30 minutes. Gasoline fumes left in gasoline tanks are extremely dangerous. When ignited, these fumes will cause a very serious explosion and fire, resulting in personal injury and property damage.

WARNING: Always store gasoline drained from a fuel tank in approved gasoline safety cans.

WARNING: Gasoline fumes are extremely dangerous! If ignited, they will cause a very serious explosion and fire, resulting in personal injury and property damage. While removing and replacing a gasoline tank and handling gasoline, be sure there are no sources of ignition in the area. Do not smoke while performing these operations, and be sure nobody else smokes in the area. Do not drag the gasoline tank on the floor.

WARNING: If gasoline is spilled on the shop floor, place approved absorbent material on it immediately. Dispose of the absorbent material in approved waste containers.

Engine fuel leaks are expensive and dangerous, and they should be corrected immediately when they are detected. Fuel leaks are expensive because they waste fuel by leaking it, rather than burning the fuel in the engine. Since gasoline is very explosive, fuel leaks are extremely dangerous because any spark near a fuel leak may start a fire or cause an explosion. If gasoline odor occurs inside or near a vehicle, it should be inspected for fuel leaks immediately. To locate a fuel leak, inspect these fuel system components:

1. Fuel tank. If a fuel tank is leaking, it must be removed and repaired. Many fuel tanks do not have a drain plug, and so the gasoline must be pumped from the tank with a hand-operated pump.
2. Fuel tank filler cap.
3. Fuel lines and filter.
4. Mechanical fuel pump (carbureted engines).
5. Vapor recovery system lines.
6. Carburetor.
7. Pressure regulator, fuel rail, and injectors (fuel injected engines).

Engine Oil Leak Diagnosis

Engine oil leaks may cause an engine to run out of oil, resulting in serious engine damage. Therefore, oil leaks should not be ignored. If an engine has an oil leak, the oil level should be checked often, and oil should be added as required until the necessary repairs are completed. When it is difficult to locate the exact cause of an oil leak, the engine should be cleaned with a steam cleaner or high-pressure washer. Then the engine may be operated while the technician watches for the source of the oil leak. Oil leaks may be difficult to locate because the oil runs down the engine surfaces. For example, the oil from a leak at a rocker arm cover may run down the side of the engine and appear as an oil pan leak. However, closer examination of the engine will indicate that the oil leak is coming from the rocker arm cover. The possible causes of oil leaks are:

1. Rear main bearing seal
2. Expansion plug in rear camshaft bearing
3. Rear oil gallery plug
4. Oil pan
5. Oil filter
6. Rocker arm covers
7. Intake manifold front and rear gaskets, (V-type engines)
8. Mechanical fuel pump gasket or worn fuel pump pivot pin, (carbureted engines)
9. Timing gear cover or seal
10. Front main bearing, (engines with cogged timing belt)
11. Oil pressure sending unit
12. Distributor O-ring or gasket
13. Engine casting porosity
14. Oil cooler or lines (where applicable)

Engine Coolant Leak Diagnosis

When a coolant leak causes low coolant level, the engine overheats in a very short time, and severe engine damage may occur. Therefore, coolant leaks must not be ignored. Coolant leaks may occur at these locations:

1. Upper radiator hose
2. Lower radiator hose
3. Heater hoses
4. By-pass hose
5. Water pump
6. Engine expansion plugs or block heater
7. Radiator
8. Thermostat housing
9. Heater core
10. Internal engine leaks caused by leaking head gasket or cracked engine components

CAUTION: Never loosen a radiator cap on a warm or hot engine. When the cap is loosened, the pressure on the coolant is released, and the coolant suddenly boils. This action may cause a technician or anyone standing nearby to be seriously burned.

Special Tools

Cooling system pressure tester

A pressure tester may be used to determine if a coolant leak is present and to locate the source of the leak. Remove the radiator cap and install the pressure tester on the the radiator filler neck (Figure 3-1). Operate the pump on the tester until the rated pressure on the radiator cap appears on the tester gauge. Leave this pressure applied to the cooling system for 15 minutes, and then check the gauge pressure. When there is no drop in gauge pressure during this time, the cooling system does not have a leak. If the pressure indicated on the tester gauge drops during the 15 minutes, the cooling system has a leak. Inspect the entire cooling system, and repair any leaks as required. Check the floor of the vehicle for any indication of coolant leaking out of the heater core and dripping onto the floor. Heater core leaks usually cause an odor inside the passenger compartment and fogging of the windshield.

The radiator pressure cap may be connected to the cooling system pressure tester with a special adapter. When the tester pump is operated, the gauge on the tester indicates the pressure required to open the cap pressure relief valve. This pressure should be the same as the pressure rating stamped on the cap.

If there are no external leaks after the cooling system has been pressure tested, the spark plugs may be removed to check for coolant in the cylinders. After the spark plugs are removed, disable the injection and ignition systems. Have an assistant crank the engine while checking for any sign of coolant discharging from the spark plug openings. When coolant is discharged from a

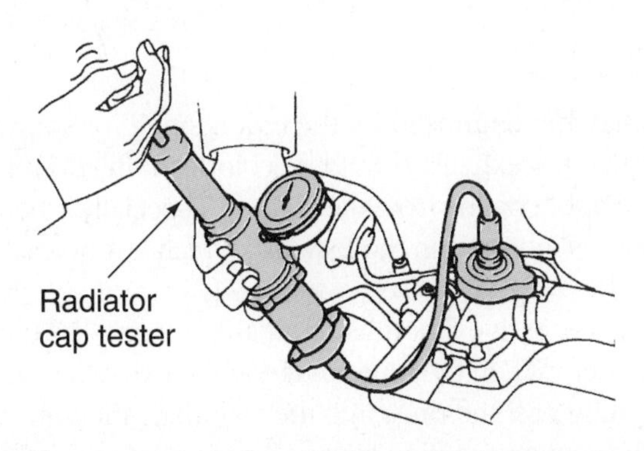

Figure 3-1 Cooling system pressure tester (Courtesy of Toyota Corporation)

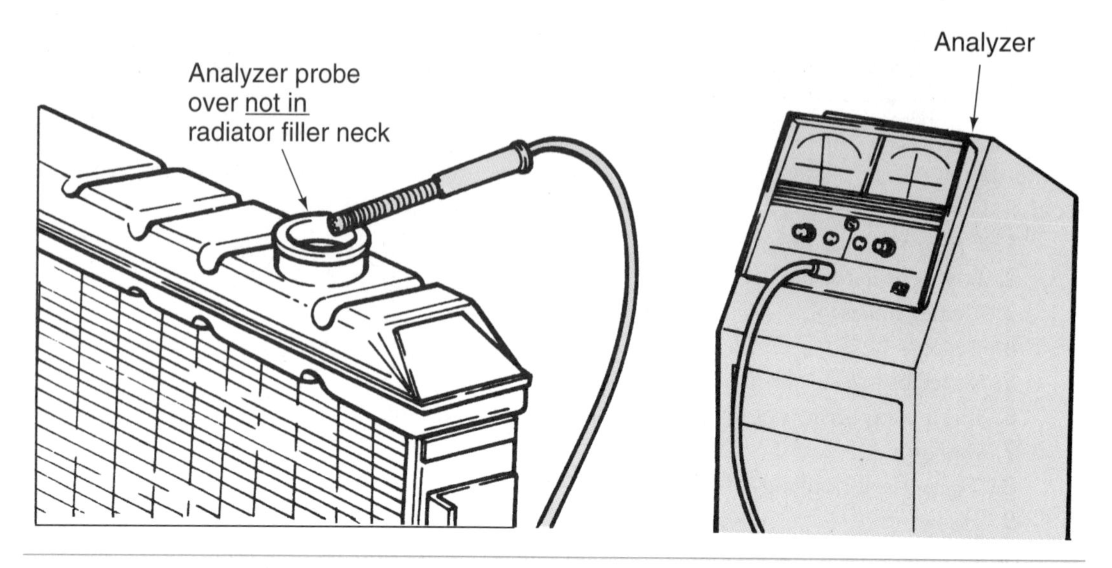

Figure 3-2 Testing combustion chamber coolant leaks with an infrared emission analyzer (Courtesy of Chrysler Corporation)

Special Tools

Infrared analyzer

spark plug opening, the head gasket, cylinder head, or cylinder wall is leaking or cracked. Coolant leaks at these locations inside the combustion chamber usually cause engine overheating.

When there are coolant leaks in the combustion chamber, bubbles appear in the coolant at the radiator filler neck with the engine running. A two-gas or four-gas infrared emissions analyzer probe may be positioned over the top of the radiator filler neck with the engine running (Figure 3-2). Do not immerse the probe in the coolant. Operate the engine at 2,000 rpm, and observe the hydrocarbon meter on the analyzer. If the combustion chamber gases are leaking into the cooling system, the HC meter provides an HC reading.

If the coolant level continues to go down with no indication of external leaks, the coolant may be leaking through a cracked block into the oil pan. If this type of leak is suspected, drain the engine oil and check for signs of coolant in the oil. The oil may be sent to a laboratory for chemical analysis of coolant content in the oil.

Engine Exhaust Diagnosis

Exhaust Color

CAUTION: Vehicle exhaust contains carbon monoxide (CO), which is a poisonous gas. Breathing CO results in health problems, and a concentrated mixture of CO and air is fatal! Do not operate a vehicle in the shop unless a shop exhaust hose is connected to the tailpipe.

Some engine problems may be diagnosed by the color, smell, or sound of the exhaust. If the engine is operating normally, the exhaust should be colorless. In severely cold weather, it is normal to see a swirl of white vapor coming from the tailpipe, especially when the engine and exhaust system are cold. This vapor is moisture in the exhaust, which is a normal by-product of the combustion process.

If the exhaust is blue, excessive amounts of oil are entering the combustion chamber, and this oil is burned with the fuel. When the blue smoke in the exhaust is more noticeable on deceleration, the oil is likely getting past the rings into the cylinder. Vacuum in the cylinder is high on deceleration, and this high vacuum is pulling oil past the rings into the combustion chamber. If the blue smoke appears in the exhaust immediately after a hot engine is restarted, the oil is likely leaking down the valve guides. In this case, the seals, guides, or valve stems may be worn.

If black smoke appears in the exhaust, the air-fuel mixture is too rich. This condition may be caused by a defect in the carburetor or fuel injection system. When a rich air-fuel mixture occurs with a carburetor system, check the choke, float, float level, needle valve and seat, and power valve. In a fuel injection system, a defective pressure regulator, injectors, or input sensors may cause a rich air-fuel mixture. A restriction in the air intake such as a plugged air filter may be responsible for a rich air-fuel mixture.

Gray smoke in the exhaust may be caused by a coolant leak into the combustion chambers. If a coolant leak is the cause of gray smoke in the exhaust, this smoke is usually most noticeable when the engine is first started after it has been shut off for over half an hour.

Exhaust Odor

On catalytic converter-equipped vehicles, a strong sulphur smell in the exhaust indicates a rich air-fuel mixture. Some sulphur smell on these engines is normal, especially during engine warm-up.

When a vehicle is not equipped with a catalytic converter, excessive odor in the exhaust usually indicates a rich air-fuel mixture. You may also notice this odor causes your eyes to water.

Exhaust Noise

CAUTION: Exhaust components may be extremely hot. If you must touch them, wear recommended protective gloves to avoid burns.

When the engine is idling, the exhaust at the tailpipe should have a smooth, even, sound. If the exhaust in the tailpipe has a "puff" sound at regular intervals, a cylinder may be misfiring. When this sound is present, check the engine compression, ignition, and fuel systems.

If the vehicle has excessive exhaust noise, check the exhaust system for leaks. The noise from exhaust leaks is most noticeable when the engine is accelerated. A small exhaust leak may cause a whistling noise when the engine is accelerated.

If the exhaust system produces a rattling noise when the engine is accelerated, check the muffler and catalytic converter for loose internal components. When this problem is suspected, rap on the muffler and converter with the engine shut off. If a rattling noise is heard, replace the appropriate component. This rattling noise in the exhaust system may also be caused by loose exhaust system hangers or an exhaust system component hitting the chassis.

When the engine has a wheezing noise at idle or with the engine running at higher rpm, check for a restricted exhaust system. Connect a vacuum gauge to the intake manifold and observe the vacuum with the engine idling. Accelerate the engine to 2,500 rpm and hold this speed for two minutes. If the exhaust is restricted, the vacuum will slowly drop below the recorded vacuum at idle. When the car is driven under normal loads, this vacuum drop is even more noticeable.

Vacuum Tests

Special Tools

Vacuum gauge

The vacuum gauge should be connected directly to the intake manifold to diagnose engine and related system conditions (Figure 3-3). When a vacuum gauge is connected to the intake manifold, the reading on the gauge should provide a steady reading between 17 and 22 in Hg with the engine idling.

Abnormal vacuum gauge readings indicate the following problems:

1. A low steady reading indicates late ignition timing (Figure 3-4).
2. If the vacuum gauge reading is steady and much lower than normal, the intake manifold has a significant leak.
3. When the vacuum gauge pointer fluctuates between approximately 11 and 16 in Hg on a carbureted engine at idle speed, the carburetor idle mixture screws require adjusting. On a fuel injected engine, the injectors require cleaning or replacing.
4. Burned or leaking valves cause a vacuum gauge fluctuation between 12 and 18 in Hg.

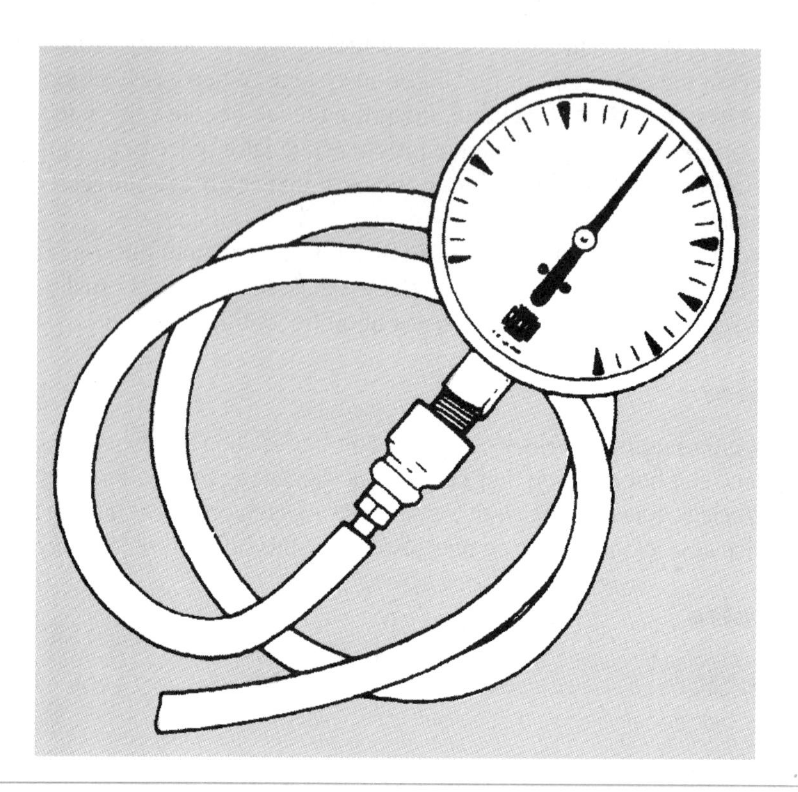

Figure 3-3 Vacuum gauge connected to the intake manifold (Reprinted with permission of Ford Motor Company)

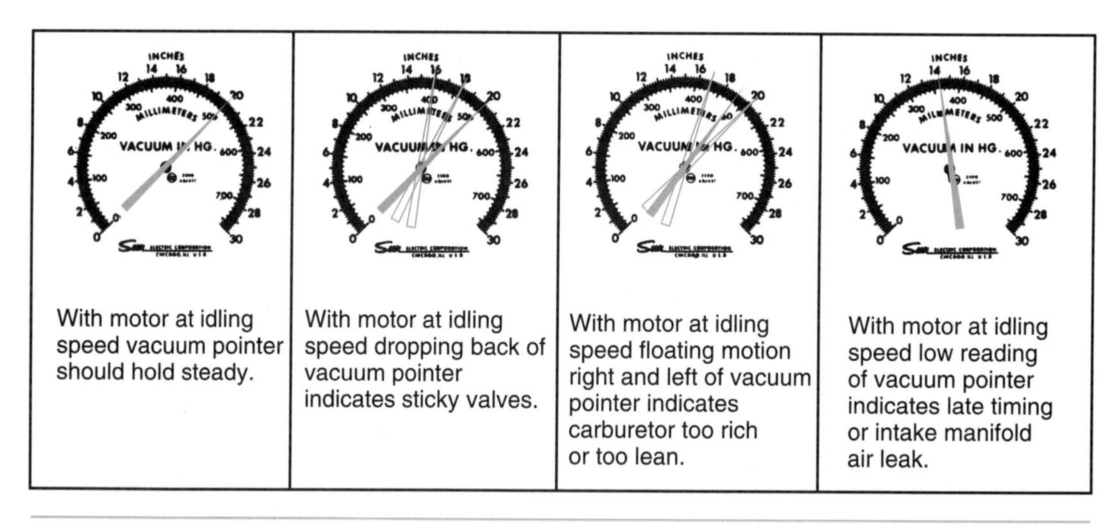

Figure 3-4 Problems indicated by abnormal vacuum gauge readings (Courtesy of Sun Electric Corporation)

5. Weak valve springs result in a vacuum gauge fluctuation between 10 and 25 in Hg.
6. A leaking head gasket may cause a vacuum gauge fluctuation between 7 and 20 in Hg.
7. If the valves are sticking, the vacuum gauge fluctuates between 14 and 18 in Hg.
8. When the engine is accelerated and held at a steady higher rpm, if the vacuum gauge pointer drops to a very low reading, the catalytic converter or other exhaust system components are restricted.

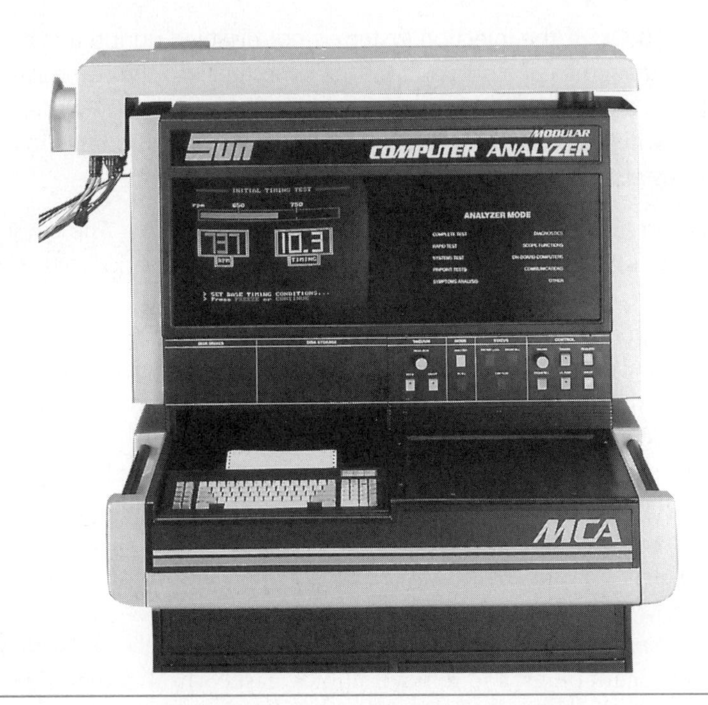

Figure 3-5 Engine analyzer (Courtesy of Sun Electric Corporation)

Engine Power Balance Test

Special Tools

Engine power balance tester

The engine power balance test is performed with an engine analyzer and with the engine at normal operating temperature (Figure 3-5). Most engine analyzers have the capability to prevent each cylinder from firing for a brief interval by stopping the ignition system from opening the primary circuit as each cylinder fires. As each cylinder misfires, the analyzer measures the amount of rpm decrease. Always follow the test equipment manufacturer's instructions when performing a cylinder balance test. These instructions may include disconnecting certain components such as an exhaust gas recirculation (EGR) valve during the test. The recommended test procedure varies depending on the model year and the emission equipment on the vehicle.

If the cylinder is working normally, a significant rpm decrease occurs when the cylinder misfires. The amount of rpm decrease depends on the type of engine. For example, if a cylinder misfires on a four-cylinder engine, the rpm decrease is higher compared to the rpm decrease when one cylinder misfires in a V-8 engine. If there is very little rpm decrease when the analyzer causes a cylinder to misfire, the cylinder is not contributing to engine power. Under this condition, the engine compression, ignition system, and fuel system should be checked to locate the cause of the problem. When all the cylinders provide the specified rpm drop, the cylinders are all contributing equally to the engine power.

Compression Test

CAUTION: If the injection system is not disabled during a compression test on a gasoline fuel injected engine, the injectors continue injecting fuel into the intake ports during the compression test. These fuel vapors are discharged from the spark plug openings and, if ignited, they may cause a serious explosion, resulting in personal injury and/or property damage.

WARNING: Since a diesel engine has much higher compression than a gasoline engine, a special compression tester is required for a diesel engine.

WARNING: If the injection system is not disabled during a compression test on a diesel engine, the injectors continue injecting fuel into the combustion chambers during the test. On a diesel engine, the compression is high enough to self-ignite the fuel in the cylinder in which the compression is tested. This self-ignition destroys the hose on the compression tester, and possibly the tester.

The compression test checks the sealing qualities of the rings, valves, and combustion chamber. Operate the engine until it reaches normal operating temperature prior to the compression test. Follow this procedure to complete the compression test:

1. Disable the ignition system by disconnecting the positive primary wire from the ignition coil. Insulate this wire with electrician's tape so it does not contact the vehicle ground. On distributorless ignition systems, disconnect the complete primary coil connector on the coil pack.
2. If the engine is fuel injected, disable the injection system by shutting off the fuel pump. On many vehicles, disconnect the wires from the fuel pump relay to shut off the fuel pump. On General Motors products, disconnect the fuel pump wire at the fuel tank. If the vehicle has a diesel engine, disconnect the fuel shut-off solenoid wire at the injection pump.
3. Loosen the spark plugs and blow any dirt from the plug recesses with an air blowgun. Remove the spark plugs, and remove all plug gaskets from the plug recesses in the cylinder heads.
4. Place a screwdriver in the throttle linkage to hold the throttle open. If the engine has a carburetor, be sure the choke is open.

Special Tools

Starter remote control switch

Special Tools

Compression gauge

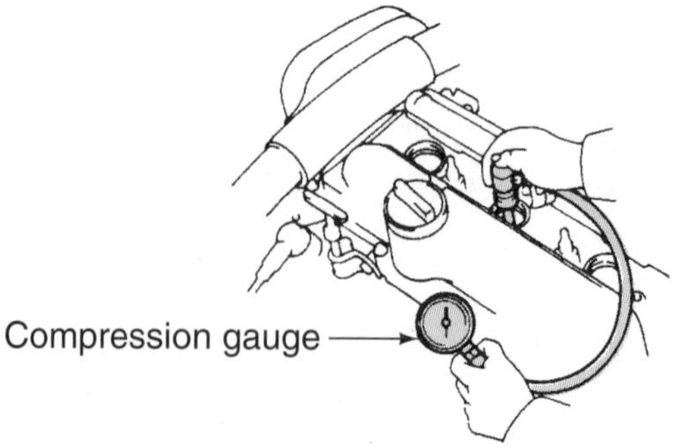

5. Install the compression tester in a spark plug hole according to the tester manufacturer's recommended procedure (Figure 3-6). Some compression testers are threaded into the spark plug openings, whereas other testers lock into position when the compression pressure is applied to them. Other testers must be held in the spark plug openings by hand.
6. Crank each cylinder through four compression strokes. A remote control switch may be connected from the starting motor main terminal to the solenoid terminal to crank the engine. The use of this switch enables the technician to perform the compression test without assistance. Since all the spark plugs are removed, the cylinder compression applied against the gauge slows the engine down to some extent, and this action can be heard. Each of the four compression strokes can be observed as a definite increase on the gauge pointer reading.
7. Release the pressure from the compression tester, and follow the same procedure to obtain the compression reading on each cylinder. Record the reading obtained on each cylinder.

Figure 3-6 Cylinder compression test (Courtesy of Toyota Motor Corporation)

Compression Test Interpretation

Compare the compression readings to the vehicle manufacturer's compression specifications in the service manual. If the compression readings on all the cylinders are equal to the specified compression, the readings are satisfactory.

If the compression on one or more cylinders is lower than the specified compression, the valves or rings are likely worn. When the compression readings on a cylinder are low on the first stroke and increase to some extent on the next three strokes, but remain below the specifications, the rings are probably worn. If the reading is low on the first compression stroke, and there is very little increase on the following three strokes, the valves are probably leaking. When all the compression readings are even, but considerably lower than the specifications, the rings and cylinders are probably worn.

When the compression readings on two adjacent cylinders are lower than specified, the cylinder head gasket is probably leaking between the cylinders. Higher than specified compression readings usually indicate excessive carbon in the combustion chamber, such as on top of the piston.

If the compression reading on a cylinder is zero, there may be a hole in the piston, or an exhaust valve may be severely burned. When the zero compression reading is caused by a hole in a piston, there is excessive blow-by from the crankcase. This blow-by is visible at the positive crankcase ventilation (PCV) valve opening in the rocker cover, if this valve is removed.

Wet Compression Test

SERVICE TIP: If you are attempting to squirt engine oil through a spark plug opening into the cylinder to perform a wet compression test, and this opening is difficult to access, place a length of vacuum hose on the end of the squirt can nozzle. Install the other end of this hose in the spark plug opening.

If a cylinder compression reading is below specifications, a wet test may be performed to determine if the valves or rings are the cause of the problem. Squirt about two or three teaspoons of engine oil through the spark plug opening into the cylinder with the low compression reading. Crank the engine for several revolutions to distribute the oil around the cylinder wall. Repeat the compression test on the cylinder with the oil added in the cylinder. If the compression reading improves considerably, the rings, or cylinders, are worn. When there is very little change in the compression reading, one of the valves is leaking.

Cylinder Leakage Test

The cylinder leakage test may be used in place of the compression test, or the leakage test may be used to determine if a low compression reading is caused by worn rings, burned valves, or other combustion chamber leaks. During this test, a regulated amount of air from the shop air supply is forced into the cylinder with both exhaust and intake valves closed. The gauge on the leakage tester indicates the percentage of leakage in the cylinder. A gauge reading of 0% indicates no cylinder leakage; if the reading is 100%, the cylinder is not holding any air.

Special Tools

Cylinder leakage tester

The spark plug must be removed from the cylinder to be leak tested. Do not remove the other spark plugs, or the air forced into the cylinder will rotate the engine until one of the valves in the cylinder opens, which causes a leak and ruins the leakage tester reading. The engine must be at normal operating temperature prior to the cylinder leakage test. There are many differences in engines, ignition systems, and fuel systems at present. A high percentage of engines produced in recent years are equipped with distributorless ignition. Other engines have an electronic ignition system with computer-controlled spark advance, and older engines have electronic ignition with mechanical distributor advances. These variations in engines and systems complicate the diagnostic procedures, and in some cases, many different procedures are recommended because of these system differences. The following cylinder leakage test procedure may be used on most engines regardless of the type of ignition system or fuel system:

1. With the pressure regulator valve in the tester in the off position, connect the shop air supply to the tester. Adjust the pressure regulator valve to obtain a zero gauge reading, if necessary.
2. Using the same methods described previously in the compression test, disable the ignition system and the fuel injection system. If the engine is equipped with a carburetor, there is no need to disable the fuel system.
3. Loosen the spark plug in the cylinder to be tested, and use an air gun to blow any dirt from the plug recess. Remove the spark plug from the cylinder to be tested.
4. Place your thumb over the spark plug hole in the cylinder to be tested. If the spark plug hole is not accessible, install a compression gauge in the spark plug hole of the cylinder to be tested.
5. Connect a remote control switch to the starter solenoid, and crank the engine a very small amount at a time. When compression is available at the spark plug hole, stop cranking the engine.
6. Connect the spark plug removed from the cylinder to the end of the disconnected spark plug wire. Attach a jumper wire from the spark plug case to ground on the engine.

Special Tools

Timing light

7. Reconnect the coil primary wires that were disconnected in step 2. Connect a timing pickup light to the spark plug wire on the cylinder being tested, and attach the positive and negative timing light leads to the battery terminals with the correct polarity.

CAUTION: When a remote control switch is used to crank the engine slowly with the ignition system and the fuel system in operation, the engine may start. Therefore, keep tools and hands away from the cooling fan and belts to avoid personal injury.

CAUTION: Never attempt to rotate the engine by installing a wrench on the crankshaft pulley bolt with the ignition system and fuel system in operation. Under this condition, the engine may start and rotate the wrench with the crankshaft. This action could result in serious personal injury and expensive vehicle damage.

8. Crank the engine a very small amount at a time with the remote control switch. When the timing light fires, stop cranking the engine. The piston is now at, or near, TDC on the compression stroke with both valves closed.
9. Connect the leakage tester air hose to the cylinder (Figure 3-7). Various adapters are available to fit different spark plug threads.
10. Rotate the pressure regulator knob on the tester until the gauge pointer reaches its highest reading and stabilizes.
11. If the reading exceeds 20%, check for air escaping from the tailpipe, positive crankcase valve (PCV) opening, and the top of the throttle body or carburetor. Air escaping from the tailpipe indicates an exhaust valve leak. When the air is coming out of the PCV valve opening, the piston rings are leaking. An intake valve is leaking if air is escaping from the top of the throttle body or carburetor.
12. Follow the same procedure in steps 2 through 11 to perform a leakage test on other cylinders.

The term *blow-by* refers to excessive combustion chamber gases escaping past the piston rings into the crankcase.

Engine Noise Diagnosis

Long before a serious engine failure occurs, there are usually warning noises from the engine. If the customer hears any abnormal noises, he or she should have these noises diagnosed by a technician. When engine problems are detected early, the engine repairs may be much less expensive. Engine defects such as damaged pistons, worn rings, loose piston pins, worn crankshaft bearings, worn camshaft lobes, and loose and worn valve train components usually produce their own peculiar noises. Certain engine defects also cause a noise under specific engine operating conditions.

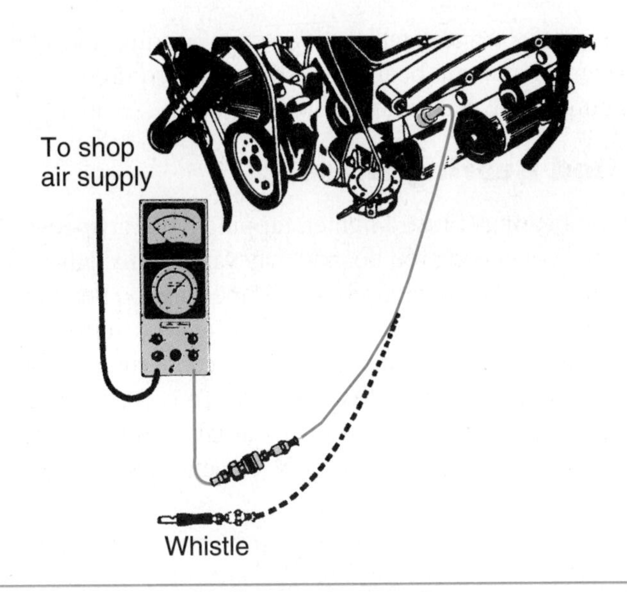

Figure 3-7 Cylinder leakage tester connected to spark plug hole (Courtesy of Sun Electric Corporation)

The experienced technician must be able to diagnose engine noises accurately. This accurate diagnosis of engine noises avoids unnecessary engine repairs. An engine noise diagnosis should be performed before doing engine repair work.

Stethoscope

CAUTION: When placing a stethoscope probe in various locations on a running engine, be careful not to catch the probe or your hands in moving components such as cooling fan blades and belts.

Since it is sometimes difficult to determine the exact location of an engine noise, a stethoscope may be useful in diagnosing noise locations. The stethoscope probe is placed on, or near, the suspected component, and the ends of the stethoscope are installed in your ears. The stethoscope amplifies sound to assist in noise location. When the stethoscope probe is moved closer to the source of the noise, the sound is louder in your ears. If a stethoscope is not available, a length of hose placed from the suspected component to your ear amplifies sound and helps to locate the cause of the noise.

Special Tools
Stethoscope

Engine Noise Diagnostic Procedure

Since lack of lubrication is a common cause of engine noise, always check the engine oil level and condition prior to noise diagnosis. Carefully observe the oil for contamination by coolant or gasoline. During the diagnosis of engine noises, always operate the engine under the conditions when the noise occurs. Remember that aluminum engine components such as pistons expand more when heated than cast iron alloy components. Therefore, a noise caused by a piston defect may occur when the engine is cold but disappear when the engine reaches normal operating temperature.

Main Bearing Noise

Loose crankshaft main bearings cause a heavy thumping knock for a brief time when the engine is first started after it has been shut off for several hours. This noise may also be noticeable on hard acceleration depending on the amount of bearing wear. Worn main bearings or crankshaft journals, or lack of lubrication cause this problem. When the thrust surfaces of a crankshaft bearing are worn, the crankshaft has excessive end play. Under this condition, a heavy thumping knock may occur at irregular intervals on acceleration. The main bearings must be replaced to correct this problem, and the crankshaft main bearing journals may require machining. If the journals are severely scored or burned, the crankshaft must be replaced.

Connecting Rod Bearing Noise

Loose connecting rod bearings cause a lighter, rapping noise at speeds above 35 mph (21 kph). The noise from a loose connecting rod bearing may vary from a light to a heavier rapping sound depending on the amount of bearing looseness. Connecting rod bearing noise may occur at idle if the bearing is very loose. When the spark plug is shorted out in the cylinder with the loose connecting rod bearing, the noise usually decreases. Worn connecting rod bearings or crankshaft journals, or lack of lubrication may cause this problem. To correct this noise, the connecting rod bearings require replacement, and the crankshaft journals should be machined. Severely worn or burned crankshaft journals require crankshaft replacement. The connecting rod alignment should also be checked.

Piston Slap

A piston slap causes a hollow, rapping noise that is most noticeable on acceleration with the engine cold. Depending on the piston condition, the noise may disappear when the engine reaches normal operating temperature. A collapsed piston skirt, cracked piston, excessive piston-to-cylinder wall clearance, lack of lubrication, or a misaligned connecting rod cause piston slap. Correction of this problem requires piston or connecting rod replacement and possible reboring of the cylinder.

Piston Pin Noise

A loose, worn piston pin causes a sharp, metallic, rapping noise that occurs with the engine idling. This noise may be caused by a worn piston pin, worn pin bores, cracked piston in the pin bore area, a worn rod bushing, or lack of lubrication. An oversized pin may be installed to correct the problem. Severely worn or cracked pistons must be replaced.

Piston Ring Noise

If the piston rings are excessively loose in the piston ring grooves, a high-pitched, clicking noise is noticeable in the upper cylinder area during acceleration. To correct this noise, the rings, and possibly the pistons, must be replaced. If the cylinders are severely worn, cylinder reboring is necessary.

Ring Ridge Noise

When the top piston rings are striking the ring ridge on the cylinder wall, a high-pitched clicking noise is heard. This noise intensifies when the engine is accelerated. Ring ridge noise may be caused by the installation of new rings without removing the ring ridge at the top of the cylinder. This noise may also be caused by a loose connecting rod bearing or piston pin, which allows the piston to move above its normal travel in the cylinder. To correct this problem, remove the ring ridge, and check the connecting rod bearing and piston pin.

Valve Train Noise

Valve train noise such as valve lifters or rocker arms appears as a light clicking noise with the engine idling. Valve train noise is slower than piston or connecting rod noise, because the camshaft is turning at one-half the crankshaft speed. This noise is less noticeable when the engine is accelerated. Sticking or worn valve lifters may provide an irregular clicking noise that is more likely to occur when the engine is first started. Valve train noise may be caused by lack of lubrication or contaminated oil. To correct valve train noise, check the valve adjustment where applicable, and check the valve lifters, pushrods, and rocker arms.

Camshaft Noise

If one lobe on a camshaft is worn, a heavy clicking noise is heard with the engine running at 2,000 to 3,000 rpm. When several camshaft lobes and lifter bottoms are scored, a continuous, heavy clicking noise is evident at idle speed. To correct this problem, the camshaft must be machined or replaced, and the valve lifter condition should be checked.

A worn, loose timing chain, sprockets, or chain tensioners causes a whirring and light rattling noise when the engine is accelerated and decelerated. Severely worn timing chains, sprockets, and tensioners may cause these noises at idle speed. The timing chain, sprockets, and tensioners must be checked and replaced as necessary to correct this noise.

Combustion Noises

The most common abnormal combustion noise is caused by detonation in the cylinders. Detonation occurs in a cylinder when the air-fuel mixture suddenly explodes rather than burning smoothly. This action drives the piston against the cylinder wall and forces the connecting rod insert suddenly against the crankshaft journal. Under this condition, a noise occurs that is similar to marbles rattling inside a metal can. Thus, the shop term *pinging* is applied to this noise. The pinging noise usually occurs when the engine is accelerated. Detonation may be caused by excessive ignition spark advance, low octane fuel, higher than normal engine temperature, and hot carbon spots on top of the piston. A lean air-fuel mixture or an inoperative exhaust gas recirculation (EGR) valve contributes to detonation. Detonation at a steady cruising speed may be caused by a defective EGR valve or related controls. To correct the detonation problem, check the ignition timing and spark advance, air-fuel mixture, engine compression, engine temperature, and EGR valve.

Detonation noise in the engine cylinders may be called pinging.

Flywheel and Vibration Damper Noise

A loose flywheel causes a thumping noise at the back of the engine. This noise varies depending on whether the engine is equipped with a fly wheel or a flexplate. The noise also varies depending on the looseness of the flywheel. To correct this problem requires transmission removal and tightening of the flywheel-to-crankshaft bolts. If the bolt holes in the flywheel are damaged, flywheel replacement is necessary.

A loose vibration damper on the front of the crankshaft causes a rumbling or thumping noise at the front of the engine, and this noise may be accompanied by engine vibrations. When the engine is accelerated under load, the noise is more noticeable. The vibration damper must be replaced to correct this problem.

Classroom Manual
Chapter 3, page 47

Engine Oil Pressure Testing

If the low oil pressure warning light or pressure gauge indicates low oil pressure, the engine oil level should be checked immediately. The oil should also be checked for contamination with gasoline or coolant. If the oil is contaminated with gasoline, the oil is thinner, and oil pressure is reduced. When the oil level and quality are satisfactory, but the oil pressure warning light or gauge indicates low pressure, the oil pressure should be checked. If the engine runs for a few minutes without proper lubrication, the crankshaft bearings and cylinder walls may be severely scored. Therefore, the vehicle should not be driven if the oil pressure warning light or gauge indicates low oil pressure.

A pressure gauge must be connected to the main oil gallery to check engine oil pressure. The oil pressure sending unit may be removed from the main oil gallery to install the pressure gauge (Figure 3-8).

Special Tools
Oil pressure gauge

The engine should be at normal operating temperature before checking the oil pressure. Engine oil pressure should be checked with the engine idling and with the engine operating at 2,000 rpm. If the oil pressure is equal to the vehicle manufacturer's specifications, but the oil pressure

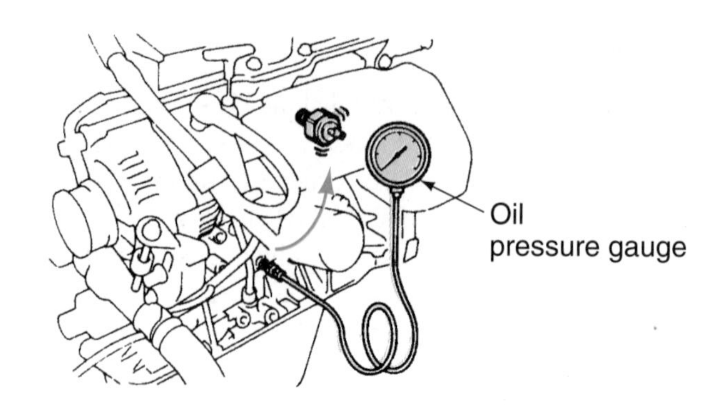

Figure 3-8 Oil pressure test gauge connected to oil pressure sending unit port in the main oil gallery (Courtesy of Toyota Motor Corporation)

warning light or gauge indicated low oil pressure, the sending unit is probably defective. When the oil pressure is lower than specified, one of these problems may exist:

1. Worn oil pump
2. Plugged oil pump pickup screen
3. Leaking oil pump pickup tube
4. Sticking pressure regulator valve in the pump
5. Broken oil pump drive
6. Loose main and connecting rod bearings
7. Worn camshaft bearings
8. Worn valve lifter bores

On many engines,when the oil pressure is lower than specified, the oil pan must be removed to check the cause of the low pressure. The oil pump and pressure regulator valve are accessible without removing the oil pan on some engines.

Photo Sequence 1 shows a typical procedure for testing engine oil pressure.

Diagnosis of Oil Consumption

With advances in piston ring technology, modern engines consume very little oil. These engines may be driven for 1,000 to 3,000 miles (1,600 to 4,800 kilometers) without adding oil to the engine. If an engine uses excessive oil, it may be leaking from the engine or burning in the combustion chambers. A plugged positive crankcase ventilation (PCV) system causes excessive pressure in the engine, and this pressure may force oil from some of the engine gaskets. If excessive amounts of oil are burned in the combustion chambers, the exhaust contains blue smoke, and the spark plugs may be fouled with oil. Excessive oil burning in the combustion chambers may be caused by worn rings and cylinders or worn valve guides and valve seals.

Classroom Manual
Chapter 3, page 49

CUSTOMER CARE: Always inform the customer if you notice problems, such as engine noises, with his or her car. Most engine failures are evidenced by unusual noises for a considerable length of time prior to the actual failure. If you inform the customer about a noise that may indicate a serious engine problem, the customer will likely consent to further accurate diagnosis of the problem. When the diagnosis is complete, the customer will likely approve the necessary repairs, and these repairs will likely be less expensive than the repairs after a complete engine failure. The alternative procedure is to let the customer drive the car until the engine noise develops into a serious engine failure. This problem may occur when the customer is miles from home. The customer will likely appreciate your diagnosis of the impending engine failure before it develops into a major problem at an inconvenient time.

Photo Sequence 1
Typical Procedure for Testing Engine Oil Pressure

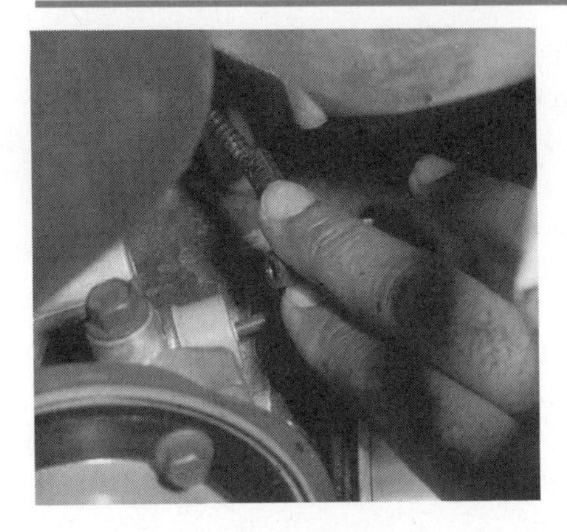

P1-1 Disconnect the wire from the oil sending unit, and remove the oil sending unit.

P1-2 Thread the fitting on the oil pressure gauge hose into the oil sending unit opening, and tighten the fitting.

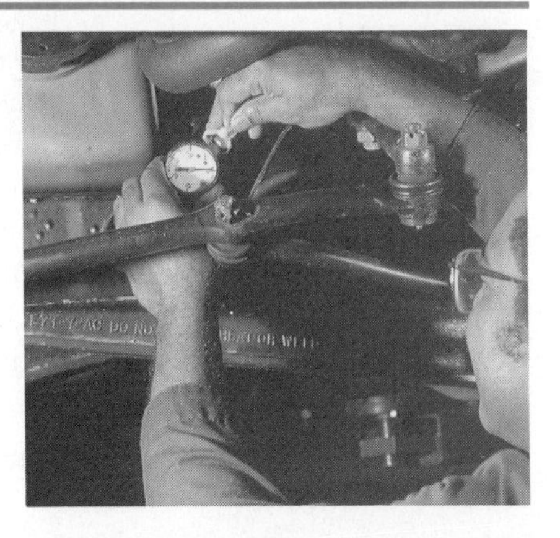

P1-3 Place the oil pressure gauge where it will not contact rotating or hot components.

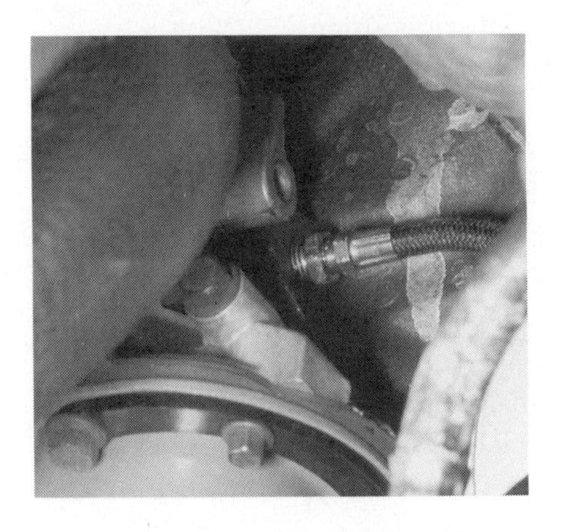

P1-4 Start the engine and check for leaks at the oil pressure gauge fittings.

P1-5 Operate the engine until it reaches normal operating temperature, and observe the oil pressure on the gauge at idle speed and 2,000 rpm.

P1-6 Compare the oil pressure gauge reading to the vehicle manufacturer's specifications.

P1-7 Remove the oil pressure gauge from the oil sending unit opening.

P1-8 Install the oil sending unit and tighten it to the specified torque, and install the sending unit wire.

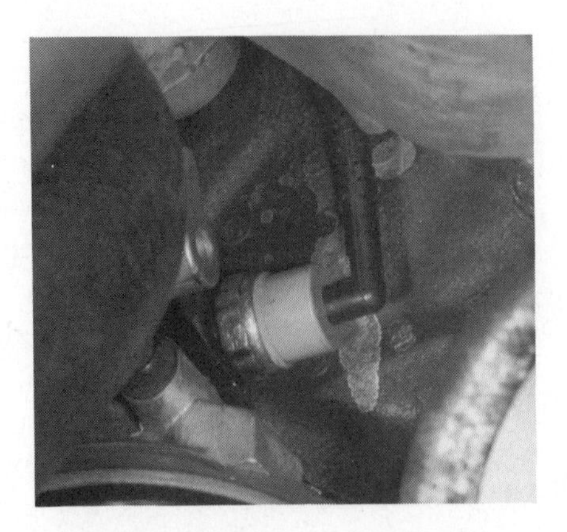

P1-9 Start the engine and check for oil leaks at the oil sending unit.

Guidelines for Diagnosing General Engine Condition

1. A general diagnostic procedure includes these steps: identify the complaint, think of the possible causes, test to locate the problem, repair the problem, and be sure the complaint is eliminated.
2. Fuel leaks in the engine compartment may be from these sources: mechanical fuel pump, carburetor, fuel rail pressure regulator, injectors, or fuel lines and filter.
3. Fuel leaks under the vehicle may come from the fuel tank, tank filler cap, fuel lines, vapor recovery system lines, or filter.
4. Engine oil leaks at the rear of the engine may be from the rear main bearing seal, expansion plug in the rear camshaft bearing, rear oil gallery plug, pan gasket, or intake manifold rear gasket on V-type engines.
5. Oil leaks at the sides of the engine may be from the rocker arm cover gaskets, oil pan, mechanical fuel pump, or oil pressure sending unit.
6. Oil leaks at the front of the engine may be from the timing gear cover and seal, front main bearing, or intake manifold front gasket.
7. An oil leak around the distributor may be caused by the distributor gasket or O-ring.
8. Engine coolant leaks at the front of the engine may be from the water pump, radiator hoses, heater hoses, by-pass hose, radiator, or thermostat housing.
9. Coolant leaks at the sides of the engine may be from the expansion plugs, block heater, or heater hoses.
10. Coolant leaks at the rear of the engine may be caused by the heater hoses.
11. If coolant leaks on the firewall, check the heater hoses and hose connections.
12. Coolant leaking on the front floor of the vehicle indicates a leaking heater core. This problem causes windshield fogging and odor inside the vehicle.
13. A pressure tester may be used to check cooling system leaks and the radiator pressure cap.
14. The hydrocarbon (HC) meter on an exhaust gas analyzer may be placed above the radiator filler neck with the engine running to check for a combustion chamber leak.
15. Blue exhaust indicates excessive amounts of engine oil entering the combustion chambers.
16. Black exhaust is caused by a rich air-fuel mixture.
17. If gray smoke is present in the exhaust immediately after the engine is started, coolant is probably entering the combustion chambers.
18. A strong sulphur smell from the exhaust of a catalytic converter-equipped vehicle is caused by a rich air-fuel ratio.
19. With the engine idling, an uneven puffing sound in the exhaust indicates a cylinder misfire.
20. Loose internal components in the muffler or catalytic converter cause a rattling noise when the engine is accelerated. This same noise may be caused by loose exhaust system hangers.
21. A vacuum gauge may be connected to the intake manifold to test for a restricted exhaust system.
22. A vacuum gauge connected to the intake manifold may be used to test for late timing, intake manifold leaks, carburetor idle mixture adjustment, burned or sticking valves, and weak valve springs.
23. A cylinder balance tester checks cylinder output by calculating the rpm drop when each cylinder is shorted out for a brief interval.
24. A cylinder compression test checks the sealing qualities of the piston rings, valves, and combustion chamber.

25. The wet compression test may be performed to determine if a low compression reading is caused by ring or valve defects.
26. A cylinder leakage test supplies a regulated amount of air into the combustion chamber to check the sealing qualities of the rings, valves, and combustion chamber.
27. A stethoscope may be used to locate the source of engine noises.
28. Loose main bearings cause a thumping noise when the engine is first started, and possibly during engine acceleration.
29. Loose connecting rod bearings result in a lighter rapping noise at vehicle speeds above 35 mph (21 kph).
30. Piston slap causes a hollow rapping noise that is most noticeable during acceleration on a cold engine.
31. Worn or loose piston pins result in a sharp, metallic rapping noise when the engine is idling.
32. If the rings are loose in the piston ring grooves, a high-pitched clicking noise occurs during acceleration, and possibly at idle. When the top ring is striking the ring ridge at the top of the cylinder, a high-pitched clicking noise is very evident on acceleration.
33. Valve train noise, such as worn lifters and rocker arms, produces a light clicking noise at idle and low speeds.
34. When one lobe on a camshaft is worn, a clicking noise is heard with the engine running at 2,000 to 3,000 rpm. Worn, scored camshaft lobes and valve lifter bottoms cause a continuous clicking noise with the engine idling.
35. Cylinder detonation, or pinging, causes a noise similar to marbles rattling in a metal can, and this noise usually occurs during acceleration. Pinging may also occur at normal cruising speed.
36. A loose flywheel causes a thumping noise at the rear of the engine at idle and lower speeds.
37. A loose vibration damper causes a rumbling, thumping noise at the front of the engine, and this noise is most noticeable when the engine is accelerated under load. If the vibration damper is loose or worn, engine vibrations may occur.
38. Low engine oil pressure may be caused by a worn oil pump, plugged pickup screen, leaking pickup pipe, loose main bearings, loose connecting rod bearings, worn camshaft bearings, and worn valve lifter bores.

CASE STUDY

A customer complained about lack of power and reduced top speed on a RWD Oldsmobile with a 307 CID V-8 engine. The customer said the engine ran smoothly and never misfired. A road test proved the customer's description of the problem to be accurate. During the road test, the technician discovered the car had a top speed of 55 mph, but the engine never misfired. The technician also noticed the exhaust had a wheezing noise, especially during hard acceleration.

The technician connected a vacuum gauge to the intake manifold and found the vacuum at idle speed was steady at 19 in Hg. When the engine was accelerated to 2,500 rpm and this speed maintained for two minutes, the vacuum dropped to 12 in Hg. The technician concluded the car must have a restricted exhaust system or air intake system. A quick check of the air filter and air cleaner did not reveal any sign of restriction. The technician decided to double-check the diagnosis by road testing the car with the vacuum gauge connected to the intake manifold. A longer hose was connected from the intake manifold through the front window to the vacuum gauge in the passenger compartment. Each time the engine was accelerated moderately during the road test, the vacuum gauge reading dropped to almost zero. This vacuum gauge reading confirmed the technician's diagnosis that the exhaust must be restricted.

The technician inspected the exhaust system, but there was no indication of damaged or collapsed components. The exhaust pipe was disconnected from the catalytic converter, and the technician discovered the inner layer of the two-ply exhaust pipe had collapsed, and the pipe was almost completely plugged. After a new exhaust pipe was installed, a road test indicated normal performance.

Terms to Know

Cooling system pressure tester	Revolutions per minute (rpm) drop	Stethoscope
Exhaust gas analyzer	Wet compression test	Pinging noise
Power balance tester	Cylinder leakage tester	Ring ridge

ASE Style Review Questions

1. While discussing engine oil leaks:
   Technician A says a leak at the rear of a V-8 engine may be caused by a leaking rear gasket on the intake manifold.
   Technician B says an oil leak may occur at the oil pressure sending unit.
   Who is correct?
   A. A only **C.** Both A and B
   B. B only **D.** Neither A nor B

2. While discussing engine coolant leaks:
   Technician A says a leaking head gasket lowers engine operating temperature.
   Technician B says a leaking heater core may cause windshield fogging.
   Who is correct?
   A. A only **C.** Both A and B
   B. B only **D.** Neither A nor B

3. While discussing the diagnosis of engine exhaust color:
   Technician A says blue smoke in the exhaust during deceleration may be caused by a rich air-fuel ratio.
   Technician B says blue smoke in the exhaust during deceleration may be caused by loose camshaft bearings.
   Who is correct?
   A. A only **C.** Both A and B
   B. B only **D.** Neither A nor B

4. While discussing exhaust color and odor diagnosis:
   Technician A says black smoke in the exhaust during acceleration indicates a lean air-fuel mixture.
   Technician B says a strong sulphur smell from a catalytic converter-equipped vehicle may be caused by a rich air-fuel mixture.
   Who is correct?
   A. A only **C.** Both A and B
   B. B only **D.** Neither A nor B

5. While discussing vacuum gauge diagnosis of engine condition:
   Technician A says a low steady vacuum reading of 12 in Hg may be caused by late ignition timing.
   Technician B says when the engine is accelerated and held at 2,500 rpm for two minutes and the intake manifold vacuum drops below the vacuum recorded at idle, the exhaust system is restricted.
   Who is correct?
   A. A only **C.** Both A and B
   B. B only **D.** Neither A nor B

6. While discussing an engine power balance test:
   Technician A says if the rpm drop on a cylinder is considerably less than the other cylinders, the cylinder may have a burned valve.
   Technician B says if the rpm drop on a cylinder is considerably less than the other cylinders, the cylinder may have a lean air-fuel mixture.
   Who is correct?
   A. A only **C.** Both A and B
   B. B only **D.** Neither A nor B

7. While discussing an engine compression test:
Technician A says the engine should be cranked until four compression strokes occur during the compression test on each cylinder.
Technician B says the gasoline fuel injection system should be left in operation during a compression test.
Who is correct?
A. A only **C.** Both A and B
B. B only **D.** Neither A nor B

8. While discussing the cylinder leakage test:
Technician A says during a cylinder leakage test, if the air escapes from the tailpipe, an intake valve is leaking.
Technician B says during a cylinder leakage test the piston must be at top dead center (TDC) on the exhaust stroke.
Who is correct?
A. A only **C.** Both A and B
B. B only **D.** Neither A nor B

9. While discussing engine noise diagnosis:
Technician A says loose main bearings cause a light rapping noise during acceleration.
Technician B says piston slap causes a hollow rapping noise that is most noticeable on acceleration with a cold engine.
Who is correct?
A. A only **C.** Both A and B
B. B only **D.** Neither A nor B

10. While discussing engine oil pressure:
Technician A says low oil pressure may be caused by a leaking oil pump pickup tube.
Technician B says low oil pressure may be caused by loose camshaft bearings.
Who is correct?
A. A only **C.** Both A and B
B. B only **D.** Neither A nor B

Table 3-1 ASE TASK

Verify driver's complaint and/or road test the vehicle; determine needed repairs.

Problem Area	Symptoms	Possible Causes	Classroom Manual	Shop Manual
ANY AUTO SYSTEM OR COMPONENT	Various problems	Diagnosed and corrected with general diagnostic procedure	37	27

Table 3-2 ASE TASK

Inspect engine assembly for fuel leaks; determine needed repairs.

Problem Area	Symptoms	Possible Causes	Classroom Manual	Shop Manual
FUEL ECONOMY	High fuel consumption, fuel odor	1. Leaking fuel tank or filler cap	43	27
		2. Leaking fuel lines or filter	43	28
		3. Leaking mechanical fuel pump or carburetor	43	28
		4. Leaking pressure regulator, rail, injectors	43	28

Table 3-3 ASE TASK

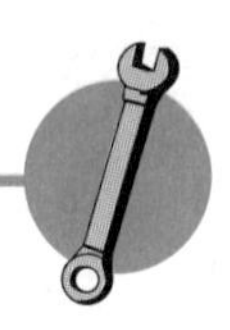

Inspect engine assembly for oil and coolant leaks; determine needed repairs.

Problem Area	Symptoms	Possible Causes	Classroom Manual	Shop Manual
OIL CONSUMPTION	Oil leaks	1. Rear main bearing seal, rear cam bearing plug, rear oil gallery plug	58	28
		2. Oil pan, rocker arm covers, oil sending unit, oil filter	58	28
		3. Intake front and rear gaskets, front main bearing, timing gear cover and seal, distributor gasket or seal	58	28
LOSS OF COOLANT	Coolant leaks	1. Radiator or heater hoses, radiator, by-pass hose, water pump	44	29
		2. Heater core, hose connections	44	29
		3. Engine expansion plugs	44	29
		4. Head gasket, cracked cylinder head or block	44	29

Table 3-4 ASE TASK

Diagnose the cause of unusual engine noise and/or vibration problems; determine needed repairs.

Problem Area	Symptoms	Possible Causes	Classroom Manual	Shop Manual
NOISE	Thumping noise on start-up and acceleration	Loose main bearings, lack of lubrication	48	37
	Rapping noise above 35 mph	Loose connecting rod bearings, lack of lubrication	48	38
	Hollow rapping noise on acceleration with a cold engine	Loose damaged pistons	48	38
	Sharp metallic rapping noise at idle	Worn, loose piston pins	48	38
	High-pitched clicking noise on acceleration	1. Worn, loose piston rings	48	38
		2. Top ring striking ring ridge	48	38
	Clicking noise at idle and higher speed	Worn camshaft lobes, lifter bottoms	50	38
	Light clicking noise at idle	Defective valve lifters, rocker arms, pushrods	50	39

Table 3-4 ASE TASK (continued)

Diagnose the cause of unusual engine noise and/or vibration problems; determine needed repairs

Problem Area	Symptoms	Possible Causes	Classroom Manual	Shop Manual
	Pinging noise on acceleration or at cruising speed	1. Excessive spark advance	43	39
		2. Overheated engine	44	39
		3. Excessive carbon in combustion chambers	51	39
		4. Inoperative EGR valve	43	39
	Thumping noise at rear of engine	Loose flywheel or flexplate	47	39
	Thumping noise with vibration at front of engine, worse during acceleration	Loose, worn vibration damper	47	39

Table 3-5 ASE TASK

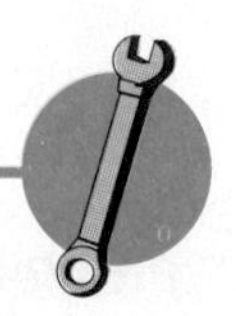

Diagnose the cause of unusual exhaust color, odor, and sound; determine needed repairs.

Problem Area	Symptoms	Possible Causes	Classroom Manual	Shop Manual
EXHAUST COLOR	Blue exhaust	Worn piston rings, valve guides, or seals	48	30
	Black exhaust	Rich air-fuel mixture	43	31
	Gray exhaust	Coolant leaks into combustion chamber	51	31
EXHAUST ODOR	Excessive sulphur smell, catalytic converter vehicles	Rich air-fuel mixture	43	31
	Strong odor causes eye to water, non-catalytic converter vehicles	Rich air-fuel mixture	43	31
EXHAUST NOISE	Loud noise on acceleration	Exhaust system leak	43	31
	Rattling noise on acceleration	1. Loose components in muffler or converter	43	31
		2. Loose exhaust hangers	43	31
		3. Exhaust component hitting chassis	43	31
	Wheezing on acceleration and higher speed	Restricted exhaust	43	31

Table 3-6 ASE TASK

Perform engine vacuum/boost (manifold absolute pressure) tests; determine needed repairs.

Problem Area	Symptoms	Possible Causes	Classroom Manual	Shop Manual
ENGINE OPERATION	Steady low vacuum	1. Late ignition timing	43	31
		2. Intake manifold leak	55	31
	Fluctuating vacuum at idle	1. Improper idle mixture adjustment, defective injectors	55	32
		2. Burned or leaking valves	50	32
		3. Sticking valves	50	32
	Severe vacuum gauge fluctuations	1. Weak valve springs	50	32
		2. Leaking head gasket	51	32
	Lower vacuum at 2,500 rpm compared to idle	Restricted exhaust	43	32

Table 3-7 ASE TASK

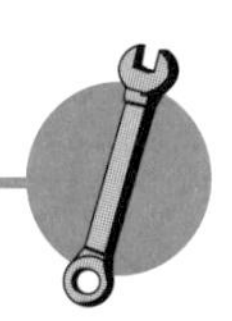

Perform cylinder power balance test; determine needed repairs.

Problem Area	Symptoms	Possible Causes	Classroom Manual	Shop Manual
ENGINE OPERATION	Rough idle, cylinder misfire, lack of rpm drop during test	1. Burned valves	50	33
		2. Worn rings, cylinders	48	33
		3. Leaking head gasket, cracked head	51	33

Table 3-8 ASE TASK

Perform cylinder compression test; determine needed repairs.

Problem Area	Symptoms	Possible Causes	Classroom Manual	Shop Manual
ENGINE OPERATION	Rough idle, cylinder misfire, compression below specifications	1. Burned valves	51	34
		2. Worn rings, cylinders	48	34
		3. Leaking head gasket, cracked head	51	34

Table 3-9 ASE TASK

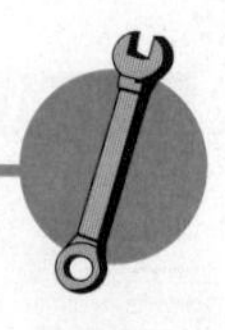

Perform cylinder leakage test; determine needed repairs.

Problem Area	Symptoms	Possible Causes	Classroom Manual	Shop Manual
ENGINE OPERATION	Rough idle, cylinder misfire, above 20% leakage on tester	1. Burned valves	51	35
		2. Worn ring, cylinders	49	35
		3. Leaking head gasket, cracked head	51	35

Basic Engine and Cooling System Service

Upon completion and review of this chapter, you should be able to:

- ❑ Perform valve adjustments on mechanical and hydraulic lifters.
- ❑ Check valve timing.
- ❑ Perform an engine temperature check.
- ❑ Diagnose engine overheating problems.
- ❑ Inspect and diagnose coolant hoses.
- ❑ Inspect and diagnose radiators and shrouds.
- ❑ Inspect and diagnose radiator pressure caps.
- ❑ Inspect coolant condition, and test coolant antifreeze protection.
- ❑ Inspect and diagnose belt condition.
- ❑ Adjust belt tension.
- ❑ Inspect and diagnose water pump and fan blade condition.
- ❑ Inspect and diagnose coolant recovery system.
- ❑ Inspect and test thermostat condition.
- ❑ Flush cooling systems.
- ❑ Inspect and diagnose viscous-drive fan clutches.
- ❑ Diagnose and repair electric-drive cooling fan circuits.

Basic Tools

Basic technician's tool set

Service manual

Feeler gauge set

12-V test light

Valve Adjustment, Mechanical Valve Lifters

Some engines have mechanical valve lifters in place of hydraulic valve lifters, and these engines require a valve adjustment. Various car manufacturers recommend different valve adjustment procedures depending on the engine. For example, on some engines, the manufacturer recommends adjusting the valves with the engine cold, whereas other manufacturers recommend performing this adjustment with the engine at normal operating temperature. Always follow the vehicle manufacturer's recommended valve adjustment procedure in the service manual.

Mechanical valve lifters are steel cylinders used in place of hydraulic valve lifters. The outside of the mechanical lifter is lubricated by engine oil, but no oil enters the lifter, and the mechanical lifter is not collapsible.

Mechanical valve lifters may be called solid tappets.

Special Tools

Remote starter switch

CAUTION: When cranking the engine slowly with a remote starter switch and the ignition switch on, the engine may start. To avoid personal injury, keep hands and tools away from fan blades and belts during this operation.

The valves in each cylinder must be adjusted with the piston at top dead center (TDC) on the compression stroke and both valves closed. To position each piston at TDC on the compression stroke, connect a timing light to the spark plug wire and crank the engine slowly with the ignition switch on and a remote starter switch connected to the starter solenoid. When the timing light flashes, the piston is near TDC on the compression stroke.

Some rocker arms have an adjustment screw and a lock screw on the push rod end of the rocker arm. With the piston at TDC on the compression stroke, place a feeler gauge of the specified thickness between the rocker arm and the valve stem (Figure 4-1). If the valve adjustment is correct, the feeler gauge slides between the rocker arm and the valve stem with a light push fit. If necessary, loosen the locknut and turn the adjustment screw to obtain the specified valve clearance. Hold the adjustment screw with a slotted screwdriver and tighten the locknut. On most engines, the specified valve clearance is different on the exhaust valves than the intake valves.

In some four-valve (4-V) engines, the rocker arms are mounted under the overhead camshafts, and the camshaft lobes contact a friction pad on the top of the rocker arm. One end of the rocker arm contacts the valve stem, and the other end of the rocker arm is mounted in a pivot in the cylinder head (Figure 4-2). The manufacturer recommends measuring the valve clearance on these engines with the cylinder head temperature less than 100°F (38°C). Follow these steps for valve adjustment:

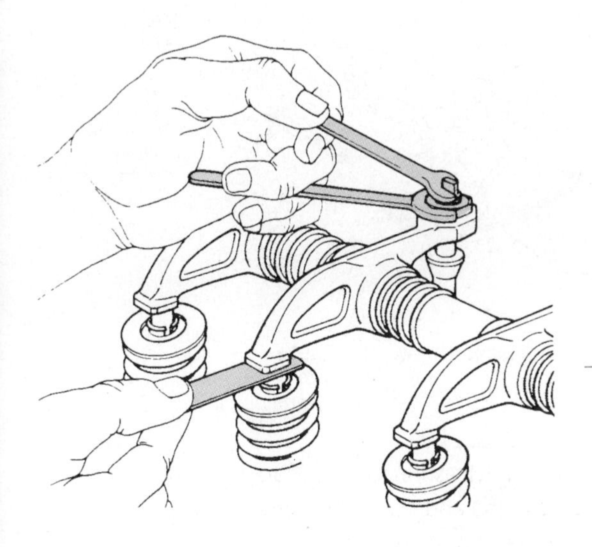

Figure 4-1 Valve adjustment with adjusting screw and locknut in the push rod end of the rocker arm (Courtesy of Chrysler Corporation)

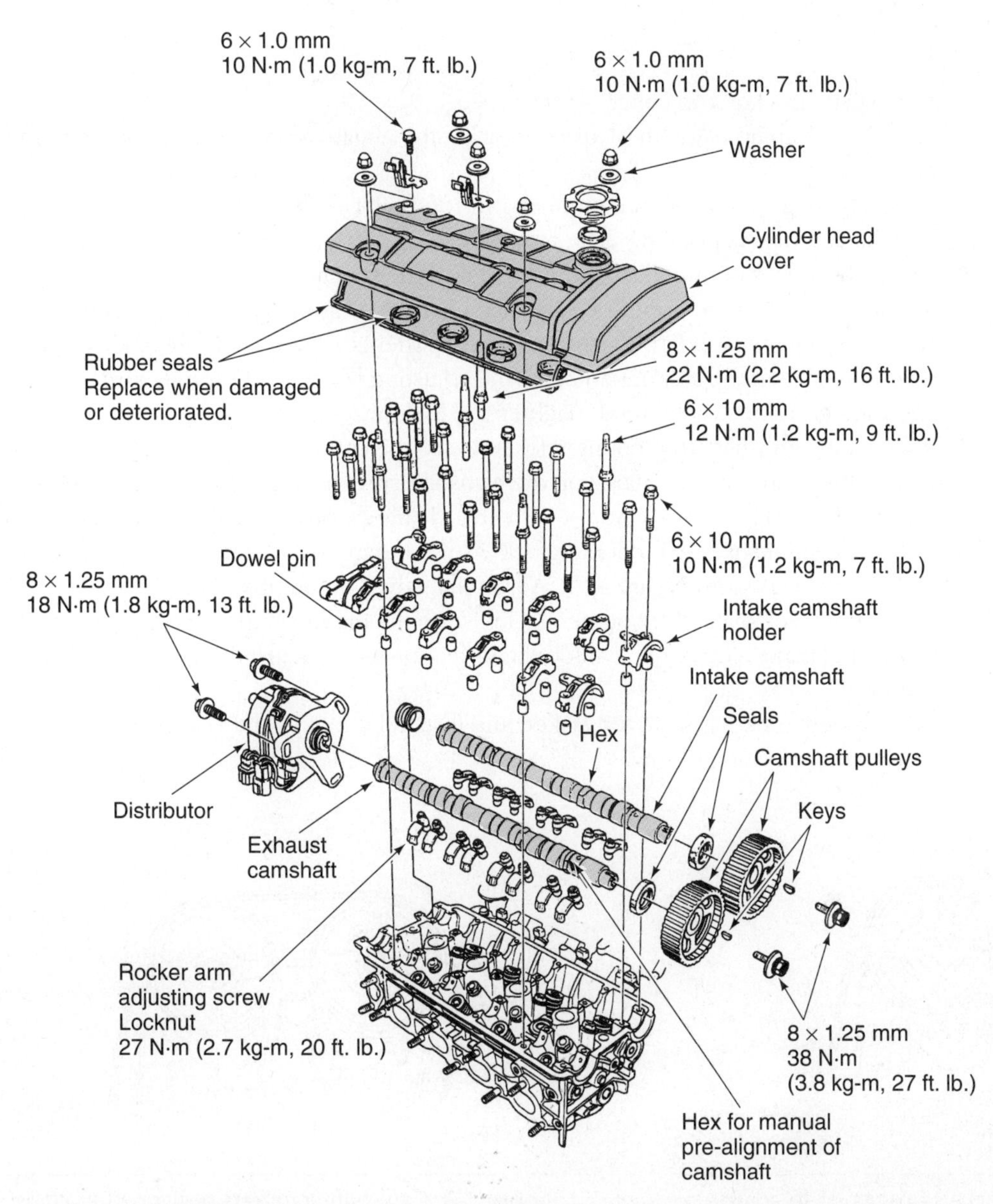

Figure 4-2 Four-valve (4-V) engine with rocker arms mounted under the camshafts (Courtesy of American Honda Motor Co., Inc.)

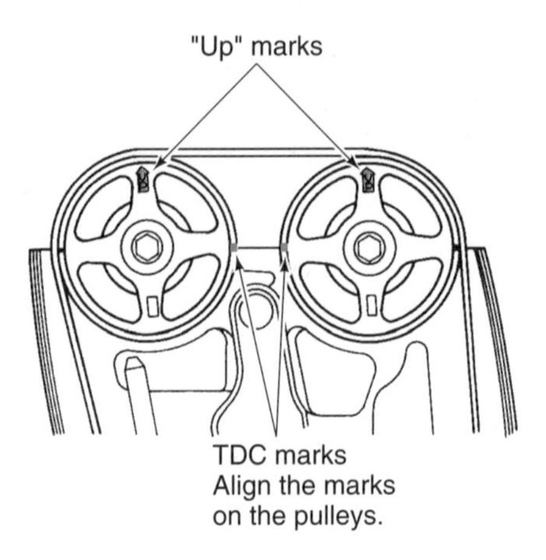

Figure 4-3 Number 1 piston at TDC and the camshaft sprocket marks facing upward (Courtesy of American Honda Motor Co., Inc.)

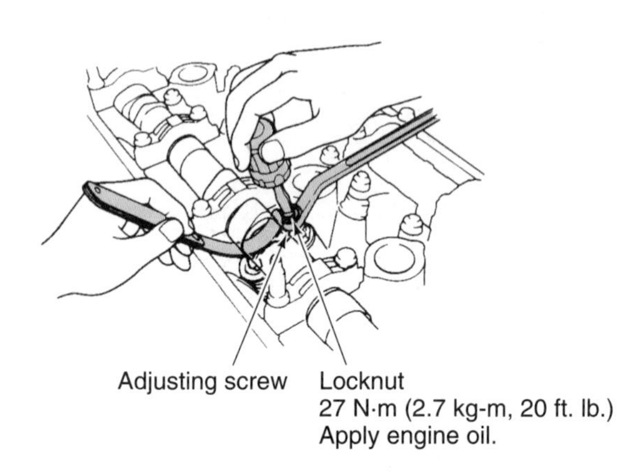

Figure 4-4 Measuring valve clearance between the camshaft lobe and the rocker arm (Courtesy of American Honda Motor Co., Inc.)

1. Remove the rocker arm cover.
2. Disable the ignition system by disconnecting the negative primary coil wire from the coil.
3. Crank the engine slowly until number 1 piston is at TDC with the marks on the camshaft sprockets pointing upward (Figure 4-3).
4. Adjust the valves on number 1 cylinder. Place the specified feeler gauge between the camshaft lobe and the upper side of the rocker arm (Figure 4-4). The feeler gauge should be a light push fit between these components. If the valve clearance requires adjusting, loosen the locknut and turn the adjusting screw in the rocker arm until the specified clearance is obtained. Tighten the locknut.
5. Rotate the crankshaft 180° counterclockwise so the camshaft sprockets move 90°. Under this condition, the marks on the camshaft sprockets face the exhaust side of the cylinder head (Figure 4-5). Adjust the valve clearance on number 3 cylinder.
6. Rotate the crankshaft 180° counterclockwise so the marks on the camshaft sprockets are facing downward (Figure 4-6). Adjust the valves on number 4 cylinder.
7. Rotate the crankshaft 180° counterclockwise so the marks on the camshaft gears are facing the intake side of the cylinder head (Figure 4-7). Adjust the valves on number 2 cylinder.
8. Install a new rocker arm cover gasket, and install the rocker arm cover. Tighten all bolts to the specified torque.

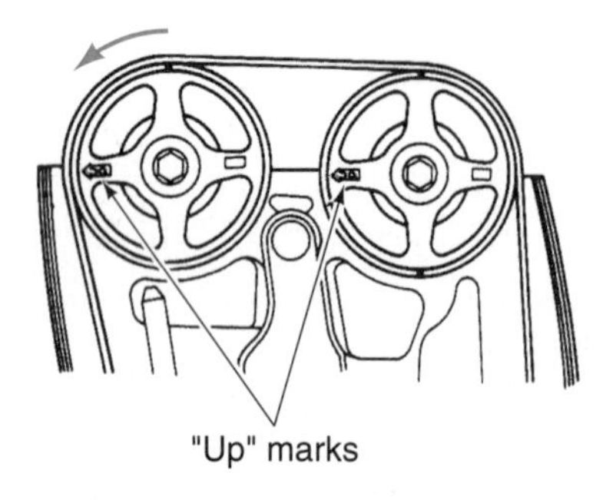

Figure 4-5 Camshaft sprockets positioned to adjust the valves in number 3 cylinder (Courtesy of American Honda Motor Co., Inc.)

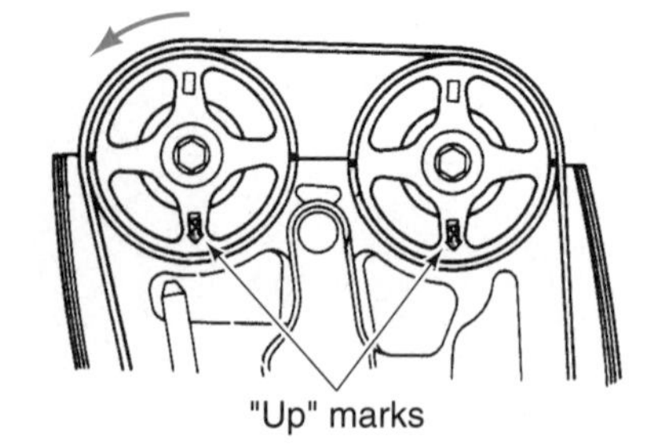

Figure 4-6 Camshaft sprockets positioned to adjust the valves in number 4 cylinder (Courtesy of American Honda Motor Co., Inc.)

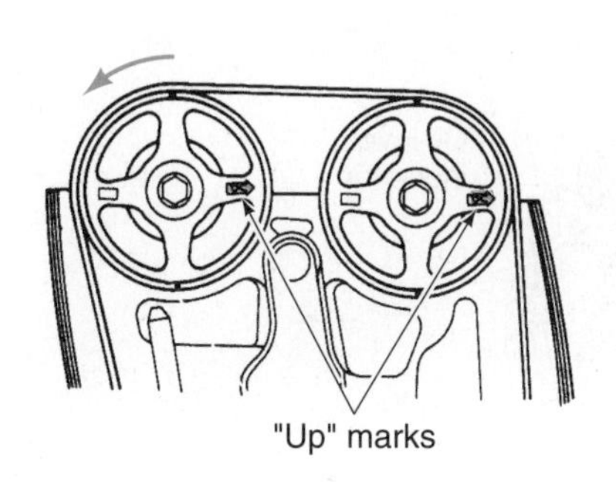

Figure 4-7 Camshaft sprockets positioned to adjust the valves in number 2 cylinder (Courtesy of American Honda Motor Co., Inc.

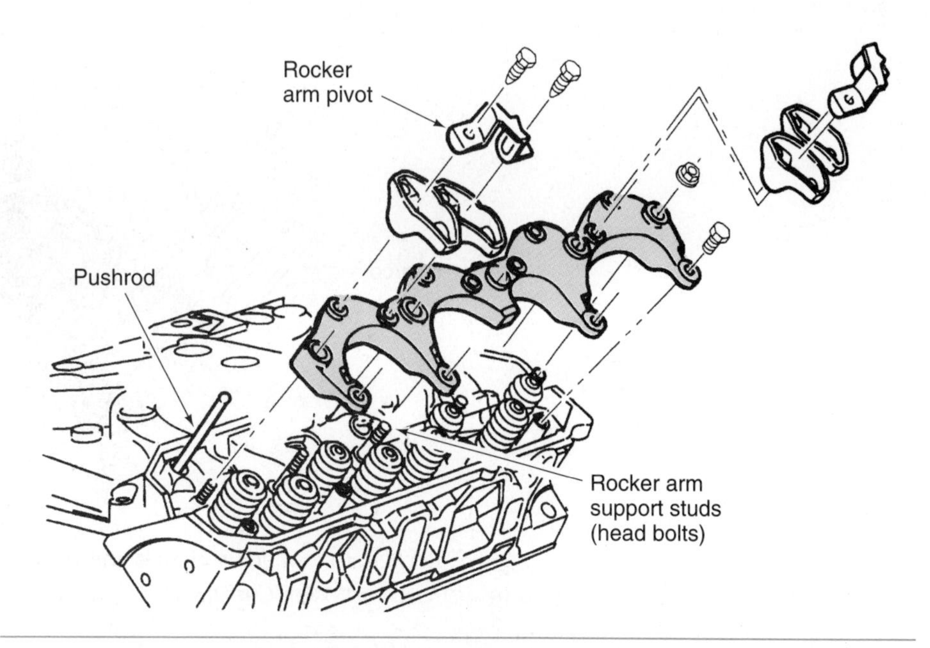

Figure 4-8 Valve train with no valve adjustment (Courtesy of Cadillac Motor Car Division, General Motors Corporation)

Valve Adjustment, Hydraulic Valve Lifters

Some engines with hydraulic valve lifters do not have a valve adjustment. In these engines, if all the valve train components are in satisfactory condition, the hydraulic valve lifters maintain zero valve clearance. However, worn or improperly serviced valve train components may affect the valve lifter operation and result in valve train noise or improper engine performance.

A valve clearance measurement with the valve lifter bottomed is recommended on many valve trains that do not have a valve adjustment (Figure 4-8). With the piston at TDC on the compression stroke, push downward on the pushrod end of the rocker arm until the valve lifter bottoms. Measure the clearance between the end of the rocker arm and the valve stem. If this clearance is more than specified, valve train components such as the rocker arm, pivot, or pushrod are worn, and this problem may cause a clicking noise at idle and low speed.

When this clearance is less than specified, the installed height of the valve in the cylinder head is too high. Excessive valve stem height is caused by grinding too much material from the valve seat, or valve face, without removing material from the end of the valve stem when the valves are reconditioned. If the valve stem height is excessively high, the valve lifter plunger may be bottomed in the lifter, and the valve may not be allowed to close. Under this condition, cylinder misfiring occurs.

Valve stem installed height is the distance from the valve spring seat to the top of the valve stem with the valve closed. Valve stem height should be measured when valves are reconditioned.

Some valve trains have hydraulic valve lifters and individual rocker arm pivots retained with self-locking nuts. These valve trains require an initial adjustment of the rocker arm nut to position the lifter plunger. With the valve closed, loosen the rocker arm nut until there is clearance between the end of the rocker arm and the valve stem. Slowly turn the rocker arm nut clockwise while rotating the pushrod (Figure 4-9). Continue rotating the rocker arm nut until the end of the rocker arm contacts the end of the valve stem and the pushrod becomes harder to turn. Continue turning the rocker arm nut clockwise the number of turns specified in the service manual. In some engines, this specification is one turn plus or minus one-quarter turn.

Valve Timing Check

If the timing belt or chain has slipped on the camshaft sprocket, the engine may fail to start because the valves are not properly timed in relation to the crankshaft. When the timing belt or chain has

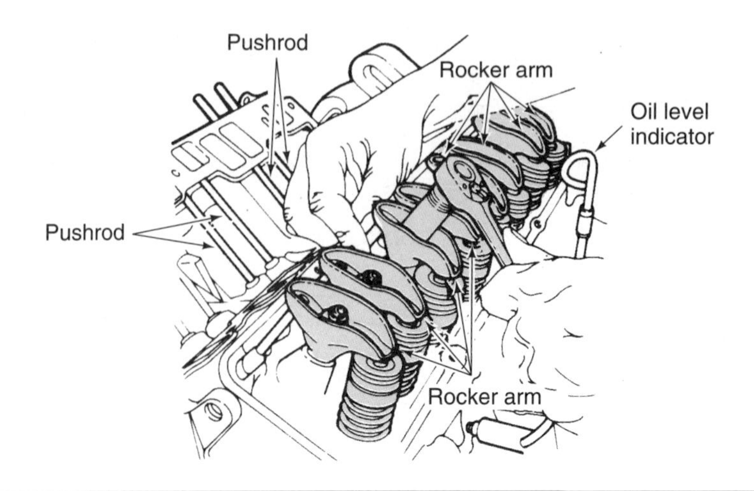

Figure 4-9 Rocker arm nut adjustment with hydraulic valve lifters and individual rocker arms (Courtesy of Cadillac Motor Car Division, General Motors Corporation)

only slipped a few cogs on the camshaft sprocket, the engine has a lack of power, and fuel consumption is excessive. Follow these steps to check the valve timing:

1. Remove the spark plug from number 1 cylinder, and place your thumb on top of the spark plug hole. If this hole is not accessible, place a compression gauge in the opening. Disconnect the positive primary wire from the ignition coil to disable the ignition system.
2. Crank the engine until compression is felt at the spark plug hole.
3. Connect a remote control switch to the starter solenoid terminal and the battery terminal on the solenoid. Slowly crank the engine until the timing mark lines up with the zero-degree position on the timing indicator. The number 1 piston is now at TDC on the compression stroke. On many engines, the timing mark is on the crankshaft pulley, and the timing indicator is mounted above the pulley.
4. Slowly crank the engine for one revolution until the timing mark lines up with the zero-degree position on the timing indicator. Number 1 piston is now at TDC on the exhaust stroke.
5. Remove the rocker arm cover and install a breaker bar and socket on the crankshaft pulley nut. Observe the valve action while rotating the crankshaft about 30° before and after TDC on the exhaust stroke. In this crankshaft position, the exhaust valve should close a few degrees after TDC on the exhaust stroke, and the intake valve should open a few degrees before TDC on the exhaust stroke. This valve position with the piston at TDC on the exhaust stroke is called valve overlap. If the valves do not open properly in relation to the crankshaft position, the valve timing is not correct. Under this condition, the timing chain or belt cover should be removed to check the position of the camshaft sprocket in relation to the crankshaft sprocket position. When the valve timing is incorrect, the timing chain or belt and/or sprockets must be replaced.

Engine Temperature Check

Lower than normal coolant temperature reduces engine efficiency and fuel economy. If the engine coolant temperature is too low, insufficient heat is discharged from the vehicle heater in cold weather. Lower than normal coolant temperature affects some computer input sensor signals, and this action may result in reduced engine performance and fuel economy. If the coolant temperature is higher than the rated temperature of the thermostat, the engine may overheat.

A thermometer may be taped to the upper radiator hose to measure coolant temperature. Be sure the temperature-sensing bulb on the thermometer is contacting the upper radiator hose. With the thermometer in place, operate the engine for 15 minutes at idle speed. The temperature indicated on the thermometer should be within a few degrees of the temperature rating of the thermostat specified in the service manual. If the temperature on the thermometer is less than the rated temperature of the thermostat, the thermostat may be opening too soon. Under this condition, the thermostat must be replaced. Photo Sequence 2 shows a typical procedure for removing, testing, and replacing a thermostat.

Special Tools

Thermometer

When the engine operating temperature is higher than the rated temperature of the thermostat, check these causes of overheating.

1. Loose or slipping fan belt
2. Electric-drive cooling fan not operating at the proper temperature
3. Defective viscous-drive fan clutch
4. Restricted air passages in radiator core
5. Partially plugged coolant passages in radiator core
6. Restricted or collapsed radiator hoses
7. Thermostat not opening at the proper temperature
8. Thermostat improperly installed
9. Leaking cylinder head gasket
10. External cooling system leaks causing low coolant level
11. Leaking radiator pressure cap
12. Defective water pump, loose impeller
13. Improper mixture of antifreeze and water
14. Excessive engine load
15. Late ignition timing and/or spark advance
16. Lean air-fuel mixture, engine vacuum leaks
17. Automatic transmission overheating
18. Dragging brakes
19. Broken, improperly positioned radiator shroud

Cooling System Inspection and Diagnosis

CAUTION: Never loosen the radiator cap on a warm or hot engine. When the cap is loosened, the reduction in pressure causes the coolant to boil, and anyone near the vehicle may be severely burned.

Visual Inspection

CAUTION: Use caution and wear protective gloves when touching cooling system components. If the engine has been running, these components may be very hot, and touching them could result in severe burns.

CAUTION: Keep hands, tools, and clothing away from electric-drive cooling fans. On many vehicles, an electric-drive fan may start turning even with the ignition switch off.

Hose Inspection. All cooling system hoses should be inspected for soft spots, swelling, hardening, chafing, leaks, and collapsing. If any of these conditions are present, hose replacement is necessary. Hose clamps should be inspected to make sure they are tight. Some radiator hoses contain a wire coil inside them to prevent hose collapse as the coolant temperature decreases. Remember to include heater hoses and the by-pass hose in the hose inspection.

Photo Sequence 2
Typical Procedure for Removing, Testing, and Replacing a Thermostat

P2-1 Drain the coolant from the radiator into an approved container.

P2-2 Remove the bolts from the thermostat housing.

P2-3 Remove the thermostat from the housing.

P2-4 Place the thermostat and a thermometer in the thermostat tester, and pour water into the tester until the thermostat is submerged in the water.

P2-5 Heat the water in the thermostat tester, and observe the temperature on the thermometer when the thermostat valve begins to open. Replace the thermostat if it does not open at the specified temperature.

P2-6 Use a scraper to remove the old gasket material from the thermostat housing and the thermostat housing mounting surface.

P2-7 Install a new gasket on the thermostat housing mounting surface.

P2-8 Install the thermostat with the temperature pellet facing toward the engine. Be sure the thermostat fits properly in the housing recess.

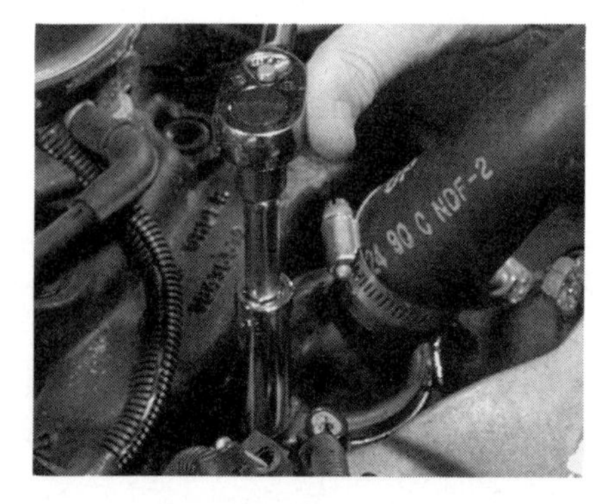

P2-9 Install and tighten the thermostat housing bolts to the specified torque, and reinstall the coolant in the radiator. Be sure the proper coolant level is present in the coolant recovery container.

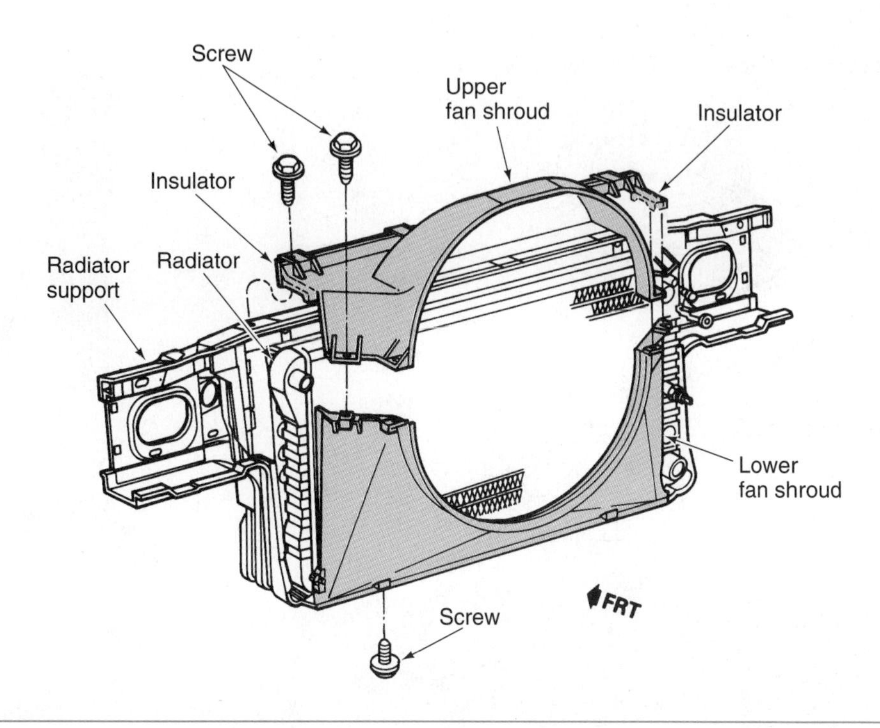

Figure 4-10 Radiator shroud concentrates air flow through the radiator core. (Courtesy of Chevrolet Motor Division, General Motors Corporation)

SERVICE TIP: If the upper radiator hose collapses when the coolant is cold, the vacuum valve may be sticking in the radiator pressure cap, or the hose to the coolant recovery container may be restricted.

Radiator and Shroud Inspection. The radiator should be inspected for restrictions such as bugs and debris in the air passages through the core. An air hose and air gun may be used to blow bugs and debris from the core. The radiator should also be inspected for leaks and loose brackets. Radiator repairs are usually done in a radiator repair shop. When the radiator cap is removed, the openings in some of the radiator tubes are visible through the filler neck. These tube openings should be free from rust and corrosion. If the tubes are restricted, cooling system flushing may remove the restriction. When flushing does not clean the radiator tubes, the radiator must be cleaned at a radiator shop. Any restrictions in the radiator air passages or coolant tubes result in higher engine operating temperature and possible overheating.

Many radiators have a shroud behind the radiator to concentrate the flow of air through the radiator core (Figure 4-10). Inspect the radiator shroud for looseness and cracks or missing pieces. If the shroud is loose, improperly positioned, or broken, air flow through the radiator is reduced, and engine overheating may result.

Radiator Cap Inspection. The radiator cap should be inspected for a damaged sealing gasket or vacuum valve (Figure 4-11). Check the pressure rating stamped on top of the cap, and be sure this is the same as the radiator cap specification in the vehicle manufacturer's service manual. Replace the cap if the pressure rating does not meet the vehicle manufacturer's specified pressure. If the sealing gasket or vacuum valve is damaged, replace the cap. Check the seat in the filler neck for damage or burrs, and remove any rough spots with fine emery paper. Visually inspect the filler neck for rust accumulation, which may indicate a contaminated cooling system. If the pressure cap sealing gasket or seat is damaged, the engine will overheat and coolant will be lost to the coolant recovery system. Under this condition, the coolant recovery container becomes over-filled with coolant.

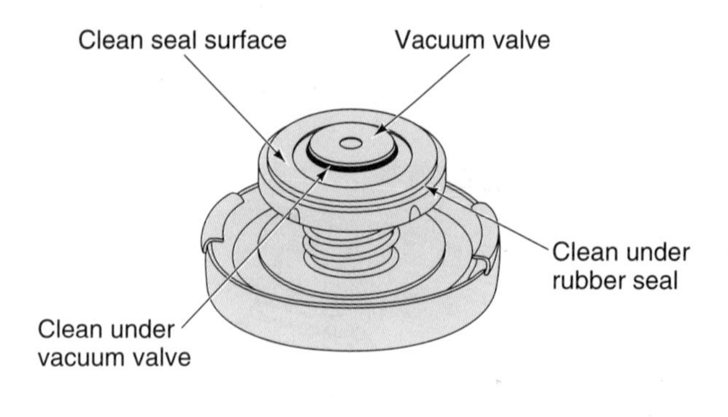

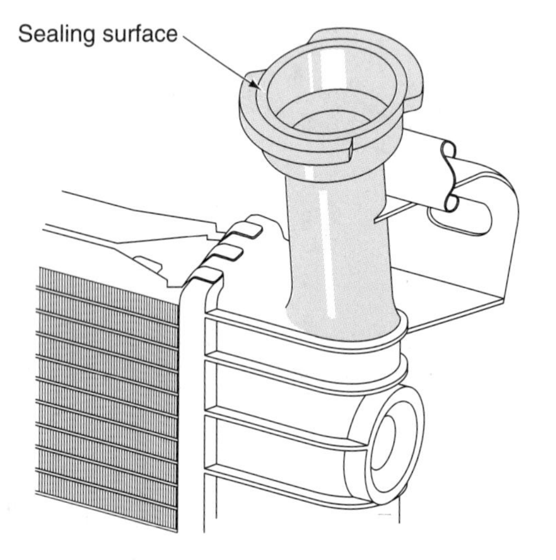

Figure 4-11 Inspecting radiator pressure cap and filler neck (Courtesy of Chrysler Corporation)

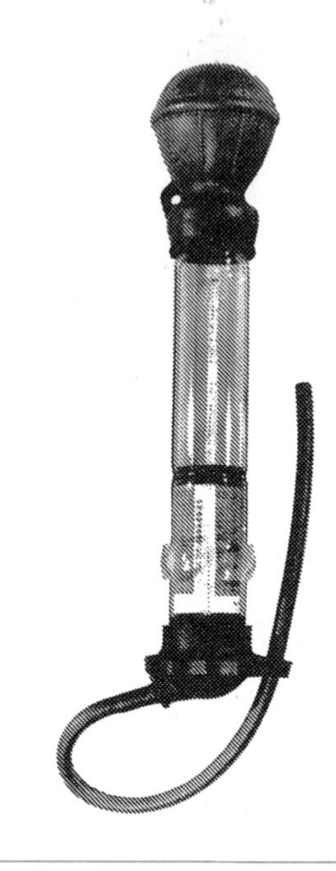

Figure 4-12 Coolant hydrometer (Courtesy of Mac Tools, Inc.)

Special Tools

Radiator pressure tester

If the cap vacuum valve is sticking, a vacuum may occur in the cooling system after the engine is shut off and the coolant temperature decreases. This vacuum may cause collapsed cooling system hoses.

A pressure tester may be used to test the pressure cap and to pressure test the entire cooling system.

Special Tools

Coolant hydrometer

A hydrometer measures the specific gravity of a liquid.

Specific gravity of a liquid is the weight of the liquid in relation to an equal volume of water.

Coolant Inspection and Testing. A coolant hydrometer pickup tube is inserted into the coolant in the radiator to test the antifreeze protection provided by the coolant. Squeeze and release the squeeze bulb on top of the hydrometer to pull coolant into the hydrometer (Figure 4-12). Fill the hydrometer chamber until the float is floating freely, but not touching the top of the chamber. When the level of coolant on the float is viewed at eye level, the marking on the float indicates the lowest temperature at which the coolant provides antifreeze protection. If the vehicle is operating in atmospheric temperatures lower than the antifreeze protection temperature provided by the coolant, some of the coolant must be drained and more antifreeze added to the cooling system.

WARNING: Used engine coolant is a hazardous waste and pollution laws prohibit the dumping of coolant in sewers. Coolant recycling equipment is available to clean contaminants from the coolant so it may be reused.

Check the condition of the coolant in the hydrometer. If rust or other contaminants appear in the coolant, the cooling system should be flushed and refilled with a new solution or recycled solution.

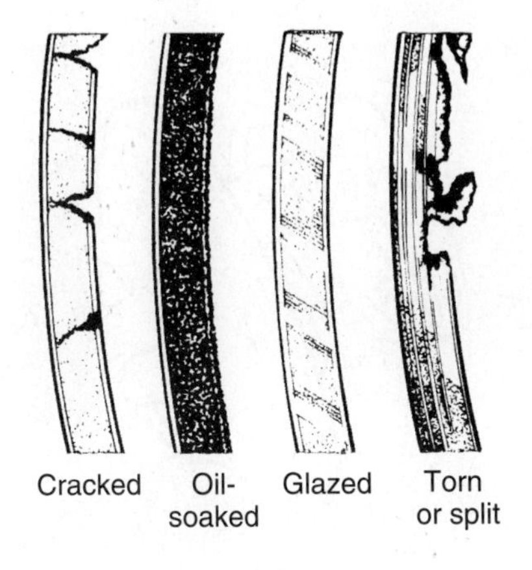

Figure 4-13 Defective belt conditions (Courtesy of Chrysler Corporation)

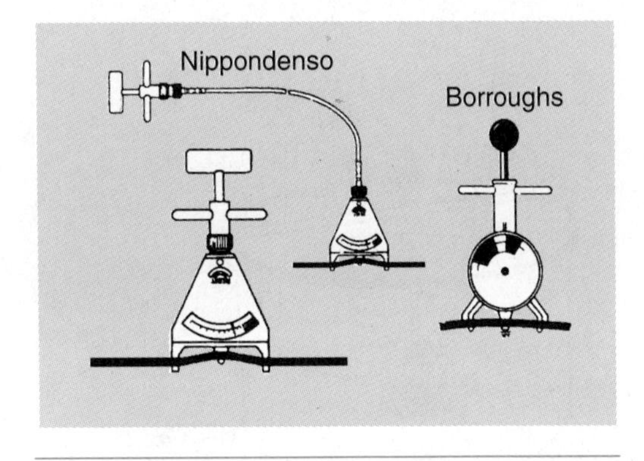

Figure 4-14 Checking belt tension with a tension gauge (Courtesy of Toyota Motor Corporation)

Belt Inspection. All the drive belts, including the water pump belt, should be inspected for cracks, oil contamination, glazing, and tears or splits (Figure 4-13). If any of these conditions are present, replace the belt or belts. Since the friction surfaces are the sides of a V-belt, the belt must be replaced if the sides are worn and the belt is contacting the bottom of the pulley.

Special Tools

Belt tension gauge

The belt tension may be checked with the engine shut off and a belt tension gauge placed over the belt at the center of the belt span (Figure 4-14). If the belt tension is less than specified, the belt must be tightened. A loose or worn belt may cause a squealing noise when the engine is accelerated. On many vehicles, the same belt drives the water pump and the alternator. If the engine has a V-belt, loosen the alternator bracket bolt and the alternator mounting bolt, and then pry on the alternator housing to tighten the belt. Ribbed V-belts usually have a spring-loaded belt tensioner. Tighten the alternator bolts, and recheck the belt tension. Since misaligned pulleys cause premature belt failure, be sure the pulleys are properly aligned.

The belt tension may also be checked by measuring the amount of belt deflection with the engine shut off. Use your thumb to depress the belt at the center of the belt span. If the belt tension is correct, the belt should have 1/2 inch deflection per foot of belt span.

Water Pump and Fan Blade Inspection. With the engine shut off, grasp the fan blades or the water pump hub and try to move the blades from side to side. This action checks for looseness in the water pump bearing. If there is any side-to-side movement in the bearing, water pump replacement is required. Check the fan blade for cracks or a bent condition. A cracked fan blade causes a clicking noise at idle speed. If the fan blades are cracked or bent, replace the fan blade assembly. Bent fan blades cause an imbalance and vibration condition, which damages the water pump bearing.

CAUTION: If a cracked, or bent, fan blade is discovered, replace the fan blade immediately. Do not start the engine. If a fan blade breaks while the engine is running, the blade becomes a very dangerous projectile, which may cause serious personal injury or vehicle damage.

CAUTION: Do not attempt to straighten bent fan blades. This action weakens the metal and may cause the blade to break suddenly, resulting in serious personal injury or vehicle damage.

If the vehicle has an electric-drive cooling fan, grasp the water pump hub to check the pump bearing for side-to-side movement. Check for coolant leaks at the water pump drain hole in the bottom of the pump and at the inlet hose connected to the pump. When coolant is dripping from

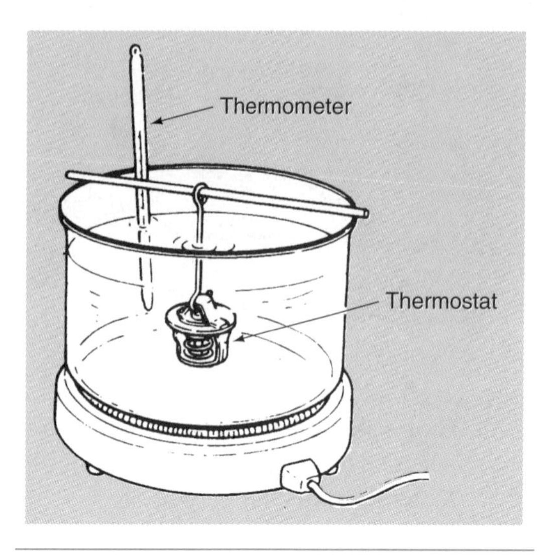

Figure 4-15 Thermostat testing (Courtesy of American Honda Motor Co., Inc.)

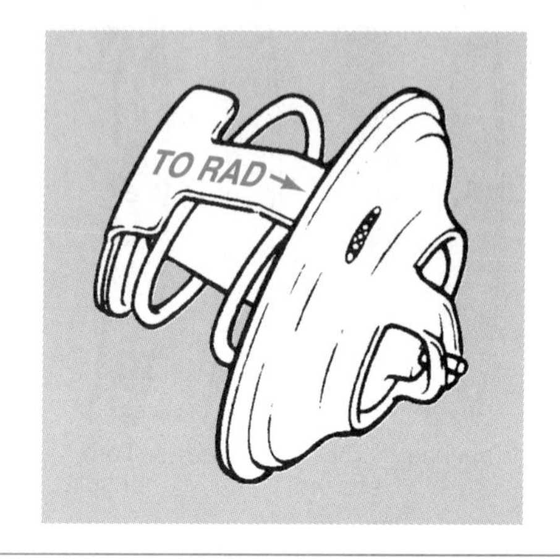

Figure 4-16 Thermostat directional marking (Courtesy of Oldsmobile Division, General Motors Corporation)

the pump drain hole, replace the pump. If coolant is leaking around the inlet hose connection on the pump, check the hose and clamp.

Coolant Recovery System Inspection. Check the coolant recovery container and hose for leaks. Observe the level of coolant in the recovery container. In cooling systems with a coolant recovery system, coolant is added to the recovery container. Most of these containers have COLD and HOT coolant level marks. The coolant level should be at the appropriate mark on the recovery container, depending on engine temperature.

Thermostat Testing and Inspection

Some of the coolant must be drained from the cooling system before the thermostat is removed. Loosen the radiator drain plug and drain the coolant into a clean coolant drain pan. Remove the thermostat housing bolts, and remove the housing. Check the thermostat valve, and replace the thermostat if the valve is stuck open. Visually inspect the thermostat housing area for rust and contaminants. If excessive rust and contaminants are present, flush the cooling system, and replace or recycle the coolant.

Special Tools

Thermostat testing container

The thermostat may be submerged with a thermometer in a container filled with water (Figure 4-15). Heat the water while observing the thermostat valve and the thermometer. The thermostat valve should begin to open when the temperature on the thermometer is equal to the rated temperature stamped on the thermostat. Replace the thermostat if it does not open at the rated temperature. Thermostat replacement is also necessary if the temperature rating on the thermostat is not the same as the thermostat temperature specified by the vehicle manufacturer.

If the opening temperature of the thermostat is higher than the rated temperature, the engine will overheat. When the thermostat opening temperature is lower than the rated temperature, the engine runs cooler than normal, providing a reduction in engine efficiency and performance. On engines with computer-controlled air-fuel mixture and spark advance, lower than normal coolant temperature changes some input sensor signals and reduces fuel economy.

Many thermostats are marked for proper installation (Figure 4-16). The pellet end of the thermostat faces toward the engine. Be sure the thermostat fits properly in the recess in the upper or lower part of the housing. Clean the mating surfaces carefully with a scraper to remove the old gasket. Install a new housing gasket and tighten the housing bolts to the specified torque.

Cooling System Flushing

CAUTION: Always wear protective clothing, including gloves and goggles, when flushing cooling systems. Some cooling system cleaning agents contain caustic chemicals, which are harmful to skin and eyes.

CAUTION: Always operate cooling system flushing equipment according to the equipment manufacturer's recommended procedure. Improper cooling system flushing procedures may cause personal injury and damage cooling system components.

WARNING: Used engine coolant and cooling system cleaning solutions must be disposed of according to state pollution laws. These solutions are considered hazardous waste.

If the radiator tubes and coolant passages in the block and cylinder head are restricted with rust and other contaminants, these components may be flushed. Cooling system flushing equipment is available for this purpose. Always operate the flushing equipment according to the equipment manufacturer's directions, and be sure that your service procedure conforms to pollution laws in your state.

Coolant reconditioning machines are available to remove harmful particles and restore corrosion additives so the coolant can be returned to the cooling system (Figure 4-17).

Special Tools

Coolant reconditioning machine

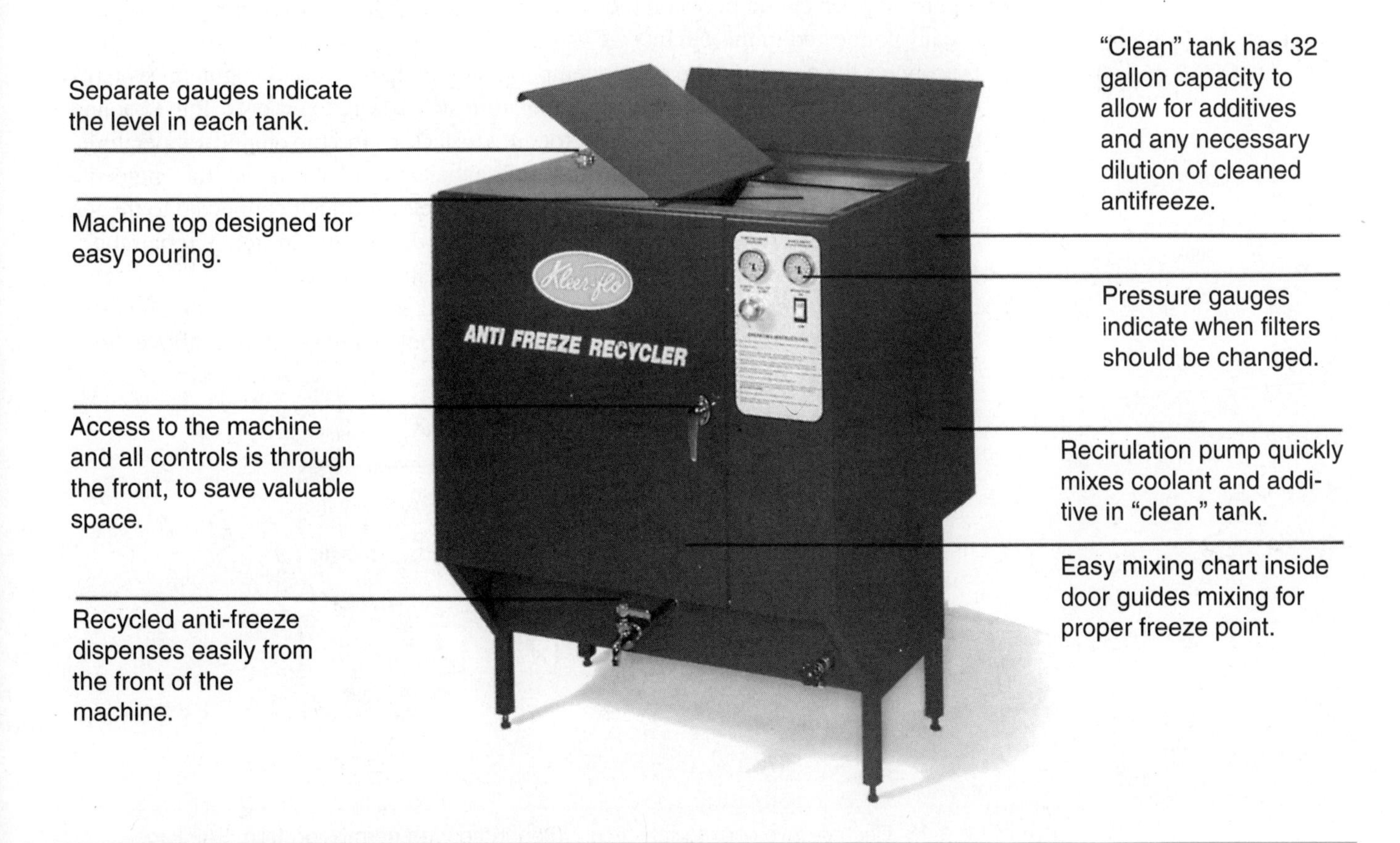

Figure 4-17 Anti-freeze recycler (Courtesy of Kleer-Flo Company)

Viscous-Drive Fan Clutch Diagnosis

The viscous-drive fan clutch should be visually inspected for leaks. If there are oily streaks radiating outward from the hub shaft, the fluid has leaked out of the clutch.

With the engine shut off, rotate the cooling fan by hand. When the engine is cold, the viscous clutch should offer a slight amount of resistance. The clutch should offer more resistance to turning when the engine is hot. If the viscous clutch allows the fan blades to rotate easily both hot and cold, the clutch should be replaced. A slipping viscous clutch results in engine overheating.

Push one of the fan blades in and out, and check for looseness in the clutch shaft bearing. If any looseness is present, replace the viscous clutch.

Electric-Drive Cooling Fan Circuit Diagnosis

Electric cooling fan circuits vary depending on the vehicle and the model year. Always follow the diagnostic procedure in the vehicle manufacturer's service manual. If the electric-drive cooling fan does not operate at the coolant temperature specified by the vehicle manufacturer, engine overheating will result, especially at idle and lower speeds when air flow through the radiator is reduced. Following is a typical electric cooling fan diagnostic procedure:

1. Connect a 12-V test light from the battery voltage supply terminal on the fan relay to ground. If 12 V is not present at this terminal, repair the open circuit between the battery positive terminal and the relay. Many circuits have a fuse link or a fuse in this circuit location.
2. With the ignition switch on, connect a 12-V test light from the ignition switch terminal on the cooling fan relay to ground (Figure 4-18). If 12 V is not present at this terminal, repair the open circuit between the ignition switch and the relay terminal. A fuse is usually connected in this circuit.
3. Disconnect the wire from the engine temperature switch, and turn the ignition switch on. Connect a jumper wire from the temperature switch wire to ground. If the cooling fan operates now, the relay and fan motor are satisfactory and the engine temperature switch or connecting wire is defective, assuming the fan did not run with the engine hot.
4. If the cooling fan did not run in step 3, ground the relay terminal connected to the temperature switch. If the fan motor operates now, but did not run in step 3, repair the open circuit in the wire from the relay to the temperature switch.
5. Leave the engine temperature switch wire disconnected, and connect an ohmmeter from this switch terminal to ground. If the engine coolant temperature is above the

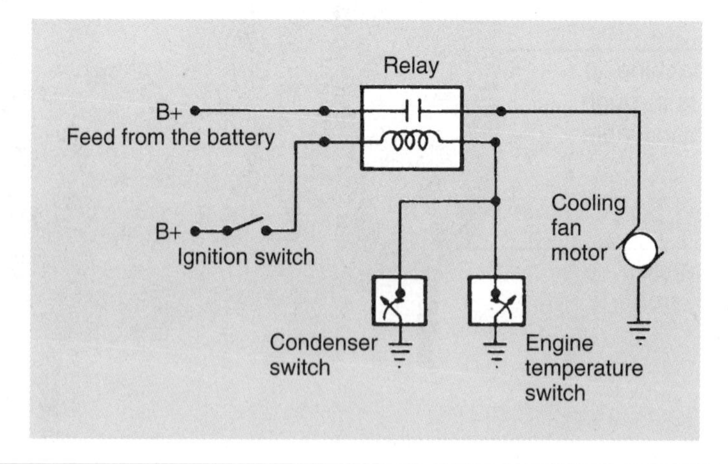

Figure 4-18 Electric-drive cooling fan circuit (Reprinted with permission from SAE Paper Number 790722 © 1979, Society of Automotive Engineers, Inc.)

vehicle manufacturer's specified temperature at which the cooling fan should operate, the engine temperature switch resistance should be very low. When the coolant temperature is below the specified cooling fan operating temperature, the engine temperature switch should provide an infinite ohmmeter reading. If the engine temperature switch ohmmeter readings are improper, replace the switch.

6. If the cooling fan did not run when the engine temperature switch wire was grounded in step 3, keep this wire grounded and connect a 12-V test light from the cooling fan motor 12-V input terminal to ground. If 12 V is not available at this terminal, connect the test light from the relay terminal connected to the fan motor and ground. If 12 V is not available at this terminal, replace the relay. When 12 V is available at the relay terminal connected to the fan motor, repair the open circuit between the relay and the fan motor.
7. Disconnect the cooling fan motor wires, and connect an ohmmeter from the motor ground wire to the battery ground. The ohmmeter reading should be very close to zero ohms (Ω). If the ohmmeter reading is above 0.5 Ω, repair the resistance problem in the motor ground wire.
8. If 12 V is supplied to the cooling fan motor in step 6, and the motor ground wire is satisfactory in step 7, replace the cooling fan motor.

Classroom Manual
Chapter 4, page 65

CUSTOMER CARE: When performing underhood service such as an engine tune-up, always make a quick visual inspection of the cooling system. Advise the customer about any unsafe conditions such as cracked radiator or heater hoses. If the customer spends a considerable amount of money on an engine tune-up, and two days later the customer's car overheats on the freeway because a radiator hose ruptured, the customer will likely be unhappy. When the customer is informed about the unsafe cooling system components, in most cases the customer will approve the installation of these cooling system components. The customer will also be thankful that your inspection may have prevented an expensive and inconvenient breakdown.

Guidelines for Lubrication System and Cooling System Diagnosis and Service

1. Most engines with mechanical valve lifters require a valve adjustment.
2. Some vehicle manufacturers recommend a valve adjustment when the engine is cold, and other manufacturers recommend this adjustment at normal engine temperature.
3. Engines with hydraulic valve lifters and no valve adjustment sometimes have a valve clearance check with the valve lifter bottomed.
4. Engines with individual rocker arms have a nut which retains the rocker arm pivot. This nut must be adjusted to position the lifter plunger properly.
5. Adjusting pads, or shims, are located in some mechanical valve lifters to provide a valve adjustment.
6. The valve timing may be checked with the crankshaft at TDC on the exhaust stroke and the valves in the overlap position.
7. There are many causes of engine overheating, and during the diagnosis of an overheating situation, all these causes should be checked until the source of the problem is found. Begin the diagnosis with the quickest tests, or checks.
8. Coolant hoses should be checked for soft spots, swelling, hardening, chafing, and leaks.
9. Radiators should be inspected for leaks, loose brackets, and restricted air and coolant passages.
10. Broken or improperly positioned radiator shrouds may cause engine overheating.

11. Defective radiator pressure caps may cause excessive coolant loss to the coolant recovery system, engine overheating, and collapsed hoses.
12. A coolant hydrometer is used to measure the antifreeze protection provided by the coolant.
13. Pollution laws prohibit the dumping of used coolant in sewers.
14. Drive belts should be inspected for cracks, oil contamination, glazing, tears, splits, and bottoming in the pulley.
15. Drive- belt tension may be measured with a belt tension gauge or by checking the belt deflection.
16. Water pumps with loose bearings or coolant leaks must be replaced.
17. Thermostat valves must open at the temperature stamped on the thermostat.
18. Most thermostats must be installed with the wax pellet toward the engine.
19. While flushing cooling systems, wear protective gloves, clothing, and goggles.
20. While flushing cooling systems, follow the directions supplied by the flushing equipment manufacturer.
21. The viscous-drive fan clutch should provide more resistance to fan blade rotation when the clutch is hot than when the clutch is cold.
22. If an electric-drive fan motor does not operate at the coolant temperature specified in the service manual, the fuse or fuse link, motor, relay, temperature switch, or connecting wires may be defective.

CASE STUDY

A customer complained about engine overheating on a Dodge Spirit with a 2.5L turbocharged engine. Further questioning of the customer revealed that the overheating problem always occurred while driving in the city at slower speeds. While driving on the freeway, the customer said that overheating had not been a problem. When asked if coolant loss had occurred, the customer replied that coolant had not been added.

A visual inspection of the cooling system components did not reveal any problems, except the coolant level was low in the coolant recovery container. The technician noticed that the electric-drive cooling fan was running when the engine was hot. A quick check of the basic ignition timing and spark advance indicated these items were satisfactory. A water pump belt tension and condition check indicated the belt was in good condition with proper tension.

After the engine cooled off, the radiator cap was removed. The coolant, radiator, and cap were inspected. Coolant level in the radiator was low, but the coolant was clean and provided adequate antifreeze protection when tested with a coolant hydrometer. The radiator coolant tubes and air passages in the radiator were not restricted. A pressure tester was used to test the pressure cap, and the cap test was normal.

The technician pressure tested the cooling system, and discovered that the tester pressure dropped off gradually while the system was pressurized for 15 minutes. A careful inspection of the entire cooling system did not indicate any external leaks. The technician placed an exhaust gas analyzer pickup in the radiator filler neck without placing the pickup in the coolant. The engine was operated at idle speed and suddenly accelerated several times. A hydrocarbon (HC) reading appeared on the analyzer HC meter indicating combustion chamber gases escaping from the coolant in the radiator. (See Chapter 3 for an explanation of this test.)

When the cylinder head was removed, the head gasket indicated signs of leaking into the number 3 cylinder. The head gasket was replaced, and the head bolts were carefully torqued. After refilling the cooling system, a pressure test indicated there was no pressure drop on the tester gauge after 15 minutes. During a road test the car performed well, and there were no heating problems.

Terms to Know

Mechanical valve lifters, or solid tappets
Hydraulic valve lifters
Adjusting pads, mechanical lifters
Valve stem installed height
Valve overlap
Viscous-drive fan clutch
Radiator shroud
Hydrometer
Specific gravity

ASE Style Review Questions

1. While discussing valve adjustments:
   Technician A says excessive valve stem installed height may cause the hydraulic valve lifter plunger to be bottomed in the lifter.
   Technician B says cylinder misfiring may be caused by excessive valve stem installed height.
   Who is correct?
   A. A only **C.** Both A and B
   B. B only **D.** Neither A nor B

2. While discussing valve timing diagnosis:
   Technician A says when the piston is at TDC on the exhaust stroke, the intake valve should be closed.
   Technician B says when the piston is at TDC on the exhaust stroke, the exhaust valve should be moving toward the open position.
   Who is correct?
   A. A only **C.** Both A and B
   B. B only **D.** Neither A nor B

3. While discussing engine overheating:
   Technician A says engine overheating may be caused by late ignition timing.
   Technician B says engine overheating may be caused by a defective viscous-drive fan clutch.
   Who is correct?
   A. A only **C.** Both A and B
   B. B only **D.** Neither A nor B

4. While discussing electric-drive cooling fan systems:
   Technician A says an electric-drive cooling fan may start turning with the ignition switch off.
   Technician B says the engine temperature switch in the electric-drive cooling fan circuit is closed when the coolant is cold.
   Who is correct?
   A. A only **C.** Both A and B
   B. B only **D.** Neither A nor B

5. While discussing radiator pressure cap diagnosis:
   Technician A says a collapsed radiator hose may be caused by a defective sealing gasket in the pressure cap.
   Technician B says a collapsed radiator hose may be caused by a defective vacuum valve in the pressure cap.
   Who is correct?
   A. A only **C.** Both A and B
   B. B only **D.** Neither A nor B

6. While discussing coolant hydrometer testing:
   Technician A says a coolant hydrometer measures the specific gravity of the coolant.
   Technician B says the specific gravity of a liquid is the weight of the liquid in relation to the weight of an equivalent volume of water.
   Who is correct?
   A. A only **C.** Both A and B
   B. B only **D.** Neither A nor B

7. While discussing drive belt diagnosis:
   Technician A says if the water pump V-belt is rubbing on the bottom of the pulley groove, the belt should be replaced.
   Technician B says a V-belt should have one inch of deflection per foot of belt span.
   Who is correct?
   A. A only **C.** Both A and B
   B. B only **D.** Neither A nor B

8. While discussing thermostat diagnosis:
   Technician A says the thermostat pellet should face toward the radiator.
   Technician B says if a defective thermostat opens at a lower temperature than the thermostat rating, this defect has no effect on computer input sensor signals.
   Who is correct?
   A. A only **C.** Both A and B
   B. B only **D.** Neither A nor B

9. While discussing viscous-drive fan clutch diagnosis:
Technician A says a viscous-drive fan clutch should offer more resistance to rotation when the engine is cold and less rotation resistance when the engine is hot.
Technician B says if there is looseness in the viscous clutch shaft bearing, the clutch assembly should be replaced.
Who is correct?
A. A only **C.** Both A and B
B. B only **D.** Neither A nor B

10. While discussing the diagnosis of an inoperative electric-drive cooling fan circuit:
Technician A says if the cooling fan motor runs with the engine temperature switch wire grounded, the defect is in the temperature switch.
Technician B says if the cooling fan motor runs with the engine temperature switch wire grounded, the cooling fan relay is defective.
Who is correct?
A. A only **C.** Both A and B
B. B only **D.** Neither A nor B

Table 4-1 ASE TASK

Adjust valves on engines with mechanical or hydraulic lifters.

Problem Area	Symptoms	Possible Causes	Classroom Manual	Shop Manual
ENGINE NOISE	Clicking at idle speed	1. Excessive valve clearance	50	50
		2. Worn valve train components	50	53
ENGINE MISFIRING	Misfiring at idle and low speeds	Insufficient valve clearance	50	53

Table 4-2 ASE TASK

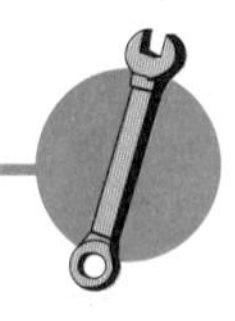

Verify correct valve timing; determine needed repairs.

Problem Area	Symptoms	Possible Causes	Classroom Manual	Shop Manual
ENGINE	Failure to start	Improper valve timing	52	53
	Reduced power and economy	Improper valve timing	52	54

Table 4-3 ASE TASK

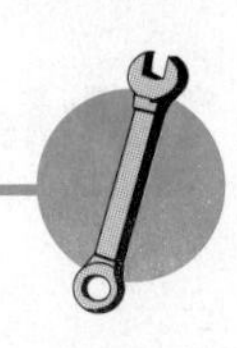

Verify engine operating temperature; determine needed repairs.

Problem Area	Symptoms	Possible Causes	Classroom Manual	Shop Manual
ENGINE TEMPERATURE	Overheating	1. Restricted radiator air or coolant passages	68	55
		2. Defective pressure cap	69	57
		3. Defective water pump	67	59
		4. Loose or worn water pump belt	67	59
		5. Coolant leaks, head gasket	66	58
		6. Broken, improperly positioned shroud	73	57
		7. Electric-drive cooling fan or viscous fan clutch	73	62
		8. Defective thermostat	72	60
		9. Restricted, collapsed radiator hoses	71	55
		10. Contaminated or improper coolant mixture	66	58
		11. Late ignition timing, excessive engine load	65	55
		12. Lean air-fuel mixture	65	55
		13. Dragging brakes	65	55
		14. Automatic transmission overheating	78	55
	Reduced engine temperature	Defective thermostat	72	60
ENGINE	Reduced power and economy	Thermostat opening too soon	72	60

Table 4-4 ASE TASK

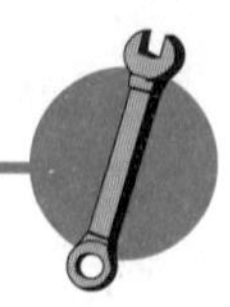

Perform cooling system pressure tests; check coolant; inspect and test radiator, pressure cap, coolant recovery tank, and hoses; determine needed repairs.

Problem Area	Symptoms	Possible Causes	Classroom Manual	Shop Manual
ENGINE TEMPERATURE	Overheating	1. Coolant leaks	71	57
		2. Radiator coolant or air passages restricted	68	57
		3. Defective pressure cap	70	57

Table 4-5 ASE TASK

Inspect, test, and replace thermostat, by-pass, and housing.

Problem Area	Symptoms	Possible Causes	Classroom Manual	Shop Manual
ENGINE TEMPERATURE	Overheating	Thermostat not properly installed	72	60
	Lack of heat from heater	Thermostat opening too soon	72	60
ENGINE PERFORMANCE	Reduced power and economy	Thermostat opening too soon	72 72	60 60

Table 4-6 ASE TASK

Inspect, test, and replace mechanical/electrical fans, fan clutch, fan shroud/ducting, and fan control devices.

Problem Area	Symptoms	Possible Causes	Classroom Manual	Shop Manual
ENGINE TEMPERATURE	Overheating	1. Inoperative electric-drive fan motor	73	62
		2. Broken, improperly positioned shroud	73	62
		3. Viscous-drive fan clutch slipping	74	62

Intake and Exhaust System Service

Upon completion and review of this chapter, you should be able to:

- ❑ Diagnose vacuum-operated components and systems with a vacuum gauge and a vacuum hand pump.
- ❑ Visually inspect the exhaust system, and determine the needed repairs.
- ❑ Diagnose intake manifold vacuum leaks.
- ❑ Service and replace air filters.
- ❑ Diagnose heated air inlet systems.
- ❑ Diagnose intake manifold temperature control systems.
- ❑ Diagnose heated grid systems.
- ❑ Inspect exhaust systems, and determine the needed repairs.
- ❑ Remove, replace, and service exhaust manifolds.
- ❑ Remove, replace, and service mufflers, pipes, and catalytic converters.
- ❑ Diagnose catalytic converters.
- ❑ Remove and replace catalytic converter pellets.
- ❑ Diagnose, service, remove, and replace heat riser valves.

Basic Tools

Basic technician's tool set

Service manual

Penetrating solvent

Trouble light

Air gun

Intake System Diagnosis and Service

Vacuum Test Equipment

Special Tools

Vacuum hand pump

The most common types of vacuum test equipment are the vacuum gauge and the vacuum hand pump. The vacuum hand pump creates vacuum when the pump is operated, and this vacuum is indicated on a gauge mounted on the pump. The hand pump may be used to check vacuum devices for leaks. For example, the hand pump hose may be connected to the air door vacuum motor in an air cleaner (Figure 5-1). Operate the hand pump until 10 in Hg (69 kPa) appears on the gauge. Under this condition, the air door should move upward to the hot air position, and the motor diaphragm should maintain the vacuum reading on the gauge for 120 seconds. If the vacuum slowly decreases on the gauge, the motor diaphragm is leaking and must be replaced.

Visual Inspection of Intake System

Visually inspect all vacuum hoses and vacuum-operated components for these defects:

1. Check all vacuum hoses to be sure they are properly routed and connected. The underhood vacuum decal indicates proper vacuum hose routing.
2. Inspect all vacuum hoses for proper tight fit between the hoses and nipples.

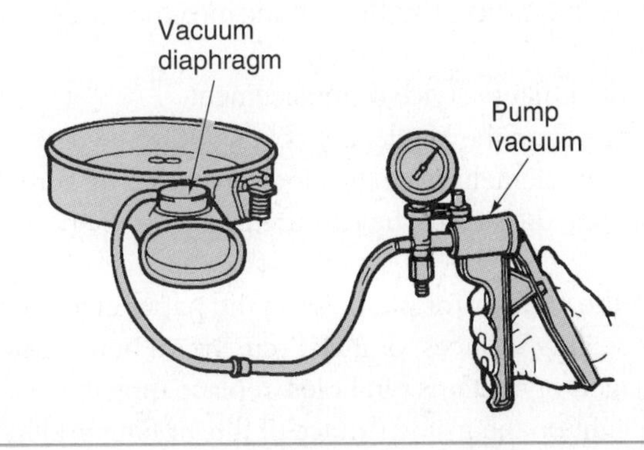

Figure 5-1 Hand vacuum pump connected to test air door vacuum motor diaphragm (Courtesy of Chrysler Corporation)

3. Inspect all vacuum hoses for kinks, breaks, and cuts.
4. Be sure vacuum hoses are not burned because they are positioned near hot components such as exhaust manifolds or exhaust gas recirculation (EGR) tubes.
5. Inspect all vacuum-operated components for damage, such as broken nipples on a thermal vacuum switch (TVS).
6. Check for evidence of oil on vacuum hose connections. Oil in a vacuum hose may contaminate vacuum-operated components or plug the vacuum hose.

Vacuum Leak Diagnosis

Broken or disconnected vacuum hoses may cause an air leak into the intake manifold, resulting in a lean air-fuel mixture. When the air-fuel mixture is leaner than normal, engine idle operation is erratic. On fuel injected engines, an intake manifold vacuum leak may cause the engine to idle faster than normal. An intake manifold vacuum leak may cause a hesitation on low-speed acceleration. If the vacuum leak is positioned in the intake manifold so it leans the air-fuel mixture on one cylinder more than the other cylinders, the cylinder may misfire at idle and lower engine speeds. Once the engine speed increases, the reduced intake manifold vacuum does not pull as much air through the vacuum leak, and the cylinder stops misfiring.

Special Tools

Vacuum gauge

A vacuum gauge connected to the intake manifold indicates a low steady reading if there is a vacuum leak into the intake manifold. (See Chapter 3 for complete vacuum gauge diagnosis.) Intake manifold mounting bolts should be checked periodically for proper torque. Loose intake manifold bolts cause leaks between the intake manifold and the cylinder head.

On some V-type engines, the intake manifold covers the valve lifter chamber. If the intake manifold gaskets are leaking on the underside of the intake manifold on these applications, oil splash from the valve lifter chamber is moved past the intake manifold gasket into the cylinders. This action results in oil consumption and blue smoke in the exhaust.

Kinked vacuum hoses shut off the vacuum to a component, making it inoperative. For example, if the vacuum hose to the EGR valve is kinked, vacuum is not supplied to this valve, and the valve remains closed. The EGR valve recirculates some exhaust into the intake manifold. Since this exhaust gas contains very little oxygen, it does not burn in the combustion chambers and combustion temperature is reduced, which lowers oxides of nitrogen (NO_x) emissions. If the EGR valve does not open, combustion chamber temperatures are higher than normal and the engine may detonate. When the EGR valve is inoperative, NO_x emissions are high.

Air Cleaner and Filter Service

The air filter should be replaced at the intervals recommended in the vehicle manufacturer's service manual. This information is also available in the owner's manual. If the vehicle is operated continually in dusty conditions, air filter replacement may be necessary at more frequent intervals. A damaged air filter may cause increased wear on cylinder walls, pistons, and piston rings. If the air filter is restricted with dirt, it restricts the flow of air into the intake manifold, and this action increases fuel consumption.

Follow these steps for air filter service or replacement:

1. Remove the wing nut on the air cleaner cover, and remove the cover.
2. Remove the air filter element from the air cleaner, and be sure that no foreign material, such as small stones, drops into the carburetor or throttle body while removing the element.
3. Visually inspect the air filter for pin holes in the paper element and damage to the paper element, sealing surfaces, or metal screens on both sides of the element. If the element is damaged or contains pin holes, replace the element.
4. Place a trouble light on the inside surface of the air filter and look through the filter element to the light. The light should be visible through the paper element, but no holes should appear in the element. If the paper element is plugged with dirt or oil, the light

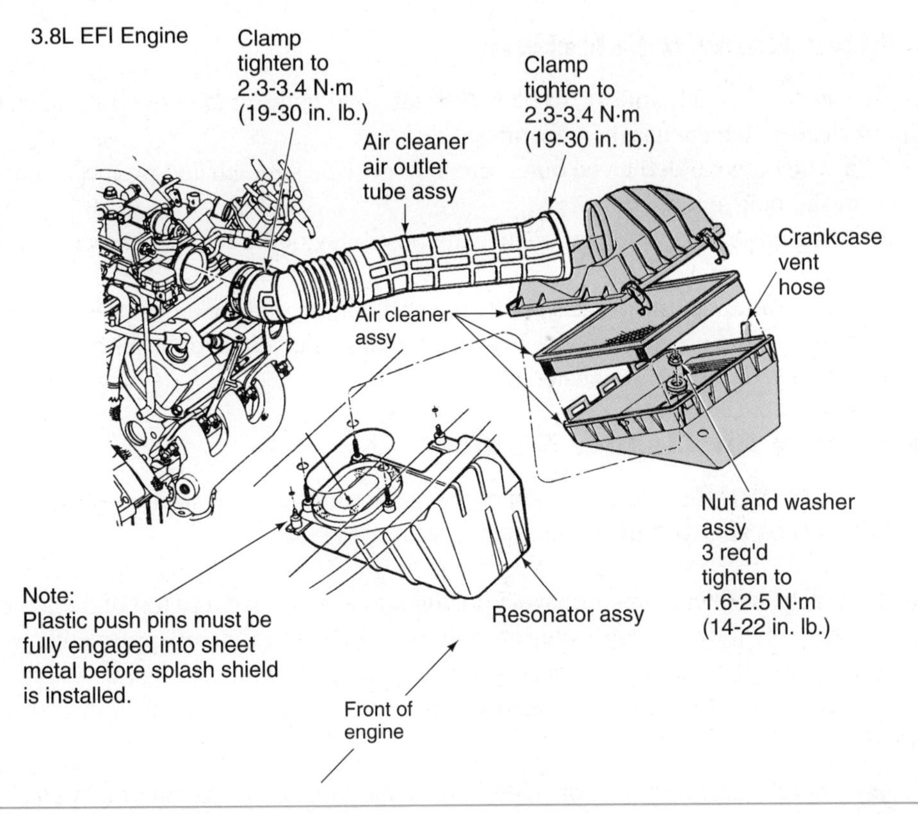

Figure 5-2 Air cleaner with positive crankcase ventilation (PVC) inlet filter (Reprinted with permission of Ford Motor Company)

is not visible through the element. When the element is plugged with dirt or oil, replace the element. If the air filter element is contaminated with oil, excessive blow-by or a defective positive crankcase ventilation (PCV) system is causing a pressure buildup in the engine. Under this condition, oil is forced up the clean air hose from the rocker arm cover into the air cleaner. Clean air should normally flow from the air cleaner through this hose to the engine.

5. If there is any dirt in the air cleaner body around the air filter, remove the air cleaner body from the engine and use a clean shop towel to remove the foreign material from the air cleaner body.
6. Check the PCV inlet filter for dirt accumulation (Figure 5-2). Many PCV inlet filters are made from foam plastic, and these filters may be washed in an approved solvent and reused. If the PCV inlet filter is damaged or plugged with dirt, replace the filter.
7. Be sure the gasket between the air cleaner and the throttle body or carburetor is in satisfactory condition, and install the air cleaner body.
8. Install the air filter element, and be sure the seal on the lower side of the element fits evenly against the matching surface on the air cleaner body.
9. Install the air cleaner cover, and be sure the cover sealing area fits against the element.

The PCV inlet filter may be called an antibackfire filter.

WARNING: Do not overtighten an air cleaner wingnut. This action may distort the air cleaner body and cause air leaks.

10. Install and tighten the wingnut.
11. Be sure the PCV hose and any other hoses or sensors are properly connected to the air cleaner.

Air Filter Element Selection

There are many different brands of air filter elements sold in the automotive aftermarket. A good quality air cleaner element has these features:

1. A wire or expanded metal outer screen to provide strength and protect against damage in shipping and handling.
2. A fine mesh inner screen to reduce the possibility of fire hazards caused by engine backfire.
3. Heat-resistant plastisol seals on the top and bottom of the element to provide a positive dust seal and prevent air flow between the filter element and the air cleaner body.
4. Oil-wetted, resin-impregnated pleated paper element to provide long filter life.

Heavy-Duty Air Cleaner Service

WARNING: Do not wring or twist a polyurethane air cleaner element cover. This action stretches the cover and makes it useless.

Some heavy-duty air filter elements have a polyurethane cover over the top of the paper air filter element. This polyurethane cover may be removed and washed in an approved solvent. After washing the cover, the excess solvent should be squeezed from the element. A light coating of ordinary crankcase oil should be placed on the polyurethane cover, and the cover may be squeezed to distribute the oil evenly.

WARNING: If air is directed from an air gun against the outside of an air filter element, the blast of air may blow dirt particles through the element and create pin holes in the element. If this element is reused in the engine, dirt will enter the engine through these pin holes, resulting in premature engine wear.

CAUTION: Do not direct air from an air gun against any part of your body. If air penetrates the skin and enters the blood stream, it will cause serious personal injury or death.

Some heavy-duty air filter elements have a heavy paper element encased between two metal end caps. This type of air filter element may be cleaned and reused. Before cleaning the element, it should be inspected for tears, punctures, pin holes, and bent end caps. If any of these conditions are present, replace the element.

An air gun on a shop air hose may be used to blow the dirt and foreign material from the air filter element. Always keep the air gun six inches away from the air filter, and direct the air against the inside of the air filter. While blowing dirt out of an air filter, do not allow the air pressure to exceed 30 psi (207 kPa). After blowing the dirt from the air filter, reinspect the air filter element for pin holes and remaining dirt with a shop light.

Some heavy-duty air filter elements may be washed in an approved filter cleaning solution for 15 minutes or more. The filter may be rinsed with low-pressure water from a water hose and then allowed to dry. Always follow the vehicle manufacturer's air filter element cleaning instructions in the service manual.

Other heavy-duty air cleaners have a precleaner containing a series of tubes and fins. These precleaners should be inspected for dirt accumulation. A brush with stiff nylon or fiber bristles may be used to clean the tubes and fins in the precleaner.

Heated Air Inlet System Diagnosis

If the air door in a heated air inlet system remains in the cold air position while the engine is cold, an acceleration stumble may occur. When the engine is hot and the air door remains in the hot air position, the engine may detonate. Follow these steps to diagnose a heated air inlet system:

Figure 5-3 A thermometer may be installed in air cleaner to check the heated air inlet system.

Special Tools

Magnetic-base thermometer

1. With the engine cold, install a thermometer inside the air cleaner near the bimetal sensor (Figure 5-3). The thermometer may be taped to the air cleaner body to keep it from being drawn into the air intake. Install the air cleaner cover and wingnut.
2. Start the engine and observe the air door in the air cleaner snorkel. This door should be lifted upward to the hot air position.
3. Observe the air door as the engine warms up. When the air door begins to move downward to a modulated position, shut the engine off and remove the air cleaner cover. The temperature on the thermometer should be 75°F to 125°F (24°C to 52°C), if the heated air inlet system is working normally.
4. Install the air cleaner cover and observe the air door. This door should move to the cool air position as the engine approaches normal operating temperature.
5. If the air door does not move to the hot air position when the engine is cold, remove the vacuum hose from the air door vacuum motor diaphragm. Install a vacuum gauge on the end of this hose with the engine idling and the engine cold. This vacuum reading should exceed 10 in Hg (69 kPa). If the vacuum reading is normal, connect a hand vacuum pump to the vacuum nipple on the air door vacuum motor. Operate the hand pump to supply 10 in Hg (69 kPa) to the vacuum motor. If this vacuum reading decreases in less than 120 seconds, replace the vacuum motor.
6. If 10 in Hg (69 kPa) is not available at the air door vacuum motor with the engine idling, check all the system hoses for vacuum leaks and replace the hoses as necessary. To test each hose for leaks, connect one end of the hose to the vacuum hand pump and plug the opposite end of the hose. Operate the hand pump and supply 20 in Hg (138 kPa) to the hose. If the gauge reading on the pump slowly decreases, the hose is leaking.
7. When the hoses are satisfactory, install a vacuum gauge in the vacuum hose from the intake manifold to the bimetal sensor. If the intake manifold vacuum is above 10 in Hg, install the vacuum gauge on the bimetal sensor nipple connected to the air door vacuum motor. If the vacuum is less than 10 in Hg (69 kPa), replace the bimetal sensor.
8. If the air door remains in the hot air position when the engine is at normal operating temperature, check all the system hoses for proper routing and connections. Check the air door to be sure it moves freely. Inspect the flexible hose and heat stove for damage and proper installation. If these system checks are satisfactory, replace the bimetal sensor.

Photo Sequence 3 shows a typical procedure for diagnosis of the heated air inlet system.

Diagnosis of Intake Manifold Temperature Control Systems

If the thermostat is sticking in the open position, engine coolant temperature is reduced. On engines with the coolant circulated through the intake manifold to provide manifold heating, a continually open thermostat reduces intake manifold temperature, resulting in decreased fuel vaporization and engine performance.

Photo Sequence 3
Typical Procedure for Diagnosis of Heated Air Inlet System

P3-1 With the engine cold remove the air cleaner cover and tape a thermometer to the air cleaner near the bimetal sensor.

P3-2 Install the air cleaner cover and start the engine.

P3-3 Observe the air door in the air cleaner snorkel. With a cold engine this door should be lifted upward to the hot air position.

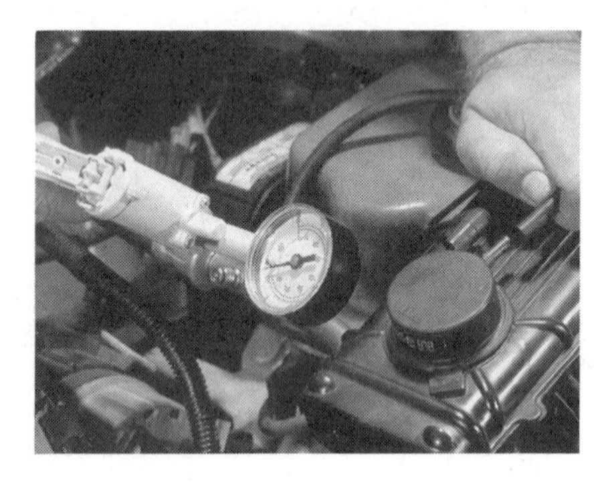

P3-4 If the air door does not move upward to the hot air position, disconnect the vacuum hose from the air door diaphragm, and connect a hand vacuum pump to this diaphragm vacuum inlet.

P3-5 Operate the vacuum hand pump and supply 10 in Hg (69 kPa) to the air door diaphragm. If the vacum reading decreases in 120 seconds, replace the diaphragm.

P3-6 Connect a vacuum gauge to the vacuum hose removed from the air door diaphragm. With the engine idling and cold the vacuum reading should exceed 10 in Hg (69 kPa). If the vacuum reading is lower than 10 in Hg (69 kPa), check the bimetal sensor and connecting hoses.

P3-7 If the air door moves upward to the hot air position in step 3, reconnect the vacuum hose to the air door diaphragm. Continue to observe the air door as the engine warms up. When the air door begins to move downward to a modulated position, shut off the engine.

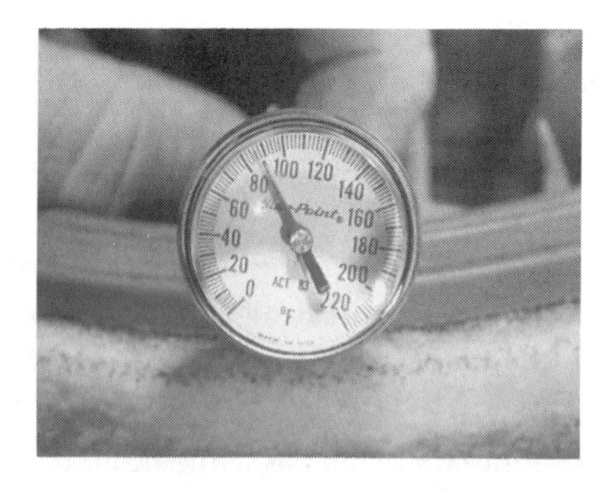

P3-8 Remove the air cleaner cover and check the thermometer reading. If the heated air inlet system is working properly, the temperature indicated on the thermometer should be 75°F to 125°F (24°C to 52°C).

P3-9 Remove the thermometer and install the air cleaner cover. Start the engine and observe the air door. As the engine warms up the air door should move to the cold air position. If the air door does not move to the cold air position, check the bimetal sensor.

Exhaust flow through the crossover passage in the intake manifold may be reduced by a heat riser valve sticking in the open position or carbon formation in the crossover passage. When the crossover passage is restricted with carbon formations, the intake manifold must be removed to clean the carbon from the crossover passage.

If the exhaust flow is reduced through the crossover passage, the engine has a lack of power and acceleration stumbles, especially when the engine is cold. Fuel consumption increases if the crossover passage does not heat the intake manifold adequately. Reduced exhaust flow through the crossover passage decreases the heat supplied to the choke spring, if this spring is located in a crossover passage well. Under this condition, the choke does not open properly, resulting in engine flooding, excessive fuel consumption, and stalling.

If the heat riser valve remains in the closed position with the engine at normal operating temperature, a loss of engine power is experienced because of the heat riser valve restriction in the exhaust system. This action is more noticeable with a vacuum-operated heat riser valve. The continually closed heat riser valve overheats the crossover passage and air-fuel mixture, resulting in engine detonation. Since the heat riser valve is part of the exhaust system, the diagnosis of this component is discussed in exhaust system diagnosis later in this chapter.

Diagnosis of Heated Grid Systems

If the heated grid does not warm the air-fuel mixture, the engine may hesitate on acceleration during cold engine operation. A defective heated grid system decreases fuel vaporization and engine performance, while increasing fuel consumption. Follow this procedure to diagnose a heated grid system:

Special Tools

Voltmeter

Ohmmeter

1. With the engine cold and the ignition switch on, disconnect the voltage supply wire from the heated grid, and connect a voltmeter from this wire to ground. The voltmeter should indicate 12 V.
2. If the voltage in step 1 is satisfactory, turn the ignition switch off and connect an ohmmeter across the grid terminals (Figure 5-4). A low reading indicates a satisfactory grid, and an infinite reading proves the grid has an open circuit. If the grid is open, replacement is necessary. Connect the ohmmeter from the grid ground wire to the battery ground. If the meter reading is nearly zero, the ground wire is satisfactory. When the meter reading exceeds 1 Ω, repair the resistance problem in the wire.
3. If the voltage reading in step 1 is less than 12 V, turn on the ignition switch and connect the voltmeter from the temperature switch terminal connected to the grid and ground. If this reading is 12 V, but the reading in step 1 is not satisfactory, repair the resistance problem in the wire from the temperature switch to the grid.

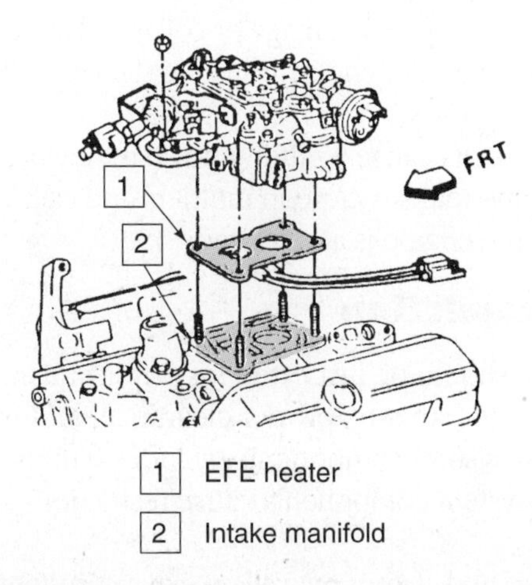

Figure 5-4 Intake manifold heater grid (Courtesy of Chevrolet Motor Division, General Motors Corporation)

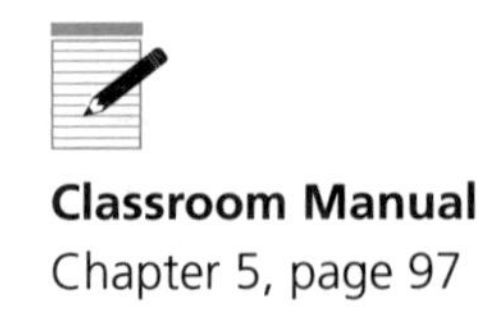

Classroom Manual
Chapter 5, page 97

4. If the voltage at the temperature switch in step 3 is not satisfactory, connect the voltmeter from the temperature switch input terminal to ground. If the voltage at this terminal is satisfactory, replace the temperature switch.
5. When the voltage at the temperature switch input terminal is not satisfactory, check the input wire and fuse.

Exhaust System Service

WARNING: Exhaust gas contains poisonous carbon monoxide (CO) gas. This gas can cause illness and death by asphyxiation. Exhaust system leaks can be dangerous for customers and for technicians, if the vehicle is running in the shop without an exhaust hose connected to the tailpipe.

CAUTION: Exhaust system components may be extremely hot if the engine has been running. Allow the engine and exhaust system to cool down prior to exhaust system service, and wear protective gloves.

CAUTION: While servicing exhaust systems, follow all safety rules regarding personal safety, vehicle hoisting, equipment safety, and shop safety. Refer to Chapters 1 and 2 for safety precautions.

Exhaust System Problems

Exhaust system components are subject to rust from inside and outside the components. Since exhaust components are exposed to road splash, they tend to rust on the outside. Stainless steel exhaust system components have excellent rust-resistant qualities. Exhaust system components are subject to large variations in temperature, which form condensation inside these components. Condensation inside the exhaust system components rusts them on the inside. A vehicle that is driven short distances and then shut off tends to get more condensation in the exhaust system than a vehicle that is driven continually for longer time periods. If a vehicle is driven for a short distance, the exhaust system heat does not have sufficient time to vaporize the condensation in the system. The rust erodes exhaust system components, causing exhaust leaks and excessive noise.

Exhaust components, particularly mufflers, catalytic converters, and pipes, may become restricted. Mufflers and catalytic converters may become restricted by loose internal components. Pipes may become restricted by physical damage or collapsed inner walls on double-walled pipes. This exhaust restriction causes a loss of engine power and increased fuel consumption. (See Chapter 3 for exhaust restriction diagnosis.)

Exhaust system components cause a rattling noise if they are touching, or almost touching, chassis components. Loose internal structure in mufflers and catalytic converters may cause a rattling noise, especially when the engine is accelerated.

Exhaust System Inspection

Exhaust system components should be inspected for physical damage, holes, rust, loose internal components, loose clamps and hangers, proper clearance, and improperly positioned shields. Physical damage on exhaust system components includes flattened areas, abnormal bends, and scrapes. Flattened exhaust system components cause restrictions. Components with this type of damage should be replaced.

Loose or broken hangers and clamps may allow exhaust system components to contact chassis components, resulting in a rattling noise. Broken hangers cause excessive movement of exhaust system components, and this action may result in premature breaking of these components.

A block of wood or a wrench may be used to tap mufflers and catalytic converters to check for loose internal components. If the internal structure is loose, a rattling noise is evident when the component is taped.

Push upward on the muffler and catalytic converter to check for weak, rusted inlet and outlet pipes. Check the entire exhaust system for leaks. With the engine running, hold a length of hose near your ear, and run the other end of the hose around the exhaust system components. The hose amplifies the sound of an exhaust leak and helps to locate the source of a leak.

Inspect all exhaust system shields for proper position, and check for the correct clearance between the exhaust system components and the shields (Figure 5-5). Check the catalytic converter for overheating, which is indicated by bluish or brownish discoloration.

Exhaust Manifold Removal and Replacement

SERVICE TIP: Some exhaust manifolds do not have a gasket between the manifold and the cylinder head. Since an exhaust manifold is subject to severe heat and temperature changes, a certain amount of manifold distortion may occur. When an exhaust manifold is reused, gaskets are available for those applications that did not have a gasket originally. The installation of an exhaust manifold gasket compensates for manifold irregularities and prevents exhaust leaks.

SERVICE TIP: Since exhaust system fasteners are subject to extreme heat, they may be difficult to loosen. A generous application of penetrating solvent on these fasteners makes removal easier.

SERVICE TIP: When exhaust system fasteners such as clamp nuts are seized, tighten the fastener. In many cases, the clamp bolt breaks off easily when you attempt to tighten it. Do not use this procedure on exhaust manifold bolts and studs.

Remove the exhaust pipe bolts at the manifold flange, and disconnect any other components in the manifold, such as an oxygen (O_2) sensor. In some applications, it is easier to disconnect the sensor wires and leave the sensor in the manifold. However, if the exhaust manifold is being replaced, the sensor must be removed and installed in the new manifold. Remove the bolts retaining the manifold to the cylinder head, and lift the manifold from the engine compartment (Figure 5-6). Remove the manifold heat shield.

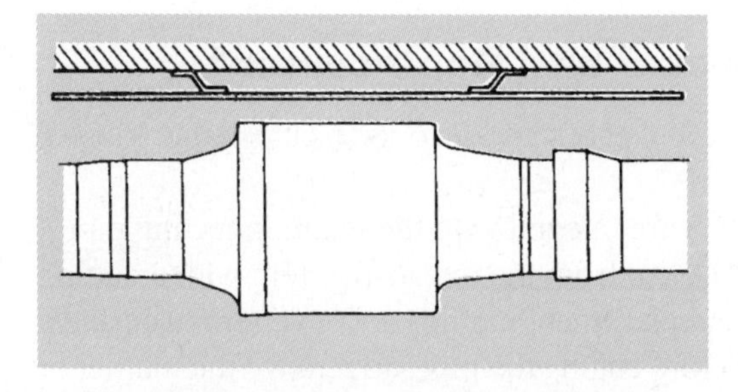

Figure 5-5 Check for proper shield position and correct clearance between the exhaust system components and the shields. (Courtesy of Toyota Motor Corporation)

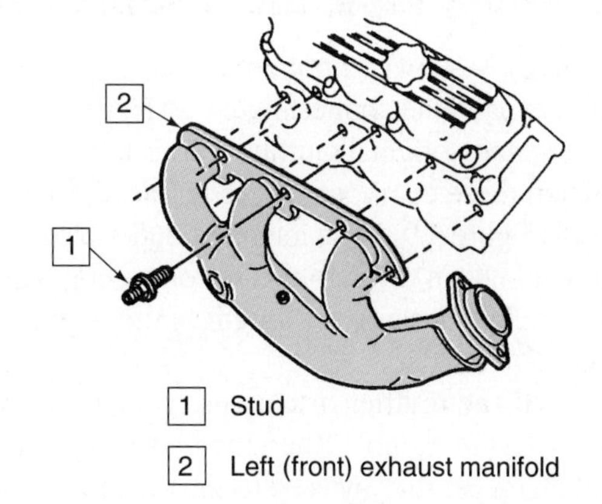

Figure 5-6 Removing exhaust manifold from the cylinder head (Courtesy of Oldsmobile Division, General Motors Corporation)

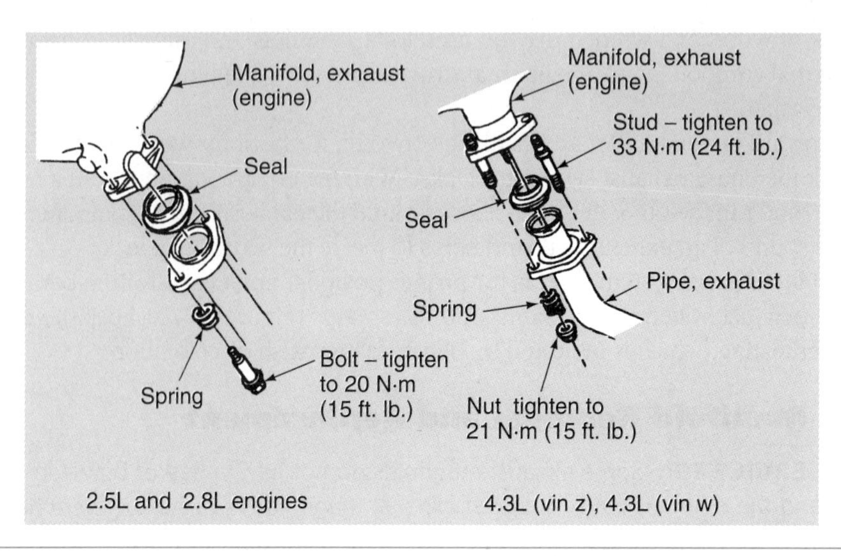

Figure 5-7 Exhaust pipe-to-manifold flange installation (Courtesy of Chevrolet Motor Division, General Motors Corporation)

Special Tools
Straightedge

Use a scraper to clean the matching surfaces on the exhaust manifold and cylinder head. Measure the exhaust manifold surface for warping with a straightedge and feeler gauge. Perform this measurement at three locations on the manifold surface. If the manifold is warped more than specified by the vehicle manufacturer, replace the manifold. Remove any gasket material from the manifold flange. If the flange has a ball connection, be sure the ball is not damaged.

If the manifold has an O_2 sensor, place some antiseize compound on the threads and install the sensor in the manifold. When this sensor is easy to access, it may be installed after the manifold is installed on the cylinder head. Install a new gasket between the cylinder head and the manifold, and install the manifold against the cylinder head. Install exhaust manifold-to-cylinder head mounting bolts, and tighten these bolts to the specified torque. Many exhaust manifold bolts, nuts, and studs are special heat-resistant fasteners. If exhaust manifold bolts or nuts must be replaced, do not substitute ordinary fasteners for heat-resistant fasteners.

Be sure the exhaust pipe mounting surface that fits against the exhaust manifold flange is in satisfactory condition. If an exhaust manifold flange gasket is required, install a new gasket and tighten the flange bolts to the specified torque (Figure 5-7).

Muffler, Pipe, and Catalytic Removal and Replacement

Exhaust system component replacement requires the use of various tools to cut, shape, loosen, and expand the connecting parts (Figure 5-8).

Special Tools
Muffler cutter

Many original mufflers and catalytic converters are integral with the interconnecting pipes. When these components are replaced, they must be cut from the exhaust system with a cutting tool (Figure 5-9). The inlet and outlet pipes on the replacement muffler or converter must have a 1.5-in (3.8-cm) overlap on the connecting pipes. Before cutting the pipes to remove the muffler or converter, measure the length of the new component and always cut these pipes to provide the required overlap.

If the muffler or converter inlet and outlet pipes are clamped to the connecting pipes, remove the clamp. When the muffler or converter is tight on the connecting pipe, a slitting tool and hammer may by used to slit and loosen the muffler pipe (Figure 5-10). A slitting tool on an air chisel may be used for this job.

Special Tools
Pipe expander

When a new muffler or converter is installed, the connecting pipes may not fit perfectly. A hydraulically operated expanding tool may be used to expand the pipe and provide the necessary fit (Figure 5-11). In some cases, adaptors may be used to provide the necessary pipe fit. A sleeve may be used to join two pieces of pipe (Figure 5-12).

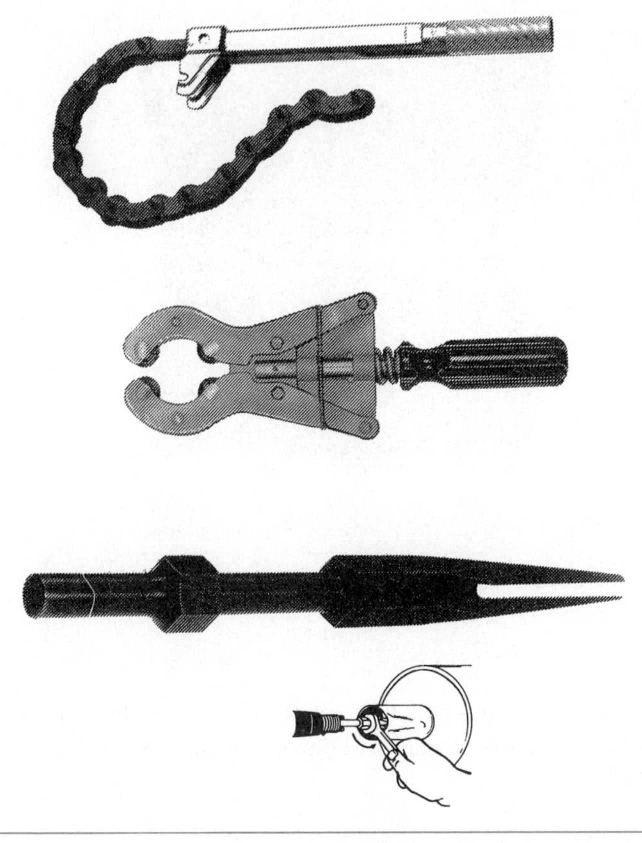

Figure 5-8 Exhaust system service tools (Courtesy of Mac Tools, Inc.)

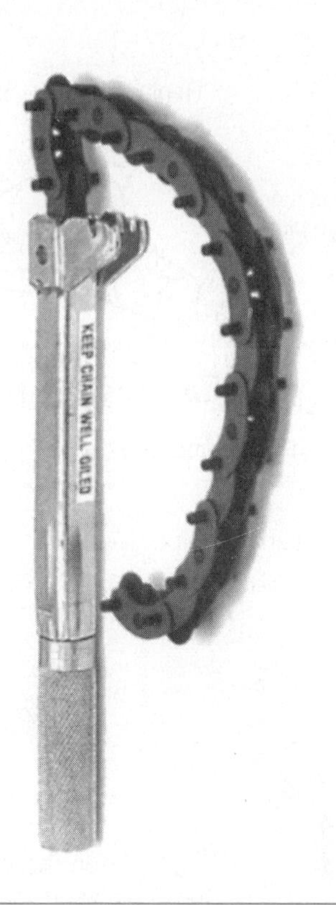

Figure 5-9 Using a cutting tool to remove a muffler (Courtesy of Snap-on Tools Corporation)

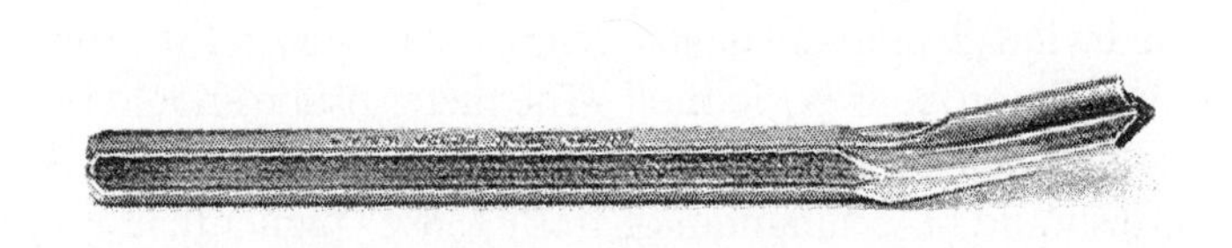

Figure 5-10 A slitting tool and a hammer may be used to slit and loosen a muffler, or converter pipe. (Courtesy of Snap-on Tools Corporation)

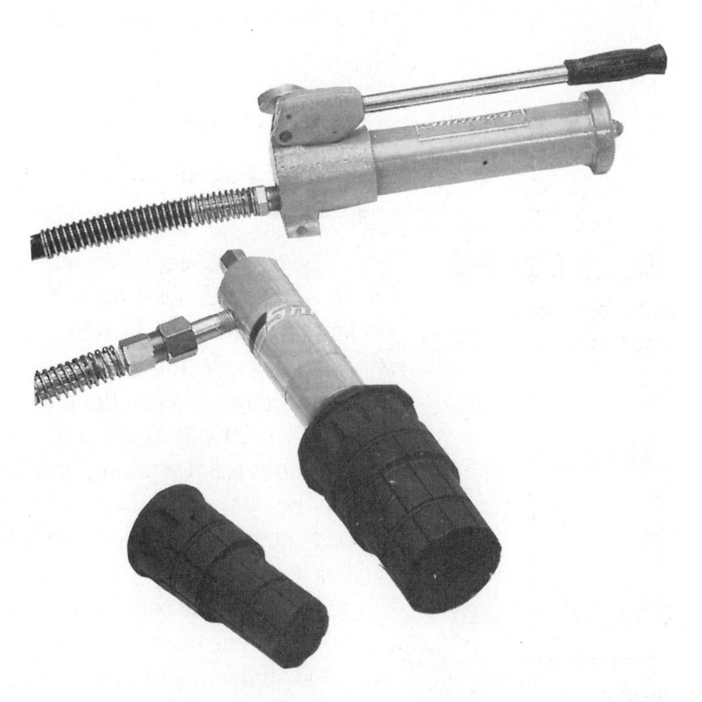

Figure 5-11 Hydraulically operated pipe expanding tool (Courtesy of Snap-on Tools Corporation)

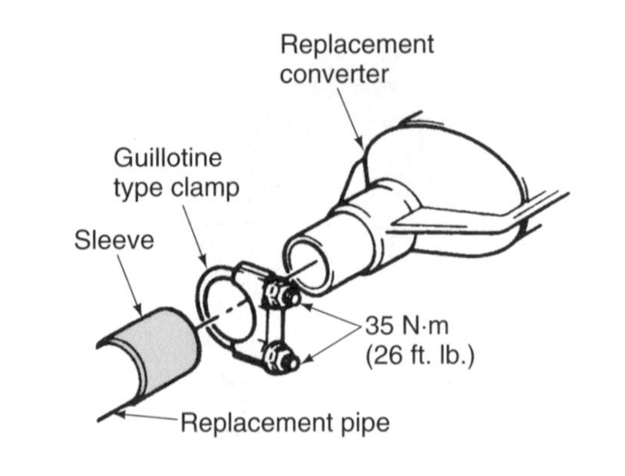

Figure 5-12 Exhaust pipe sleeve (Courtesy of Oldsmobile Division, General Motors Corporation)

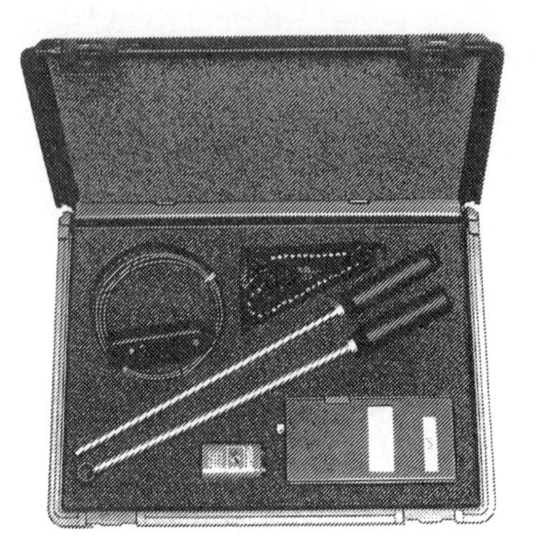

Figure 5-13 Digital pyrometer for measuring catalytic converter temperature (Courtesy of Mac Tools, Inc.)

When the new exhaust system components are installed, there must be adequate clearance between all the exhaust components and the chassis. Install and tighten all clamps and hangers securely.

Catalytic Converter Diagnosis and Service

The most accurate method of testing a catalytic converter is to measure the tailpipe emissions with a four-gas infrared exhaust analyzer. Other systems, such as the fuel, ignition, and emission systems, also affect tailpipe emissions. An understanding of these systems is necessary before we discuss exhaust gas analysis with a four-gas analyzer. All this information is provided later in the appropriate chapters.

WARNING: Discarding, disconnecting, or tampering with any emission component is a serious federal offence for automotive technicians in the United States.

Special Tools

Hand-held digital pyrometer

The catalytic converter may be tested with a hand-held digital pyrometer (Figure 5-13). This electronic device measures heat wherever the probe is positioned. The meter probe should be positioned to measure the temperature at the converter inlet and outlet. If the catalytic converter is working properly, the outlet temperature should be a minimum of 100°F (38°C) higher than the inlet temperature. When the temperature difference between the converter inlet and outlet is less than 100°F (38°C), the converter is not working properly, and it should be replaced or repaired. If the converter is not working properly, always check the air pump system to be sure it is pumping air into the converter when the engine is at normal operating temperature. When this air flow is not present, converter operation is inefficient. Never install a piece of exhaust pipe in place of the catalytic converter.

Special Tools

Converter cover sealant

Catalytic converter vibrator tool

Air-operated vacuum pump

Some catalytic converters have an inner and outer shell. If the outer shell is physically damaged on this type of converter, the outer shell may be cut apart to inspect the inner converter (Figure 5-14). If the inner part of the converter is damaged, replace the complete unit. If the damage is limited to the outer shell, a high-temperature sealant may be applied to the upper and lower shells. These shells are then retained with channels and clamps (Figure 5-15).

The pellets may be replaced in some pellet-type catalytic converters. A plug in the bottom of the converter is removed, and a vibrator tool is connected to the plug opening. The vibrator tool shakes the pellets out of the converter into a storage can on the tool. Install the new pellets in the

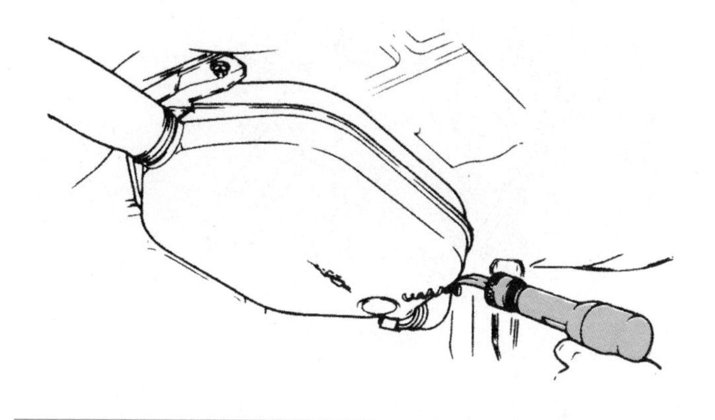

Figure 5-14 Cutting outer shell from a catalytic converter (Courtesy of Chevrolet Motor Division, General Motors Corporation)

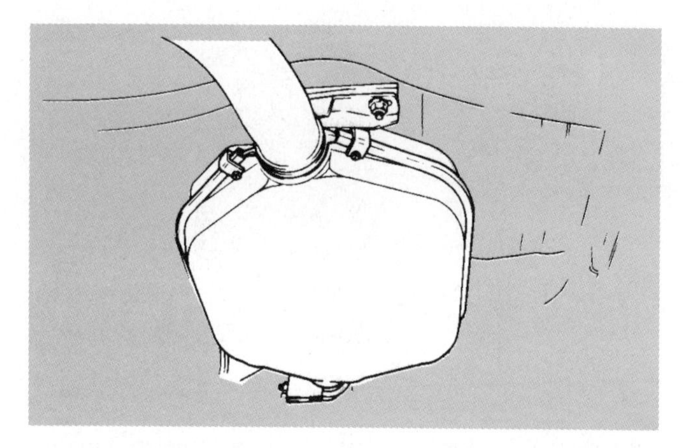

Figure 5-15 Catalytic converter outer shell reassembled with sealant, channels, and clamps (Courtesy of Chevrolet Motor Division, General Motors Corporation)

storage can on the vibrator tool, and connect a vacuum pump to the tailpipe. When a shop air hose is connected to the vacuum pump, the pellets are drawn from the storage can into the converter. If pellets come out of the tailpipe, replace the converter.

Catalytic Converter Removal and Replacement

Since there are many different catalytic converter mountings, the converter removal and replacement procedure varies depending on the vehicle. Following is a typical converter removal and replacement procedure for a converter mounted directly to the exhaust manifold:

1. Allow the engine and exhaust system time to cool down before working on the vehicle. Disconnect the negative battery terminal, and wait one minute before working on the vehicle if the car has an air bag. This gives the air bag module time to power down and prevents accidental air bag deployment.
2. Lift the vehicle on a hoist, and disconnect the two bolts and front exhaust pipe bracket. Remove the two bolts holding the front exhaust pipe to the center exhaust pipe.
3. Disconnect the three nuts holding the front exhaust pipe to the front catalytic converter (Figure 5-16).
4. Be sure the front catalytic converter is cool, and disconnect the suboxygen sensor connector.
5. Remove the bolt, nut, and number 1 manifold bracket (Figure 5-17).

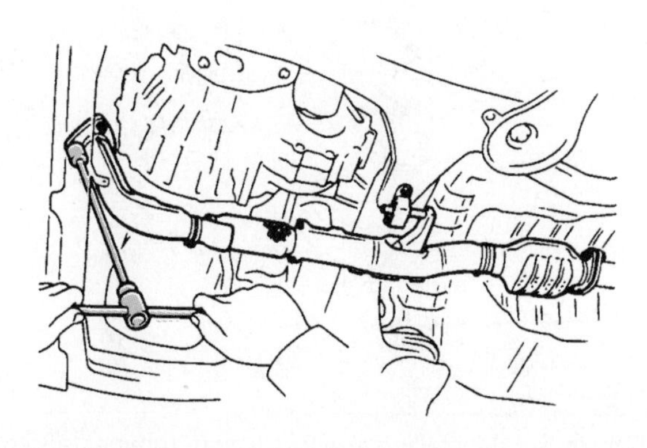

Figure 5-16 Removing three nuts from the front exhaust pipe (Courtesy of Toyota Motor Corporation)

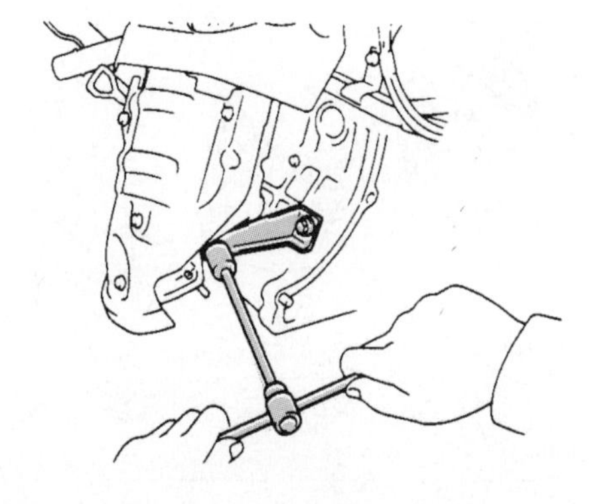

Figure 5-17 Removing bolt, nut, and number 1 manifold bracket (Courtesy of Toyota Motor Corporation)

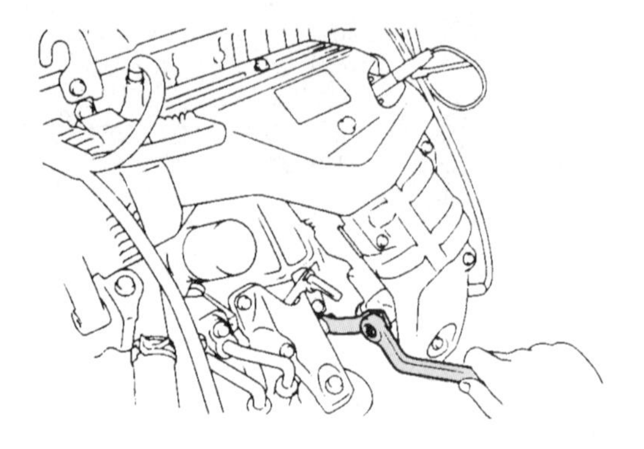

Figure 5-18 Removing bolt, nut, and number 2 manifold bracket (Courtesy of Toyota Motor Corporation)

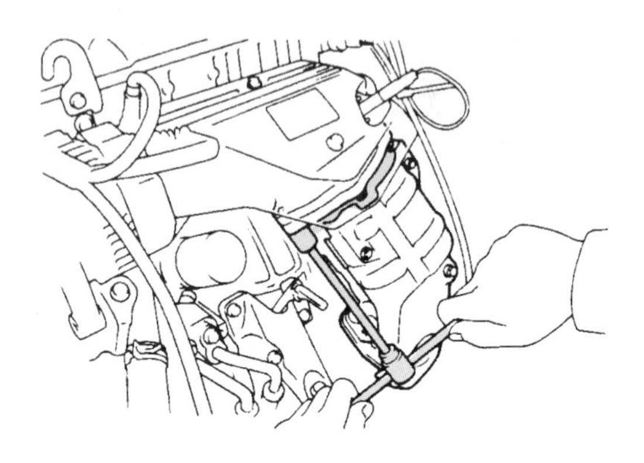

Figure 5-19 Removing two nuts and three bolts, holding the catalytic converter to the exhaust manifold (Courtesy of Toyota Motor Corporation)

6. Remove the bolt, nut, and number 2 manifold bracket (Figure 5-18).
7. Remove the two nuts and three bolts holding the catalytic converter to the manifold, and remove the converter (Figure 5-19). Remove the gasket, retainer, and cushion from the converter. Remove the eight bolts holding the outer heat insulators on the converter, and remove the heat shields (Figure 5-20).
8. Install the heat shields on the new catalytic converter and tighten the retaining bolts to the specified torque.
9. Place a new cushion, retainer, and gasket on the catalytic converter (Figure 5-21).
10. Install the catalytic converter on the exhaust manifold, and install the three new bolts and two new nuts (Figure 5-22). Tighten the bolts and nuts to the specified torque.
11. Install the number 1 and number 2 exhaust manifold support brackets, and tighten the fasteners to the specified torque.
12. Place new gaskets on the front and rear of the front exhaust pipe, and install the front exhaust pipe in its proper position. Install the two bolts and nuts holding the front exhaust pipe to the center exhaust pipe.
13. Install the three nuts holding the front exhaust pipe to the converter, and tighten these nuts to the specified torque (Figure 5-23).
14. Tighten the two bolts and nuts holding the front exhaust pipe to the center exhaust pipe. Install the front exhaust pipe bracket and tighten the fastener to the specified torque.

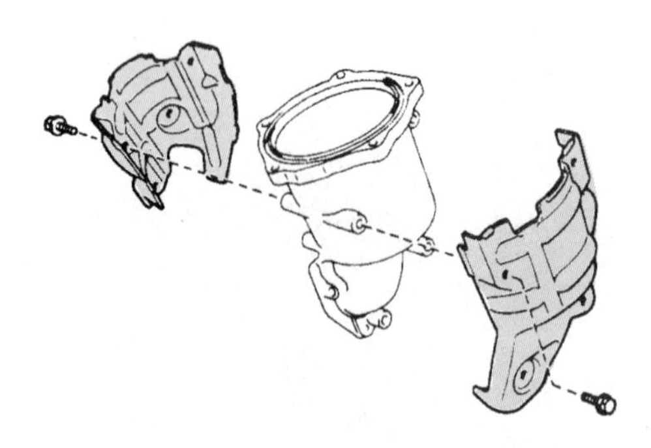

Figure 5-20 Removing outer heat shields from the catalytic converter (Courtesy of Toyota Motor Corporation)

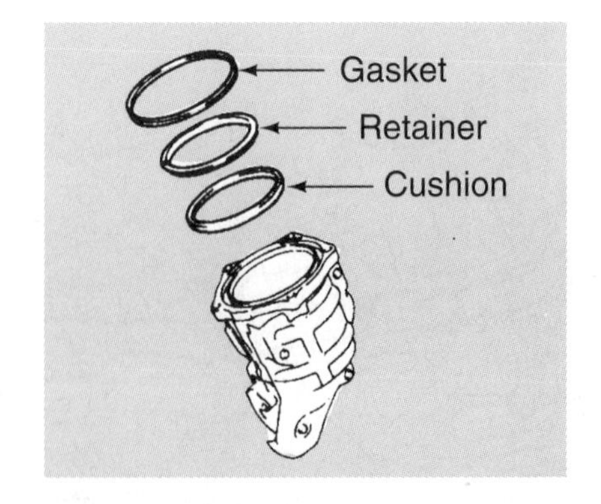

Figure 5-21 Installing a new cushion, retainer, and gasket on the catalytic converter (Courtesy of Toyota Motor Corporation)

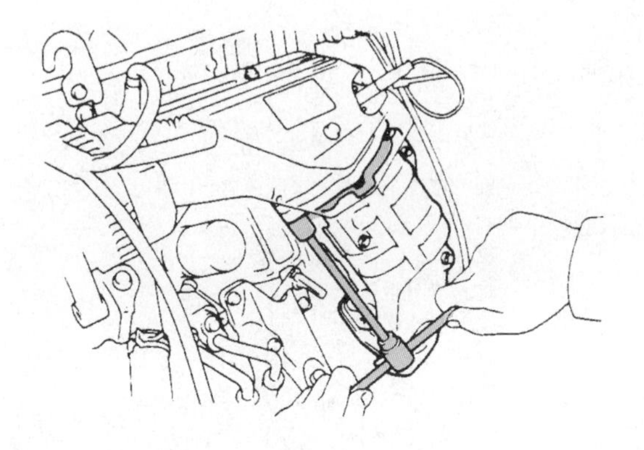

Figure 5-22 Installing the catalytic converter on the exhaust manifold (Courtesy of Toyota Motor Corporation)

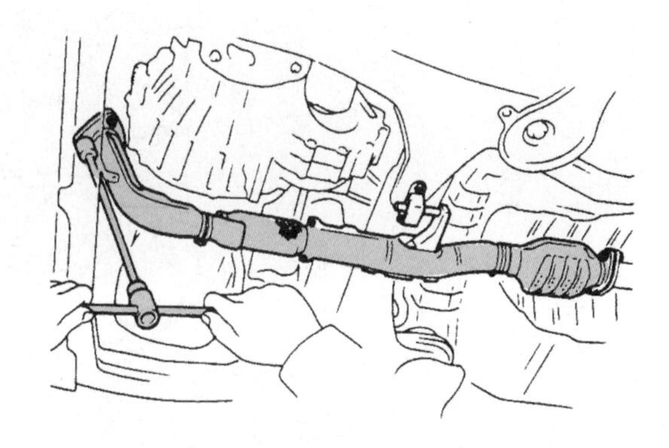

Figure 5-23 Installing the front exhaust pipe on the catalytic converter (Courtesy of Toyota Motor Corporation)

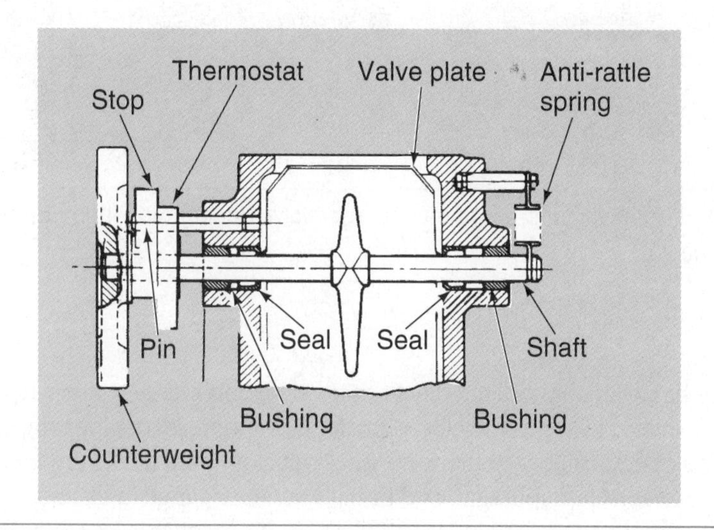

Figure 5-24 Mechanical heat riser valve (Courtesy of Chrysler Corporation)

Heat Riser Valve Diagnosis and Service

If the engine has a mechanical heat riser valve, observe the counterweight on the valve while accelerating the engine from idle speed while the engine is cold (Figure 5-24). At idle speed, the heat riser valve should be closed on a cold engine, and the valve should move toward the open position each time the engine is accelerated. If there is no movement of the heat riser valve, apply some graphite oil to each end of the shaft. Grasp the counterweight with a pair of pliers and move it back and forth until the valve moves freely. When this lubrication does not loosen the heat riser valve, replace the valve. If the heat riser valve rattles on acceleration, check the shaft for side-to-side movement in the housing. When the shaft is loose in the housing, valve replacement is necessary.

If the heat riser valve is sticking in the open position, exhaust flow is reduced through the exhaust crossover passage in the intake manifold. This action results in reduced fuel vaporization, lack of engine power, and acceleration stumbles, especially when thc engine is warming up. When the choke spring is positioned in a crossover passage well, the choke does not open properly if the exhaust flow is reduced through the crossover passage. When the heat riser valve is sticking in the closed position, this exhaust restriction causes a lack of engine power.

Graphite oil contains graphite particles. When this lubricant is applied to the heat riser valve shaft and the engine is started, the oil burns off, leaving the graphite deposited on the shaft as an excellent dry lubricant.

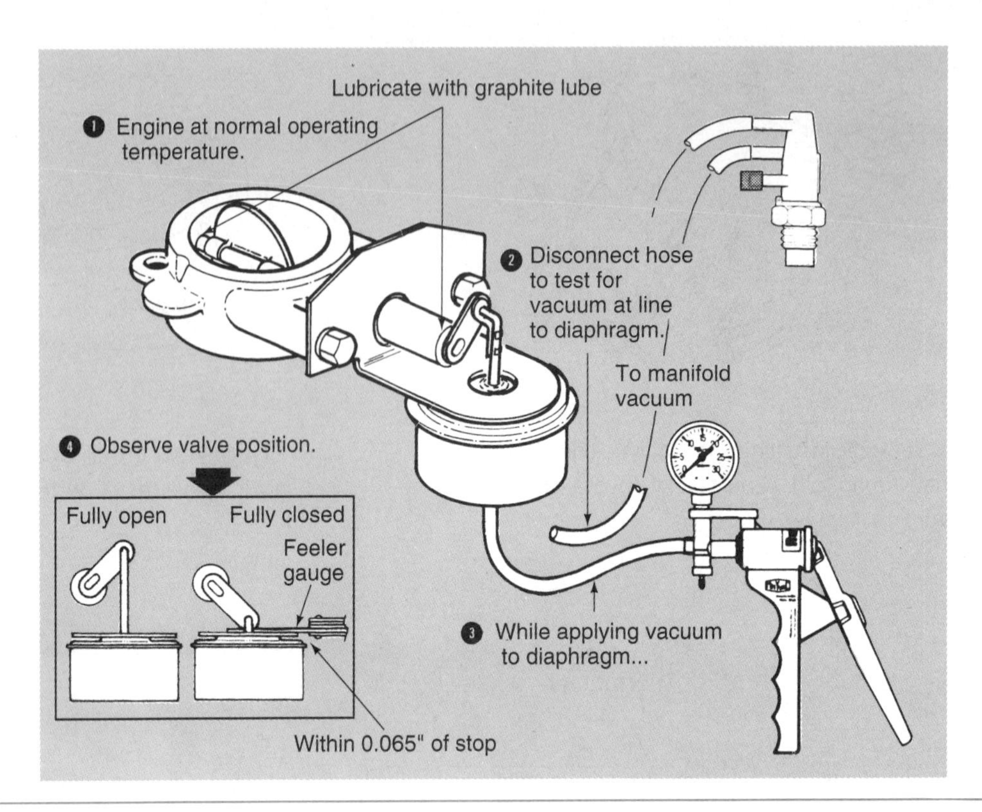

Figure 5-25 Testing vacuum-operated heat riser valve (Reprinted with permission of Ford Motor Company)

If the engine has a vacuum-operated heat riser valve, observe the valve when a cold engine is started. The valve should move to the closed position. At a preset engine temperature, the valve should move to the open position. If the valve does not close with a cold engine, remove the vacuum hose from the heat riser diaphragm, and install a vacuum gauge in this hose. With the engine idling, the vacuum should exceed 10 in Hg (69 kPa). If the vacuum is less than 10 in Hg (69 kPa), check the hoses and thermal vacuum switch (TVS). Testing of the TVS and hoses is discussed earlier in this chapter.

When the specified vacuum is available at the heat riser valve diaphragm, disconnect the diaphragm linkage from the valve shaft and try to rotate the shaft. If the shaft is seized, follow the same procedure used to loosen the mechanical heat riser valve shaft. Replace the heat riser valve assembly if the valve remains seized.

If the valve moves freely, connect a hand vacuum pump to the diaphragm port, and slowly apply vacuum to the diaphragm (Figure 5-25). Most heat riser valves should begin to close at 1 to 6 in Hg (7 to 41 kPa), and the valve should be fully closed at 10 in Hg (69 kPa). With 10 in Hg (69 kPa) applied to the diaphragm, the vacuum reading should not decrease in a two-minute interval. If the vacuum reading slowly decreases, the diaphragm is leaking. In the fully closed position, the clearance between the arm on the shaft and the stop should be 0.065 in (1.65 mm). When the heat riser valve does not operate at the specified vacuum, or the diaphragm indicates a leak, replace the heat riser assembly.

Classroom Manual
Chapter 5, page 107

CUSTOMER CARE: While discussing automotive problems with customers, remember that the average customer is not familiar with the latest automotive terminology. Always use basic terms that customers can understand when explaining complicated technical problems. Most customers appreciate a few minutes spent by service personnel to explain their automotive problems. It is not necessary to provide customers with a lesson in the latest automotive technology, but it is important that customers understand the basic cause of the problem with their car so they feel satisfied the repair expenditures are necessary. A satisfied customer is usually a repeat customer!

Guidelines for Servicing Intake and Exhaust Systems

1. Many vacuum system components such as thermal vacuum switches may be tested with a hand vacuum pump and a vacuum gauge.
2. Vacuum hoses should be inspected for cracks, proper connections, kinks, looseness, proper routing, proximity to hot components, and indication of oil inside the hoses.
3. Vacuum leaks into the intake manifold result in a lean air-fuel mixture, rough idle operation, and a low-speed acceleration stumble.
4. Intake vacuum leaks may cause misfiring on one cylinder at idle and low speeds.
5. Intake vacuum leaks on fuel injected engines may cause faster than normal idle speed.
6. An intake vacuum leak causes a low steady reading on a vacuum gauge connected to the intake manifold.
7. Intake manifold gaskets leaking on the underside of the manifold on V-type engines may result in oil consumption.
8. When using an air gun to blow dirt from an air filter, keep the gun six inches from the filter, and direct the air against the inside of the filter.
9. A trouble light may be placed inside an air filter to visually inspect the filter.
10. When servicing an air filter, always service the positive crankcase ventilation (PCV) inlet filter.
11. Polyurethane air filter covers on heavy-duty air filter elements may be washed and reoiled.
12. Some heavy-duty air filter elements may be washed in an approved cleaner and reused.
13. A heated air inlet system should maintain air cleaner temperature between 75°F and 125°F (24°C and 52°C).
14. If the air door always remains in the cold air position in a heated air inlet system, acceleration stumbles occur while the engine is warming up.
15. When the air door always remains in the hot air position in a heated air inlet system, the engine may detonate.
16. If exhaust flow is reduced through the exhaust crossover passage, acceleration stumbles, lack of power, and reduced fuel economy are experienced.
17. An inoperative heated grid in the intake system causes low-speed acceleration stumbles while the engine is cold.
18. Exhaust systems should be checked for leaks, loose internal components, loose or broken hangers and clamps, and rusted, weakened components.
19. When replacement exhaust system components are installed, the pipes should overlap 1.5 in (3.8 cm).
20. If a catalytic converter is working properly, there should be at least 100°F (38°C) temperature difference between the converter inlet and outlet temperatures.
21. On some pellet-type catalytic converters, a vibrating tool and an air-operated vacuum pump may be used to remove and replace the pellets.
22. When a cold engine is accelerated, the mechanical heat riser valve should move toward the open position.
23. A vacuum-operated heat riser valve should be closed on a cold engine, and this valve should move to the fully open position at a preset coolant temperature.

CASE STUDY

A customer brought in a 5.2L Dodge 1/2-ton truck for a valve job and a timing chain replacement. The cylinder heads were reconditioned and the timing chain and sprocket were replaced in what seemed to be a routine repair procedure. When the left cylinder head was removed, the technician noticed that the exhaust manifold was broken. The customer was informed regarding the additional cost of the exhaust manifold, and the new manifold was installed after the cylinder heads were reinstalled.

When the engine was started, the technician noticed it was misfiring on one cylinder. The technician connected the engine analyzer and discovered that number 1 cylinder was not producing any rpm drop during a cylinder power balance test. A quick check of the ignition system indicated the maximum coil voltage and normal firing voltage were satisfactory. The technician removed the spark plug and it appeared to be in normal condition. All the spark plugs were removed, and a compression test indicated the compression was within the manufacturer's specifications. The spark plugs were reinstalled and the intake manifold was checked for vacuum leaks. Since the intake manifold did not indicate any vacuum leaks, the technician removed the rocker arm cover and checked the valve lift and valve spring condition on the valves in number 1 cylinder. These valve train components appeared to be in normal condition. The technician performed a quick check of the valve timing and found it to be accurate.

While listening to the engine, the technician noticed the engine had an unusual sound, similar to a detonation noise, when the engine was accelerated. The technician began to think about the causes of this problem and realized the engine did not have a misfire on number 1 cylinder when the vehicle was brought into the shop. After logical reasoning, the technician came to the conclusion that the only thing different on the left side of the engine was the new exhaust manifold, assuming the cylinder head was properly reconditioned. The compression test and visual inspection of the valve train did not indicate any cylinder head problems.

The technician removed the new exhaust manifold and poured solvent into the manifold exhaust port for number 1 cylinder. The exhaust manifold port filled up with solvent, and the solvent did not run through the port. The technician dumped the solvent out of the manifold and attempted to push a small steel rod through the port. The steel rod struck metal a short distance inside the port, and the technician knew for sure the casting in the exhaust port was completely plugging the port. This completely restricted port was not allowing any exhaust out of the cylinder, causing the misfire on number 1 cylinder.

The new exhaust manifold was returned to the parts supplier and another new manifold supplied. The technician ran a steel rod through all the exhaust manifold ports to make sure this manifold was not restricted. When this new manifold was installed, the engine operation was normal, and a road test did not reveal any other problems. From this experience, the technician learned not to always assume that new parts are in satisfactory condition.

Terms to Know

Thermal vacuum switch (TVS)
Polyurethane air cleaner cover
Bimetal sensor
Magnetic-base thermometer
Pipe expander
Slitting tool
Hand-held digital pyrometer
Graphite oil
Catalytic converter vibrator tool
Air-operated vacuum pump

ASE Style Review Questions

1. While discussing intake manifold vacuum leaks:
   Technician A says an intake manifold vacuum leak may cause a cylinder misfire at idle and low engine speeds.
   Technician B says an intake manifold vacuum leak may cause a cylinder misfire when the throttle is nearly wide open.
   Who is correct?
   A. A only **C.** Both A and B
   B. B only **D.** Neither A nor B

2. While discussing intake manifold vacuum leaks:
   Technician A says an intake manifold vacuum leak may result in faster than normal idle speed on fuel injected engines.
   Technician B says an intake manifold vacuum leak causes a low steady reading on a vacuum gauge connected to the intake manifold.
   Who is correct?
   A. A only **C.** Both A and B
   B. B only **D.** Neither A nor B

3. While discussing air filter service:
   Technician A says while using an air gun to blow dirt from an air filter, the air gun should be held against the surface of the filter.
   Technician B says while using an air gun to blow dirt from an air filter, the air gun should be positioned on the outside of the filter.
   Who is correct?
   A. A only **C.** Both A and B
   B. B only **D.** Neither A nor B

4. While discussing heated air inlet system diagnosis:
   Technician A says if the air door remains in the hot air position while the engine is at normal operating temperature, the engine may detonate.
   Technician B says the air door should begin to modulate the air flow into the air cleaner snorkel at 175°F (79°C).
   Who is correct?
   A. A only **C.** Both A and B
   B. B only **D.** Neither A nor B

5. While discussing intake manifold temperature control systems:
   Technician A says a plugged crossover passage in the intake manifold causes cylinder misfiring.
   Technician B says a plugged crossover passage in the intake manifold results in an acceleration stumble, especially when the engine is cold.
   Who is correct?
   A. A only **C.** Both A and B
   B. B only **D.** Neither A nor B

6. While discussing heated grid systems:
   Technician A says 12 V is supplied through the temperature switch to the grid when the switch contacts are closed.
   Technician B says the temperature switch contacts should be closed when the engine is at normal operating temperature.
   Who is correct?
   A. A only **C.** Both A and B
   B. B only **D.** Neither A nor B

7. While discussing exhaust system diagnosis:
   Technician A says a rattling noise when the engine is accelerated may be caused by loose internal components in the muffler.
   Technician B says a lack of engine power may be caused by a restricted exhaust system.
   Who is correct?
   A. A only **C.** Both A and B
   B. B only **D.** Neither A nor B

8. While discussing catalytic converter diagnosis:
   Technician A says the temperature should be the same at the catalytic converter inlet and outlet, if the converter is working normally.
   Technician B says a defective belt-driven air pump may cause improper converter operation.
   Who is correct?
   A. A only **C.** Both A and B
   B. B only **D.** Neither A nor B

9. While discussing catalytic converter service and diagnosis:
Technician A says the pellets may be changed in some converters.
Technician B says a vacuum cleaner is used to remove the converter pellets.
Who is correct?
A. A only **C.** Both A and B
B. B only **D.** Neither A nor B

10. While discussing vacuum-operated heat riser valves:
Technician A says vacuum is supplied through a reducer valve to the heat riser valve diaphragm.
Technician B says vacuum should be supplied to the heat riser valve diaphragm when the engine is hot.
Who is correct?
A. A only **C.** Both A and B
B. B only **D.** Neither A nor B

Table 5-1 ASE TASK

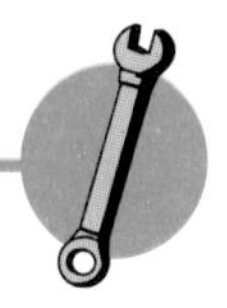

Inspect, service, and repair or replace air filtration system components.

Problem Area	Symptoms	Possible Causes	Classroom Manual	Shop Manual
FUEL ECONOMY	Excessive fuel consumption	Restricted air filter	97	70
ENGINE LIFE	Cylinder walls, pistons, rings scored prematurely	Pin holes in air filter	97	70

Table 5-2 ASE TASK

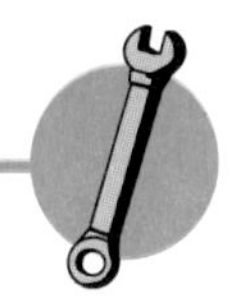

Test operation of inlet air temperature control systems.

Problem Area	Symptoms	Possible Causes	Classroom Manual	Shop Manual
ENGINE PERFORMANCE	Acceleration stumble	Inlet air door in cold air position during warmup	100	73
	Detonation	Inlet air door in hot air position with engine hot	100	73

Table 5-3 ASE TASK

Inspect, test, and replace sensors, diaphragm, and hoses of inlet air temperature systems.

Problem Area	Symptoms	Possible Causes	Classroom Manual	Shop Manual
ENGINE PERFORMANCE	Acceleration stumble	1. Defective bimetal sensor	101	73
		2. Defective air door vacuum motor diaphragm	101	73
		3. Leaking vacuum hoses	101	73

Table 5-4 ASE TASK

Inspect, test, and replace heat stove shroud, hot air pipe, and damper of inlet air temperature control systems.

Problem Area	Symptoms	Possible Causes	Classroom Manual	Shop Manual
ENGINE PERFORMANCE	Acceleration stumble	Leaking, damaged, or missing stove or hot air pipe	101	73

Table 5-5 ASE TASK

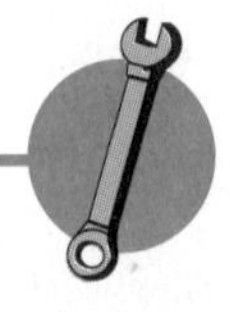

Inspect, service, and replace exhaust manifold, exhaust pipes, mufflers, resonators, tail pipes, and heat shields.

Problem Area	Symptoms	Possible Causes	Classroom Manual	Shop Manual
NOISE	Excessive exhaust noise especially on acceleration	Leaking muffler, pipes, converter, or resonator	114	77
	Rattling noise	1. Loose internal parts in muffler or converter	113	76
		2. Exhaust components hitting chassis	115	76
		3. Broken hangers, clamps	115	76
ENGINE PERFORMANCE	Lack of power	Restricted pipes, muffler, or converter	109	76

Table 5-6 ASE TASK

Inspect, test, service, and replace converter catalyst or converter(s) or catalytic converter systems.

Problem Area	Symptoms	Possible Causes	Classroom Manual	Shop Manual
EMISSIONS	High emission levels	Defective catalyst in converter	111	80
NOISE	Excessive exhaust noise, especially on acceleration	Leaking converter or pipe connections	110	77
ENGINE PERFORMANCE	Lack of power	Restricted converter	112	76

Introduction to Tune-Up Equipment

Upon completion and review of this chapter, you should be able to:

- ❑ Connect and use fuel pressure gauges properly.
- ❑ Connect and operate an injector balance tester.
- ❑ Connect and use pressurized injector cleaning containers.
- ❑ Connect and use circuit testers.
- ❑ Complete proper volt-ampere tester connections.
- ❑ Complete correct multimeter connections.
- ❑ Connect various tach-dwellmeters properly and obtain meter readings.
- ❑ Operate a photoelectric tachometer.
- ❑ Operate a magnetic probe-type tachometer.
- ❑ Check ignition timing and spark advance with an advance-type timing light.
- ❑ Operate a digital timing meter with a magnetic probe.
- ❑ Connect and operate an ignition module tester.
- ❑ Connect oscilloscope leads to the ignition system.
- ❑ Perform a correct scan tester connection to the vehicle, and program the tester for the vehicle being tested.
- ❑ Prepare an emissions analyzer to make accurate emission readings on a vehicle.
- ❑ Connect an engine analyzer and program it for a specific vehicle.

Chapter Purpose

The purpose of this chapter is to familiarize the reader with the tune-up equipment used in the remaining chapters in the book. Our purpose in this chapter is to provide a general description of the most common types of tune-up equipment and to describe briefly the basic tests that may be performed with this equipment. The use of this equipment for detailed diagnosis of various systems is provided in the appropriate following chapters.

Basic Tools

Basic technician's tool set

Service manual

Circuit testers, 12-V, and self-powered

Fuel Pressure Gauges

CAUTION: While testing fuel pressure, be careful not to spill gasoline. Gasoline spills may cause explosions and fires, resulting in serious personal injury and property damage.

WARNING: Electronic fuel injection systems are pressurized, and these systems require depressurizing prior to fuel pressure testing and other service procedures. (Refer to Chapter 11 for detailed fuel pressure testing.)

The fuel pressure gauge is used to measure the electric fuel pump pressure in throttle body injection (TBI) and port fuel injection (PFI) systems. The mechanical fuel pump pressure on a carbureted engine is also checked with a pressure gauge.

A fuel pump pressure gauge is usually sold in a kit that contains the necessary fittings and hoses to connect the gauge to the fuel system. Since fuel system design varies depending on the manufacturer and the type of fuel system, the fuel pressure gauge requires a number of adaptors. A fuel pressure gauge kit for a TBI system must have adaptors to connect the gauge at the throttle body fuel inlet line (Figure 6-1).

Special Tools

Fuel pressure gauges

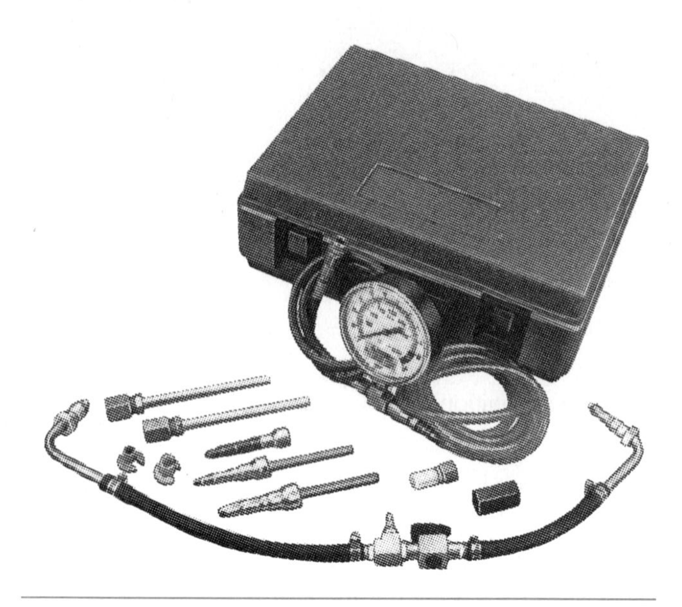

Figure 6-1 Fuel pressure gauge and adaptors for throttle body injection (TBI) systems (Courtesy of OTC Division, SPX Corp.)

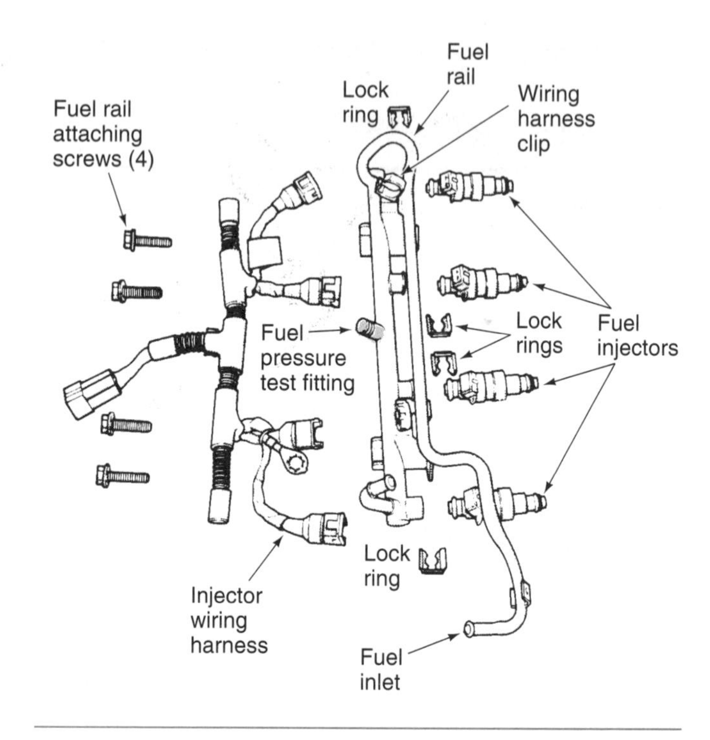

Figure 6-2 On a PFI system, the fuel pressure gauge is connected to the Schrader valve on the fuel rail to test fuel pump pressure. (Courtesy of Chrysler Corporation)

When the fuel pump pressure is tested on a PFI system, the fuel pressure gauge must be connected to the Schrader valve on the fuel rail (Figure 6-2). Since there are several different sizes of Schrader valves on various PFI systems, a number of adaptors are supplied with the PFI fuel pressure gauge kit (Figure 6-3).

If the fuel pump is defective in a PFI or TBI system, the engine usually fails to start or stops intermittently. A plugged fuel filter may cause the same symptoms. When fuel pressure is excessive in a PFI or TBI system, the air-fuel mixture is too rich. A lean air-fuel mixture in these systems may be caused by lower than specified fuel pressure.

On a carbureted engine, the fuel pressure gauge is connected at the fuel inlet fitting on the carburetor to measure the mechanical fuel pump pressure. The gauge fitting is threaded into the

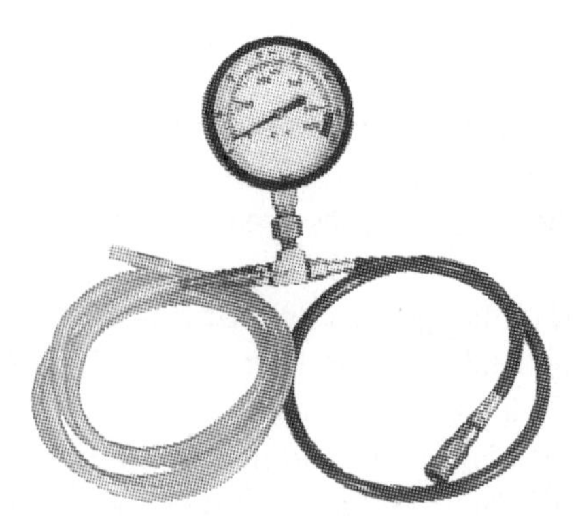

Figure 6-3 Fuel pressure gauge for port fuel injection (PFI) systems (Courtesy of OTC Division, SPX Corp.)

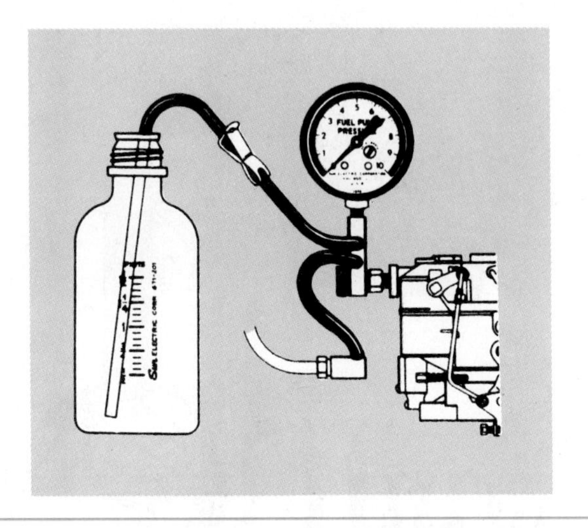

Figure 6-4 Fuel pump pressure and flow tester, carbureted engine (Courtesy of Sun Electric Corporation)

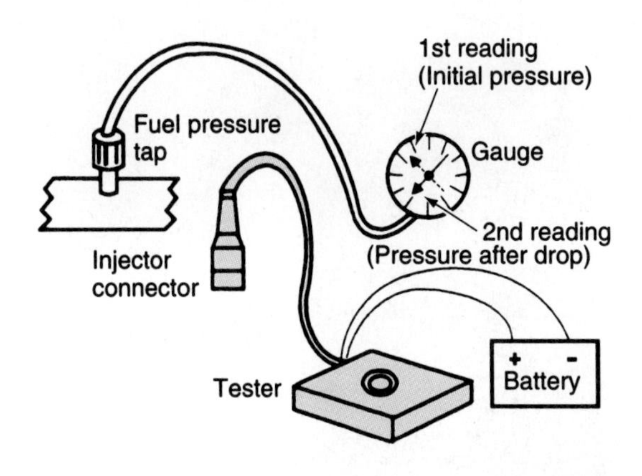

Figure 6-5 Injector balance tester (Courtesy of Oldsmobile Division General Motors Corporation)

carburetor inlet fitting, and one of the hoses on the tester is connected to the fuel inlet line. A second hose on the tester is installed in a graduated plastic container, and a special clip closes this line (Figure 6-4). With the engine idling, the clip is closed to measure the fuel pump pressure, and then the clip is released to allow fuel to flow into the plastic container to measure fuel pump volume. The fuel flow into the plastic container is timed for a specific length of time such as 30 or 45 seconds, depending on the specifications. A typical fuel pump flow specification would be 1 pint in 30 seconds.

Injector Balance Tester

Special Tools

Injector balance tester

The injector balance tester is used to test the injectors in a port fuel injected (PFI) engine for proper operation. A fuel pressure gauge is also used during the injector balance test. The injector balance tester contains a timing circuit, and some injector balance testers have an off-on switch. A pair of leads on the tester must be connected to the battery with the correct polarity (Figure 6-5). The injector terminals are disconnected, and a second double lead on the tester is attached to the injector terminals.

Before the injector test, the fuel pressure gauge is connected to the Schrader valve on the fuel rail, and the ignition switch should be cycled two or three times until the specified fuel pressure is indicated on the pressure gauge. When the tester push button is depressed, the tester energizes the injector winding for a specific length of time, and the technician records the pressure decrease on the fuel pressure gauge. This procedure is repeated on each injector.

Some vehicle manufacturers provide a specification of 3 psi (20 kPa) maximum difference between the pressure readings after each injector is energized. If the injector orifice is restricted, there is not much pressure decrease when the injector is energized. Acceleration stumbles, engine stalling, and erractic idle operation are caused by restricted injector orifices. The injector plunger is sticking open if excessive pressure drop occurs when the injector is energized. Sticking injector plungers may result in a rich air-fuel mixture.

Pressurized Injector Cleaning Container

Special Tools

Pressurized injector cleaning container

The pressurized injector cleaning container is designed for cleaning the injectors in PFI systems. The hose on this container is connected to the Schrader valve on the fuel rail, and various fittings

Figure 6-6 Pressurized injector cleaning container (Courtesy of OTC Division, SPX Corp.)

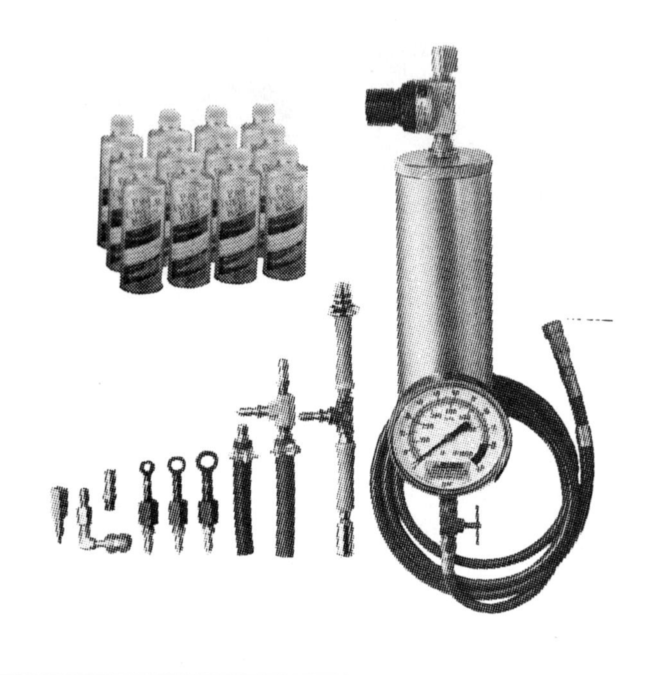

Figure 6-7 Canister pressurized with shop air supply for injector cleaning (Courtesy of OTC Division, SPX Corp.)

are available for this connection. After the lid is removed from the container, a specific quantity of unleaded gasoline and injector cleaner is placed in the container. When the lid is installed, the hand pump on the container is operated until the specified fuel system pressure is indicated on the container pressure gauge (Figure 6-6). During the injector cleaning process, the electric fuel pump must be disabled and the fuel return line plugged. Under this condition, the engine runs on the fuel and injector cleaning solution in the pressurized container.

Some automotive equipment suppliers market canister-type injector cleaning containers. These containers have a valve and a hose for connection to the Schrader valve (Figure 6-7). A premixed solution of unleaded gasoline and injector cleaner is placed in the container, and the specified fuel pressure is supplied from the shop air supply. With the return fuel line plugged and the electric fuel pump disabled, the engine runs on the solution in the canister to clean the injectors.

Circuit Testers

WARNING: Do not use a conventional 12-V test light to diagnose components and wires in computer systems. The current draw of these test lights may damage computers and computer system components. High-impedance test lights are available for diagnosing computer systems. Always be sure the test light you are using is recommended by the tester manufacturer for testing computer systems.

WARNING: Do not use any type of test light or circuit tester to diagnose automotive air bag systems. Use only the vehicle manufacturer's recommended equipment on these systems.

Circuit testers are used to diagnose open circuits, grounds, and shorts in electric circuits. There is a large variety of circuit testers available for automotive testing, but the most common circuit tester is

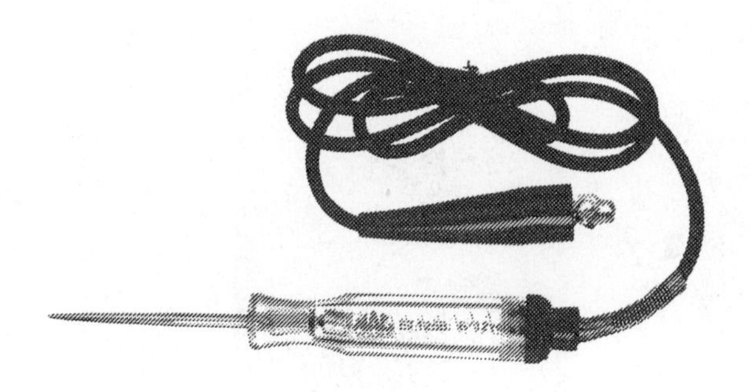

Figure 6-8 12-V test light (Courtesy of Mac Tools, Inc.)

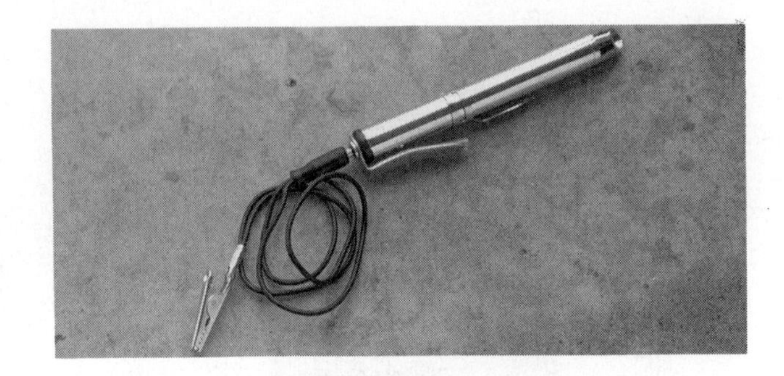

Figure 6-9 Self-powered test light

the 12-V test light. A sharp probe is molded into a handle on the 12-V test light, and the upper part of the handle is transparent. The 12-V bulb inside the tester is visible through the transparent handle (Figure 6-8). A ground clip extending from the handle is connected to one terminal on the bulb, and the other bulb terminal is connected to the probe. In most test situations, the ground clip is connected to a ground connection on the vehicle, and the probe is connected to a circuit to determine if voltage is available. With this type of test connection, the bulb is illuminated if voltage is available at the probe.

A self-powered test light is similar in appearance to a 12-V test light, but the self-powered test light contains an internal battery (Figure 6-9). One end of many automotive circuits is connected to ground, and battery voltage is supplied to the other end of the circuit. The ground side of the circuit may be called the negative side of the circuit, and the positive side of the circuit is connected to the battery positive terminal. The positive side of the circuit may be disconnected and the test light lead connected to the battery positive terminal. When the self-powered test light probe is connected to this disconnected circuit, the test light should be illuminated if the other end of the circuit is connected to ground. If the circuit is open, the light remains off.

A self-powered test light may be referred to as a continuity tester.

Volt-Ampere Tester

A volt-ampere tester is used to perform voltmeter and ammeter tests in any automotive electrical circuit. Some of the most common tests performed with a volt-ampere tester are:

1. Battery load test
2. Starter current draw test
3. Alternator maximum output test
4. Alternator normal system voltage test

Special Tools

Volt-ampere tester

The volt-ampere tester contains an ammeter, voltmeter, and a carbon pile load. A carbon pile load is a stack of heavy carbon discs. A control knob on the tester adjusts the position of these discs. When the control knob is turned all the way counterclockwise to the off position, the carbon discs are not contacting each other. As the control knob is rotated clockwise, the discs make contact, and further clockwise rotation supplies more pressure to the discs, which reduces the resistance of the carbon pile. Two heavy leads are connected to the carbon pile load, and these leads are connected across the battery terminals with the correct polarity. The red lead is connected to the positive battery terminal, and the negative lead is attached to the negative battery terminal. Many volt-ampere testers have digital ammeter and voltmeter readings (Figure 6-10), but some of these testers are equipped with analog meters.

Some ammeters have a pair of leads that are connected in series in the circuit to be tested, but many ammeters have an inductive clamp that clips over the wire in which the current is to be measured. The inductive clamp and the ammeter read the amount of current flow from the strength of the magnetic field surrounding the wire.

An inductive clamp is a type of ammeter pickup that fits over the wire in which the current is to be measured. The inductive clamp senses the amount of current flowing in the wire from the magnetic field surrounding the wire.

Figure 6-10 Volt-ampere tester with digital ammeter and voltmeter readings (Courtesy of Snap-on Tools Corporation)

Figure 6-11 The control knob is rotated to select the desired reading and scale on a digital multimeter (Courtesy of OTC Division, SPX Corp.)

Multimeter

WARNING: A high-impedance digital multimeter must be used to test the voltage of some components and systems such as an oxygen (O_2) sensor circuit. If a low-impedance analog meter is used in this type of circuit, the current flow through the meter is high enough to damage the sensor. Always use the type of meter specified by the vehicle manufacturer.

WARNING: Always be sure the proper scale is selected on the multimeter, and the correct lead connections are completed for the component or system being tested. Improper multimeter lead connections or scale selections may blow the internal fuse in the meter or cause meter damage.

Special Tools

Multimeter

Multimeters are small hand-held meters that provide these readings on several different scales:

1. DC volts
2. AC volts
3. Ohms
4. Amperes
5. Milliamperes

Analog meters have a pointer, and a scale to indicate a specific reading.

Since multimeters do not have heavy leads, the highest ammeter scale on this type of meter is sometimes 10 amperes. A control knob on the front of the multimeter must be rotated to the desired reading and scale (Figure 6-11). Some multimeters are autoranging, which means the meter automatically switches to a higher scale if the reading goes above the value of the scale being used. For example, if the meter is reading on the 10-V scale and the leads are connected to a 12-V battery, the meter automatically changes to the next highest scale. If the multimeter is not autoranging, the technician must select the proper scale for the component or circuit being tested.

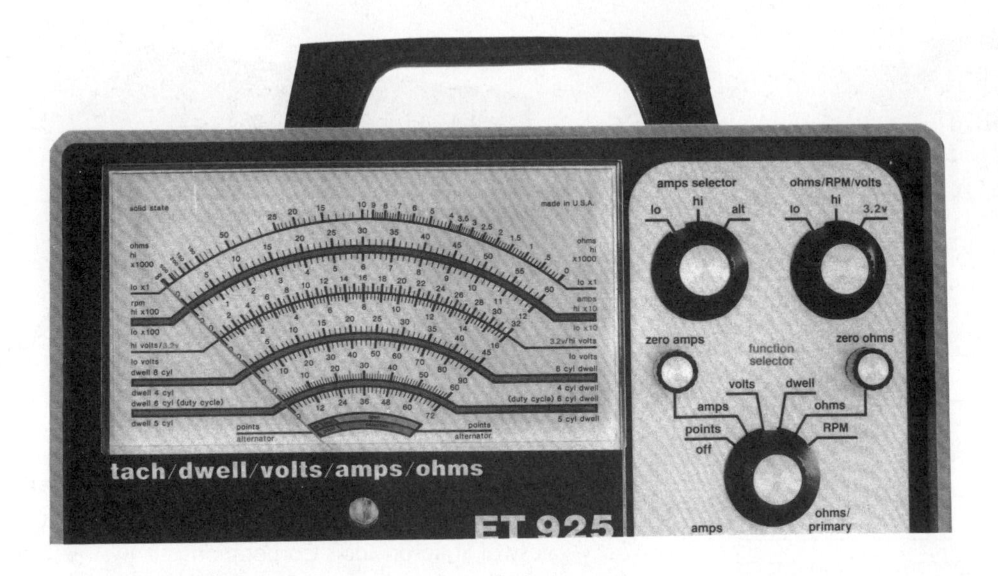

Figure 6-12 Analog multimeter (Courtesy of Mac Tools, Inc.)

Multimeters usually have an internal fuse which blows if the meter is connected improperly. Multimeters usually have digital readings, but analog multimeters are available (Figure 6-12). Digital multimeters have high impedance compared to analog multimeters, but digital multimeters have different impedances depending on the meter design. Always use the type of multimeter recommended by the vehicle manufacturer. The leads are plugged into terminals in the front of the multimeter. Some of these meters have several terminals. The reading provided by the terminal position is indicated beside the terminal. The technician must plug the leads into the correct terminal to obtain the desired reading. Some multimeters have a common (com) terminal. The negative, black meter lead must be plugged into this terminal.

Some multimeters have additional capabilities such as testing of diode condition, frequency, temperature, engine rpm, ignition dwell, and distributor condition. This type of multimeter has more switch positions and more lead terminals (Figure 6-13).

Digital meters have a digital reading to indicate a specific value.

Meter impedance refers to the internal resistance inside the meter. Digital multimeters usually have higher impedance than analog meters.

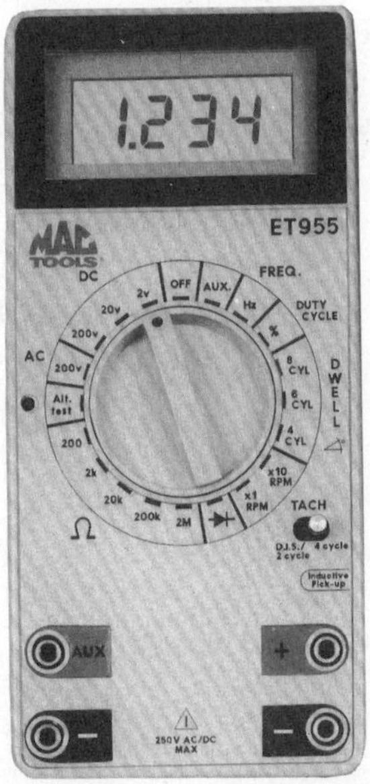

Figure 6-13 Multifunctional digital multimeter (Courtesy of Mac Tools, Inc.)

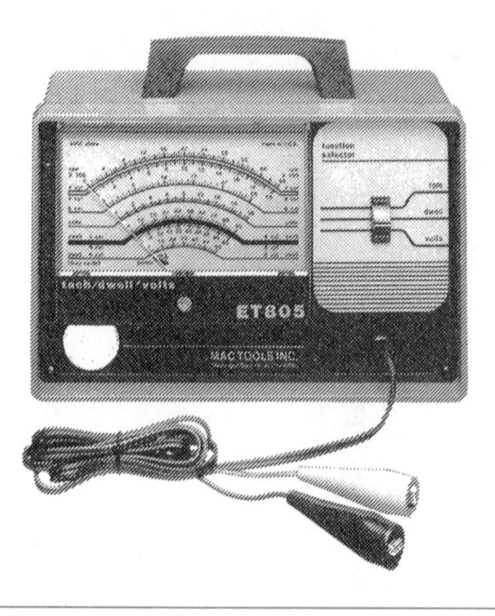

Figure 6-14 Tach-dwellmeter for checking engine rpm and dwell (Courtesy of Mac Tools Inc.)

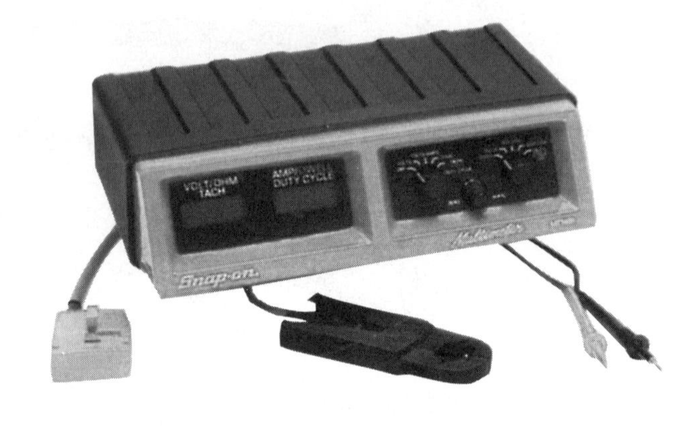

Figure 6-15 Digital tachometer with an inductive clamp that fits over number 1 spark plug wire (Courtesy of Snap-on Tools Corporation)

Special Tools

Tach-dwellmeter

Digital tachometer

Magnetic probe-type tachometer

Ignition Test Equipment

Tach-dwellmeter

A tachometer is designed to read engine revolutions per minute (rpm).

The tach-dwellmeter is one of the commonly used pieces of tune-up equipment. A small internal dry-cell battery powers most tach-dwellmeters. The red tach-dwellmeter lead is connected to the negative primary coil terminal, and the black meter lead is connected to ground. A switch on the meter must be set in the rpm, or dwell, position (Figure 6-14). On distributorless ignition systems a special tachometer lead may be provided in the ignition system for the tach-dwellmeter connection.

Digital Tachometers

A dwellmeter is designed to read dwell on the ignition system.

Since the dwell is controlled by the module in electronic ignition systems, the dwell reading is not used for diagnostic purposes. Therefore, the dwell function is eliminated on some meters, and the tachometer is available by itself. Digital tachometers are now available with an inductive pickup that is clamped over the number one spark plug wire. These meters provide an rpm reading from the speed of spark plug firings (Figure 6-15). This type of tachometer is suitable for distributorless ignition systems.

Special Tools

Timing lights, advance-type, and digital advance-type

Photoelectric tachometers are available to read engine rpm or the rpm of other rotating components. An internal light source is powered by a battery in the tachometer or by the car battery. A piece of reflecting tape is applied to the crankshaft pulley, and the light in the photoelectric tachometer is pointed at this tape when the engine is running. The photoelectric cell in the tachometer senses the reflected light each time the reflective tape rotates through the meter light beam. An engine rpm calculation is made by the tachometer from these reflected light pulses.

On a point-type ignition system, dwell is the number of degrees the points remain closed on each distributor cam lobe.

Many engines manufactured in recent years have a magnetic probe receptacle mounted above the crankshaft pulley (Figure 6-16). Digital tachometers are available with a magnetic pickup that fits in this magnetic probe receptacle. Each timing mark rotation past the pickup sends a pulse signal to the meter (Figure 6-17). The digital tachometer provides an rpm reading from these pulses. This type of digital tachometer may be used on diesel engines. The magnetic probe receptacle may also be used for engine timing purposes.

Timing Light

Classroom Manual
Chapter 6, page 119

WARNING: Never pierce the number one spark plug wire to complete a timing light connection. This action results in high-voltage leakage from the spark plug wire.

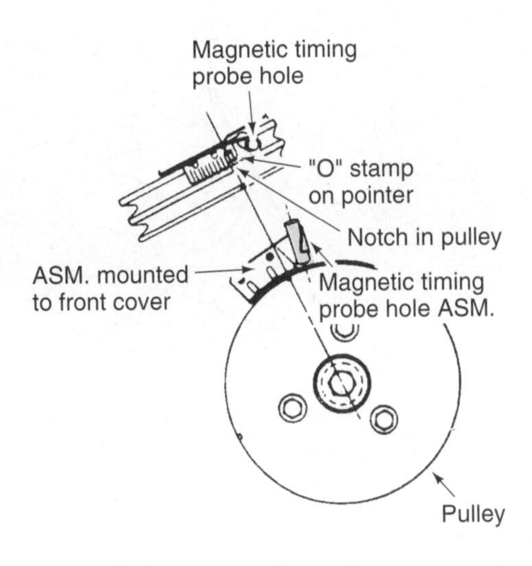

Figure 6-16 Magnetic probe receptacle mounted above the crankshaft pulley (Courtesy of Chevrolet Motor Division, General Motors Corporation)

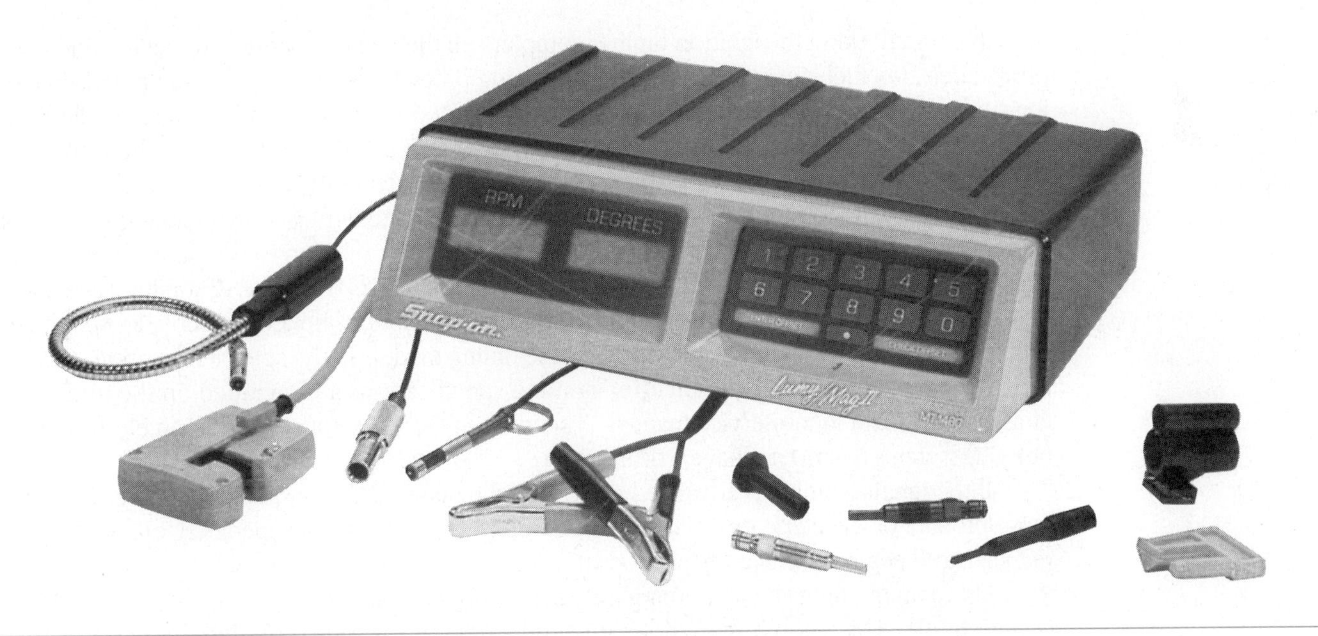

Figure 6-17 Digital tachometer with magnetic probe-type pickup (Courtesy of Snap-on Tools Corporation)

In an electronic distributor ignition system, dwell is the degrees of distributor rotation the primary circuit is turned on prior to each cylinder firing. Since the dwell is determined electronically on these systems, it cannot be adjusted.

A timing light is essential for checking the ignition timing in relation to crankshaft position. Two leads on the timing light must be connected to the battery terminals with the correct polarity. Most timing lights have an inductive clamp that fits over the number one spark plug wire (Figure 6-18). Older timing lights have a lead that goes in series between the number one spark plug wire and the spark plug. A trigger on the timing light acts as an off/on switch. When the trigger is pulled with the engine running, the timing light emits a beam of light each time the spark plug fires.

The timing marks are usually located on the crankshaft pulley or on the flywheel. A stationary pointer, line, or notch is positioned above the rotating timing marks. The timing marks are lines on the crankshaft pulley or flywheel that represent various degrees of crankshaft rotation when the number one piston is before top dead center (BTDC) on the compression stroke. The TDC crankshaft position and the degrees are usually identified in the group of timing marks. Some timing marks include degree lines representing the after top dead center (ATDC) crankshaft position (Figure 6-19).

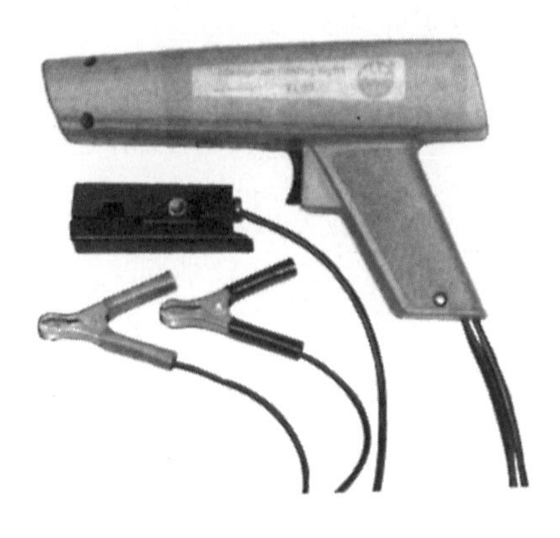

Figure 6-18 Timing light with inductive clamp that fits over the number one spark plug wire (Courtesy of Mac Tools, Inc.)

Figure 6-19 Various timing marks on the crankshaft pulley (Courtesy of Chevrolet Motor Division, General Motors Corporation)

Before checking the ignition timing, complete all the vehicle manufacturer's recommended procedures. On fuel injected engines, special timing procedures are required, such as disconnecting a timing connector to be sure the computer does not provide any spark advance while checking basic ignition timing. These timing instructions are usually provided on the underhood emission label. (Detailed timing instructions are provided later in the appropriate chapter.)

With the engine running at the specified idle speed, the light is aimed at the timing marks. The timing mark should appear at the specified position in relation to the timing pointer. For example, if the vehicle manufacturer's timing specification is 12° BTDC, the 12° timing mark should be directly under the timing pointer. If the timing mark is not positioned properly, loosen the distributor clamp and rotate the distributor until the timing mark is in the specified location, and then retighten the clamp. The manufacturer's timing specifications are included on the underhood emission label and in the service manual. The ignition timing is not adjustable on electronic ignition (EI) systems that do not have a distributor.

If the ignition timing is advanced more than the manufacturer's specified timing, the engine may detonate on acceleration. When the ignition timing is later than specified, engine power is reduced, and engine overheating may occur.

Many timing lights have a timing advance knob that may be used to check spark advance (Figure 6-20). This knob has an index line and a degree scale surrounding the knob. Prior to checking the spark advance, the basic timing should be checked. Accelerate the engine to 2,500 rpm or the speed recommended by the vehicle manufacturer. While maintaining this rpm, slowly

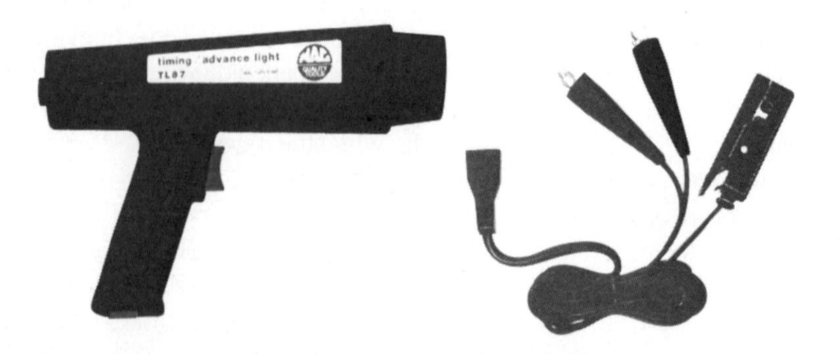

Figure 6-20 Timing light with advance knob (Courtesy of Mac Tools Inc.)

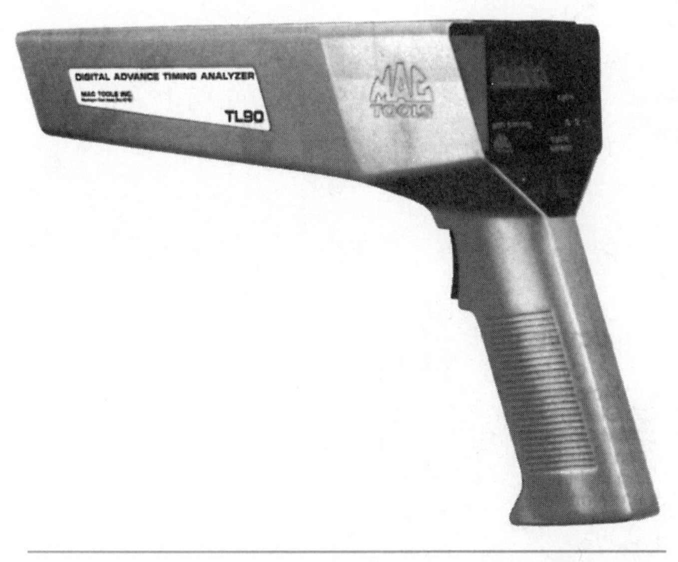

Figure 6-21 Digital advance-type timing light provides spark advance or rpm reading (Courtesy of Mac Tools, Inc.)

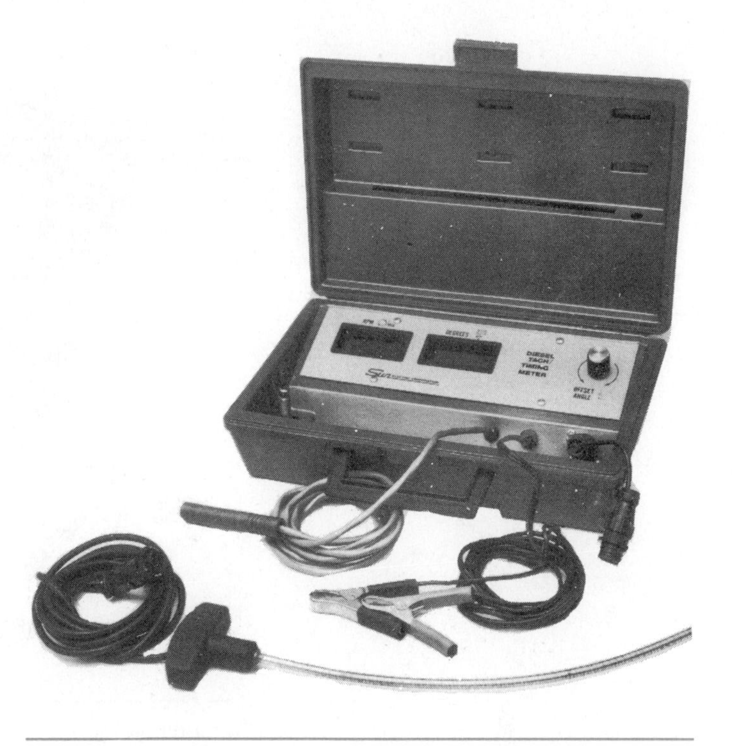

Figure 6-22 Digital timing meter with magnetic probe (Courtesy of Sun Electric Corporation)

rotate the advance knob toward the advanced position until the timing marks come back to the basic timing position. Under this condition, the index mark on the advance knob is pointing to the number of degrees advance provided by the computer or distributor advances. The reading on the degree scale may be compared to the vehicle manufacturer's specifications to determine if the spark advance is correct.

Some timing lights have a digital reading in the back of the light that displays the number of degrees advance as the engine is accelerated (Figure 6-21). When the trigger on the light is squeezed, the light flashes and the digital display reads the degrees of spark advance. If the trigger is released, the digital reading indicates engine rpm.

Magnetic Timing Probe

Digital timing meters are available with a magnetic probe that fits into the magnetic timing receptacle above the crankshaft pulley. Some of these timing meters also have an inductive clamp for the number one spark plug wire (Figure 6-22). Therefore, this type of meter may be used with either pickup. The magnetic probe should be pushed into the timing receptacle until it lightly touches the crankshaft pulley. With the engine running, the probe senses the timing mark rotating past the probe to provide a reading on the digital display.

Special Tools

Magnetic probe-type digital timing meter

Since the magnetic timing receptacle is not located exactly above the number one TDC position on the crankshaft pulley, an offset adjustment on the meter must be completed before the timing is checked with the magnetic timing probe. The amount of offset varies depending on the engine. Always set the timing offset according to the equipment manufacturer's instructions. If the timing offset is not adjusted properly, the basic timing reading is not accurate.

Ignition Module Tester

Without an ignition module tester, the ignition module is usually tested by the process of elimination. The procedure is to test all the other components in the electronic ignition system, and if they are satisfactory, the module must be the cause of the performance problem.

Special Tools

Ignition module tester

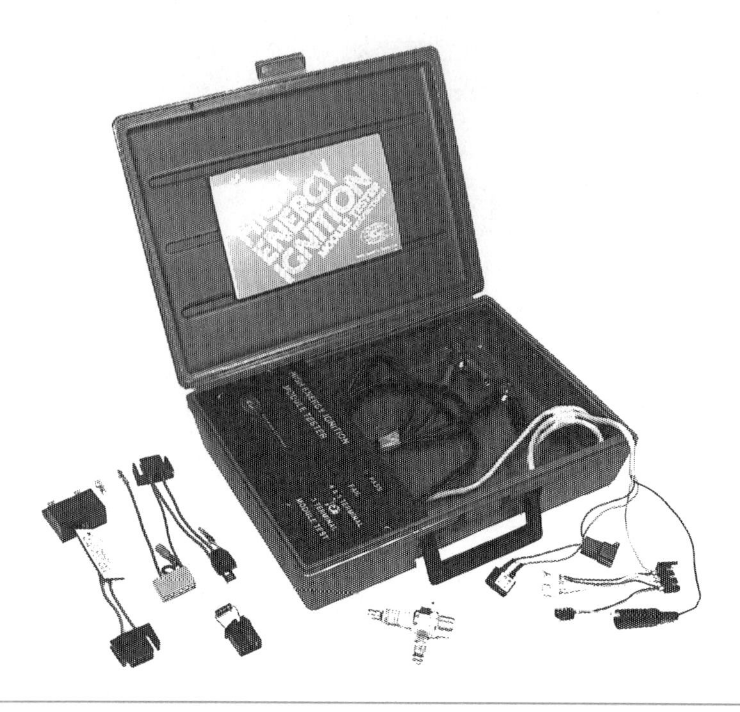

Figure 6-23 Ignition module tester (Courtesy of Automotive Group, Kent-Moore Division SPX Corp.)

The ignition module tester has a number of lead wires, and separate jumper wires are supplied with the tester (Figure 6-23). This type of tester will test a number of different ignition systems. It must always be connected according to the equipment manufacturer's recommendations. Buttons on the tester are pushed to initiate specific tests, and the tester indicates a pass or fail condition for each test.

In many cases, a defective ignition module causes a no-start problem. Sometimes a defective ignition module causes the ignition system to intermittently stop firing. Typical tests performed by an ignition module tester are:

1. Key on/engine off test
2. Cranking current test
3. Idle current test
4. Cruise current test
5. Shorted module test
6. Cranking primary voltage test
7. Idle primary voltage test
8. Cruise primary voltage test

Oscilloscope

Special Tools

Oscilloscope

An oscilloscope may be called a scope.

The oscilloscope is very useful in diagnosing ignition problems quickly and accurately. Digital and analog voltmeters do not react fast enough to read secondary ignition voltages. Each time a spark plug fires, the voltage and current are only present at the spark plug electrodes for approximately 1.5 milliseconds. The oscilloscope may be considered as a very fast reacting voltmeter that reads and displays voltages in the ignition system. These voltage readings appear as a voltage trace on the oscilloscope screen (Figure 6-24). An oscilloscope screen is a cathode ray tube (CRT), which is very similar to the picture tube in a television set. High voltage from an internal source is supplied to an electron gun in the back of the CRT when the oscilloscope is turned on. This electron gun emits a continual beam of electrons against the front of the CRT. The external leads on the oscillo-

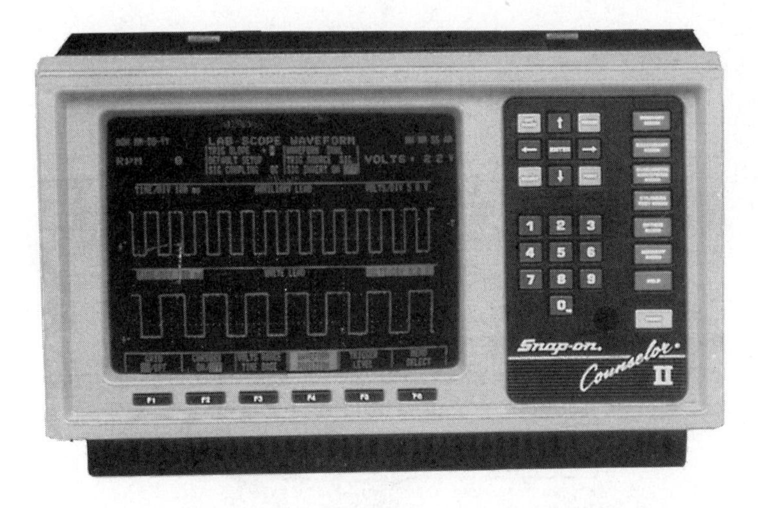

Figure 6-24 Oscilloscope (Courtesy of Snap-on Tools Corporation)

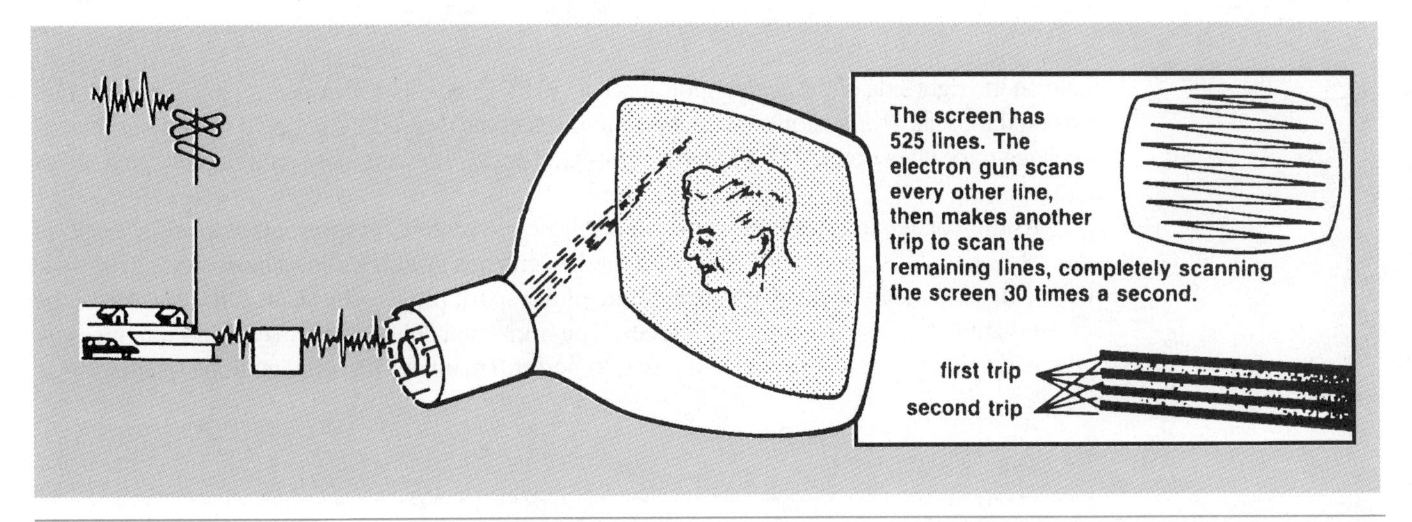

Figure 6-25 Cathode ray tube (CRT) operation (Reprinted with permission of Ford Motor Company of Canada Limited)

scope are connected to deflection plates above and below and on each side of the electron beam (Figure 6-25). When a voltage signal is supplied from the external leads to the deflection plates, the electron beam is distorted and strikes the front of the screen in a different location to indicate the voltage signal from the external leads.

One inductive clamp on the oscilloscope fits over the secondary coil wire, and a second inductive clamp is placed on the number one spark plug wire. The primary scope leads are connected from the negative primary coil terminal to ground with the correct polarity. The black primary lead wire is always connected to ground. Most oscilloscopes have a pair of voltmeter lead wires that are usually connected to the battery terminals with the correct polarity.

The negative primary coil terminal may be referred to as a "tach" terminal.

Oscilloscope Scales and Tests

An upward movement of the voltage trace on an oscilloscope screen indicates an increase in voltage, and a downward movement of this trace represents a decrease in voltage. The voltage scale on the left side of the screen reads secondary ignition voltages from 0 to 25 kilovolts (kV), and the

One kilovolt (kV) is equal to 1,000 volts.

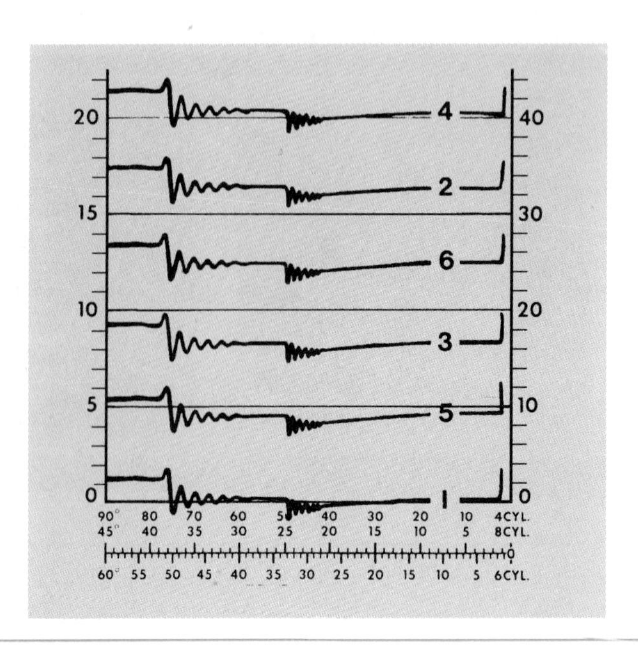

Figure 6-26 Typical scales on an oscilloscope screen (Courtesy of Sun Electric Corporation)

scale on the right side of the screen indicates 0 to 50 kV (Figure 6-26). A control knob or push buttons on the oscilloscope allows the technician to select the desired kV scale. If the highest part of a voltage trace goes upward to 40 on the right-hand scale, the secondary ignition voltage is 40 kV at that instant.

As the voltage trace moves across an oscilloscope screen, it represents a specific length of time. A horizontal scale about halfway up on the screen is graduated in milliseconds. This scale may be used to measure the actual firing time of the spark plugs. The scale at the bottom of the screen is graduated in percentage or degrees. The horizontal degree scale may be used to measure ignition dwell. An oscilloscope may be used to perform many tests on the ignition system.

Scan Testers

Special Tools

Scan tester

Scan testers are used to test automotive computer systems. These testers will retrieve fault codes from the onboard computer memory and display these codes in the digital reading on the tester. The scan tester will perform many other diagnostic functions depending on the year and make of the vehicle. Most scan testers have removable modules that are updated each year. These modules are designed to test the computer systems on various makes of vehicles. For example, some scan testers have a 3-in-1 module that tests the computer systems on Chrysler, Ford, and General Motors vehicles. A 10-in-1 module is also available to diagnose computer systems on vehicles imported by 10 different manufacturers. These modules plug into the scan tester (Figure 6-27).

Scan testers are capable of testing many onboard computer systems such as engine computers, antilock brake computers, air bag computers, and suspension computers, depending on the year and make of the vehicle and the type of scan tester. In many cases, the technician must select the computer system to be tested with the scan tester after the tester is connected to the vehicle.

The scan tester is connected to specific diagnostic connectors on various vehicles. Some manufacturers have one diagnostic connector, and they connect the data wire from each onboard computer to a specific terminal in this connector. Other vehicle manufacturers have several different diagnostic connectors on each vehicle, and each of these connectors may be connected to one or more onboard computers. A set of connectors is supplied with the scan tester to allow tester connection to various diagnostic connectors on different vehicles (Figure 6-28).

Figure 6-27 Scan tester with various modules (Courtesy of Snap-on Tools Corporation)

Figure 6-28 Scan tester connectors to fit various diagnostic connectors on different vehicles (Courtesy of OTC Division, SPX Corp.)

The scan tester must be programmed for the model year, make of vehicle, and type of engine. With some scan testers, this selection is made by pressing the appropriate buttons on the tester, as directed by the digital tester display. On other scan testers, the appropriate memory card must be installed in the tester for the vehicle being tested (Figure 6-29). Some scan testers have a built-in printer to print test results, while other scan testers may be connected to an external printer.

As automotive computer systems become more complex, the diagnostic capabilities of scan testers continue to expand. Many scan testers now have the capability to store, or "freeze," data into the tester during a road test, and then play back this data when the vehicle is returned to the shop.

Some scan testers now display diagnostic information based on the fault code in the computer memory. Service bulletins published by the scan tester manufacturer may be indexed by the tester after the vehicle information is entered in the tester. Other scan testers will display sensor specifications for the vehicle being tested.

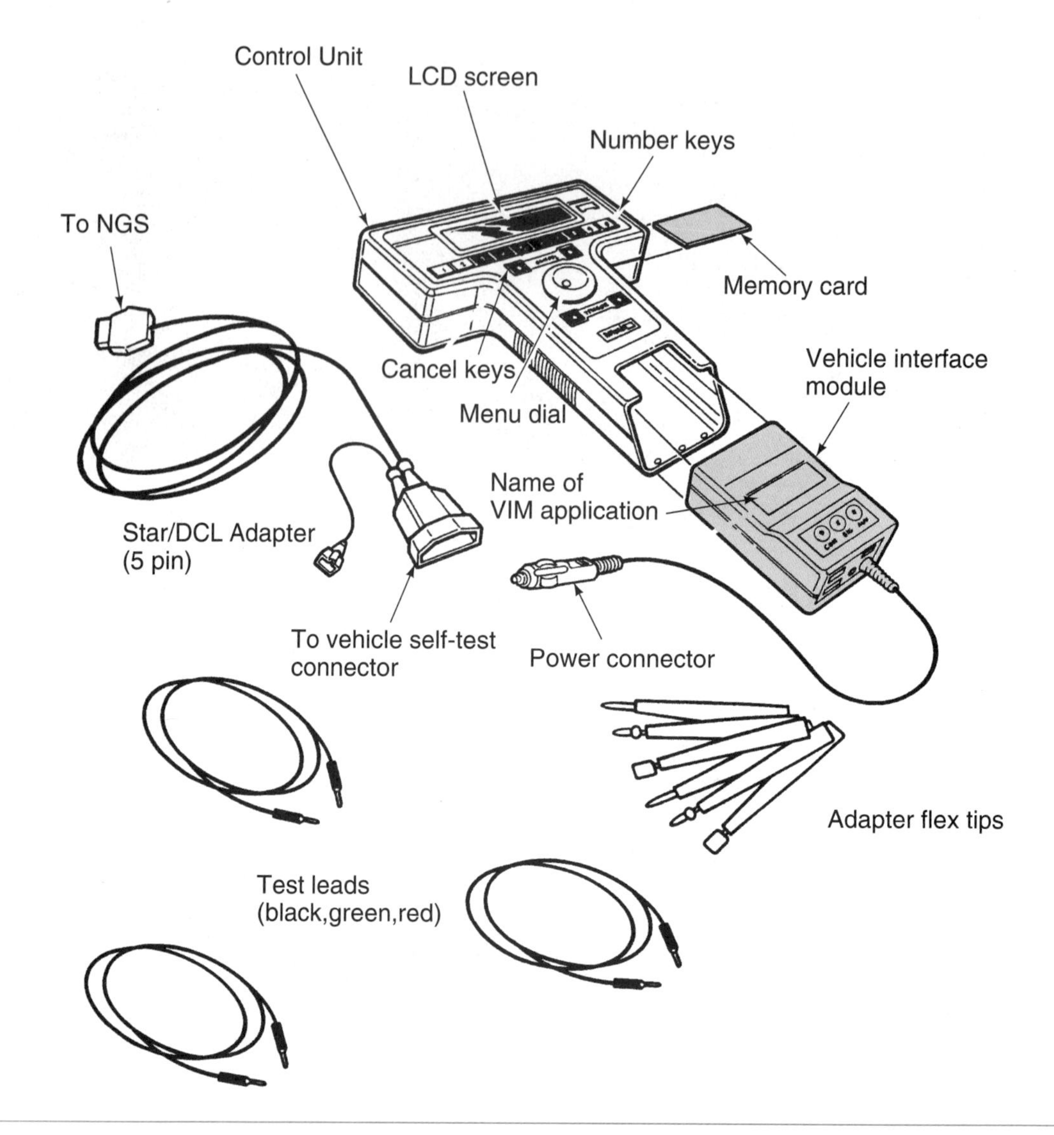

Figure 6-29 Scan tester with memory card and vehicle interface module (Reprinted with permission of Ford Motor Company)

Emissions Analyzer

Special Tools
Emissions analyzer

Vehicles produced in the United States have been required to meet specific emission standards for many years. Many states have emission inspection and maintenance (I/M) programs that require car owners to maintain their vehicles to meet emission standards. The emissions analyzer is used to measure tail pipe emissions.

Older two-gas emissions analyzers measured hydrocarbons (HC) and carbon monoxide (CO). Newer four-gas emissions analyzers measure HC and CO, plus oxygen (O_2) and carbon dioxide (CO_2) (Figure 6-30). A catalytic converter affects HC and CO readings. If a two-gas analyzer is used to measure the actual emissions levels coming out of the engine, the analyzer pickup must be installed ahead of the converter. Some exhaust pipes have a removable plug for this purpose. If the exhaust pipe does not have a plug, the pipe may be drilled and threaded, and a brass plug may be installed. If a four-gas analyzer pickup is installed in the tailpipe, O_2 and CO_2 readings are not affected by the catalytic converter.

Two-gas emissions analyzers measure hydrocarbons (HC) and carbon monoxide (CO).

Emissions analyzers require a warm-up and calibration interval of approximately 15 minutes. Some emissions analyzers perform this warm-up mode automatically, and the technician is reminded on the digital reading that the tester is in the warm-up mode. Other emissions testers do

Figure 6-30 Four-gas emissions analyzer (Courtesy of Bear Automotive Service Equipment Company)

not have an automatic warm-up mode. The technician must turn on the tester and wait the proper length of time. Modern four-gas emissions analyzers have an automatic calibration function, but older analog emissions analyzers had to be calibrated manually with calibration knobs.

Emissions analyzers have a filter or filters to remove water and other contaminants from the exhaust sample. These filters must be replaced periodically. A warning light on the tester is illuminated if the filter or anything else is restricting the exhaust flow through the tester. An emissions analyzer may be used to check the emission levels on a vehicle to determine if these levels meet state and federal emission standards. (Four-gas emission analysis and emission standards are explained later in the appropriate chapter.) These items may also be checked with a four-gas analyzer:

Four-gas analyzers measure the same emissions as a two-gas analyzer plus oxygen (O_2) and carbon dioxide (CO_2).

1. Air-fuel mixture, rich or lean
2. Air pump malfunctions
3. Defective injectors or carburetor
4. Cylinder misfiring
5. Intake manifold vacuum leaks
6. Catalytic converter defects
7. Air pump defects
8. Leaking head gaskets
9. Defective EGR valve
10. Leaking or restricted exhaust system
11. Excessive spark advance

Engine Analyzers

SERVICE TIP: When diagnosing any engine performance or economy problem, always test the basic items first. For example, always be sure the engine has satisfactory compression, ignition, and emission component operation before diagnosing the computer system.

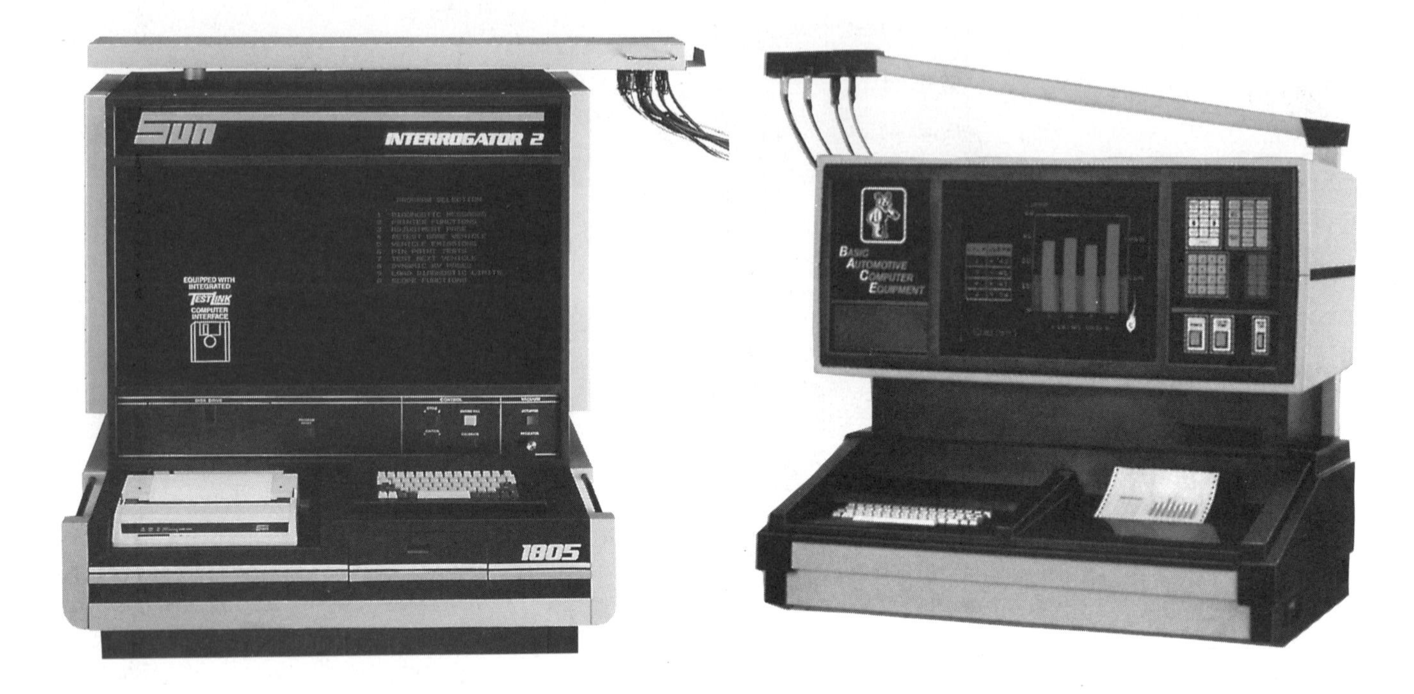

Figure 6-31 An engine analyzer is a combination of many automotive testers. (Courtesy of Sun Electric Corporation)

Figure 6-32 Engine analyzer with key board, printer, and graphic display (Courtesy of Bear Automotive Service Equipment Company)

Special Tools

Engine analyzer

Engine analyzers are actually a combination of different automotive testers (Figure 6-31). Many engine analyzers have an oscilloscope screen that displays ignition voltages and many other readings. Engine analyzers have many leads which must be connected according to the analyzer manufacturer's instructions. A typical engine analyzer will perform the same diagnostic functions as these individual testers:

1. Vacuum gauge
2. Pressure gauge
3. Vacuum pump
4. Compression gauge (power balance tester)
5. Voltmeter
6. Ammeter
7. Ohmmeter
8. Advance-type timing light
9. Oscilloscope
10. Scan tool
11. Emissions analyzer

The engine analyzer contains a keyboard that is used to enter vehicle data, commands, and specifications (Figure 6-32). A printer in the engine analyzer provides a printout of the test results for the customer. Some engine analyzers display the tests results graphically on the screen.

The normal specified sensor readings are programmed into the computer in the engine analyzer for each year and model of vehicle. During a test procedure, this computer compares the vehicle data to the specifications programmed into the analyzer computer. The engine analyzer identifies readings that are not within specifications and provides diagnostic information to find the exact cause of the defective reading.

Engine analyzers usually have manual and automatic test modes. If the technician selects the automatic mode, the analyzer automatically performs a complete series of tests. In the manual mode, the technician selects the specific tests he or she wants to perform. During the series of tests in the automatic mode, the analyzer performs tests to prove the condition of the ignition system, starting system, charging system, cylinder compression, fuel system, and emission systems.

Classroom Manual
Chapter 6, page 120

CUSTOMER CARE: During many automotive service operations such as fuel system service, the technician literally has the customer's life in his or her hands! Always perform automotive service, including fuel system repairs, carefully and thoroughly. Always watch for unsafe conditions such as fuel leaks or damaged fuel lines, and report these problems to the customer. When you prove to the customer that you are concerned about vehicle safety, you will probably have a steady customer.

Guidelines for Using Automotive Tune-Up Equipment

1. Electronic fuel injection (EFI) systems must be depressurized before connecting a fuel gauge or disconnecting fuel system components.
2. On a port fuel injection (PFI) system, the pressure gauge is connected to the Schrader valve on the fuel rail.
3. On throttle body injection (TBI) or carburetor systems, the fuel pressure gauge must be connected in series at the fuel inlet line.
4. The injector balance tester is a timing device that opens each injector for a specific length of time.
5. During an injector balance test, the fuel system pressure drop is recorded when each injector is opened for a specific length of time.
6. Injectors are cleaned with a solution of unleaded gasoline and injector cleaner in a pressurized container connected to the Schrader valve on the fuel rail.
7. During the injector cleaning procedure, the return fuel line must be blocked and the electric fuel pump must be disabled.
8. Conventional 12-V or self-powered test lights should not be used to test computer circuits. High-impedance test lights should be used for this purpose.
9. Test lights of any type must not be used to diagnose air bag circuits. Only the vehicle manufacturer's recommended tools should be used on these systems.
10. Volt-ampere testers are commonly used to perform battery load tests, starter current draw tests, alternator output tests, and alternator normal system voltage tests.
11. Multimeters read ac volts, dc volts, milliamperes, amperes, and ohms on various scales.
12. Digital multimeters have higher impedance than analog multimeters.
13. Only high-impedance digital multimeters should be used to test computer system components such as oxygen (O_2) sensors.
14. Tach-dwellmeters read engine rpm and ignition dwell.
15. A piece of reflective tape attached to a rotating component such as the crankshaft pulley provides a signal for a photoelectric tachometer.
16. The probe on a magnetic probe-type digital tachometer is installed in the magnetic timing receptacle above the crankshaft pulley.
17. The control knob on an advance-type timing light may be used to check spark advance.
18. When a magnetic probe-type digital timing meter is used, the offset on the tester must be adjusted for the engine being tested.
19. Many ignition module testers check the ignition module and primary ignition circuit voltage and current under various engine operating conditions.
20. An oscilloscope contains a cathode ray tube (CRT), which provides a voltage trace of the ignition system voltage much like a very fast reacting voltmeter.
21. Scan testers retrieve fault codes from the computer memory and perform many other diagnostic functions.

22. The scan tester must be programmed for the vehicle make, model year, and type of engine.
23. Since most scan testers have the capability to test various computer systems on the vehicle, the tester must be programmed for the computer system being tested.
24. The scan tester must be connected to the appropriate diagnostic connector on the vehicle.
25. A two-gas emissions analyzer reads carbon monoxide (CO) and hydrocarbon (HC) emission levels.
26. A four-gas emissions analyzer reads the same emission levels as a two-gas analyzer, plus oxygen (O_2) and carbon dioxide (CO_2). These two latter emissions are not affected by the catalytic converter.
27. Emission analyzers require a 15-minute warm-up and calibration period when they are turned on.
28. An engine analyzer contains a combination of many different automotive testers.

CASE STUDY

A customer brought a 5.2 L Dodge 1/2 ton into the shop with an engine misfiring complaint. When the technician listened to the engine running, he found the engine was misfiring at idle speed, and the misfire became worse during acceleration.

The technician connected the scope to the engine, and discovered the spark plug firing voltages from cylinders 5 and 7 were very low. The other spark plug firing voltages were normal. After thinking about this problem for a few minutes, the technician concluded that cylinders 5 and 7 are beside each other at the rear of the left bank; thus, a compression problem on both of these cylinders could be the cause of the low firing voltages. The technician also realized that a high-voltage leakage problem between the number 5 and 7 spark plug wires could be the cause of the problem.

The technician shut off the engine and removed the number 5 and 7 wires from the distributor cap. When the engine was restarted, the firing voltages from the number 5 and 7 cylinders still remained very low. This action proved that the problem had to be in the distributor cap. If the problem was in the cylinders, the firing voltages would have gone very high when the spark plug wires were removed from the distributor cap.

The distributor cap was removed. A close examination of the cap revealed a small crack in the corner of the cap between the number 5 and 7 spark plug wire terminals. After a new distributor cap was installed, the spark plug firing voltages were normal, and the engine misfiring problem disappeared. From this experience, the technician learned these facts:

1. The oscilloscope is very useful in diagnosing ignition problems quickly and accurately.
2. The technician's understanding of automotive systems is also extemely important when diagnosing automotive problems.
3. The technician's ability to think about the causes of automotive problems and arrive at some logical steps to pinpoint the trouble is also very important.

Terms to Know

Injector balance tester
Pressurized injector cleaning container
Canister-type pressurized injector cleaning container
Self-powered test light
Volt-ampere tester
Analog meter
Digital meter
Meter impedance
Tach-dwellmeter
Photoelectric tachometer
Magnetic probe-type digital tachometer
Advance-type timing light
Magnetic probe-type digital timing meter
Ignition module tester
Oscilloscope
Scan tester
Two-gas emissions analyzer
Four-gas emissions analyzer
Engine analyzer

ASE Style Review Questions

1. While discussing electric fuel pump pressure testing:
   Technician A says the fuel pressure gauge should be connected to the return fuel line to test the fuel pressure in a port fuel injection (PFI) system.
   Technician B says the fuel pressure gauge should be connected to the Schrader valve on the fuel rail to test fuel pressure in a PFI system.
   Who is correct?
   A. A only **C.** Both A and B
   B. B only **D.** Neither A nor B
2. While discussing electric fuel pump pressure testing:
   Technician A says the fuel pressure gauge should be connected in series at the fuel inlet line on throttle body injection (TBI) systems.
   Technician B says the Schrader valves on the fuel rails in various fuel systems are all the same size.
   Who is correct?
   A. A only **C.** Both A and B
   B. B only **D.** Neither A nor B
3. While discussing injector balance testers:
   Technician A says the injector balance tester contains a polarity sensor.
   Technician B says the injector balance tester contains a timing circuit.
   Who is correct?
   A. A only **C.** Both A and B
   B. B only **D.** Neither A nor B
4. While discussing an injector balance test:
   Technician A says the differences between the pressured drops on the injectors should not exceed 10 psi (69 kPa).
   Technician B says if there is excessive pressure drop when an injector is energized, the injector orifice is restricted.
   Who is correct?
   A. A only **C.** Both A and B
   B. B only **D.** Neither A nor B
5. While discussing injector cleaning:
   Technician A says the fuel return line must be blocked during the injector cleaning process.
   Technician B says the electric fuel pump should be disabled during the injector cleaning process.
   Who is correct?
   A. A only **C.** Both A and B
   B. B only **D.** Neither A nor B
6. While discussing circuit testers:
   Technician A says a conventional 12-V test light may be used to diagnose automotive computer circuits.
   Technician B says a self-powered test light may be used to diagnose an air bag circuit.
   Who is correct?
   A. A only **C.** Both A and B
   B. B only **D.** Neither A nor B

7. While discussing multimeters:
Technician A says an analog multimeter should be used to test the oxygen (O_2) sensor voltage.
Technician B says a high-impedance digital multimeter should be used to test the O_2 sensor voltage.
Who is correct?
A. A only **C.** Both A and B
B. B only **D.** Neither A nor B

8. While discussing tach-dwellmeters:
Technician A says many tach-dwellmeter leads are connected from the negative primary coil terminal to ground.
Technician B says many tach-dwellmeter leads are connected from the positive primary coil terminal to ground.
Who is correct?
A. A only **C.** Both A and B
B. B only **D.** Neither A nor B

9. While discussing timing lights and timing meters:
Technician A says if a timing meter is used with a magnetic probe, the offset must be adjusted on the meter.
Technician B says while using a magnetic timing probe, the offset adjustment is the same on all engines.
Who is correct?
A. A only **C.** Both A and B
B. B only **D.** Neither A nor B

10. While discussing oscilloscopes:
Technician A says the upward voltage traces on an oscilloscope screen indicate a specific length of time.
Technician B says the cathode ray tube (CRT) in an oscilloscope is like a very fast reacting voltmeter.
Who is correct?
A. A only **C.** Both A and B
B. B only **D.** Neither A nor B

Table 6-1 ASE TASK

Diagnose engine mechanical, electrical, and fuel problems with an oscilloscope and/or engine analyzer; determine needed repairs.

Problem Area	Symptoms	Possible Causes	Classroom Manual	Shop Manual
ENGINE PERFORMANCE	Detonation	Ignition timing too far advanced	120	107
	Lack of power	Ignition timing retarded	120	107
	Overheating	Ignition timing retarded	120	107

Table 6-2 ASE TASK

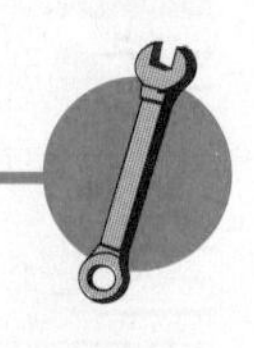

Inspect, test, and replace electronic ignition system control unit (module).

Problem Area	Symptoms	Possible Causes	Classroom Manual	Shop Manual
ENGINE PERFORMANCE	Failure to start	Defective ignition module	120	102
	Intermittent ignition operation	Defective ignition module	120	102

Table 6-3 ASE TASK

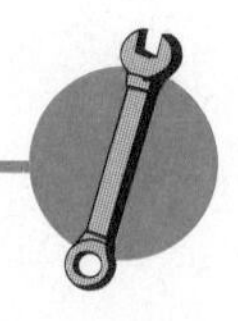

Inspect, test, and repair or replace fuel pressure regulation system and components of a fuel injection system.

Problem Area	Symptoms	Possible Causes	Classroom Manual	Shop Manual
ENGINE	Failure to start	1. Defective fuel pump	121	91
	2.	Plugged fuel filter	121	91
	Rich air-fuel mixture	Excessive fuel pressure	121	91
	Lean air-fuel mixture	Low fuel pressure	121	91

Table 6-4 ASE TASK

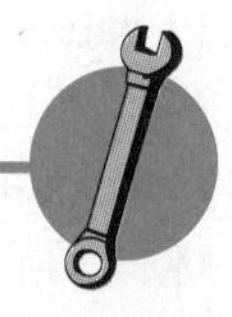

Inspect, test, clean, and replace fuel injectors.

Problem Area	Symptoms	Possible Causes	Classroom Manual	Shop Manual
ENGINE PERFORMANCE	Acceleration stumble	Injector orifices restricted	122	93
	Rich air-fuel mixture	Injector plungers sticking	122	93
	Stalling, erratic idle operation	Injector orifices restricted	122	93

Battery, Starter, and Charging System Service and Diagnosis

Upon completion and review of this chapter, you should be able to:

- ❑ Perform general battery service such as cleaning battery terminals and a battery case, and removing and replacing a battery.
- ❑ Perform battery hydrometer test, and determine battery condition.
- ❑ Perform battery capacity test, and determine battery condition.
- ❑ Perform battery charging procedures.
- ❑ Connect booster battery cables properly.
- ❑ Measure battery drain accurately.
- ❑ Perform starter current draw test.
- ❑ Perform voltage drop tests on starter insulated-side and ground-side circuit.
- ❑ Perform voltage drop tests on starter control circuit.
- ❑ Inspect alternator belt condition.
- ❑ Check belt tension on V-belts and ribbed V-belts.
- ❑ Adjust belt tension.
- ❑ Perform an alternator output test, and determine alternator condition.
- ❑ Check alternator regulator voltage.
- ❑ Diagnose undercharged and overcharged battery conditions.
- ❑ Inspect and repair wiring harness and connectors.

Basic Tools

Technician's tool set

Service manual

Battery terminal puller

Battery terminal cleaner

12-gauge jumper wire

Splicing tape

Mylar tape

Resin-core solder

Battery Service and Diagnosis

General Battery Service

WARNING: Battery electrolyte contains sulfuric acid, which is a strong, corrosive acid. Electrolyte is very damaging to vehicle paint, upholstery, and clothing. Do not allow the electrolyte to contact these items.

CAUTION: Battery electrolyte is very harmful to human skin and eyes. Always wear face and eye protection, protective gloves, and clothing when handling batteries or electrolyte. If electrolyte contacts your skin or eyes, flush with clean water immediately and obtain medical help.

CAUTION: Never smoke or allow other sources of ignition near a battery. Hydrogen gas discharged while charging the battery is explosive (Figure 7-1)! A battery explosion may cause personal injury or property damage. Even when the battery has not been charged for several hours, hydrogen gas still may be present under the battery cover.

A battery should be checked for terminal corrosion, loose or broken holddown, frayed or corroded cables, cracked case, and electrolyte level (Figure 7-2). If the battery has removable vent caps, the electrolyte level may be checked through the vent cap openings. When the vent caps are not removable, the electrolyte level may be checked by observing the level through the translucent plastic case. The battery may be removed from the vehicle, washed with a baking soda and water solution, and rinsed with clean water.

WARNING: On air-bag equipped vehicles, the negative battery cable should be disconnected and the technician should wait one minute before working on the vehicle electrical system to avoid accidental air bag deployment. The wait time may vary depending on the vehicle manufacturer. Always follow the instructions in the vehicle manufacturer's service manual.

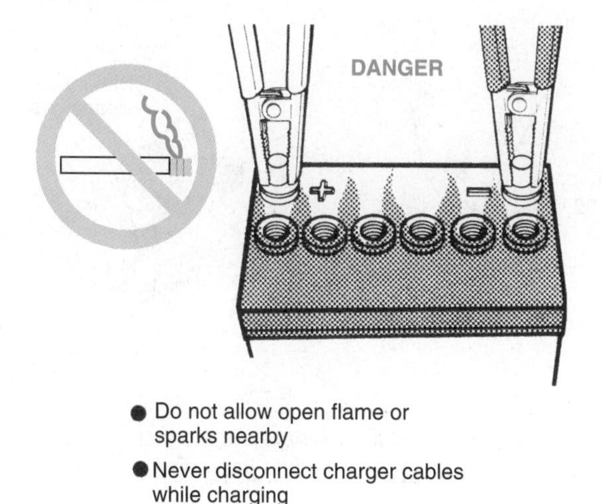

Figure 7-1 Do not smoke or allow other sources of ignition near a battery. (Courtesy of Toyota Motor Corporation)

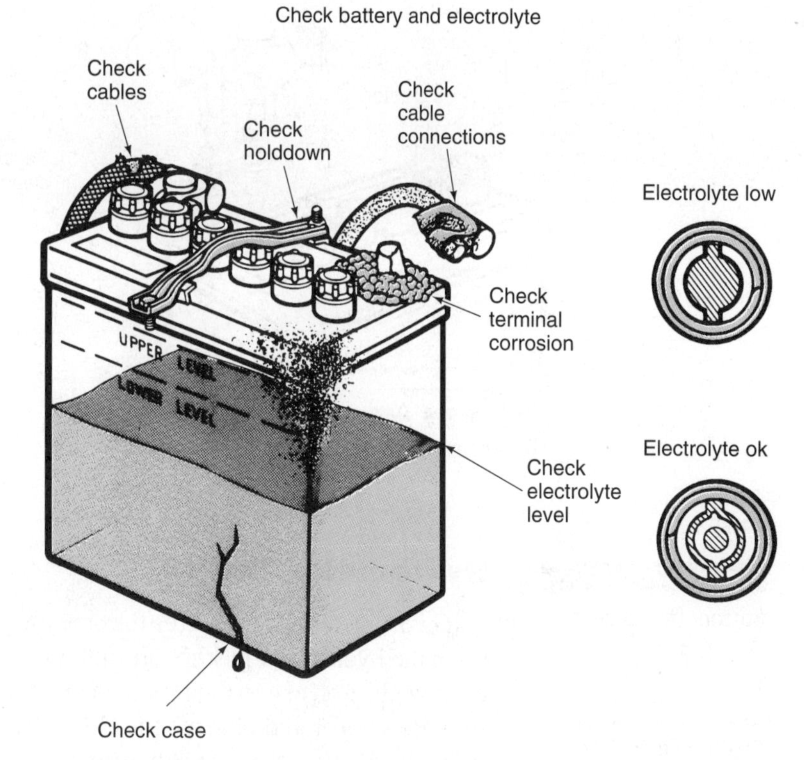

Figure 7-2 General battery service (Courtesy of Toyota Motor Corporation)

SERVICE TIP: Always disconnect the negative battery cable before the positive cable. If the positive cable is disconnected first, the wrench may slip and ground this cable to the chassis. This action may result in severe burns or a battery explosion.

Disconnecting a battery in a computer-equipped vehicle causes these problems:

1. Erases adaptive memories in various computers.
2. Erases memory in memory seats and memory mirror systems.
3. Erases station memory programmed in the radio.
4. On some cars such as Chrysler products, the theft deterrent computer will not allow the vehicle to restart after a battery disconnection until one of the door lock cylinders is cycled from locked to unlocked.

To avoid these problems, some tool manufacturer's supply a dry cell voltage source that may be plugged into the cigarette lighter prior to disconnecting the battery.

WARNING: Do not hammer or twist battery cable ends to loosen them. This action will loosen the terminal in the battery cover and cause electrolyte leaks.

After the battery terminal nuts have been loosened, the battery terminals may be removed with a terminal puller (Figure 7-3). Battery terminals and cable ends may be cleaned with a battery terminal cleaner (Figure 7-4).

WARNING: Always tighten the battery holddown bolts to the specified torque. Overtightening these bolts may damage the battery cover and case.

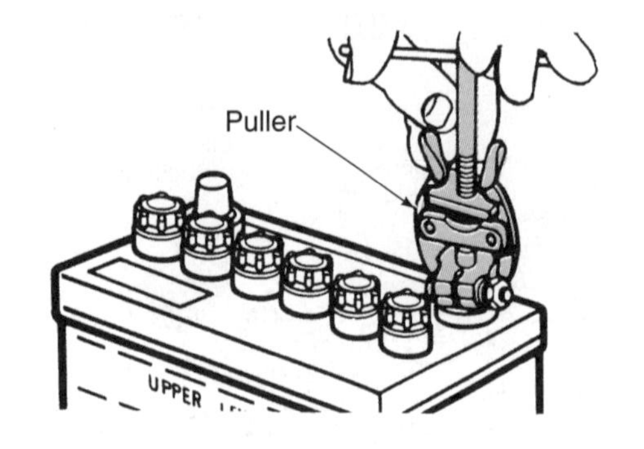

Figure 7-3 Battery terminal puller (Courtesy of Toyota Motor Corporation)

Figure 7-4 Using a terminal cleaner to clean battery terminals and cable ends (Courtesy of Toyota Motor Corporation)

Special Tools

Battery hydrometer

The specific gravity of a liquid is the weight of the liquid in relation to the weight of an equal volume of water. Water has a specific gravity of 1.000, sulfuric acid has a specific gravity of 1.835, and the electrolyte in a fully charged battery has a specific gravity of 1.270.

Hydrometer Testing

The hydrometer test measures the specific gravity of the electrolyte. Remove the vent caps and insert the hydrometer pickup tube into the electrolyte. Squeeze the hydrometer bulb and release it to draw electrolyte into the hydrometer until the float is floating freely (Figure 7-5).

Bend over and observe the electrolyte level on the hydrometer float. Read the electrolyte specific gravity on the float at the electrolyte level. A fully charged battery has a specific gravity of 1.270. If the battery specific gravity is less than 1.270, the battery requires charging. A battery with a 1.155 specific gravity is 50% charged (Figure 7-6). A voltmeter may be connected across the battery terminals to check the open-circuit voltage.

Since the specific gravity is affected by temperature, some correction of the specific gravity reading in relation to temperature is required. A thermometer in the lower part of some hydrometers measures electrolyte temperature. For every 10°F below 80°, subtract 4 points from the hydrometer reading, and add 4 points to the reading for each 10° above 80°F (Figure 7-7).

Some maintenance-free batteries have a built-in hydrometer in the cover. When a green dot appears in the center of the hydrometer, the battery is ready for testing. If the hydrometer appears dark with no green dot, the battery and charging circuit should be tested. A light yellow or bright hydrometer indicates the electrolyte is below the bottom of the hydrometer (Figure 7-8). Under this condition, the battery replacement is required, and the battery should not be tested or charged.

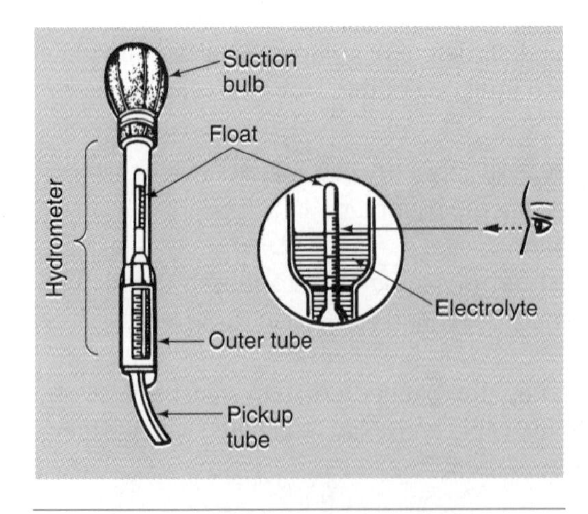

Figure 7-5 Battery hydrometer (Courtesy of Toyota Motor Corporation)

STATE OF CHARGE	SPECIFIC GRAVITY	OPEN-CIRCUIT VOLTAGE
100%	1.265	12.6
75%	1.225	12.4
50%	1.190	12.2
25%	1.155	12.0
DEAD	1.120	11.9

Figure 7-6 Battery specific gravity reading in relation to state of charge and open-circuit voltage (Courtesy of Toyota Motor Corporation)

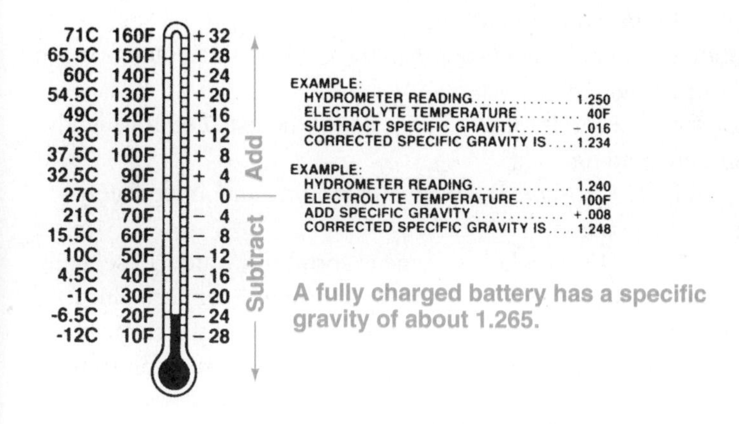

Figure 7-7 Correcting specific gravity reading in relation to temperature (Courtesy of Toyota Motor Corporation)

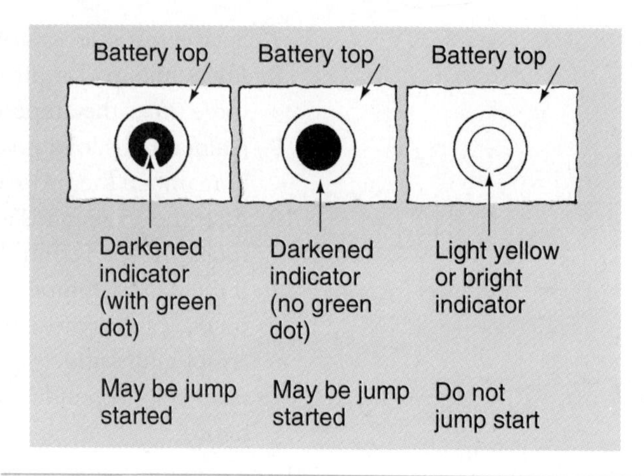

Figure 7-8 Built-in hydrometer in battery cover (Courtesy of Oldsmobile Division, General Motors Corporation)

Battery Capacity Test

A battery must have a minimum specific gravity of 1.190 prior to a capacity test, and the battery cables should be removed. The battery capacity test procedure may vary depending on the type of test equipment and whether the tester has digital or analog meters. Always follow the test equipment manufacturer's instructions.

Follow these steps for the capacity test with an analog-type capacity tester:

1. Rotate the capacity tester load control to the off position, and set the voltmeter control to the 18-V position.
2. Check the mechanical zero on each meter and adjust if necessary.

SERVICE TIP: On automotive electrical test equipment such as battery testers and chargers, the positive cable clamp is red and the negative cable clamp is black.

3. Connect the tester leads to the battery terminals with the proper polarity. The red lead is always connected to the positive terminal, and the black lead is always connected to the negative terminal.
4. Observe the battery open-circuit voltage, which should be above 12.2 V.
5. Set the test selector to the #2 charging position, and adjust the ammeter pointer to the zero position with the electrical zero adjust control.
6. Connect the ammeter inductive clamp over the negative tester cable (Figure 7-9).

Special Tools

Volt-ampere tester

Battery open-circuit voltage is the voltage with no electrical load on the battery.

A battery capacity test may be called a load test.

A small screw or thumb screw on the front of an analog meter may be rotated to adjust the pointer to the zero reading on the scale with the meter disconnected. This is called the mechanical zero.

ELECTROLYTE TEMPERATURE	MINIMUM VOLTAGE UNDER LOAD
70F (21C) & above	9.6 volts
60F (16C)	9.5
50F (10C)	9.4
40F (4C)	9.3
30F (-1C)	9.1
20F (-7C)	8.9
10F (-12C)	8.7
0F (-18C)	8.5

Figure 7-9 Volt-ampere tester connections to the battery (Courtesy of Toyota Motor Corporation)

7. Set the test selector to the #1 starting position.
8. Rotate the load control until the ammeter reads one-half the cold cranking rating or three times the ampere-hour rating.
9. Maintain this load position for 15 seconds. If necessary, adjust the load control slightly to maintain the proper ampere reading.
10. After the 15 seconds are completed, observe the voltmeter reading and immediately rotate the load control to the off position.
11. If the battery temperature is 70°F (21°C) and the voltage is above 9.6 V after 15 seconds on the capacity test, the battery is satisfactory and may be returned to service. Since temperature affects battery capacity, the minimum voltage at the end of the capacity test varies depending on temperature as indicated in the volt-ampere tester connection figure.

If the battery voltage is less than the minimum load voltage at the end of the load test, the battery should be charged and the load test repeated. After charging and a repeat load test, if the battery voltage is less than the minimum load voltage, the battery should be replaced.

Photo Sequence 4 shows a typical procedure for testing battery capacity.

Battery Charging

Low-Maintenance Battery Charging

Fast Charging. Before charging the battery, the battery case and terminals should be cleaned and the vent caps should be removed. Check the electrolyte level and add distilled water as required. If the battery is in the vehicle, disconnect both battery cables. Determine the charging rate and time (Figure 7-10).

CAUTION: Never connect or disconnect battery charger cables from the battery terminals with the charger in operation. This procedure causes sparks at the battery terminals, which may result in a battery explosion, personal injury, and/or property damage.

Special Tools

Battery charger

Follow these steps for battery charging:

1. Be sure the charger timer switch and main switch are off.
2. Connect the negative charger cable to the negative battery terminal, and the positive charger cable to the positive battery terminal.
3. Connect the charger power cable to an electrical outlet.
4. Set the charger voltage switch to the proper battery voltage, and turn on the main switch.
5. Set the timer to the required battery charging time.

Reserve Capacity Rating	20-Hour Rating	5 Amperes	10 Amperes	20 Amperes	30 Amperes	40 Amperes
75 Minutes or less	50 Ampere-Hours or less	10 Hours	5 Hours	2 1/2 Hours	2 Hours	
Above 75 To 115 Minutes	Above 50 To 75 Ampere-Hours	15 Hours	7 1/2 Hours	3 1/4 Hours	2 1/2 Hours	2 Hours
Above 115 To 160 Minutes	Above 75 To 100 Ampere-Hours	20 Hours	10 Hours	5 Hours	3 Hours	2 1/2 Hours
Above 160 To 245 Minutes	Above 100 To 150 Ampere-Hours	30 Hours	15 Hours	7 1/2 Hours	5 Hours	3 1/2 Hours

Figure 7-10 Battery charging time in relation to the battery rating (Courtesy of Toyota Motor Corporation)

Photo Sequence 4
Typical Procedure for Testing Battery Capacity

P4-1 Observe the hydrometer in the battery top. If the hydrometer is green, proceed with the load test. A black hydrometer indicates battery charging is necessary, and a yellow hydrometer indicates battery replacement is required.

P4-2 Remove the battery cables and install adaptors in the battery terminals.

P4-3 Be sure the load control is in the off position, and set the voltmeter control to 18 V.

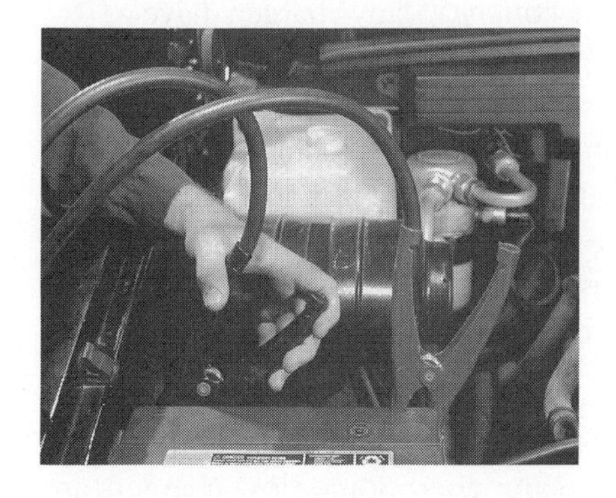

P4-4 Connect the volt-ampere tester cables to the battery terminals with the correct polarity.

P4-5 Connect the ammeter clamp over the negative tester cable.

P4-6 Set the tester control to the battery test position.

P4-7 Rotate the load control until the ammeter reads one-half the cold cranking ampere rating of the battery.

P4-8 Hold the load control in this position for 15 seconds, and read the battery voltage with the load still applied. If the battery temperature is 70°F, the voltage should be above 9.6 V.

P4-9 Immediately rotate the load control to the off position, and disconnect the tester cables.

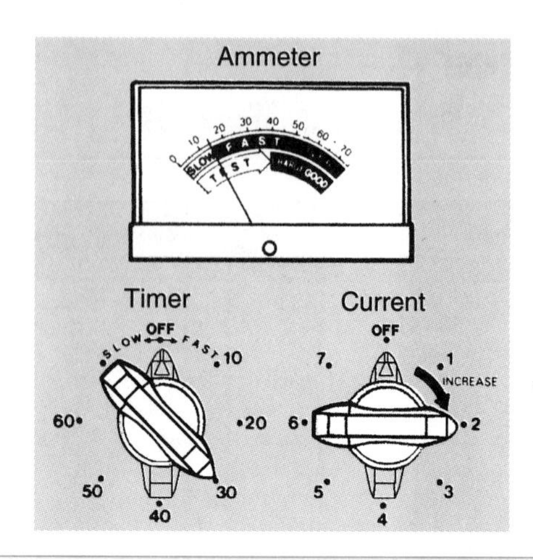

Figure 7-11 Battery charger ammeter and timer control (Courtesy of Toyota Motor Corporation)

6. Adjust the current control to obtain the proper charging rate (Figure 7-11). Some battery chargers have a reverse polarity protector, which prevents charger operation if the charger cables are connected to a battery with reversed polarity. If the charger cables are connected properly and the battery voltage is very low, the reverse polarity protector may not allow the charger to begin charging the battery. Many chargers have a bypass button that must be pressed to bypass the reverse polarity protector when this condition occurs.
7. Shut off the charger while testing the battery. Continue charging the battery until the specific gravity is above 1.250 and the open-circuit voltage is above 12.6 V.
8. When the battery is above 1.225 specific gravity, a high charging rate causes excessive gassing. Reduce the charging rate as required to prevent excessive gassing.
9. Shut off the timer, current control, and main switch, and disconnect the charger cables from the battery.

Slow Charging. A battery may be slow charged at one-tenth of the ampere-hour rating. For example, an 80 ampere-hour battery should be charged at 8 amperes. Some slow chargers are capable of charging several batteries connected in series. A current control on the charger is rotated until the desired charging rate is obtained on the charger ammeter. The battery should be charged until the specific gravity is above 1.250, or until there is no further increase in specific gravity or battery voltage in one hour.

Charging Maintenance-Free Batteries

Maintenance-free batteries require different charging rates compared to low-maintenance batteries. Some vehicle manufacturers recommend using the reserve capacity rating of the battery to determine the ampere-hours charge required by the battery. For example, a battery with a reserve capacity of 75 minutes requires 25 amperes × 3 hours charge = 75 ampere-hours. Continue charging the battery until the green dot appears in the hydrometer, and the battery passes a capacity test. Always use the vehicle or battery manufacturer's recommended charging rate.

Battery Boosting

WARNING: When jump starting a vehicle, always turn off all electrical accessories in the vehicle being boosted and the boost vehicle. Electrical accessories left on during the boost procedure may be damaged.

WARNING: A booster battery must be connected with the proper polarity and the correct procedure. Reversed battery polarity or improper procedure may result in severe electrical system damage on the boost vehicle and the vehicle being boosted. This procedure may also result in battery explosion, personal injury, and property damage.

CAUTION: Do not boost a vehicle if the battery built-in hydrometer indicates clear or light yellow. This action may result in battery explosion with resulting personal injury and property damage.

Always follow the battery boost procedure in the vehicle manufacturer's service manual. Following is a typical battery boost procedure:

1. Apply the parking brake and place the transmission in park.
2. Be sure all electrical accessories on both vehicles are turned off, and check the built-in hydrometer. It must indicate green or black.
3. Connect one end of the positive booster cable to the positive cable end in the boost vehicle.
4. Connect the other end of the positive cable to the positive battery cable end in the vehicle being boosted.
5. Connect one end of the negative booster cable to the negative battery cable end in the boost vehicle.
6. Connect the other end of the negative booster cable to an engine ground on the vehicle being boosted (Figure 7-12). This ground connection may be an alternator bracket or A/C compressor bracket, and this connection should be at least 18 inches from the battery in the vehicle being boosted.
7. Start the engine in the boost vehicle, and then start the engine in the vehicle being boosted.
8. Reverse steps 3 through 6 to disconnect the booster cables. When disconnecting the booster cables, the negative booster cable on the vehicle being boosted must be disconnected first.

Special Tools

Booster cables

Battery boosting may be called jump starting.

Battery Drain Testing

Most computers have a few milliamperes of current flow when they are not in operation. This current flow is called battery drain. Since many vehicles today have several computers, this current flow may discharge the battery if the vehicle is not driven for several weeks. Many computers have a higher drain for a short period of time after the ignition switch is turned off, and then the drain is reduced. It is very important for technicians to understand how to perform an accurate battery drain test and determine if the drain is normal or excessive.

Battery drain refers to the current flow out of the battery through some of the electrical or electronic components on the vehicle with the ignition switch off.

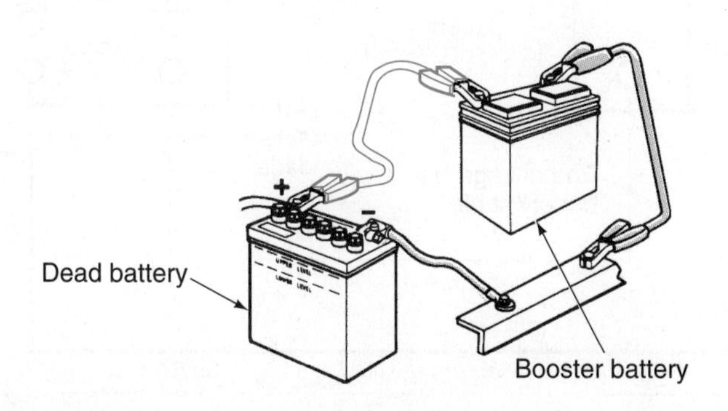

Figure 7-12 Booster battery connections (Courtesy of Toyota Motor Corporation)

Battery drain on a computer-equipped vehicle may be called parasitic drain.

Special Tools

Multimeter

Classroom Manual
Chapter 7, page 135

Follow these steps for battery drain testing:

1. Be sure the ignition switch is off. Make sure the glove compartment and trunk lights are off and the doors are closed.
2. Disconnect the negative battery cable, and connect a 12-gauge jumper wire between the cable end and the negative battery terminal. Do not attempt to start the engine once this jumper is connected.
3. Connect the ammeter on a multimeter in series between the negative battery cable end and the negative battery terminal (Figure 7-13). Set the ammeter scale on 20 amperes.
4. Turn on the ignition switch, and turn on such computer-controlled accessories as the A/C and radio/stereo.
5. Turn off all electrical accessories and the ignition switch.
6. Disconnect the 12-gauge jumper wire.
7. If the ammeter reading is very low, switch the ammeter scale to the lowest milliampere scale and observe the reading.
8. Check the vehicle manufacturer's specifications for the required time for the computers to reduce the amount of drain (Figure 7-14).
9. Compare the drain on the ammeter to the vehicle manufacturer's specifications. If the drain is excessive, disconnect the fuses or circuit breakers in various circuits to locate the source of the high drain.
10. Disconnect the multimeter, and tighten the negative battery cable.

Starter Diagnosis and Service

Starter Current Draw Test

The battery specific gravity must be 1.190 or above before the starter current draw test is performed. Always follow the starter current draw test procedure in the vehicle manufacturer's service manual. Following is a typical starter current draw test procedure:

1. Rotate the load control on the tester to the off position.
2. If an analog tester is used, check the mechanical zero on each meter and adjust as necessary.
3. Set the voltmeter selector to the internal (INT) 18-V position.
4. Connect the positive tester cable to the positive battery cable, and connect the negative tester cable to the negative battery cable.

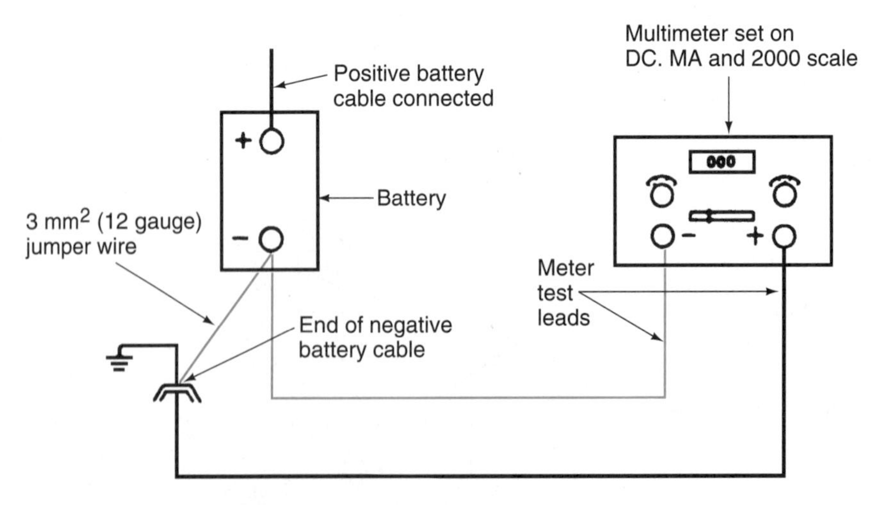

Figure 7-13 Jumper wire and ammeter connected to test battery drain (Courtesy of Oldsmobile Division, General Motors Corporation)

COMPONENT OR MODULE	CURRENT DRAW (MA)		SERVICE MANUAL SECTION	CORRESPONDING FUSE/POWER CIRCUIT
	TYPICAL	MAX		
Radio U1X CD U1B	7.0 1.8	8.5 3.5	8A-150 8A-150	11 11
ELC (After 7 Minute Time Out)		1.0	8A-42	12
Multi-Function Chime Module Key Removed From Ignition	1.0	1.0	8A-76	11
PCM	5.0	7.0	6E	
Generator Heated Windshield Control Module	2.0 0.3	2.0 0.4	8A-30 8A-69	N/A 8
RAC (Theft Deterrent) (Illuminated Entry) (Auto Door Locks)		3.8	8A-15 8A-133 8A-115 8A-131	11
Pass Key Decoder Module	.75	1.0	8A-133	4
Instrument Panel Digital Cluster	4.0	6.0	8A-82	14
Oil Level Module		0.1	8A-81	11
Light Control Module	0.5	1.0	8A-15	14
Heated Seat Control Modules		0.5	8A-148	Circuit Breaker 4

Figure 7-14 Computer drain specifications (Courtesy of Oldsmobile Division, General Motors Corporation)

5. Set the test selector to the #2 charging position.
6. Adjust the ammeter to read zero with the zero adjust control.
7. Connect the ammeter inductive clamp on the negative battery cable or cables (Figure 7-15).
8. Be sure all the electrical accessories on the vehicle are off and the doors are closed.
9. Set the test selector to the #1 starting position.

WARNING: Always follow the vehicle manufacturer's recommended procedure for disabling the ignition system. If these instructions are not followed, electronic components may be damaged.

10. Disable the ignition system by disconnecting the positive primary coil wire. Always follow the vehicle manufacturer's recommended procedure for disabling the ignition system.
11. Crank the engine and observe the ammeter and voltmeter readings. The starter current draw should equal the manufacturer's specifications, and the voltage should be above the minimum cranking voltage.
12. Disconnect the tester cables, and reconnect the positive primary coil wire.

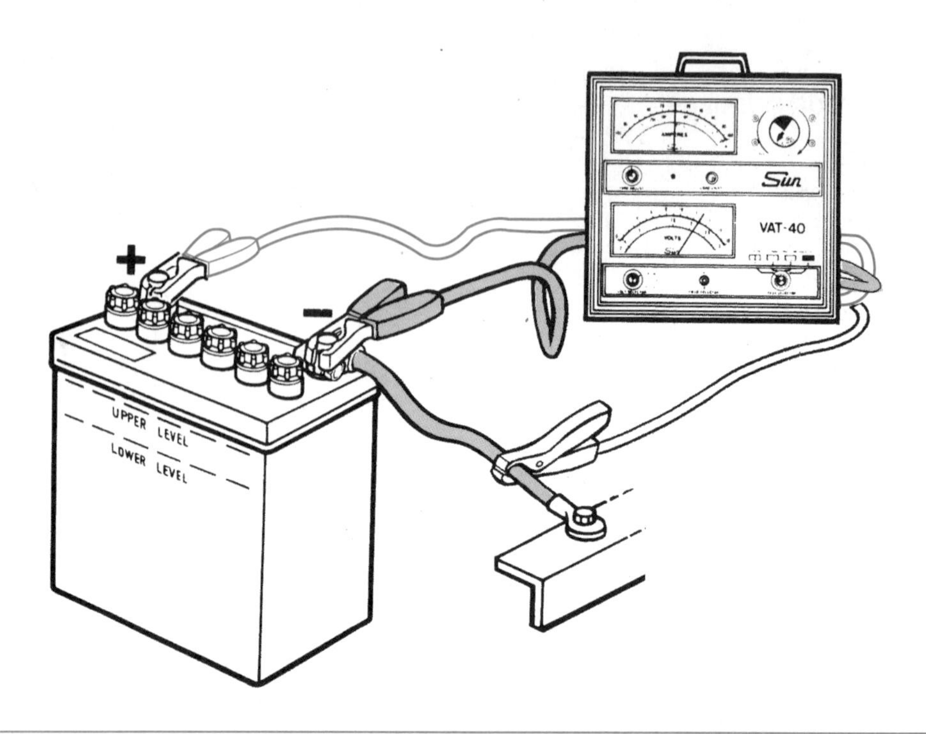

Figure 7-15 Volt-ampere tester connections for starter current draw test (Courtesy of Toyota Motor Corporation)

Starter Current Draw Test Results

A starter that is drawing too much current and cranking slowly may be called a dragging starter.

High current draw and low cranking speed usually indicate a defective starter. High current draw may also be caused by internal engine problems, such as partial bearing seizure.

A low cranking speed and low current draw with high cranking voltage usually indicate excessive resistance in the starter circuit, such as cables and connections. Always remember that the battery and battery cable connections must be in satisfactory condition to obtain accurate starter current draw test results.

Voltage Drop Test, Starting Motor Circuit Insulated Side

The resistance in an electrical wire may be checked by measuring the voltage drop across the wire with normal current flow in the wire. To measure the voltage drop across the positive battery cable, connect the positive voltmeter lead to the positive battery cable at the battery, and connect the negative voltmeter lead to the other end of the positive battery cable at the starter solenoid (Figure 7-16). Set the voltmeter selector switch to the lowest scale, and disable the ignition system. Crank the engine. The voltage drop indicated on the meter should not exceed 0.5 V. If the voltage reading is above this figure, the cable has excessive resistance. If the cable ends are clean and tight, replace the cable.

Connect the positive voltmeter lead to the positive battery cable on the starter solenoid, and connect the negative voltmeter lead to the starting motor terminal on the other side of the solenoid. Leave the voltmeter on the lowest scale and crank the engine. If the voltage drop exceeds 0.3 V, the solenoid disc and terminals have excessive resistance.

Voltage Drop Test, Starting Motor Ground Side

Connect the positive voltmeter lead to the starter case and connect the negative voltmeter lead to the negative battery cable on the battery. Leave the ignition system disabled and place the voltmeter selector on the lowest scale. If the voltage drop reading exceeds 0.2 V while cranking the engine, the negative battery cable or ground return circuit has excessive resistance.

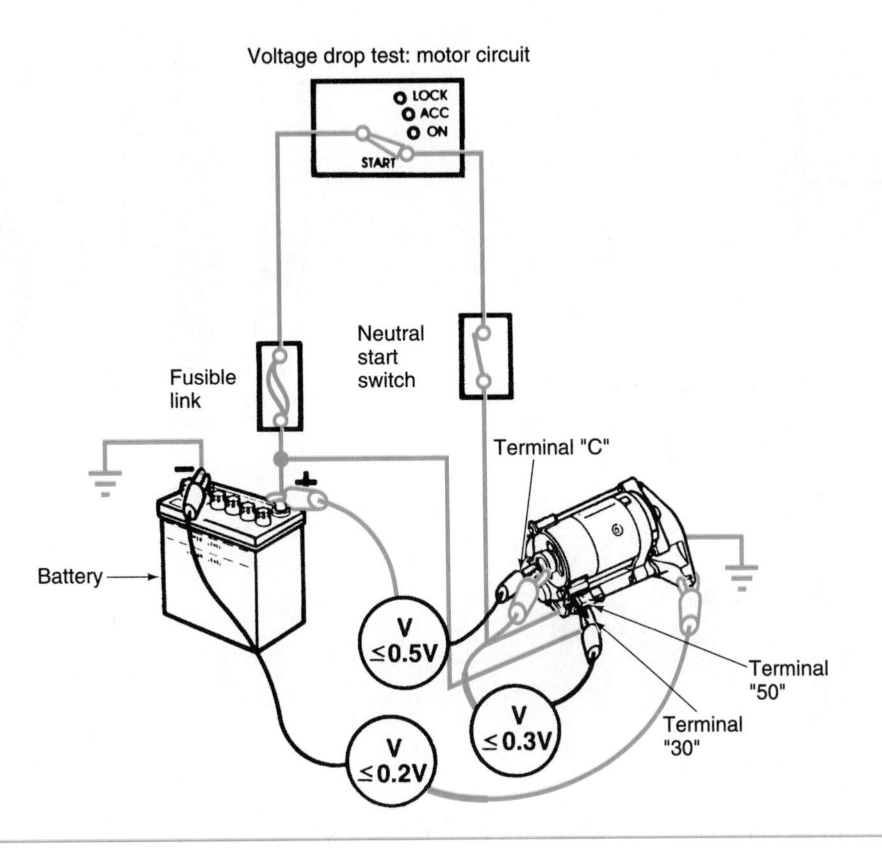

Figure 7-16 Voltmeter connections for starter insulated-side and ground-side voltage drop tests (Courtesy of Toyota Motor Corporation)

Voltage Drop Tests, Control Circuit

Connect the positive voltmeter lead to the positive battery cable at the battery, and connect the negative voltmeter lead to the solenoid winding terminal on the solenoid (Figure 7-17). Leave the ignition system disabled, and place the voltmeter selector on the lowest scale. If the voltage drop across the control circuit exceeds 1.5 V while cranking the engine, individual voltage drop tests on control circuit components are necessary to locate the high-resistance problem.

Connect the voltmeter leads across individual control circuit components such as the ignition switch, neutral safety switch, and starter relay contacts to measure the voltage drop across these components while cranking the engine. In most starting motor control components, if the voltage drop exceeds 0.2 V, the component is defective. The voltmeter leads may also be connected across individual wires in the starter control circuit to test voltage drop in the wires. Before any starting motor circuit component is replaced, always remove the negative battery cable. High resistance in starting motor control components and wires may cause the starting circuit to be inoperative. High resistance in the solenoid disc and terminals may cause a clicking action when the ignition switch is turned to the start position.

Classroom Manual
Chapter 7, page 139

Alternator Service and Diagnosis

Alternator Belt Service and Diagnosis

Alternator belt condition and tension are extremely important for satisfactory alternator operation. A loose belt causes low alternator output and a discharged battery. A loose, dry, or worn belt may cause squealing and chirping noises, especially during engine acceleration and cornering.

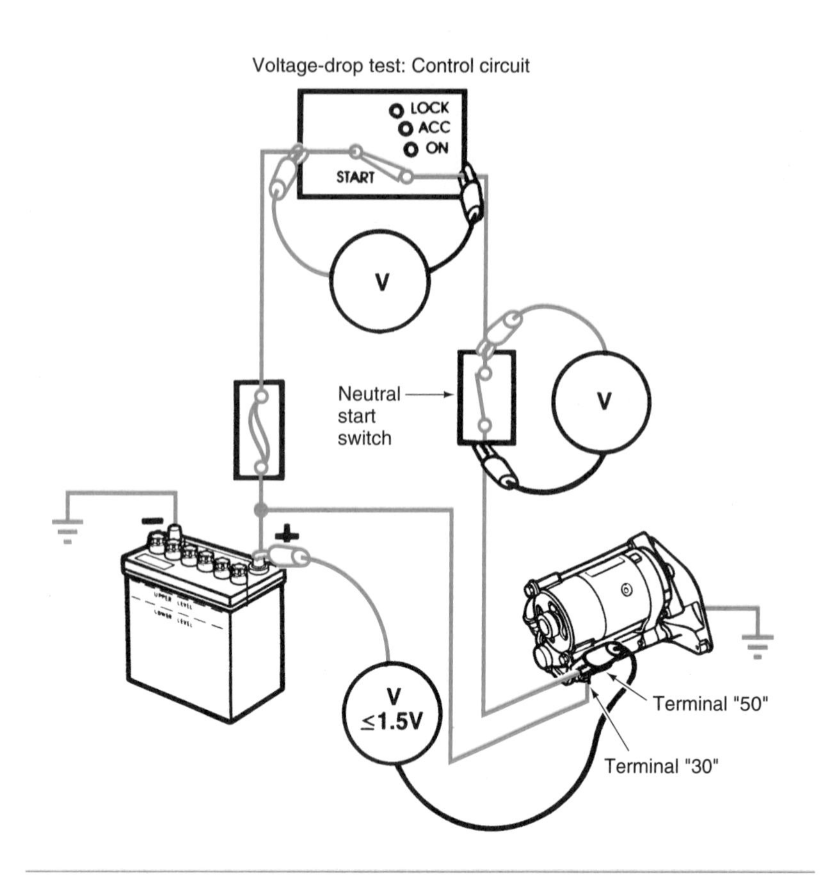

Figure 7-17 Voltmeter connections to test voltage drop in the starter control circuit (Courtesy of Toyota Motor Corporation)

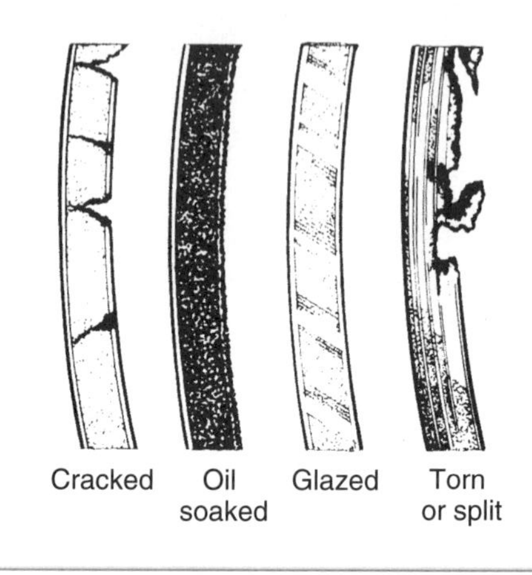

Figure 7-18 Defective belt conditions (Courtesy of Chrysler Corporation)

Special Tools

Belt tension gauge

The alternator belt should be checked for cracks, oil soaking, worn or glazed edges, tears, and splits (Figure 7-18). If any of these conditions are present, belt replacement is necessary.

Since the friction surfaces are on the sides of a V-type belt, wear occurs in this area. If the belt edges are worn, the belt may be rubbing on the bottom of the pulley. This condition requires belt replacement. Belt tension may be checked by measuring the belt deflection. Press on the belt with the engine stopped to measure the belt deflection, which should be 1/2 inch per foot of free span. The belt tension may be checked with a belt tension gauge placed over the belt (Figure 7-19). The tension on the gauge should equal the vehicle manufacturer's specifications. If a V-belt requires tightening, follow this procedure:

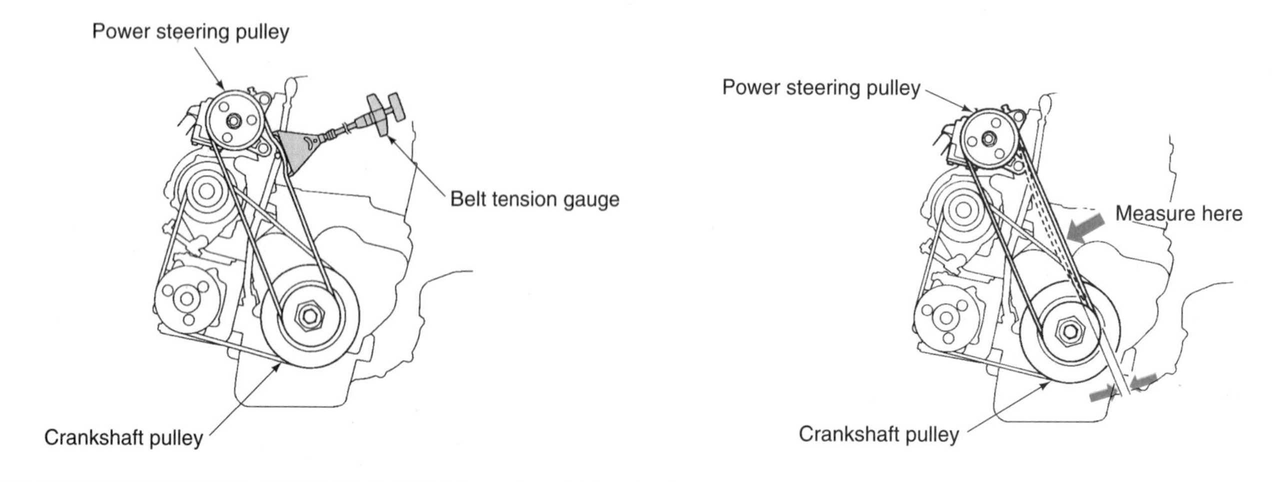

Figure 7-19 Methods of checking belt tension (Courtesy of American Honda Motor Co., Inc)

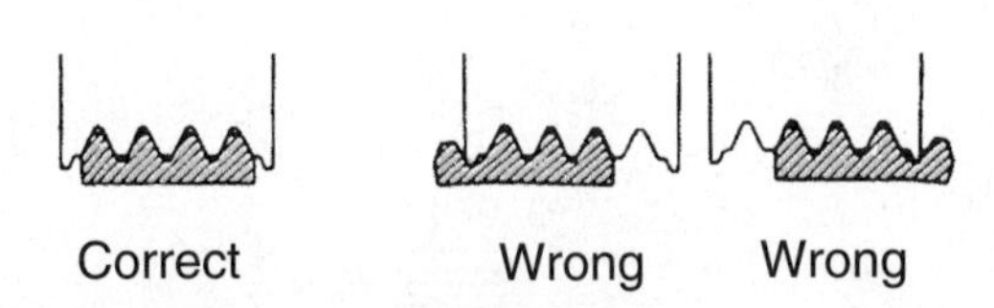

Figure 7-20 Proper and improper installation of ribbed V-belt on a pulley (Courtesy of Toyota Motor Corporation)

1. Loosen the alternator bracket bolt.
2. Loosen the alternator mounting bolt.
3. Check the bracket and alternator mounting bolts for wear. If these bolts or bolt openings in the bracket or alternator housing are worn, replacement is necessary.
4. Pry against the alternator housing with a pry bar to tighten the belt.
5. Hold the alternator in the position described in step 4, and tighten the bracket bolt.
6. Recheck the belt tension with the tension gauge. If the belt does not have the specified tension, repeat steps 1 through 5.
7. Tighten the bracket bolt and the mounting bolt to the manufacturer's specified torque.

Some alternators have a ribbed V-belt. Many ribbed-V belts have an automatic tensioning pulley; therefore, a tension adjustment is not required. The ribbed V-belt should be checked to make sure it is installed properly on each pulley in the belt drive system (Figure 7-20). The tension on a ribbed V-belt may be checked with a belt tension gauge in the same way as the tension on a V-belt.

A ribbed V-belt may be referred to as a serpentine belt.

Many ribbed V-belts have a spring-loaded tensioner pulley that automatically maintains belt tension. As the belt wears or stretches, the spring moves the tensioner pulley to maintain the belt tension. Some of these tensioners have a belt-length scale that indicates new belt range and used belt range (Figure 7-21). If the indicator on the tensioner is out of the used belt length range, belt replacement is required. Many belt tensioners have a one-half inch drive opening in which a ratchet or flex handle may be installed to move the tensioner pulley off the belt during belt replacement (Figure 7-22).

Belt pulleys must be properly aligned to minimize belt wear. The edges of the pulleys must be in line when a straightedge is placed on the pulleys (Figure 7-23).

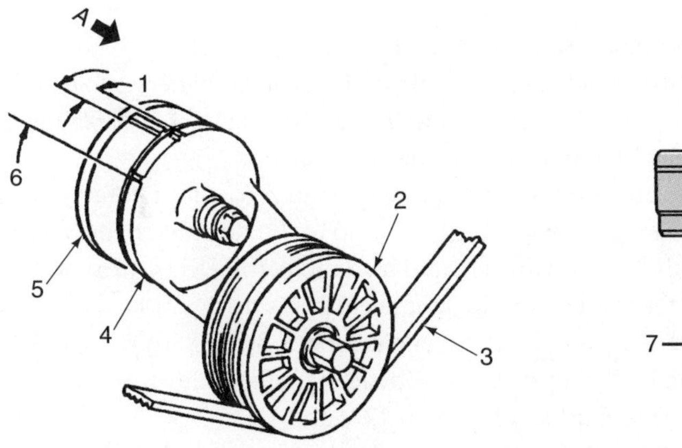

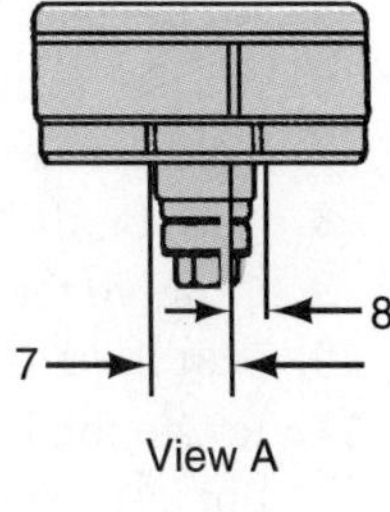

1. New belt range
2. Pulley
3. Belt
4. Arm
5. Spindle
6. Used belt acceptable wear range
7. Used belt length range
8. New belt length range

Figure 7-21 Belt tensioner scale (Courtesy of Chevrolet Motor Division, General Motors Corporation)

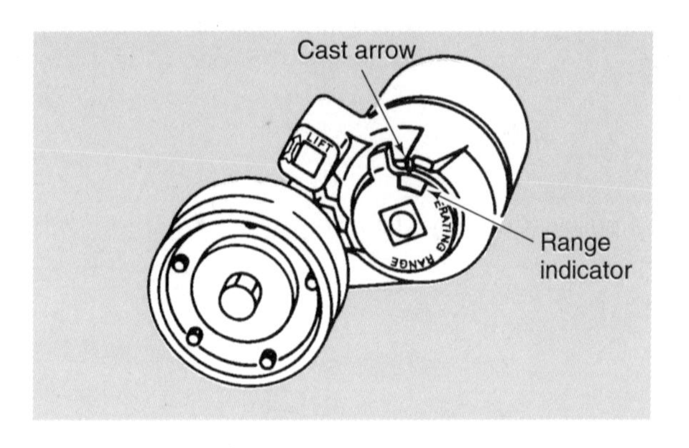

Figure 7-22 One-half inch drive opening in the tensioner pulley (Courtesy of Cadillac Motor Car Division, General Motors Corporation)

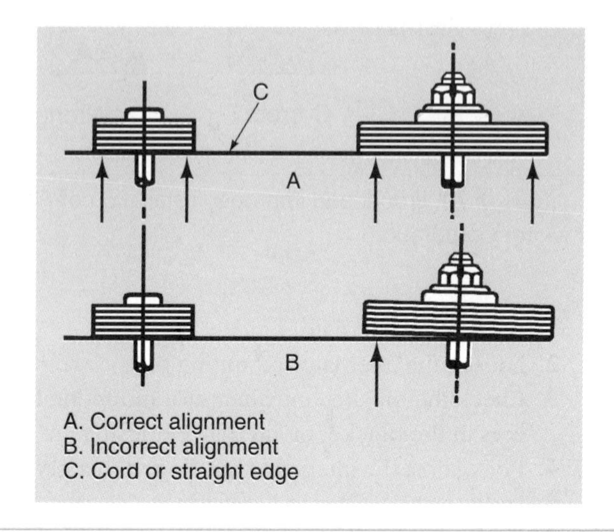

Figure 7-23 Checking pulley alignment (Courtesy of Chevrolet Motor Division, General Motors Corporation)

Alternator Output Test

The alternator belt must be in satisfactory condition before the output test is performed. Always follow the vehicle manufacturer's recommended procedure for testing alternator output. Following is a typical alternator output test procedure:

WARNING: Never allow the alternator voltage to go above 15.5 V. High voltage may damage computers and other electrical and electronic equipment on the vehicle.

WARNING: Never disconnect the circuit between the alternator and the battery with the engine running. This action may cause extremely high voltage momentarily, which causes electronic component damage.

1. Be sure the tester load control is in the off position.
2. If the tester has analog meters, check the mechanical zero on the meters, and adjust as required.
3. Set the voltmeter selector to the INT 18-V position.
4. Connect the positive tester cable to the positive battery cable at the battery, and connect the negative tester cable to the negative battery cable at the battery.
5. Set the test selector to the #2 charging position.
6. Adjust the ammeter to read zero using the zero adjust control, if necessary.
7. Connect the ammeter inductive clamp to the negative battery cable.
8. Turn on the ignition switch, and turn off all the electrical accessories.
9. Observe the reading on the ammeter; this is the ignition and alternator field draw.
10. Start the engine and hold the engine speed at 2,000 rpm.
11. Rotate the load control to the vehicle manufacturer's recommended voltage for the output test. This is usually between 12 V and 13 V.
12. Observe the ammeter reading, and rotate the load control to the off position.
13. Obtain the alternator output by adding the ammeter reading in step 9 to the reading in step 12.

Since the ammeter inductive pickup is installed on the negative battery cable, the ammeter only indicates the alternator current flow through the battery during the output test. The current is supplied directly from the alternator to the ignition system and alternator field, and so the current draw for these two systems must be added to the alternator output to obtain an accurate output reading. If the alternator output is not within 10% of the vehicle manufacturer's specified output, further diagnosis is required.

Alternator Regulator Voltage Test

To check the alternator for proper regulator voltage, leave the volt-ampere tester connected as in the output test. Operate the engine at 2,000 rpm until the battery is sufficiently charged to provide a charging rate of 10 amperes or less. Observe the voltmeter reading. The voltage should be at the vehicle manufacturer's specified regulator voltage.

Alternator Diagnosis

If the alternator output is zero, the alternator field circuit may have an open circuit. The most likely place for an open circuit in the alternator field is at the slip rings and brushes. When the alternator voltage is normal, but the amps output is zero, the fuse link between the alternator battery terminal and the positive battery cable may be open. With the ignition switch off, battery voltage should be available on each end of the fuse link. If the voltage is normal on the battery side of the fuse link and zero on the alternator side, the fuse link is open and must be replaced.

If the alternator output is less than specified, always be sure that the belt and belt tension are satisfactory. When the belt and belt tension are satisfactory and the alternator output is less than specified, the alternator is defective.

Battery overcharging causes excessive gassing of the battery. Overcharging may be caused by a defective voltage regulator, which allows a higher-than-specified charging voltage.

Battery undercharging eventually results in a discharged battery and a no-start problem. Undercharging may be caused by a loose or worn belt, a defective alternator, low regulator voltage, high resistance in the battery wire between the alternator and the battery, or an open fuse link.

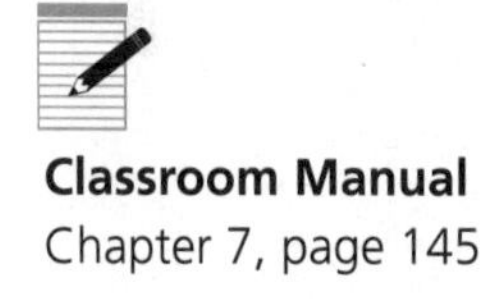

Classroom Manual
Chapter 7, page 145

Wiring Harness and Terminal Diagnosis and Repair

Terminal and Wire Problems

WARNING: Always follow the vehicle manufacturer's wiring and terminal repair procedure in the service manual. On some components and circuits, manufacturers recommend component replacement rather than wiring repairs. For example, some manufacturers recommend replacing air bag system components such as sensors, if the wiring or terminals are damaged on these components.

Wires should be checked for a burned or melted condition (Figure 7-24). Connector ring terminals should be checked for loose retaining nuts, which cause high resistance or intermittent open circuits (Figure 7-25). An open circuit may be caused by a terminal that is backed out of the connector (Figure 7-26). Terminals that are bent or damaged may cause shorts or open circuits (Figure 7-27). An open circuit occurs when the terminal is crimped over the insulation instead of the wire core (Figure 7-28). A greenish-white corrosion on terminals results in high resistance or an open circuit (Figure 7-29).

Splicing Copper Wire With Splice Clips

Follow this procedure to splice copper wire with splice clips:

1. Cut the tape off the harness if necessary. Be careful not to cut the wire insulation.
2. Cut the wire to be repaired as necessary. Do not cut more than necessary off the wire.
3. Use the appropriate stripper opening in a pair of wire strippers to strip the proper amount of insulation off the wire ends. Enough insulation must be stripped off the

Special Tools

Wire crimping tool

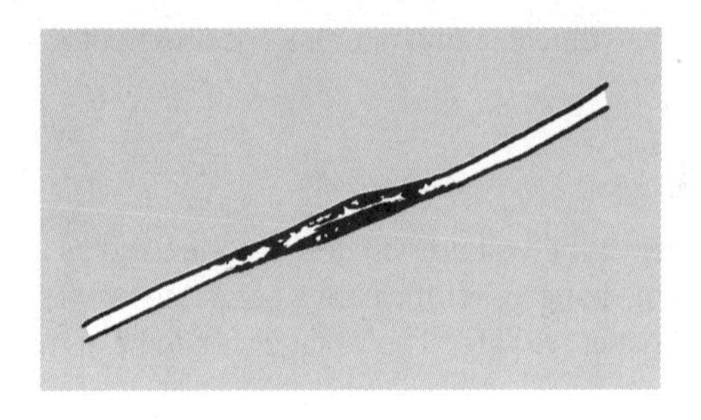

Figure 7-24 Burned, melted wiring insulation (Courtesy of Cadillac Motor Car Division, General Motors Corporation)

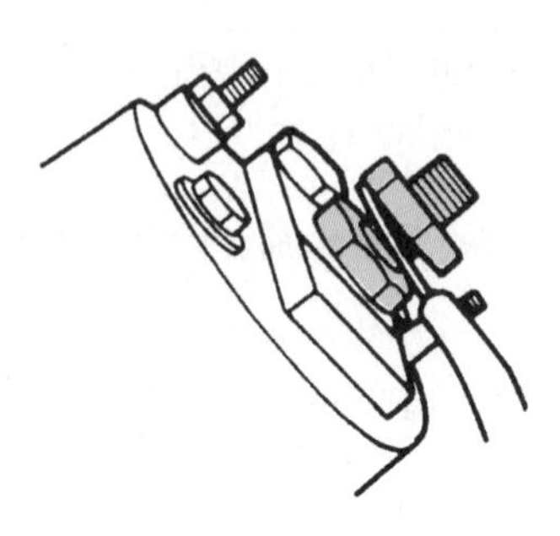

Figure 7-25 Loose retaining nuts on ring-type terminals cause high resistance or an intermittent open circuit. (Courtesy of Cadillac Motor Car Division, General Motors Corporation)

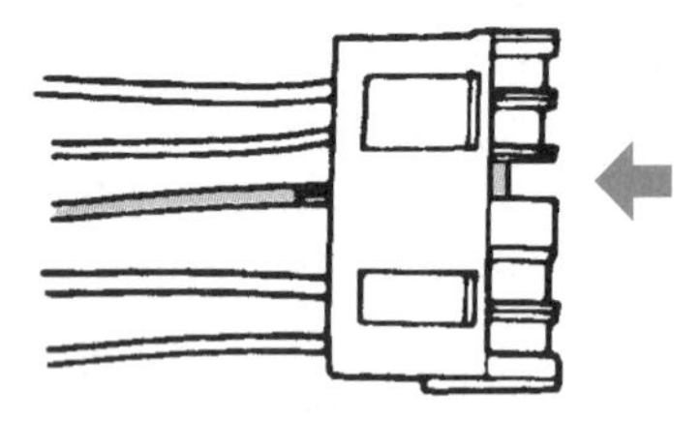

Figure 7-26 Terminals that are backed out of connectors cause open circuits. (Courtesy of Cadillac Motor Car Division, General Motors Corporation)

Figure 7-27 Bent or damaged component terminals may result in shorted or open circuits. (Courtesy of Cadillac Motor Car Division, General Motors Corporation)

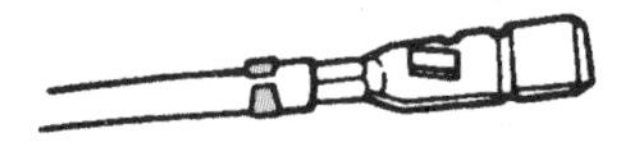

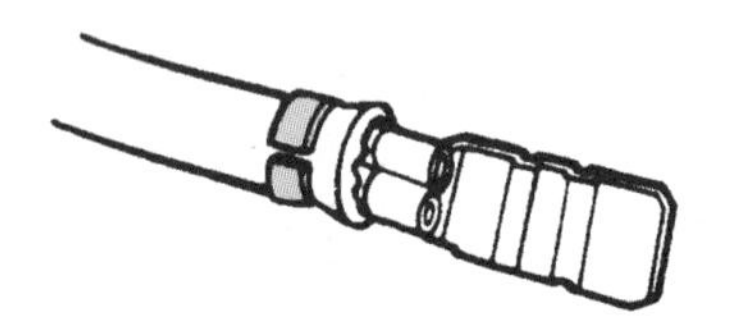

Figure 7-28 An open circuit occurs when a terminal is crimped over the insulation rather than the wire core. (Courtesy of Cadillac Motor Car Division, General Motors Corporation)

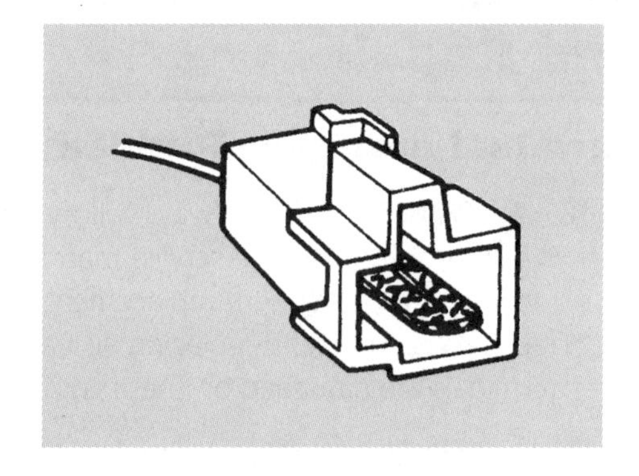

Figure 7-29 A greenish-white corrosion on terminals causes high resistance or an open circuit. (Courtesy of Cadillac Motor Car Division, General Motors Corporation)

wire ends to allow the wire ends to fit into the splice clip, and a small amount of wire should be visible on each side of the clip.

4. Place the wire ends in the splice clip so the clip is centered on the wire ends (Figure 7-30). Be sure all the wire strands are in the clip.
5. Place the proper size opening in the crimping tool over the splice clip so this tool is even with one edge of the clip (Figure 7-31). Apply steady pressure on the crimping tool to crimp the clip onto the wire.
6. Repeat step 5 on the opposite end of the clip.

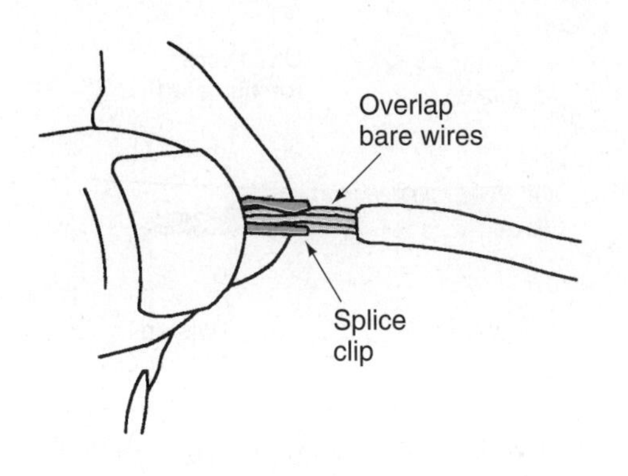

Figure 7-30 Wire ends placed in splice clip (Courtesy of Cadillac Motor Car Division, General Motors Corporation)

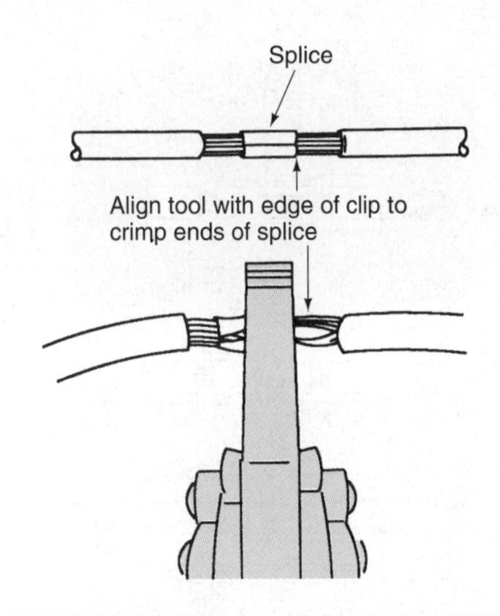

Figure 7-31 Wire crimping tool placed on splice clip (Courtesy of Cadillac Motor Car Division, General Motors Corporation)

7. Heat the splice clip with a soldering gun and melt resin-core solder into the clip (Figure 7-32).
8. Use splicing tape to cover the splice, and do not flag the tape (Figure 7-33).
9. If the wire is installed in a harness, tape the complete harness in the area where the tape was removed. When the wire remains by itself, wrap splicing tape over the spliced area and the wire insulation near the spliced area.

The term *tape flagging* refers to a large piece of excess splicing tape hanging from one side of a wire.

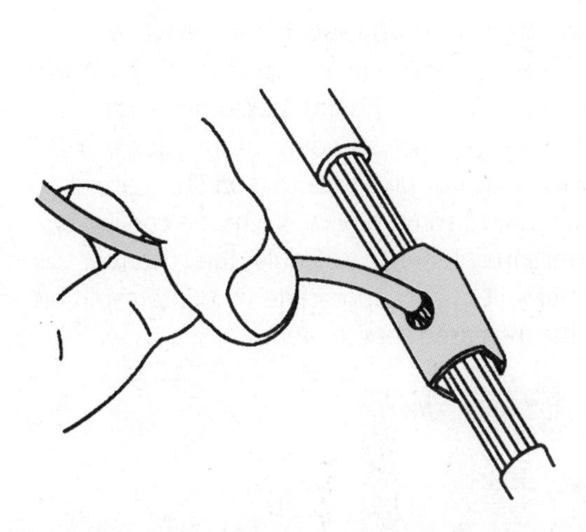

Figure 7-32 Soldering the splice clip (Courtesy of Cadillac Motor Car Division, General Motors Corporation)

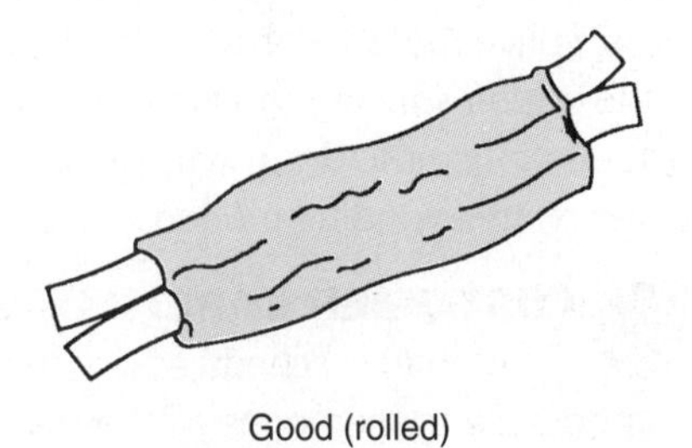

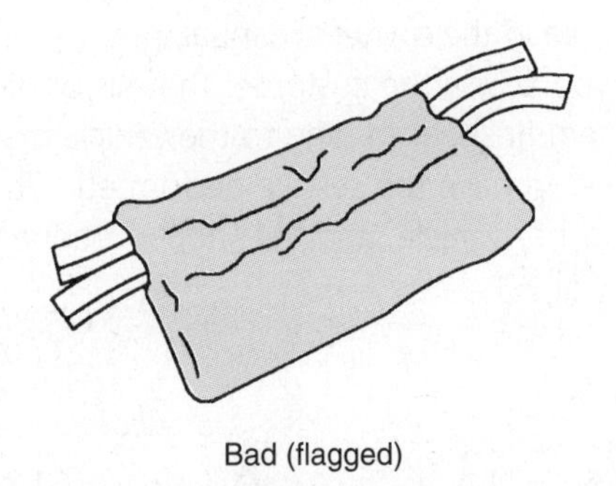

Figure 7-33 Proper and improper splice clip taping (Courtesy of Cadillac Motor Car Division, General Motors Corporation)

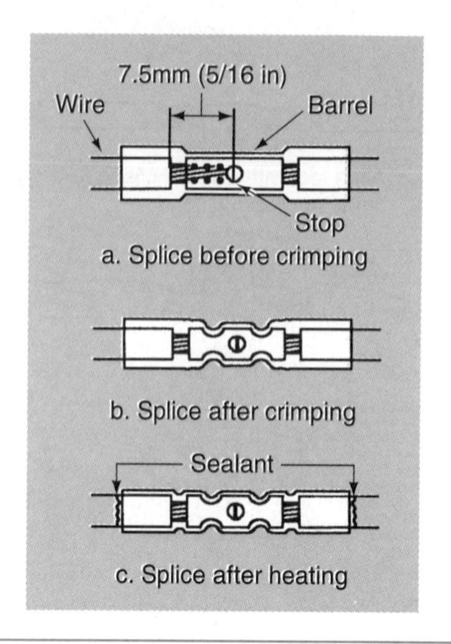

Figure 7-34 Crimp and seal splice sleeve after crimping and heating. (Courtesy of Cadillac Motor Car Division, General Motors Corporation)

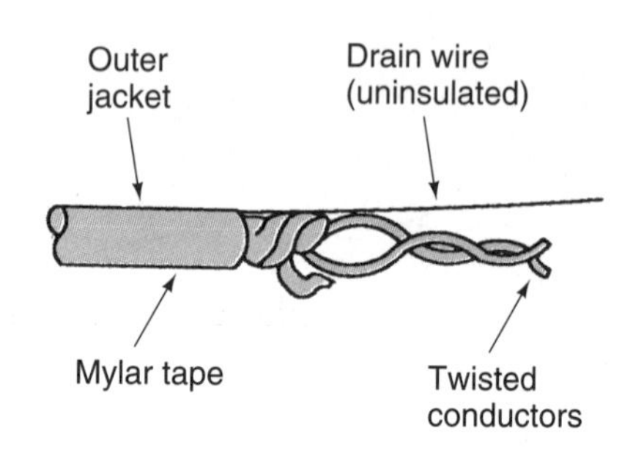

Figure 7-35 Wiring harness with drain wire (Courtesy of Cadillac Motor Car Division, General Motors Corporation)

Splicing Copper Wire With Crimp and Seal Splice Sleeves

Crimp and seal splice sleeves may be used to splice copper wire. These sleeves have a heat-shrink sleeve over the outside of the sleeve. The stripped wire ends are placed in the sleeve until they contact the stop in the center of the sleeve. A wire crimping tool is used to crimp the sleeve near each end. After the crimping procedure, the sleeve is heated slightly with a heat torch to shrink the insulating heat shrink sleeve onto the terminal. When the shrinking process is complete, sealant should appear at each end of the sleeve (Figure 7-34).

Some wiring harnesses contain a drain wire (Figure 7-35). When repairing wires in this type of harness, splice the wire or wires as explained previously. If more than one splice is required, stagger the wiring splices, and tape the splices with mylar tape. Do not tape the drain wire onto the harness. Splice the drain wire separately (Figure 7-36) and then place mylar tape over the spliced area with a winding motion.

CUSTOMER CARE: As an automotive technician, you should be familiar with the maintenance schedules recommended by various vehicle manufacturers. Of course, it is impossible to memorize all the maintenance schedules on different makes of vehicles, but maintenance schedule books are available. This maintenance schedule information is available in the owner's manual, but the vehicle owner may not take time to read this manual. If you advise the customer that his or her vehicle requires some service, such as a cooling system flush, according to the vehicle manufacturer's maintenance schedule, the customer will often have the service performed. The customer will usually appreciate your interest in his or her vehicle, and the shop will benefit from the increased service work.

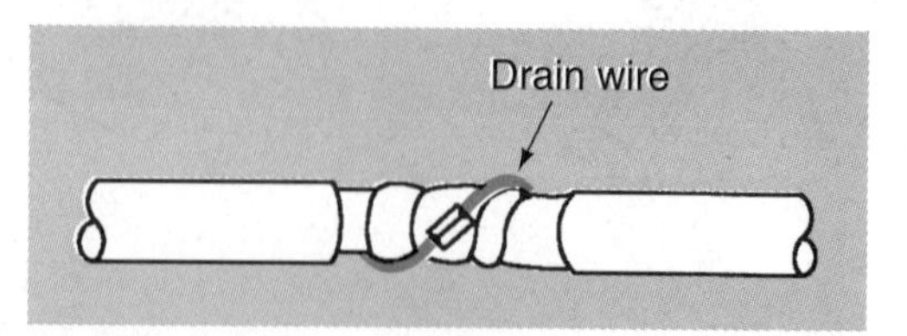

Figure 7-36 Separate splice for drain wire (Courtesy of Cadillac Motor Car Division, General Motors Corporation)

Guidelines for Battery, Starting System, and Charging System Service and Diagnosis

1. Battery electrolyte is very harmful to human skin, eyes, vehicle paint, and upholstery.
2. While charging a battery, explosive hydrogen gas is given off. Never smoke or allow other sources of ignition near a battery.
3. Always remove the negative battery cable first when disconnecting battery cables.
4. Disconnecting a battery erases computer adaptive memories, memory seat and mirror positions, and programmed radio station memory.
5. A hydrometer test indicates the specific gravity of the battery electrolyte.
6. During the battery capacity test, the battery is discharged at one-half the cold cranking amperes rating or three times the ampere-hour rating for 15 seconds. If the battery is satisfactory, the battery voltage will remain above 9.6 V, if the battery temperature is 70°F (21°C).
7. Connecting a booster battery with reversed polarity may result in severe electrical system damage.
8. Computers have a very low electrical drain. When the ignition switch is turned off, this drain is higher, and the drain is gradually reduced within a few minutes.
9. High current draw and low cranking speed usually indicate a defective starter or internal engine problems.
10. A low cranking speed, low current draw, and high cranking voltage usually indicate excessive resistance in the starter circuit.
11. Voltage drop tests may be performed to measure resistance in the starter circuit and starter control circuit.
12. Many ribbed V-belts have a spring-loaded tensioner pulley, which automatically maintains belt tension.
13. Many belt tension pulleys have a belt length scale.
14. If high voltage is allowed during the alternator output test, electrical component damage may occur.
15. Battery overcharging may be caused by high regulator voltage.
16. Battery undercharging may be caused by a loose or worn belt, low regulator voltage, defective alternator, high resistance in the alternator circuit, or an open fuse link.
17. Resin-core solder should be used to solder splice clips.
18. After a crimp and seal splice sleeve is heated, sealant should appear at each end of the sleeve.

CASE STUDY

A customer complained about a no-start problem on an Oldsmobile Cutlass Calais. When the service writer questioned the customer about the problem, the customer indicated the battery seemed to be discharged and would not crank the engine. The customer also indicated this problem occurred about once a month.

When the technician lifted the hood, she noticed the battery, alternator, and starting motor had been replaced, which indicated that someone else had likely been trying to correct this problem. The technician checked the belt and all wiring connections in the starting and charging systems without finding any problems. Next, the technician performed starter current draw and alternator output tests, and these systems performed satisfactorily. The technician checked the regulator voltage and found it was within specifications. After the battery was charged, the technician performed a capacity test, and the battery performance was satisfactory. Since the battery, starting system, and charging system tests were normal, the technician concluded there must be a battery drain problem.

The technician connected an ammeter and jumper wire in series between the negative battery cable and terminal, and proceeded with the drain test. When the jumper wire was disconnected, the ammeter indicated a continual 1-ampere drain. The technician began removing the fuses one at a time and checking the ammeter reading. When the courtesy lamp fuse was disconnected, the drain disappeared.

The technician checked the wiring diagram for the vehicle and found the vehicle had a remote keyless entry module. When the courtesy lamp fuse was installed and the remote keyless entry module disconnected, the drain disappeared on the ammeter. This module was draining current through it to ground, but the current flow was not high enough to illuminate the courtesy lights.

A new remote keyless entry module was installed, and the drain was rechecked. The drain was now within the vehicle manufacturer's specifications.

Terms to Know

Specific gravity
Hydrometer
Capacity test
Mechanical zero
Jump starting
Battery boosting
Parasitic battery drain
Starter current draw test
Voltage drop tests
Tensioner pulley
Tape flagging
Splice clip
Crimp and seal splice sleeves
Drain wire

ASE Style Review Questions

1. While discussing battery service:
Technician A says if a battery is disconnected, computer adaptive memories are erased.
Technician B says if a battery is disconnected, the memory seat and memory mirror positions are erased.
Who is correct?
A. A only **C.** Both A and B
B. B only **D.** Neither A nor B

2. While discussing battery service:
Technician A says the positive battery cable should be disconnected first.
Technician B says the negative battery cable should be disconnected first.
Who is correct?
A. A only **C.** Both A and B
B. B only **D.** Neither A nor B

3. While discussing the battery capacity test:
Technician A says the battery should be discharged at one-third of the cold cranking ampere rating.
Technician B says a satisfactory battery at 70°F has 10 V at the end of the 15-second capacity test.
Who is correct?
A. A only **C.** Both A and B
B. B only **D.** Neither A nor B

4. While discussing battery boosting:
Technician A says the negative booster cable should be connected to the negative battery terminal on the vehicle being boosted.
Technician B says the positive booster cable should be connected to the positive battery terminal on the vehicle being boosted.
Who is correct?
A. A only **C.** Both A and B
B. B only **D.** Neither A nor B

5. While discussing battery drain testing:
Technician A says computers have a low milliampere drain with the ignition off.
Technician B says the fuses may be disconnected one at a time to locate the cause of a battery drain.
Who is correct?
A. A only **C.** Both A and B
B. B only **D.** Neither A nor B

6. While discussing the starter current draw test:
Technician A says the battery should be at least 50% charged before the starter draw test is performed.
Technician B says the electrical accessories on the vehicle should be on during the starter draw test.
Who is correct?
A. A only **C.** Both A and B
B. B only **D.** Neither A nor B

7. While discussing the starter current draw test results:
Technician A says low current draw and low cranking speed with high cranking voltage usually indicate a defective starter.
Technician B says high current draw and low cranking speed usually indicate a defective starter.
Who is correct?
A. A only **C.** Both A and B
B. B only **D.** Neither A nor B

8. While discussing alternator output:
Technician A says zero alternator output may be caused by an open field circuit.
Technician B says zero alternator output may be caused by an open alternator fuse link.
Who is correct?
A. A only **C.** Both A and B
B. B only **D.** Neither A nor B

9. While discussing battery condition:
Technician A says an undercharged battery may be caused by a low regulator voltage.
Technician B says an undercharged battery may be caused by high regulator voltage.
Who is correct?
A. A only **C.** Both A and B
B. B only **D.** Neither A nor B

10. While discussing wiring harness and terminal repairs:
Technician A says acid-core solder should be used for soldering electrical terminals.
Technician B says sealant should appear at both ends of a crimp and seal sleeve after it is heated.
Who is correct?
A. A only **C.** Both A and B
B. B only **D.** Neither A nor B

Table 7-1 ASE TASK

Inspect, service, and replace battery, cables, clamps, and holddown devices.

Problem Area	Symptoms	Possible Causes	Classroom Manual	Shop Manual
ENGINE STARTING	No-start	1. Discharged battery	135	114
		2. Corroded battery terminals and cables	136	114

Table 7-2 ASE TASK

Perform battery capacity (load, high-rate discharge) test; determine needed repairs.

Problem Area	Symptoms	Possible Causes	Classroom Manual	Shop Manual
ENGINE STARTING	No-start	Discharged battery	137	117

Table 7-3 ASE TASK

Slow and fast charge conventional/maintenance-free batteries.

Problem Area	Symptoms	Possible Causes	Classroom Manual	Shop Manual
ENGINE STARTING	No-start	Discharged battery	138	118

Table 7-4 ASE TASK

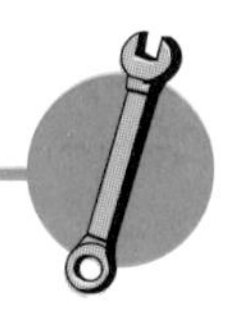

Perform starter current draw test; determine needed repairs.

Problem Area	Symptoms	Possible Causes	Classroom Manual	Shop Manual
ENGINE STARTING	High current draw, slow cranking speed	Defective starter	139	122

Table 7-5 ASE TASK

Perform starter circuit voltage drop tests; determine needed repairs.

Problem Area	Symptoms	Possible Causes	Classroom Manual	Shop Manual
ENGINE STARTING	Low current draw, low cranking speed, with cranking voltage	High resistance in starter control circuit	141	124

Table 7-6 ASE TASK

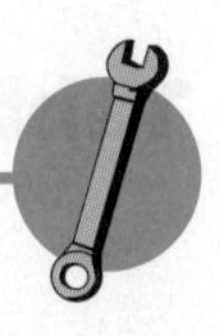

Inspect, test, and repair or replace components and wires in the starter control circuit.

Problem Area	Symptoms	Possible Causes	Classroom Manual	Shop Manual
ENGINE STARTING	Slow cranking or no cranking	High resistance in starter control circuit wires or components	140	125

Table 7-7 ASE TASK

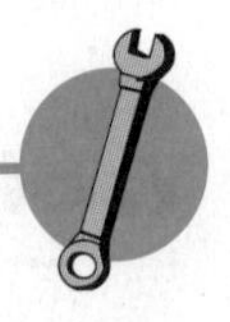

Inspect, test, and replace starter relays and solenoids.

Problem Area	Symptoms	Possible Causes	Classroom Manual	Shop Manual
ENGINE STARTING	Clicking action, no cranking	Defective solenoid or starter relay	141	125

Table 7-8 ASE TASK

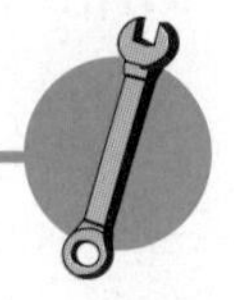

Diagnose charging system problems that cause an undercharge, overcharge, or a no-charge condition.

Problem Area	Symptoms	Possible Causes	Classroom Manual	Shop Manual
ENGINE STARTING	Undercharged battery	1. Defective alternator	146	128
		2. Low regulator voltage	147	129
		3. Loose alternator belt	146	126
		4. Open alternator fuse link	148	129
		5. High resistance in alternator circuit	148	129
	Overcharged battery	High regulator voltage	148	129

Table 7-9 ASE TASK

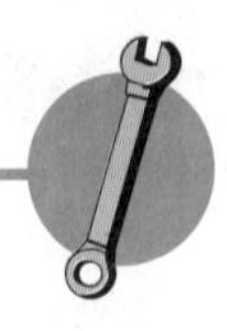

Inspect, adjust, and replace alternator drive belts, pulleys, and fans.

Problem Area	Symptoms	Possible Causes	Classroom Manual	Shop Manual
ENGINE STARTING	No-start, discharged battery	Loose, worn alternator belt	146	126
NOISE	Squealing on acceleration	Loose, worn alternator belt	146	126

Table 7-10 ASE TASK

Inspect and repair or replace charging circuit connectors and wires.

Problem Area	Symptoms	Possible Causes	Classroom Manual	Shop Manual
ENGINE STARTING	Undercharged battery, no-start	High resistance in alternator wires or terminals	148	130

Input Sensor Diagnosis and Service

Upon completion and review of this chapter, you should be able to:

- ❑ Diagnose computer voltage supply wires.
- ❑ Diagnose computer ground wires.
- ❑ Test and diagnose oxygen sensors.
- ❑ Test and diagnose engine coolant temperature sensors.
- ❑ Test and diagnose air charge temperature sensors.
- ❑ Test, diagnose, and adjust throttle position sensors.
- ❑ Test and diagnose different types of manifold absolute pressure sensors.
- ❑ Test and diagnose various types of mass air flow sensors.
- ❑ Test and diagnose knock sensors.
- ❑ Test and diagnose exhaust gas recirculation valve position sensors.
- ❑ Test and diagnose vehicle speed sensors.
- ❑ Test and diagnose park/neutral switches.

Diagnosis of Computer Voltage Supply and Ground Wires

Basic Tools

Basic technician's tool set

Service manual

Electrical probes

Propane torch

Clear, heat-resistant water container

Computer Voltage Supply Wire Diagnosis

SERVICE TIP: Never replace a computer unless the ground wires and voltage supply wires are proven to be in satisfactory condition.

WARNING: Always observe the correct meter polarity when connecting test meters. Since nearly all cars and light trucks have negative battery ground, the negative meter lead is connected to ground for many tests. Connecting a meter with incorrect polarity will damage the meter.

A computer will never operate properly unless it has proper ground connections and a satisfactory voltage supply at the required terminals. A computer wiring diagram for the vehicle being tested must be available for these tests. Backprobe the BATT terminal on the computer and connect a pair of digital voltmeter leads from this terminal to ground (Figure 8-1). Always ground the black meter lead.

The voltage at this terminal should be 12 V with the ignition switch off. If 12 V are not available at this terminal, check the computer fuse and related circuit. Turn on the ignition switch and connect the red voltmeter lead to the +B and +B1 terminals with the black lead still grounded. The voltage measured at these terminals should be 12 V with the ignition switch on. When the specified voltage is not available, test the voltage supply wires to these terminals. These terminals may be connected through fuses, fuse links, or relays. Always refer to the vehicle manufacturer's wiring diagram for the vehicle being tested.

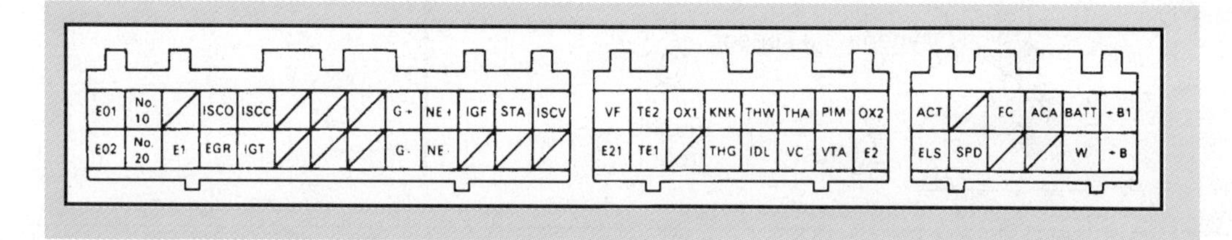

Figure 8-1 Typical computer terminals (Courtesy of Toyota Motor Corporation)

Computer Ground Wire Diagnosis

SERVICE TIP: When diagnosing computer problems, it is usually helpful to ask the customer about service work that has been performed lately on his or her vehicle. If service work has been performed in the engine compartment, it is possible that a computer ground wire may be loose or disconnected.

Classroom Manual
Chapter 8, page 163

Computer ground wires usually extend from the computer to a ground connection on the engine or battery. With the ignition switch on, connect a pair of digital voltmeter leads from the E1 and E2 computer terminals to ground. The voltage drop across the ground wires should be 0.2 V or less. If the voltage reading is more than specified, repair the ground wires.

Input Sensor Diagnosis and Service

Oxygen Sensor Diagnosis

Voltage Signal Diagnosis. The engine must be at normal operating temperature before the oxygen (O_2) sensor is tested. Always follow the test procedure in the vehicle manufacturer's service manual, and use the specifications supplied by the manufacturer.

WARNING: An oxygen sensor must be tested with a digital voltmeter. If an analog meter is used for this purpose, the sensor may be damaged.

SERVICE TIP: A contaminated oxygen sensor may provide a continually high voltage reading because the oxygen in the exhaust stream does not contact the sensor.

SERVICE TIP: If the insulation on a wire must be probed to connect a meter to the wire, place a small amount of silicon sealant over the wire puncture after the test procedure.

Special Tools

Digital multimeter with frequency capabilities

A digital voltmeter is connected from the O_2 sensor wire to ground to test this sensor (Figure 8-2). Use an electric probe to backprobe the connector near the O_2 sensor to connect the voltmeter to the sensor signal wire. If possible, avoid probing the insulation to connect a meter to a wire. With the engine idling, if the O_2 sensor and the computer system are working properly, the sensor voltage should be cycling from low voltage to high voltage. The signal from many O_2 sensors varies between 0 V and 1 V.

WARNING: When working on an engine with an O_2 sensor, always use the room temperature vulcanizing (RTV) sealant recommended by the vehicle manufacturer. The use of other RTV sealants may contaminate the O_2 sensor.

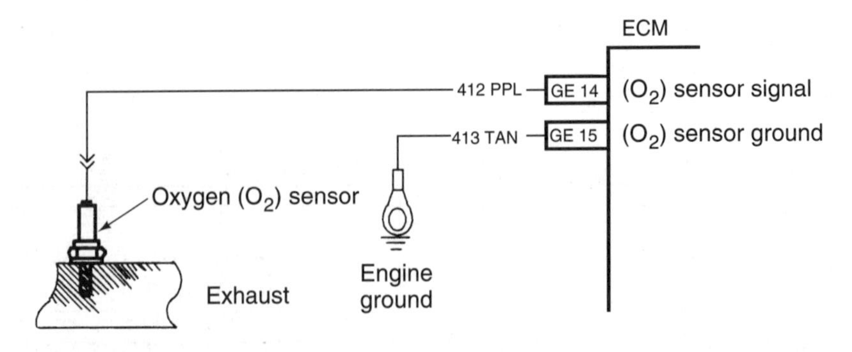

Figure 8-2 Oxygen (O_2) sensor wiring to the computer (Courtesy of Oldsmobile Division, General Motors Corporation)

If the voltage is continually high, the air-fuel ratio may be rich or the sensor may be contaminated. The O_2 sensor may be contaminated with room temperature vulcanizing (RTV) sealant, antifreeze, or lead from leaded gasoline.

When the O_2 sensor voltage is continually low, the air-fuel ratio may be lean, the sensor may be defective, or the wire between the sensor and the computer may have a high-resistance problem. If the O_2 sensor voltage signal remains in a mid-range position, the computer may be in open loop or the sensor may be defective.

When the O_2 sensor is removed from the engine, a digital voltmeter may be connected from the signal wire to the sensor case, and the sensor element may be heated in the flame from a propane torch. The propane flame keeps the oxygen in the air away from the sensor element, causing the sensor to produce voltage. While the sensor element is in the flame, the voltage should be nearly 1 V. The voltage should drop to zero immediately when the flame is removed from the sensor. If the sensor does not produce the specified voltage, it should be replaced.

Oxygen Sensor Wiring Diagnosis. If a defect in the O_2 sensor signal wire is suspected, backprobe the sensor signal wire at the computer and connect a digital voltmeter from the signal wire to ground with the engine idling. The difference between the voltage readings at the sensor and at the computer should not exceed the vehicle manufacturer's specifications. A typical specification for voltage drop across the average sensor wire is 0.2 V.

With the engine idling, connect a digital voltmeter from the sensor case to the sensor ground wire on the computer. A typical maximum voltage drop reading across the sensor ground circuit is 0.2 V. Always use the vehicle manufacturer's specifications. If the voltage drop across the sensor ground exceeds specifications, repair the ground wire or the sensor ground in the exhaust manifold.

Oxygen Sensor Heater Diagnosis. If the O_2 sensor heater is not working, the sensor warm-up time is extended, and the computer stays in open loop longer. In this mode, the computer supplies a richer air-fuel ratio, and the fuel economy is reduced. Disconnect the O_2 sensor connector and connect a digital voltmeter from the heater voltage supply wire and ground. With the ignition switch on, 12 V should be supplied on this wire. If the voltage is less than 12 V, repair the fuse in this voltage supply wire, or repair the wire itself.

With the O_2 sensor wire disconnected, connect an ohmmeter across the heater terminals in the sensor connector (Figure 8-3). If the heater does not have the specified resistance, replace the sensor.

Open loop occurs during engine warm-up. In this mode, the computer program controls the air-fuel ratio while ignoring the O_2 sensor signal.

Closed loop occurs when the engine is at, or near, normal operating temperature. In this mode, the computer uses the O_2 sensor signal to control the air-fuel ratio.

Engine Coolant Temperature (ECT) Ohmmeter Diagnosis

WARNING: Never apply an open flame to an engine coolant temperature (ECT) sensor or air charge temperature (ACT) sensor for test purposes. This action will damage the sensor.

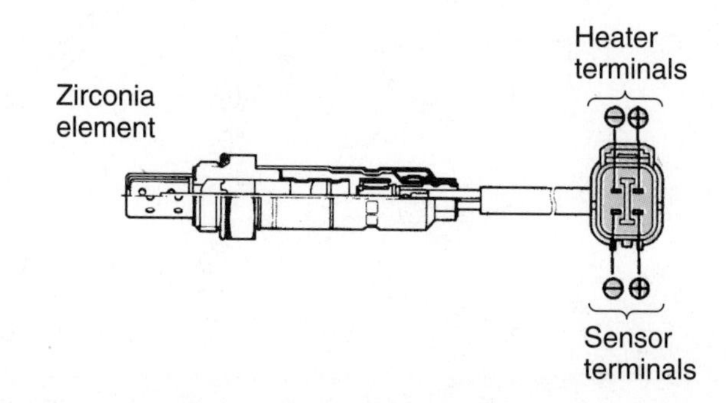

Figure 8-3 Heater terminals in the oxygen (O_2) sensor (Courtesy of Honda Motor Co., Ltd.)

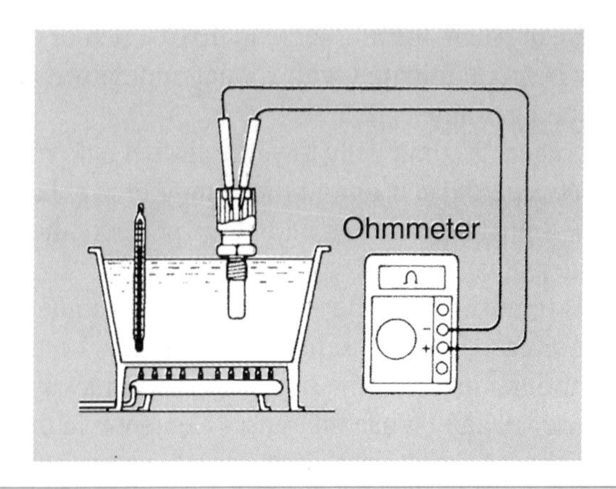

Figure 8-4 Engine coolant temperature (ECT) test connections with sensor removed and placed in hot water (Courtesy of Toyota Motor Corporation)

A defective ECT sensor may cause some of the following problems:

1. Hard engine starting
2. Rich or lean air-fuel ratio
3. Improper operation of emission devices
4. Reduced fuel economy
5. Improper converter clutch lockup
6. Hesitation on acceleration
7. Engine stalling

The ECT sensor may be removed and placed in a container of water with an ohmmeter connected across the sensor terminals (Figure 8-4). A thermometer is also placed in the water. When the water is heated, the sensor should have the specified resistance at any temperature (Figure 8-5). Always use the vehicle manufacturer's specifications. If the sensor does not have the specified resistance, replace the sensor.

Engine Coolant Temperature Sensor Wiring Diagnosis

WARNING: Before disconnecting any computer system component, be sure the ignition switch is turned off. Disconnecting components may cause high induced voltages and computer damage.

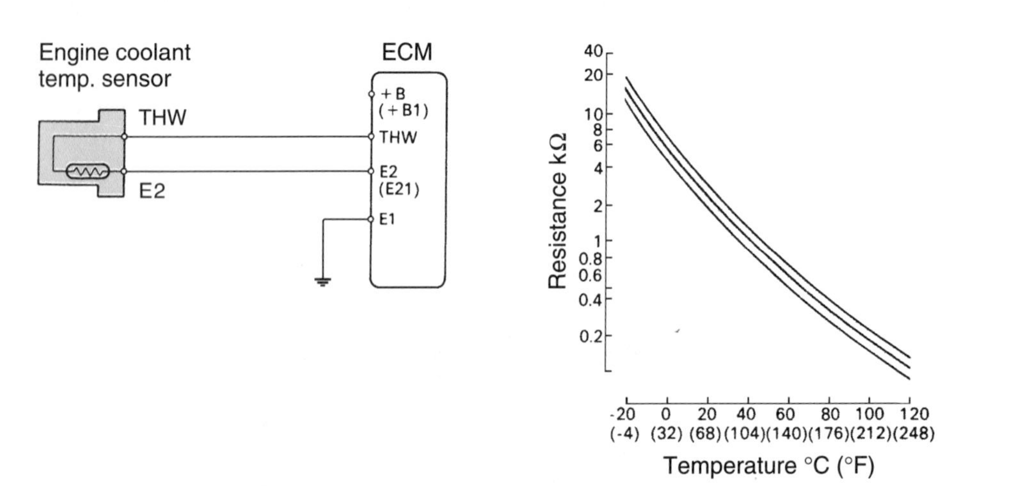

Figure 8-5 Engine coolant temperature (ECT) sensor wiring and specifications (Courtesy of Toyota Motor Corporation)

COLD CURVE 10,000-OHM RESISTOR USED		HOT CURVE CALCULATED RESISTANCE OF 909 OHMS USED	
–20°F	4.70 V	110°F	4.20 V
–10°F	4.57 V	120°F	4.00 V
0°F	4.45 V	130°F	3.77 V
10°F	4.30 V	140°F	3.60 V
20°F	4.10 V	150°F	3.40 V
30°F	3.90 V	160°F	3.20 V
40°F	3.60 V	170°F	3.02 V
50°F	3.30 V	180°F	2.80 V
60°F	3.00 V	190°F	2.60 V
70°F	2.75 V	200°F	2.40 V
80°F	2.44 V	210°F	2.20 V
90°F	2.15 V	220°F	2.00 V
100°F	1.83 V	230°F	1.80 V
110°F	1.57 V	240°F	1.62 V
120°F	1.25 V	250°F	1.45 V

Figure 8-6 Voltage drop specifications for the engine coolant temperature (ECT) sensor (Courtesy of Chrysler Corporation)

With the wiring connectors disconnected from the ECT sensor and the computer, connect an ohmmeter from each sensor terminal to the computer terminal to which the wire is connected. Both sensor wires should indicate less resistance than specified by the vehicle manufacturer. If the wires have higher resistance than specified, the wires or wiring connectors must be repaired.

Engine Coolant Temperature Sensor Voltmeter Diagnosis

With the sensor installed in the engine, the sensor terminals may be backprobed to connect a digital voltmeter to the sensor terminals. The sensor should provide the specified voltage drop at any coolant temperature (Figure 8-6).

Some computers have internal resistors connected in series with the ECT sensor. The computer switches these resistors at approximately 120°F (49°C). This resistance change inside the computer causes a significant change in voltage drop across the sensor as indicated in the specifications. This is a normal condition on any computer with this feature. This change in voltage drop is always evident in the vehicle manufacturer's specifications.

Air Charge Temperature Sensor Diagnosis

The results of a defective air charge temperature (ACT) sensor may vary depending on the vehicle make and year. A defective air charge temperature sensor may cause these problems:

1. Rich or lean air-fuel ratio
2. Hard engine starting
3. Engine stalling or surging
4. Acceleration stumbles
5. Excessive fuel consumption

The ACT sensor may be removed from the engine and placed in a container of water with a thermometer (Figure 8-7). When a pair of ohmmeter leads is connected to the sensor terminals

Special Tools

Thermometer

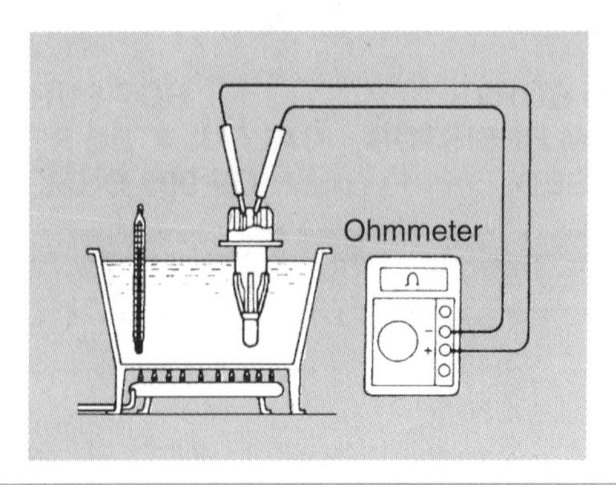

Figure 8-7 Air charge temperature (ACT) sensor placed in a container of water with an ohmmeter connected across the sensor terminals (Courtesy of Toyota Motor Corporation)

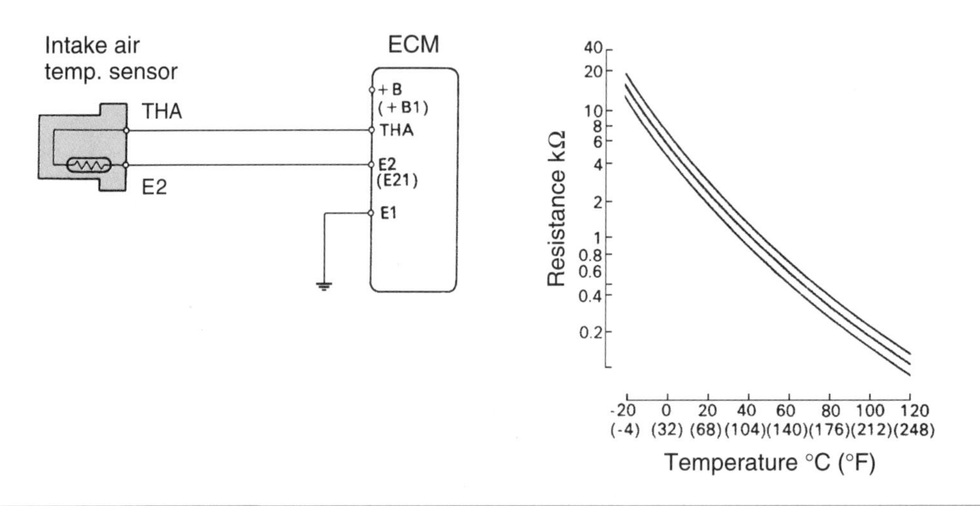

Figure 8-8 Air charge temperature (ACT) sensor wiring and resistance specifications (Courtesy of Toyota Motor Corporation)

and the water in the container is heated, the sensor should have the specified resistance at any temperature (Figure 8-8). If the sensor does not have the specified resistance, sensor replacement is required.

With the ACT sensor installed in the engine, the sensor terminals may be backprobed and a voltmeter may be connected across the sensor terminals. The sensor should have the specified voltage drop at any temperature (Figure 8-9). The wires between the air charge temperature sensor and the computer may be tested in the same way as the ECT wires.

Throttle Position Sensor Diagnosis, Three-Wire Sensor

A defective throttle position sensor (TPS) may cause acceleration stumbles, engine stalling, and improper idle speed. Backprobe the sensor terminals to complete the meter connections. With the ignition switch on, connect a voltmeter from the 5-V reference wire to ground (Figure 8-10). The voltage reading on this wire should be approximately 5 V. Always refer to the vehicle manufacturer's specifications.

If the reference wire is not supplying the specified voltage, check the voltage on this wire at the computer terminal. If the voltage is within specifications at the computer, but low at the sensor, repair the 5-V reference wire. When this voltage is low at the computer, check the voltage supply wires and ground wires on the computer. If these wires are satisfactory, replace the computer.

CHARGED TEMPERATURE SENSOR TEMPERATURE VS. VOLTAGE CURVE	
Temperature	**Voltage**
–20°F	4.81 V
0°F	4.70 V
20°F	4.47 V
40°F	4.11 V
60°F	3.67 V
80°F	3.08 V
100°F	2.51 V
120°F	1.97 V
140°F	1.52 V
160°F	1.15 V
180°F	0.86 V
200°F	0.65 V
220°F	0.48 V
240°F	0.35 V
260°F	0.28 V

Figure 8-9 Air charge temperature (ACT) sensor voltage drop specifications (Courtesy of Chrysler Corporation)

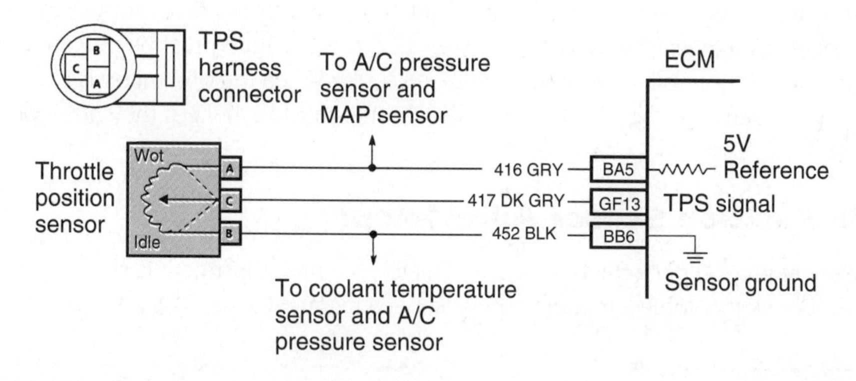

Figure 8-10 Throttle position sensor (TPS) and related wiring (Courtesy of Oldsmobile Division, General Motors Corporation)

With the ignition switch on, connect the voltmeter from the sensor ground wire to the battery ground. If the voltage drop across this circuit exceeds specifications, repair the ground wire from the sensor to the computer.

SERVICE TIP: When testing the throttle position sensor voltage signal, use an analog voltmeter, because the gradual voltage increase on this wire is quite visible on the meter pointer. If the sensor voltage increase is erratic, the voltmeter pointer fluctuates.

Special Tools

Analog voltmeter

SERVICE TIP: When the throttle is opened gradually to check the throttle position sensor voltage signal, tap the sensor lightly and watch for fluctuations on the voltmeter pointer, indicating a defective sensor.

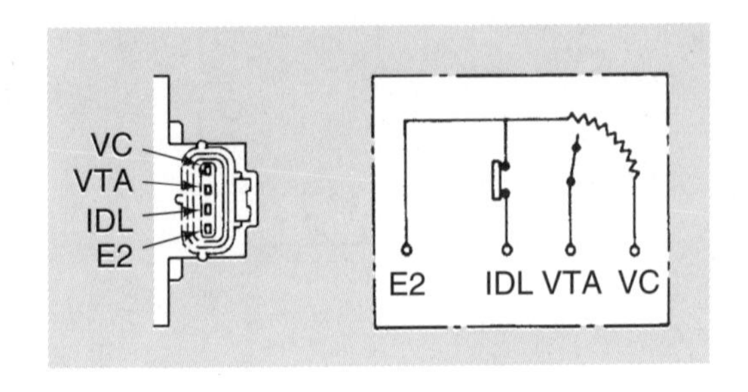

Figure 8-11 Four-wire throttle position sensor (TPS) with idle switch (Courtesy of Toyota Motor Corporation)

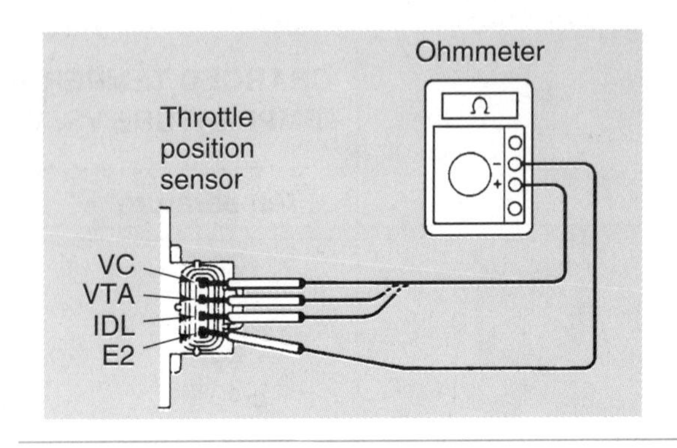

Figure 8-12 Ohmmeter test connections from the ground terminal to each of the other terminals on the throttle position sensor (TPS) (Courtesy of Toyota Motor Corporation)

With the ignition switch on, connect a voltmeter from the sensor signal wire to ground. Slowly open the throttle and observe the voltmeter. The voltmeter reading should increase smoothly and gradually. Typical TPS voltage readings would be 0.5 V to 1 V with the throttle in the idle position, and 4 V to 5 V at wide-open throttle. Always refer to the vehicle manufacturer's specifications. If the TPS does not have the specified voltage or if the voltage signal is erratic, replace the sensor.

Throttle Position Sensor Diagnosis, Four-Wire Sensor

Some TPSs contain an idle switch that is connected to the computer. These sensors have the same wires as a three-wire TPS and an extra wire for the idle switch (Figure 8-11).

The four-wire TPS is tested with an ohmmeter connected from the sensor ground terminal to each of the other terminals (Figure 8-12). A specified feeler gauge must be placed between the throttle lever and the stop for some of the ohmmeter tests. When the ohmmeter is connected from the ground (E2) terminal to the VTA terminal, the throttle must be held in the wide-open position (Figure 8-13).

Throttle Position Sensor Adjustment

A TPS adjustment may be performed on some vehicles, but this adjustment is not possible on other applications. Check the vehicle manufacturer's service manual for the TPS adjustment procedure.

CLEARANCE BETWEEN LEVER AND STOP SCREW	BETWEEN TERMINALS	RESISTANCE
0 mm (0 in.)	VTA – E2	0.28 – 6.4 kΩ
0.35 mm (0.014 in.)	IDL – E2	0.5 kΩ or less
0.70 mm (0.028 in.)	IDL – E2	Infinity
Throttle valve fully open	VTA – E2	2.0 – 11.6 kΩ
–	VC – E2	2.7 – 7.7 kΩ

Figure 8-13 Specifications for throttle position sensor (TPS) ohmmeter tests (Courtesy of Toyota Motor Corporation)

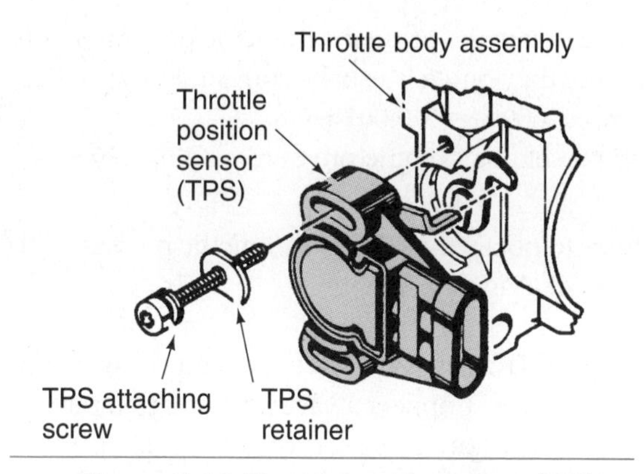

Figure 8-14 Throttle position sensor (TPS) with elongated slots for sensor adjustment (Courtesy of Chevrolet Motor Division, General Motors Corporation)

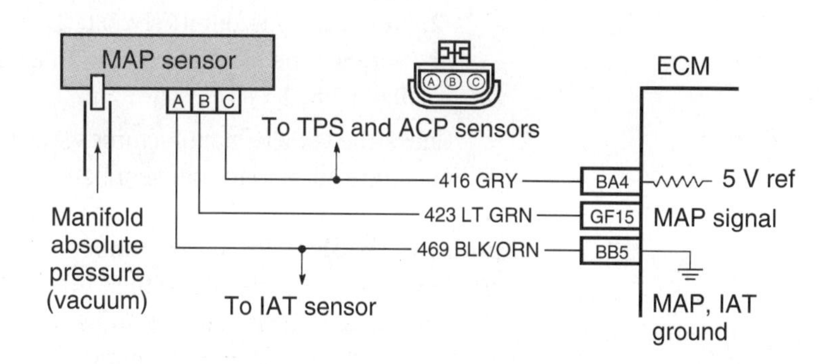

Figure 8-15 Manifold absolute pressure (MAP) sensor and connecting wires (Courtesy of Oldsmobile Division, General Motors Corporation)

An improper TPS adjustment may cause inaccurate idle speed, engine stalling, and acceleration stumbles. Follow these steps for a typical TPS adjustment:

1. Backprobe the TPS signal wire and connect a voltmeter from this wire to ground.
2. Turn on the ignition switch and observe the voltmeter reading with the throttle in the idle position.
3. If the TPS does not provide the specified signal voltage, loosen the TPS mounting bolts and rotate the sensor housing until the specified voltage is indicated on the voltmeter (Figure 8-14).
4. Hold the sensor in this position and tighten the mounting bolts to the specified torque.

Manifold Absolute Pressure (MAP) Sensor Diagnosis

Barometric Pressure Voltage Signal Diagnosis. A defective manifold absolute pressure sensor may cause a rich or lean air-fuel ratio, excessive fuel consumption, and engine surging. This diagnosis applies to MAP sensors that produce an analog voltage signal. With the ignition switch on, backprobe the 5-V reference wire and connect a voltmeter from the 5-V reference wire to ground (Figure 8-15).

SERVICE TIP: Manifold absolute pressure sensors have a much different calibration on turbocharged engines than on nonturbocharged engines. Be sure you are using the proper specifications for the sensor being tested.

If the reference wire is not supplying the specified voltage, check the voltage on this wire at the computer. If the voltage is within specifications at the computer, but low at the sensor, repair the 5-V reference wire. When this voltage is low at the computer, check the voltage supply wires and ground wires on the computer. If these wires are satisfactory, replace the computer.

With the ignition switch on, connect the voltmeter from the sensor ground wire to the battery ground. If the voltage drop across this circuit exceeds specifications, repair the ground wire from the sensor to the computer.

Backprobe the MAP sensor signal wire and connect a voltmeter from this wire to ground with the ignition switch on. The voltage reading indicates the barometric pressure signal from the MAP sensor to the computer. Many MAP sensors send a barometric pressure signal to the computer each time the ignition switch is turned on and each time the throttle is in the wide-open position. If the voltage supplied by the barometric pressure signal in the MAP sensor does not equal the vehicle manufacturer's specifications, replace the MAP sensor.

The barometric pressure voltage signal varies depending on altitude and atmospheric conditions. Follow this calculation to obtain an accurate barometric pressure reading:

1. Phone your local weather or TV station and obtain the present barometric pressure reading; for example, 29.85 inches. The pressure they quote is usually corrected to sea level.
2. Multiply your altitude by 0.001; for example, 600 feet X 0.001 = 0.6.
3. Subtract the altitude correction from the present barometric pressure reading: 29.85 - 0.6 = 29.79.

Check the vehicle manufacturer's specifications to obtain the proper barometric pressure voltage signal in relation to the present barometric pressure (Figure 8-16).

Special Tools

Vacuum hand pump

Manifold Absolute Pressure Sensor Voltage Signal Diagnosis. Leave the ignition switch on and the voltmeter connected to the MAP sensor signal wire. Connect a vacuum hand pump to the MAP sensor vacuum connection and apply 5 inches of vacuum to the sensor. On some MAP sensors, the sensor voltage signal should change 0.7 V to 1.0 V for every 5 inches of vacuum change applied to the sensor. Always use the vehicle manufacturer's specifications. If the barometric pressure voltage signal was 4.5 V with 5 inches of vacuum applied to the MAP sensor, the voltage should be 3.5 V to 3.8 V. When 10 inches of vacuum is applied to the sensor, the voltage signal should be 2.5 V to 3.1 V. Check the MAP sensor voltage at 5-inch intervals from 0 to 25 inches. If the MAP sensor voltage is not within specifications at any vacuum, replace the sensor.

Diagnosis of Manifold Absolute Pressure Sensor with Voltage Frequency Signal. If the MAP sensor produces a digital voltage signal of varying frequency, check the 5-V reference wire and the ground wire with the same procedure used on other MAP sensors. This sensor diagnosis is based on the use of a MAP sensor tester that changes the MAP sensor varying frequency voltage to an analog voltage. Follow these steps to test the MAP sensor voltage signal:

Special Tools

MAP sensor tester

1. Turn off the ignition switch, and disconnect the wiring connector from the MAP sensor.
2. Connect the connector on the MAP sensor tester to the MAP sensor (Figure 8-17).
3. Connect the MAP sensor tester battery leads to a 12-V battery.
4. Connect a pair of digital voltmeter leads to the MAP tester signal wire and ground.

Absolute Baro Reading	Lowest Allowable Voltage at –40°F	Lowest Allowable Voltage at 257°F	Lowest Allowable Voltage at 77°F	TBI MAP Sensor Designed Output Voltage	Highest Allowable Voltage at 77°F	Highest Allowable Voltage at 257°F	Highest Allowable Voltage at –40°F
31.0″	4.548 V	4.632 V	4.716 V	4.800 V	4.884 V	4.968 V	5.052 V
30.9″	4.531 V	4.615 V	4.699 V	4.783 V	4.867 V	4.951 V	5.035 V
30.8″	4.514 V	4.598 V	4.682 V	4.766 V	4.850 V	4.934 V	5.018 V
30.7″	4.497 V	4.581 V	4.665 V	4.749 V	4.833 V	4.917 V	5.001 V
30.6″	4.480 V	4.564 V	4.648 V	4.732 V	4.816 V	4.900 V	4.984 V
30.5″	4.463 V	4.547 V	4.631 V	4.715 V	4.799 V	4.883 V	4.967 V
30.4″	4.446 V	4.530 V	4.614 V	4.698 V	4.782 V	4.866 V	4.950 V
30.3″	4.430 V	4.514 V	4.598 V	4.682 V	4.766 V	4.850 V	4.934 V
30.2″	4.413 V	4.497 V	4.581 V	4.665 V	4.749 V	4.833 V	4.917 V
30.1″	4.396 V	4.480 V	4.564 V	4.648 V	4.732 V	4.816 V	4.900 V
30.0″	4.379 V	4.463 V	4.547 V	4.631 V	4.715 V	4.799 V	4.883 V

Figure 8-16 Barometric pressure voltage signal specifications at various barometric pressures (Courtesy of Chrysler Corporation)

Figure 8-17 Manifold absolute pressure (MAP) sensor tester (Courtesy of Thexton Manufacturing Company)

5. Turn on the ignition switch and observe the barometric pressure voltage signal on the meter. If this voltage signal does not equal the manufacturer's specifications, replace the sensor.
6. Supply the specified vacuum to the MAP sensor with a hand vacuum pump.
7. Observe the voltmeter reading at each specified vacuum. If the MAP sensor voltage signal does not equal the manufacturer's specifications at any vacuum, replace the sensor.

Photo Sequence 5 shows a typical procedure for testing a Ford MAP sensor.

Mass Air Flow Sensor Diagnosis

Voltmeter Diagnosis of Vane-Type Mass Air Flow Sensor. Always check the voltage supply wire and the ground wire to the MAF module before checking the sensor voltage signal. Always follow the recommended test procedure in the manufacturer's service manual and use the specifications supplied by the manufacturer. The following procedure is based on the use of a Fluke multimeter. Follow these steps to measure the MAF sensor voltage signal:

1. Set the multimeter on the VDC scale, and connect the black meter lead to the COM terminal in the meter while the red meter lead is installed in the V/RPM meter connection.
2. Connect the red meter lead to the MAF signal wire with a special piercing probe, and connect the black meter lead to ground (Figure 8-18).
3. Turn on the ignition switch and press the min/max button to activate the min/max feature (Figure 8-19).

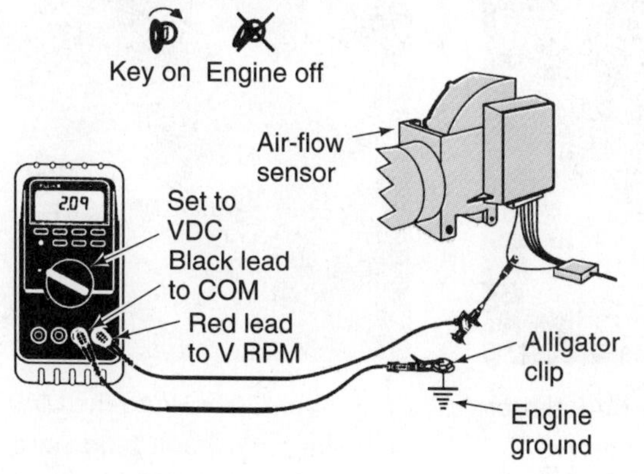

Figure 8-18 Voltmeter connected to measure mass air flow sensor (MAF) voltage signal (Courtesy of John Fluke Mfg. Co., Inc.)

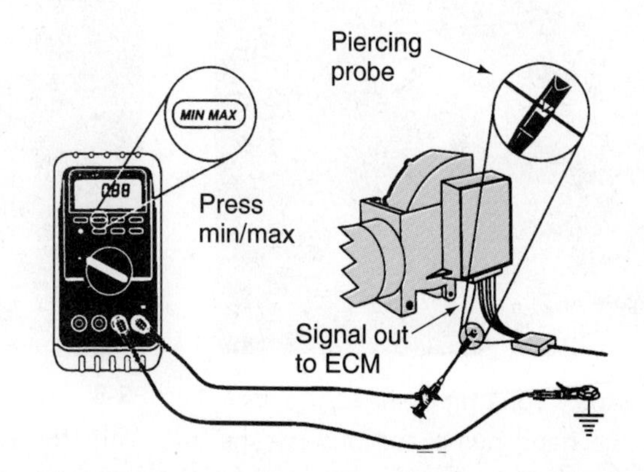

Figure 8-19 Press the min/max button to engage the min/max test mode. (Courtesy of John Fluke Mfg. Co., Inc)

Photo Sequence 5
Typical Procedure for Testing a Ford MAP Sensor

P5-1 Remove the MAP sensor wiring connector and vacuum hose.

P5-2 Connect the MAP sensor tester to the MAP sensor.

P5-3 Connect the digital voltmeter leads from the proper MAP sensor tester lead to ground.

P5-4 Connect the MAP sensor tester leads to the battery terminals with the proper polarity.

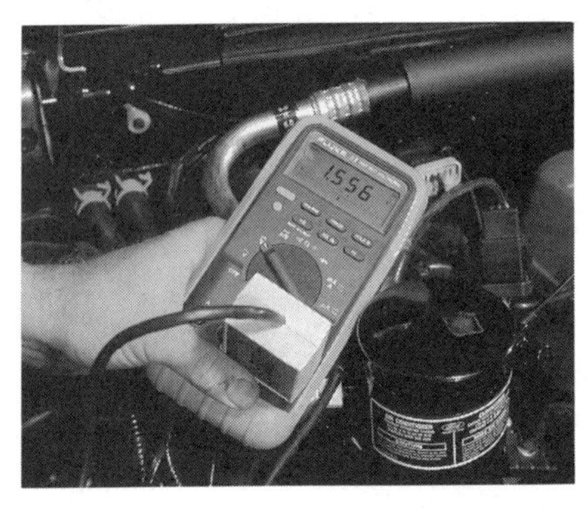

P5-5 Observe the MAP sensor barometric pressure (Baro) voltage reading on the voltmeter and compare this reading to specifications.

P5-6 Connect a vacuum hand pump hose to the MAP sensor and apply 5 in Hg to the MAP sensor. Observe the MAP sensor voltage signal on the voltmeter. Compare this voltmeter reading to specifications.

P5-7 Apply 10 in Hg to the MAP sensor with the hand pump and observe the voltmeter reading. Compare this reading to specifications.

P5-8 Apply 15 in Hg to the MAP sensor with the hand pump and observe the voltmeter reading. Compare this reading to specifications.

P5-9 Apply 20 in Hg to the MAP sensor with the hand pump and observe the voltmeter reading. Compare this reading to specifications. If any of the MAP sensor readings do not meet specifications, replace the MAP sensor.

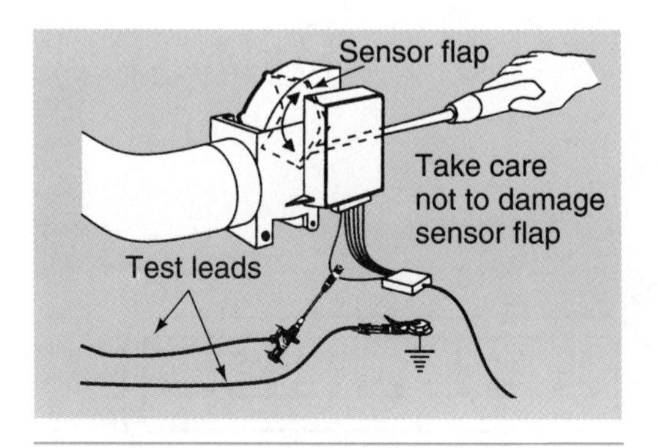

Figure 8-20 Pushing the mass air flow (MAF) sensor vane from the open to the closed position (Courtesy of John Fluke Mfg. Co., Inc.)

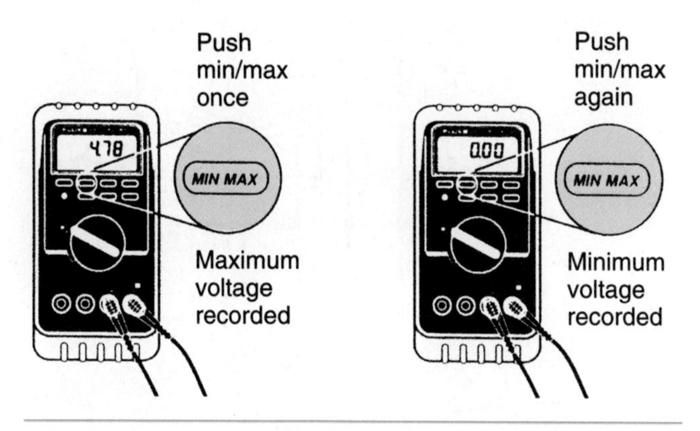

Figure 8-21 Pressing min/max button to read minimum and maximum sensor voltage signals (Courtesy of John Fluke Mfg. Co., Inc.)

WARNING: While pushing the mass air flow sensor vane open and closed, be careful not to mark or damage the vane or sensor housing.

4. Slowly push the MAF vane from the closed to the wide-open position, and allow the vane to slowly return to the closed position (Figure 8-20).
5. Touch the min/max button once to read the maximum voltage signal recorded, and press this button again to read the minimum voltage signal (Figure 8-21). If the minimum voltage signal is zero, there may be an open circuit in the MAF sensor variable resistor. When the voltage signal is not within the manufacturer's specifications, replace the sensor.

Vane-Type Mass Air Flow Ohmmeter Tests. Some vehicle manufacturers specify ohmmeter tests for the MAF sensor. With the MAF sensor removed, connect the ohmmeter to the E2 and VS MAF sensor terminals (Figure 8-22). The resistance at these terminals should be 200 Ω to 600 Ω (Figure 8-23).

Connect the ohmmeter leads to the other recommended MAF sensor terminals, and record the resistance readings (Figure 8-24). Since the THA and E2 sensor terminals are connected internally to the thermistor, temperature affects the ohm reading at these terminals as indicated in the specifications. If the specified resistance is not available in any of the test connections, replace the MAF sensor.

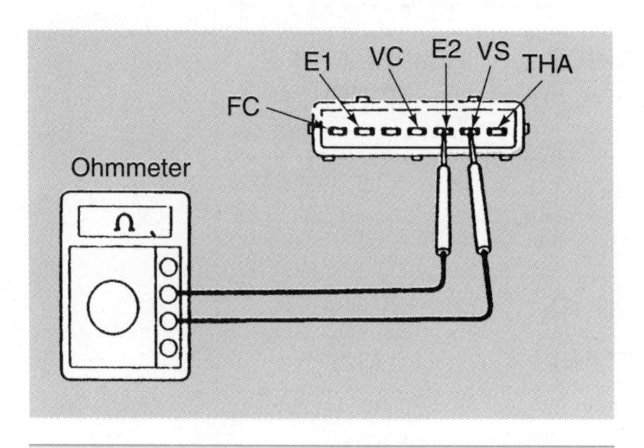

Figure 8-22 Ohmmeter connected to mass air flow (MAF) E2 and VS terminals (Courtesy of Toyota Motor Corporation)

BETWEEN TERMINALS	RESISTANCE (Ω)	MEASURING PLATE OPENING
FC – E1	Infinity	Fully closed
FC – E1	Zero	Other than closed
VS – E2	200 – 600	Fully closed
VS – E2	20 – 1,200	Fully open

Figure 8-23 Ohm specifications at mass air flow (MAF) sensor terminals (Courtesy of Toyota Motor Corporation)

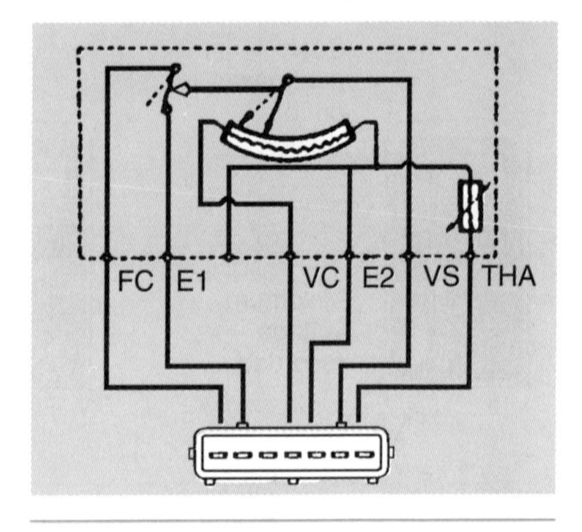

Figure 8-24 Mass air flow (MAF) sensor terminals and internal electric circuit (Courtesy of Toyota Motor Corporation)

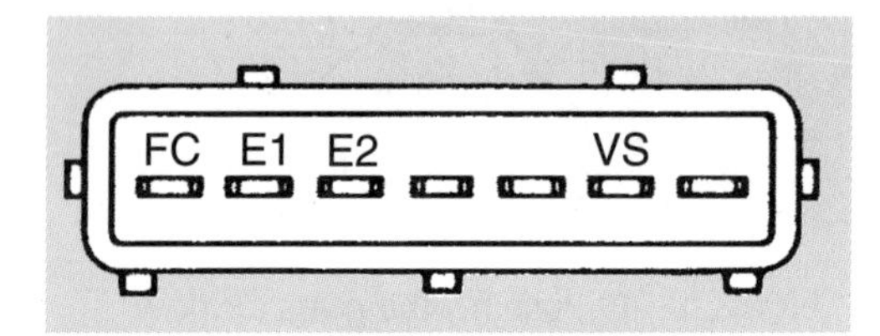

Figure 8-25 Mass air flow (MAF) sensor terminals for ohmmeter tests at specific vane positions (Courtesy of Toyota Motor Corporation)

Connect the ohmmeter leads to the specified MAF sensor terminals (Figure 8-25), and move the vane from the fully closed to the fully open position. With each specified meter connection and vane position, the ohmmeter should indicate the specified resistance (Figure 8-26). When the ohmmeter leads are connected to the E2 and VS sensor terminals, the ohm reading should increase smoothly as the sensor vane is opened and closed.

Heated Resistor or Hot Wire Mass Air Flow Sensor Frequency Test. The test procedure for heated resistor and hot wire MAF sensors varies depending on the vehicle make and year. Always follow the test procedure in the vehicle manufacturer's service manual. A frequency test may be performed on some MAF sensors, such as the AC Delco MAF on some General Motors products. The following test procedure is based on the use of a Fluke multimeter. Follow these steps to check the MAF sensor voltage signal and frequency:

1. Place the multimeter on the V/RPM scale and connect the meter leads from the MAF voltage signal wire to the ground wire (Figure 8-27).
2. Start the engine and observe the voltmeter reading. On some MAF sensors, this reading should be 2.5 V. Always refer to the manufacturer's specifications.
3. Lightly tap the MAF sensor housing with a screwdriver handle and watch the voltmeter pointer. If the pointer fluctuates or the engine misfires, replace the MAF sensor. Some

BETWEEN TERMINALS	RESISTANCE (Ω)	TEMPERATURE °C (°F)
VS – E2	200 – 600	–
VC – E2	200 – 400	–
THA – E2	10,000 – 20,000	−20 (−4)
	4,000 – 7,000	0 (32)
FC – E1	Infinity	–

Figure 8-26 Mass air flow (MAF) sensor ohm specifications at various vane positions (Courtesy of Toyota Motor Corporation)

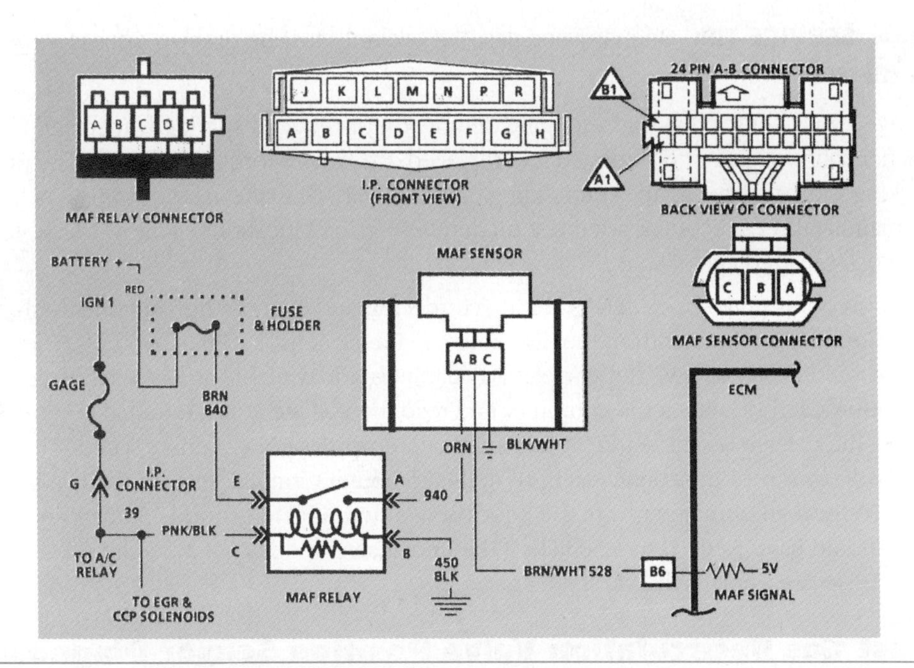

Figure 8-27 Mass air flow (MAF) sensor wiring diagram (Courtesy of Chevrolet Motor Division, General Motors Corporation)

MAF sensors have experienced loose internal connections, which cause erratic voltage signals and engine misfiring and surging.

4. Be sure the meter dial is on dc volts, and press the RPM button three times so the meter displays voltage frequency. The meter should indicate about 30 hertz (Hz) with the engine idling.
5. Increase the engine speed, and record the meter reading at various speeds.
6. Graph the frequency readings. The MAF sensor frequency should increase smoothly and gradually in relation to engine speed. If the MAF sensor frequency reading is erratic, replace the sensor (Figure 8-28).

Knock Sensor Diagnosis

SERVICE TIP: If a knock sensor is overtorqued, it may become too sensitive and cause reduced spark advance and fuel economy.

SERVICE TIP: When the knock sensor torque is less than specified, the sensor may have reduced sensitivity, and engine detonation may occur.

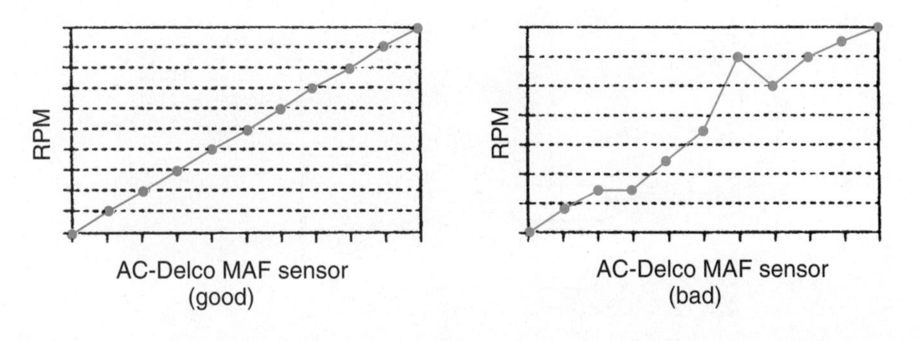

Figure 8-28 Satisfactory and unsatisfactory mass air flow (MAF) sensor frequency readings (Courtesy of John Fluke Mfg. Co., Inc.)

SERVICE TIP: On many engines, the coolant must be drained prior to knock sensor removal.

A defective knock sensor may cause engine detonation or reduced spark advance and fuel economy. When a knock sensor is removed and replaced, the sensor torque is critical. The procedure for checking a knock sensor varies depending upon the vehicle make and year. Always follow the vehicle manufacturer's recommended test procedure and specifications. Follow these steps for a typical knock sensor diagnosis:

1. Disconnect the knock sensor wiring connector, and turn on the ignition switch.
2. Connect a voltmeter from the disconnected knock sensor wire to ground. The voltage should be 4 V to 6 V. If the specified voltage is not available at this wire, backprobe the knock sensor wire at the computer and read the voltage at this terminal (Figure 8-29). If the voltage is satisfactory at this terminal, repair the knock sensor wire. When the voltage is not within specifications at the computer terminal, replace the computer.
3. Connect an ohmmeter from the knock sensor terminal to ground. Some knock sensors should have 3,300 Ω to 4,500 Ω. If the knock sensor does not have the specified ohms resistance, replace the sensor.

Exhaust Gas Recirculation Valve Position Sensor Diagnosis

CAUTION: The EGR valve and EVP sensor may be very hot if the engine has been running. Wear protective gloves if it is necessary to service these components.

Many exhaust gas recirculation valve position (EVP) sensors have a 5-V reference wire, a voltage signal wire, and a ground wire. The 5-V reference wire and the ground wire may be checked using the same procedure explained previously on TPS and MAP sensors. Connect a pair of voltmeter leads from the voltage signal wire to ground, and turn on the ignition switch. The voltage signal should be approximately 0.8 V. Connect a vacuum hand pump to the vacuum fitting on the EGR valve, and slowly increase the vacuum from zero to 20 in Hg. The EVP sensor voltage signal should gradually increase to 4.5 V at 20 in Hg. Always use the EVP test procedure and specifications supplied by the vehicle manufacturer. If the EVP sensor does not have the specified voltage, replace the sensor.

Vehicle Speed Sensor Diagnosis

A defective vehicle speed sensor may cause different problems depending on the computer output control functions. A defective vehicle speed sensor (VSS) may cause these problems:

1. Improper converter clutch lockup
2. Improper cruise control operation
3. Inaccurate speedometer operation

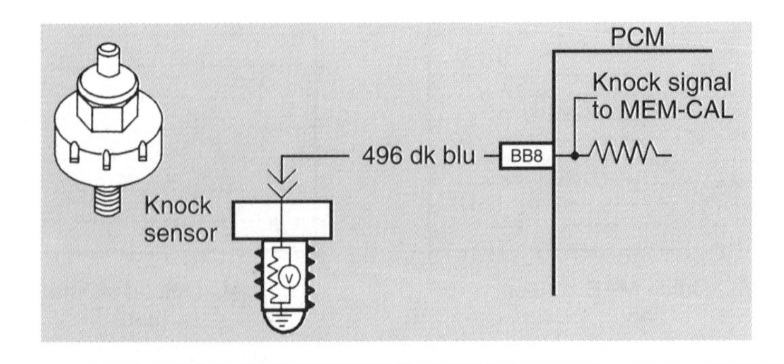

Figure 8-29 Knock sensor wiring diagram (Courtesy of Oldsmobile Division, General Motors Corporation)

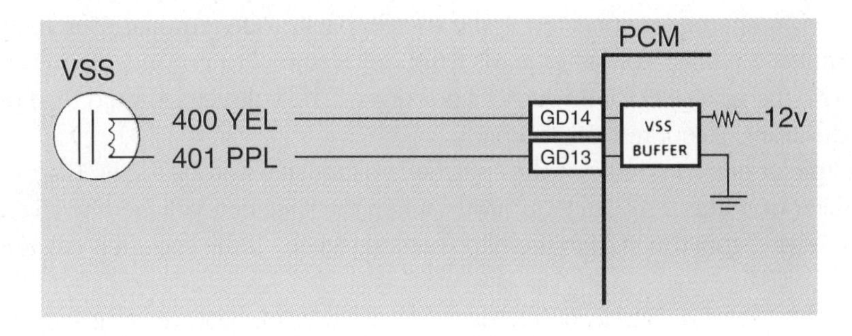

Figure 8-30 Vehicle speed sensor (VSS) wiring diagram (Courtesy of Oldsmobile Division, General Motors Corporation)

Prior to VSS diagnosis, the vehicle should be lifted on a hoist so the drive wheels are free to rotate. Backprobe the VSS yellow wire, and connect a pair of voltmeter leads from this VSS wire to ground. Then start the engine (Figure 8-30).

Select the 20 V ac scale on the voltmeter. Place the transaxle in drive and allow the drive wheels to rotate. If the VSS voltage signal is not 0.5 V, or more, replace the sensor. When the VSS provides the specified voltage signal, backprobe the GD 14 PCM terminal and repeat the voltage signal test with the drive wheels rotating. If 0.5 V is available at this terminal, the trouble may be in the PCM.

When 0.5 V is not available at this terminal, turn the ignition switch off and disconnect the VSS terminal and the PCM terminals. Connect the ohmmeter leads from the 400 VSS terminal to the GD 14 PCM terminal. The meter should read zero ohms. Repeat the test with the ohmmeter leads connected to the 401 VSS terminal and the GD 13 PCM terminal. This wire should also have zero ohms resistance. If the resistance in these wires is more than specified, repair the wires.

Park/Neutral Switch Diagnosis

A defective park/neutral switch may cause improper idle speed and failure of the starting motor circuit. Always follow the test procedure in the vehicle manufacturer's service manual. Disconnect the park/neutral switch wiring connector and connect a pair of ohmmeter leads from the B terminal in the wiring connector to ground (Figure 8-31). If the ohmmeter does not indicate less than 0.5 Ω, repair the ground wire.

A park/neutral switch may be referred to as a neutral drive switch.

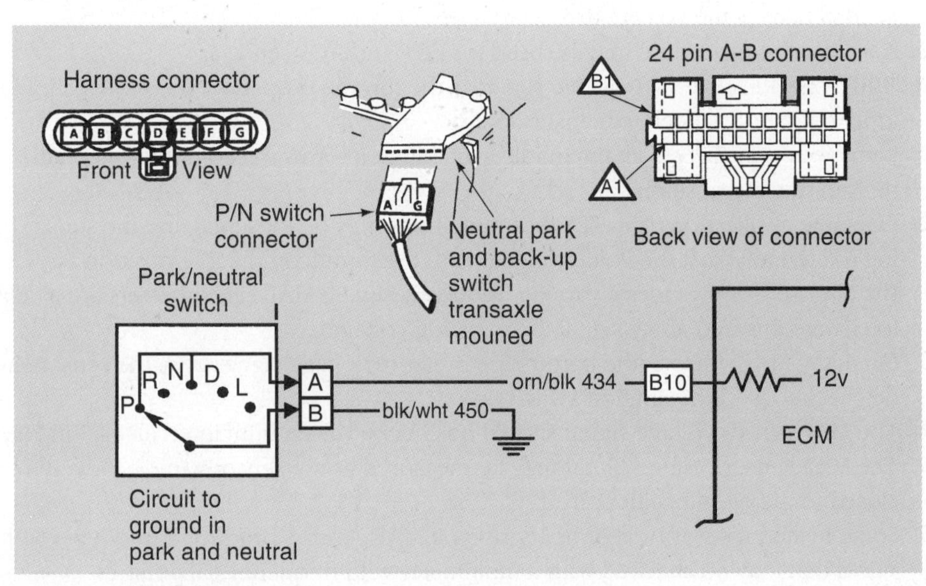

Figure 8-31 Park/neutral switch wiring diagram (Courtesy of Chevrolet Motor Division, General Motors Corporation)

With the wiring harness connected to the switch, backprobe terminal A on the park/neutral switch and connect a pair of voltmeter leads from this terminal to ground. Turn on the ignition switch, and move the gear selector through all positions. The voltmeter should read over 5 V in all gear selector positions except neutral and park.

If the voltmeter does not indicate the specified voltage, backprobe the PCM terminal B10 and connect the meter from this terminal to ground. When the specified voltage is available at this terminal, repair the wire from the PCM to the park/neutral switch. If the specified voltage is not available, replace the PCM.

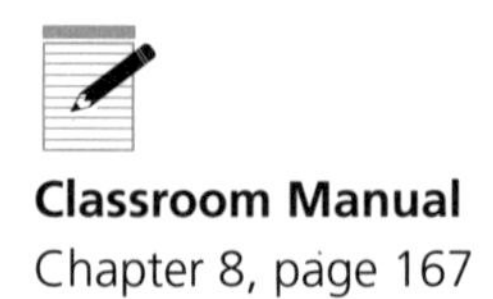

Classroom Manual
Chapter 8, page 167

When the gear selector is placed in neutral or park, the voltmeter reading should be less than 0.5 V. If the specified voltage is not available in these gear selector positions, replace the park/neutral switch.

CUSTOMER CARE: Like everyone else, individuals involved in the automotive service industry do make mistakes. If you make a mistake that results in a customer complaint, always be willing to admit your mistake and correct it. Do not try to cover up the mistake or blame someone else. Customers are usually willing to live with an occasional mistake that is corrected quickly and efficiently.

Guidelines for Input Sensor Diagnosis

1. Defective computer voltage supply wires or ground wires may cause defects in engine performance and economy. These wires must always be tested before the computer is replaced.
2. If the O_2 sensor voltage signal is higher than specified, the air-fuel ratio may be rich or the sensor may be contaminated.
3. When the O_2 sensor voltage signal is lower than specified, the air-fuel ratio may be lean or the sensor may be defective.
4. An O_2 sensor must be tested with a digital voltmeter.
5. A defective O_2 sensor heater may cause extended open loop time and reduced fuel economy.
6. An ECT sensor may be checked with an ohmmeter or a voltmeter.
7. Some computers switch an internal resistor in the ECT circuit at 120°F (49°C), and this action changes the voltage drop across the sensor.
8. A similar test procedure may be used for ECT and ACT sensors.
9. While checking the TPS voltage signal as the throttle is opened, the sensor should be tapped lightly to check for sensor defects.
10. Many four-wire TPSs contain an idle switch that informs the computer when the throttle is in the idle position.
11. On some applications, the TPS mounting bolts may be loosened and the sensor housing rotated to adjust the voltage signal with the throttle in the idle position.
12. If a MAP sensor provides a varying frequency signal, MAP sensor testers are available to change this frequency signal to an analog voltage.
13. On many MAP sensors, the barometric pressure voltage signal from the sensor should be checked with the ignition switch on.
14. The MAP sensor voltage signal should be checked at vacuum intervals of 5 in Hg.
15. The MAF sensor voltage signal may be measured as the sensor vane is moved from the closed to the open position.
16. Some heated resistor-type or hot wire-type MAF sensors produce a frequency voltage signal that may be checked with a multimeter with frequency capabilities.
17. The EVP sensor voltage signal should change from about 0.8 V with the EGR valve closed to 4.5 V with the EGR valve wide open.

18. The VSS signal is used by the computer to control the torque converter clutch lockup, cruise control, and the electronic speedometer.
19. The voltage signal at the park/neutral switch should be low with the gear selector in park or neutral, and high in other gear selector positions.

CASE STUDY

A customer complained about hard starting on a Chevrolet Celebrity with a 2.5 L four-cylinder engine. The technician asked the customer when the hard starting problem occurred. The customer indicated that the problem occurred after the car had been parked all night. The customer also revealed that the problem did not occur if the block heater was plugged in. In the cold climate where this customer lives, many vehicles have block heaters to provide easier starting in cold weather. Further questioning of the customer revealed there was evidence of black smoke from the tailpipe after a cold start. The technician informed the customer that a check of the computer input sensors would be required. Since the computer supplies air-fuel ratio enrichment in response to the ECT signal in this system, the technician suspected a problem in this sensor. The technician checked the vehicle manufacturer's specifications for the vehicle and found an approximate ohm value for various sensor temperatures (Figure 8-32).

The technician tested the sensor and found the resistance of the sensor did vary about 200 Ω from the specifications at some temperatures. Since the specifications indicated the specified ohm values were approximate, the technician had to decide if this variation from specifications was enough to cause the problem.

The technician obtained a new ECT sensor and compared the ohm reading on the new sensor to the reading on the old sensor. He found the new sensor had the specified resistance at various temperatures within 10 ohms. The new ECT sensor was installed and a quick test was performed on the other sensors without finding any problems. Since the vehicle was needed for commuting purposes, the car was returned to the customer.

On the following day, the service manager phoned the owner of the vehicle to find out the results of the sensor installation. The vehicle owner was very pleased that the car started easily and ran well when it was started with the engine cold.

COOLANT SENSOR TEMPERATURE TO RESISTANCE VALUES (APPROXIMATE)		
°F	°C	OHMS
210	100	185
160	70	450
100	38	1,800
70	20	3,400
40	4	7,500
20	–7	13,500
0	–18	25,000
–40	–40	100,700

Figure 8-32 Engine coolant temperature (ECT) sensor resistance specifications in relation to temperature (Courtesy of Chevrolet Motor Division, General Motors Corporation)

Terms to Know

Open loop
Closed loop
Room temperature vulcanizing (RTV) sealant
Oxygen (O_2) sensor
Engine coolant temperature (ECT) sensor
Air charge temperature (ACT) sensor
Throttle position sensor (TPS)
Reference voltage
Manifold absolute pressure (MAP) sensor
Barometric (Baro) pressure sensor
Mass air flow (MAF) sensor
Vane-type MAF sensor
Heated resistor-type MAF sensor
Hot wire-type MAF sensor
Knock sensor
Vehicle speed sensor (VSS)
Exhaust gas recirculation valve position (EVP) sensor
Park/neutral switch
Neutral/drive switch (NDS)

ASE Style Review Questions

1. While discussing O_2 sensor diagnosis:
   Technician A says the voltage signal on a satisfactory O_2 sensor should always be cycling between 0.5 V and 1 V.
   Technician B says a contaminated O_2 sensor provides a continually low voltage signal.
   Who is correct?
   A. A only **C.** Both A and B
   B. B only **D.** Neither A nor B

2. While discussing ECT sensor diagnosis:
   Technician A says a defective ECT sensor may cause hard cold engine starting.
   Technician B says a defective ECT sensor may cause improper operation of emission devices.
   Who is correct?
   A. A only **C.** Both A and B
   B. B only **D.** Neither A nor B

3. While discussing ECT sensor diagnosis:
   Technician A says the ECT sensor resistance should increase as the sensor temperature increases.
   Technician B says some computers have internal resistors connected in series with the ECT sensor, and the computer switches these resistors at 120°F (49°C).
   Who is correct?
   A. A only **C.** Both A and B
   B. B only **D.** Neither A nor B

4. While discussing ACT sensor diagnosis:
   Technician A says the ACT sensor resistance should decrease as the sensor temperature decreases.
   Technician B says the ACT sensor resistance should increase as the sensor temperature decreases.
   Who is correct?
   A. A only **C.** Both A and B
   B. B only **D.** Neither A nor B

5. While discussing TPS diagnosis:
   Technician A says the TPS voltage signal should increase smoothly from 1 V at idle to 6 V at wide-open throttle.
   Technician B says a defective TPS may cause improper idle speed.
   Who is correct?
   A. A only **C.** Both A and B
   B. B only **D.** Neither A nor B

6. While discussing TPS diagnosis:
   Technician A says a four-wire TPS contains an idle switch.
   Technician B says in some applications the TPS mounting bolts may be loosened and the TPS housing rotated to adjust the voltage signal with the throttle in the idle position.
   Who is correct?
   A. A only **C.** Both A and B
   B. B only **D.** Neither A nor B

7. While discussing MAP sensor diagnosis:
   Technician A says with the ignition switch on, the MAP sensor produces a barometric pressure voltage signal.
   Technician B says the MAP sensor reference voltage and the ground wire should be checked with the ignition switch on.
   Who is correct?
   A. A only **C.** Both A and B
   B. B only **D.** Neither A nor B

8. While discussing MAP sensor diagnosis:
   Technician A says on some MAP sensors, the voltage signal should increase 2 V for each 5 inches of vacuum applied to the sensor.
   Technician B says on some MAP sensors, the voltage signal should decrease 0.7 V to 1 V for every 5 inches of vacuum increase applied to the sensor.
   Who is correct?
   A. A only **C.** Both A and B
   B. B only **D.** Neither A nor B

9. While discussing MAF sensor diagnosis:
   Technician A says on a vane-type MAF sensor, the voltage signal should be checked as the vane is moved from fully closed to fully open.
   Technician B says on a vane-type MAF sensor, the voltage signal should decrease as the vane is opened.
   Who is correct?
   A. A only **C.** Both A and B
   B. B only **D.** Neither A nor B

10. While discussing knock sensor service and diagnosis:
   Technician A says if the knock sensor torque is more than specified, the sensor has reduced sensitivity.
   Technician B says if a knock sensor torque is more than specified, the sensor may be too sensitive, resulting in reduced spark advance.
   Who is correct?
   A. A only **C.** Both A and B
   B. B only **D.** Neither A nor B

Table 8-1 ASE Task

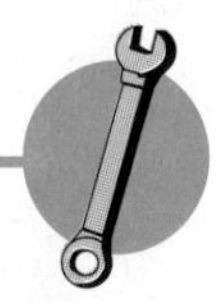

Inspect, test, adjust, and replace sensor and actuator components and circuits of electronic/computer-controlled systems.

Problem Area	Symptoms	Possible Causes	Classroom Manual	Shop Manual
ENGINE PERFORMANCE	Acceleration stumbles	1. Defective throttle position sensor (TPS)	171	144
		2. Improper throttle position sensor (TPS) adjustment	172	146
		3. Defective engine coolant temperature (ECT) sensor	170	141
		4. Defective air change temperature (ACT) sensor	171	143
		5. Defective oxygen (O_2) sensor	167	140
	Improper idle speed, engine stalling	1. Defective engine coolant temperature (ECT) sensor	170	141
		2. Defective throttle position sensor (TPS)	171	144
		3. Improper throttle position sensor (TPS) adjustment	172	146
		4. Defective air charge temperature (ACT) sensor	171	143

Table 8-1 ASE Task (continued)

Inspect, test, adjust, and replace sensor and actuator components and circuits of electronic/computer-controlled systems.

Problem Area	Symptoms	Possible Causes	Classroom Manual	Shop Manual
ENGINE PERFORMANCE	Engine surging	1. Intake manifold vacuum leak	172	147
		2. Defective manifold absolute pressure (MAP) sensor	172	147
		3. Defective mass air flow (MAF) sensor	173	149
	Detonation	1. Defective knock sensor	176	153
		2. Insufficient knock sensor torque	176	153
	Hard starting	Defective engine coolant temperature (ECT) sensor	170	141
FUEL ECONOMY	Reduced fuel mileage	1. Defective oxygen (O_2) sensor	167	140
		2. Defective engine coolant temperature (ECT) sensor	167	141
		3. Defective air charge temperature (ACT) sensor	171	143
		4. Defective manifold absolute pressure (MAP) sensor	172	147
		5. Defective mass air flow (MAF) sensor	173	149
		6. Defective knock sensor	176	153
		7. Excessive knock sensor torque	176	153

Ignition System Service and Diagnosis

Upon completion and review of this chapter, you should be able to:

- ❑ Diagnose ignition system problems.
- ❑ Perform a no-start diagnosis and determine the cause of the no-start condition.
- ❑ Diagnose primary circuit wiring.
- ❑ Remove, inspect, and service distributors.
- ❑ Inspect, service, and adjust ignition points.
- ❑ Install and time distributors to the engine.
- ❑ Inspect and test secondary ignition system wires.
- ❑ Inspect distributor caps and rotors.
- ❑ Inspect, service, and test ignition coils.
- ❑ Inspect, service, and test pickup coils.
- ❑ Inspect, service, and test ignition modules.
- ❑ Check and adjust ignition timing.
- ❑ Check ignition spark advance.
- ❑ Remove, service, and replace spark plugs.
- ❑ Perform pickup tests on distributor ignition (DI) systems.
- ❑ Perform tests on optical-type pickups.
- ❑ Perform no-start ignition tests on the cam and crankshaft sensors on electronic ignition (EI) systems.
- ❑ Perform no-start ignition tests on the coil and power train control module (PCM) on EI systems.
- ❑ Replace cam and crankshaft sensors on EI systems.
- ❑ Perform coil tests on EI systems.
- ❑ Adjust crankshaft sensors on EI systems.
- ❑ Perform no-start diagnoses on EI systems.
- ❑ Install and time the cam sensor on an EI system, 3.8 L turbocharged engine.
- ❑ Perform magnetic sensor tests on EI systems.
- ❑ Perform no-start tests on EI systems with magnetic sensors.
- ❑ Diagnose engine misfiring on EI-equipped engines.

Ignition System Diagnosis

Basic Tools

Basic technician's tool set

Service manual

12-V test light

Test spark plug

Jumper wires

Measuring tape

No-Start Diagnosis

These ignition defects may cause a no-start condition or hard starting:

1. Defective coil
2. Defective cap and rotor
3. Defective pickup coil
4. Open secondary coil wire
5. Low or zero primary voltage at the coil
6. Fouled spark plugs
7. Burned ignition points (point-type ignition)

Engine Misfiring Diagnosis

If engine misfiring occurs, check these items:

1. Engine compression
2. Intake manifold vacuum leaks
3. High resistance in spark plug wires, coil secondary wire, or cap terminals
4. Electrical leakage in the distributor cap, rotor, plug wires, coil secondary wire, or coil tower
5. Defective coil
6. Defective spark plugs

7. Low primary voltage and current
8. Improperly routed spark plug wires
9. Insufficient dwell (point-type ignition)
10. Worn distributor bushings

Power Loss

Check these items to diagnosis a power loss condition:

1. Engine compression
2. Restricted exhaust or air intake
3. Late ignition timing
4. Insufficient centrifugal or vacuum advance
5. Cylinder misfiring

Engine Detonation, Spark Knock

If the engine detonates, check for:

1. Higher than specified engine compression
2. Ignition timing too far advanced
3. Excessive centrifugal or vacuum advance
4. Spark plug heat range too hot
5. Improperly routed spark plug wires

Reduced Fuel Mileage

When the fuel consumption is excessive, check these components:

1. Engine compression
2. Late ignition timing
3. Lack of spark advance
4. Cylinder misfiring

No-Start Ignition Diagnosis, Primary Ignition Circuit

The same no-start diagnosis may be performed on most electronic ignition systems. Follow these steps for the no-start diagnosis:

1. Connect a 12-V test lamp from the coil tachometer (tach) terminal to ground, and turn on the ignition switch (Figure 9-1). On high energy (HEI) systems, the test light should be on because the module primary circuit is open. If the test light is off, there is an open circuit in the coil primary winding or in the circuit from the ignition switch to the coil battery terminal. On Chrysler electronic or Ford Dura-Spark II systems, the test light should be off because the module primary circuit is closed. Since there is primary current flow, most of the voltage is dropped across the primary coil winding. This action results in very low voltage at the tach terminal, which does not illuminate the test light. On these systems, if the test light is illuminated, there is an open circuit in the module or in the wire between the coil and the module.
2. Crank the engine and observe the test light. If the test light flutters while the engine is cranked, the pickup coil signal and the module are satisfactory. When the test lamp does not flutter, one of these components is defective. The pickup coil may be tested with an ohmmeter. If the pickup coil is satisfactory, the module is defective.

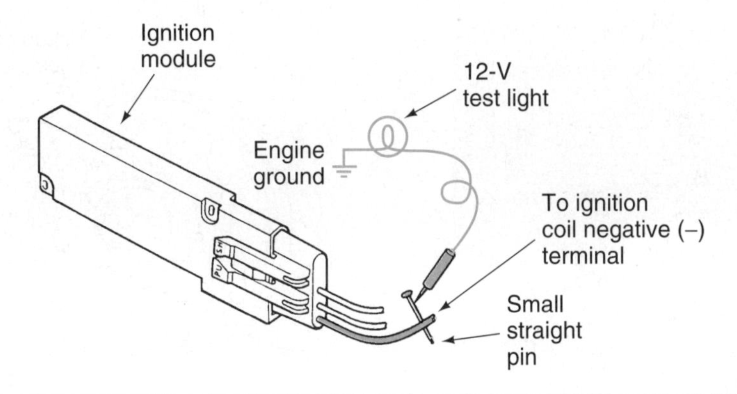

Figure 9-1 Test light connected to the negative primary coil terminal and ground (Reprinted with the permission of Ford Motor Company)

No-Start Ignition Diagnosis, Secondary Ignition Circuit

1. If the test light flutters in the primary circuit no-start diagnosis, connect a test spark plug to the coil secondary wire and ground the spark plug case (Figure 9-2). The test spark plug must have the correct voltage requirement for the ignition system being tested. For example, test spark plugs for HEI systems have a 25,000-V requirement compared to a 20,000-V requirement for many other test spark plugs. A short piece of vacuum hose may be used to connect the test spark plug to the center distributor cap terminal on HEI systems with an integral coil in the distributor cap.
2. Crank the engine and observe the spark plug. If the test spark plug fires, the ignition coil is satisfactory. If the test spark plug does not fire, the coil is likely defective because the primary circuit no-start test proved the primary circuit is triggering on and off.
3. Connect the test spark plug to several spark plug wires and crank the engine while observing the spark plug. If the test spark plug fired in step 2 but does not fire at some of the spark plugs, the secondary voltage and current is leaking through a defective distributor cap, rotor, or spark plug wires, or the plug wire is open. If the test spark plug fires at all the spark plugs, the ignition system is satisfactory.

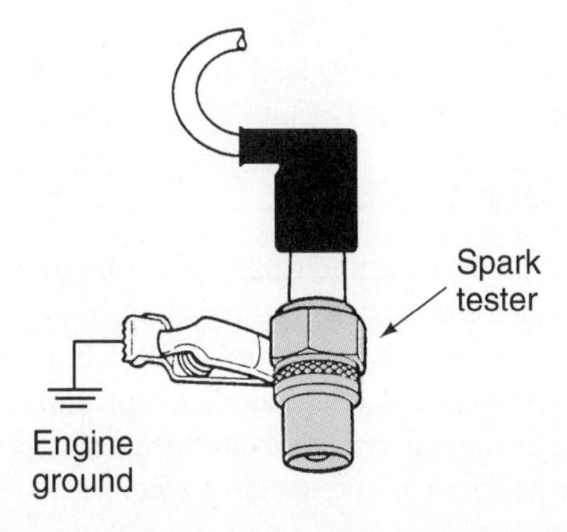

Figure 9-2 Test spark plug connected to coil secondary wire (Reprinted with the permission of Ford Motor Company)

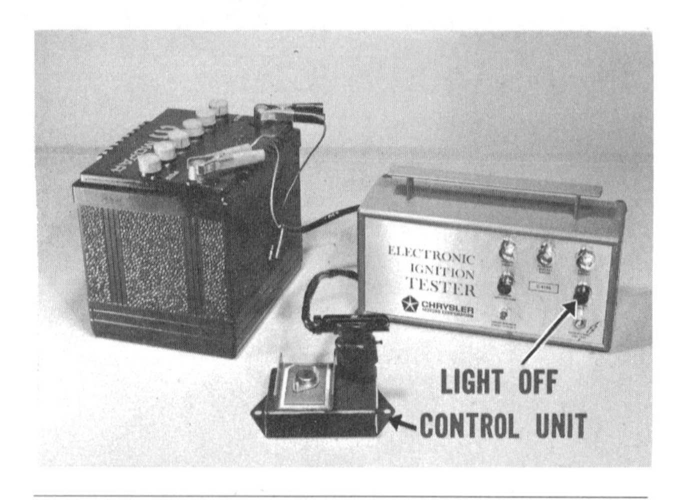

Figure 9-3 If the ignition module tester light remains off, the module is defective. (Courtesy of Chrysler Corporation)

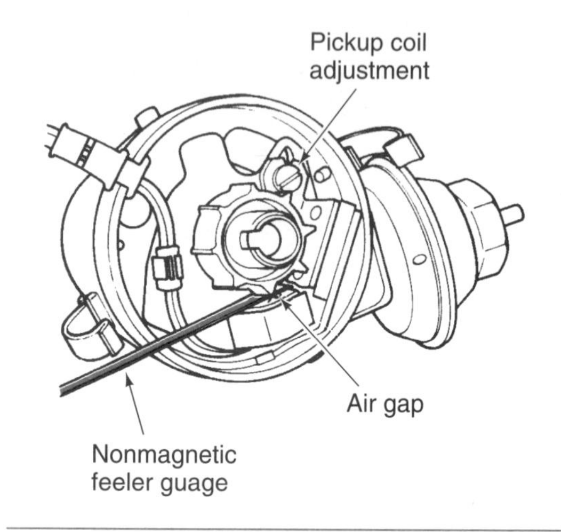

Figure 9-4 Pickup coil air gap adjustment (Courtesy of Chrysler Corporation)

Ignition Module Testing

Special Tools
Ignition module tester

A variety of ignition module testers are available from vehicle and test equipment manufacturers. These ignition module testers check the module's capability to switch the primary ignition circuit on and off. Always follow the instructions published by the vehicle or test equipment manufacturer. The module tester leads are connected to the module, and the power supply wires are connected to the battery terminals with the correct polarity. On some testers, a green light is illuminated if the module is satisfactory, and this light remains off when the module is defective (Figure 9-3).

Pickup Coil Adjustment and Tests

Pickup Gap Adjustment

Special Tools
Nonmagnetic feeler gauge

When the pickup coil is bolted to the pickup plate, such as on Chrysler distributors, the pickup air gap may be measured with a nonmagnetic copper feeler gauge positioned between the reluctor high points and the pickup coil (Figure 9-4).

If a pickup gap adjustment is required, loosen the pickup mounting bolts and move the pickup coil until the manufacturer's specified air gap is obtained. Retighten the pickup coil retaining bolts to the specified torque. Some pickup coils are riveted to the pickup plate. A pickup gap adjustment is not required for these pickup coils.

When checking the pickup coil, always check the distributor bushing for horizontal movement, which changes the pickup gap and may cause engine misfiring.

Pickup Coil Ohmmeter Tests

Special Tools
Ohmmeter

Remove the pickup leads from the module and calibrate an ohmmeter on the X10 scale. Connect the ohmmeter to the pickup coil terminals to test the pickup coil for an open circuit or a shorted condition. While the ohmmeter leads are connected, pull on the pickup leads and watch for an erratic meter reading, indicating an intermittent open in the pickup leads. Most pickup coils have 150 Ω to 900 Ω resistance, but the manufacturer's exact specifications must be used. If the pickup coil is open, the ohmmeter provides an infinite reading, whereas a meter reading below the specified resistance indicates a shorted pickup coil.

Connect an ohmmeter from one of the pickup leads to ground to test the pickup coil for a grounded condition. If the pickup coil is not grounded, the ohmmeter provides an infinite reading.

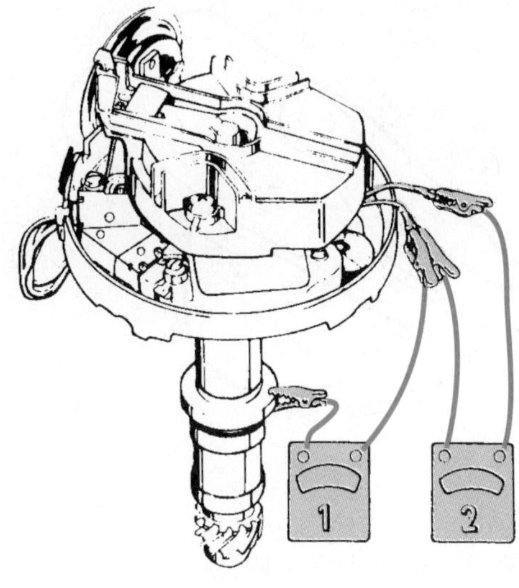

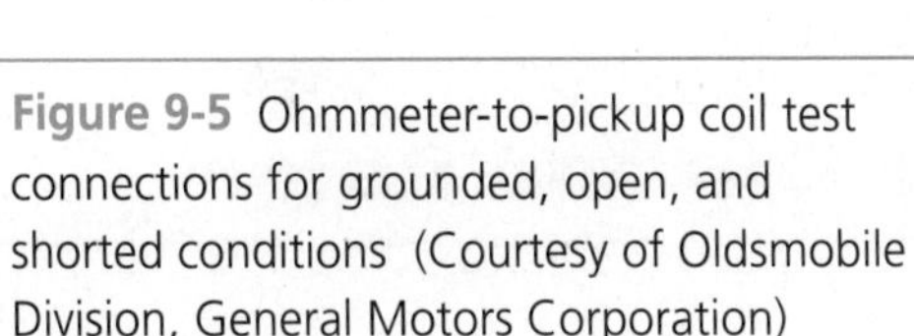

Figure 9-5 Ohmmeter-to-pickup coil test connections for grounded, open, and shorted conditions (Courtesy of Oldsmobile Division, General Motors Corporation)

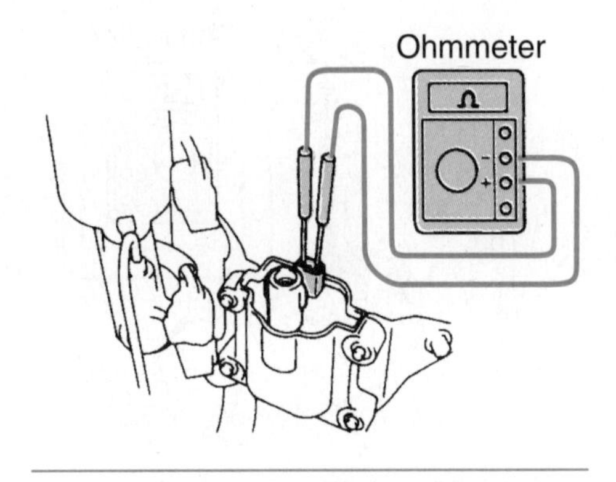

Figure 9-6 Ohmmeter connected to primary coil terminals (Courtesy of Toyota Motor Corporation)

A grounded pickup coil gives a low meter reading. Ohmmeter 1 (Figure 9-5) is connected to test the pickup coil for a grounded condition. Ohmmeter 2 in the figure illustrates the connection to test the pickup coil for an open circuit or a shorted condition.

Ignition Coil Inspection and Tests

The ignition coil should be inspected for cracks or any evidence of leakage in the coil tower. The coil container should be checked for oil leaks. If the oil is leaking from a coil, air space is present in the coil, which allows condensation to form internally. Condensation in an ignition coil causes high voltage leaks and engine misfiring.

Calibrate an ohmmeter on the X1 scale and connect the meter leads to the primary coil terminals to test the primary winding for an open circuit or a shorted condition (Figure 9-6). An infinite ohmmeter reading indicates an open winding. The winding is shorted if the meter reading is below the specified resistance. Most primary windings have a resistance of 0.5 Ω to 2 Ω, but the exact manufacturer's specifications must be compared to the meter readings.

An ohmmeter must be calibrated on the X1,000 scale and then connected from the coil secondary terminal to one of the primary terminals to test the secondary winding for a shorted or open circuit (Figure 9-7). A meter reading below the specified resistance indicates a shorted secondary winding, and an infinite meter reading proves that the winding is open.

In some HEI integral coils, the secondary winding is connected from the secondary terminal to the coil frame. When the secondary winding is tested in these coils, the ohmmeter must be connected from the secondary coil terminal to the coil frame or to the ground wire terminal extending from the coil frame to the distributor housing.

Many secondary windings have 8,000 Ω to 20,000 Ω resistance, but the meter readings must be compared to the manufacturer's specifications. The ohmmeter tests on the primary and secondary windings indicate satisfactory, open, or shorted windings. However, the ohmmeter tests do not indicate such defects as defective insulation around the coil windings, which causes high voltage leaks. Therefore, an accurate indication of coil condition is the coil maximum voltage output test with a test spark plug connected from the coil secondary wire to ground as explained in the no-start diagnosis.

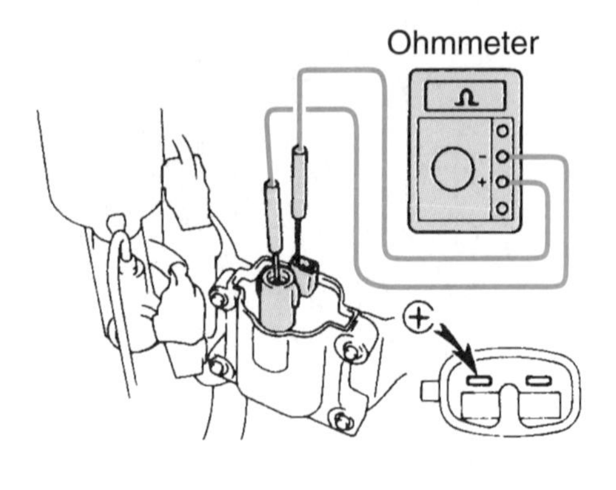

Figure 9-7 Ohmmeter connected from one primary terminal to the coil tower to test the secondary winding (Courtesy of Toyota Motor Corporation)

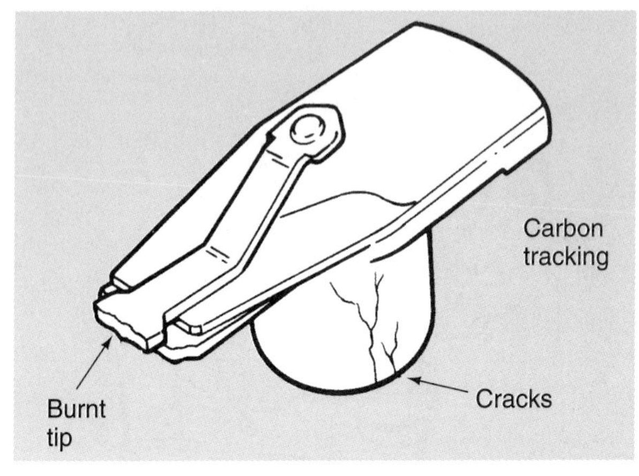

Figure 9-8 Inspecting rotor for cracks and evidence of high voltage leaks (Courtesy of Chrysler Corporation)

Inspection of Distributor Cap and Rotor

WARNING: Avoid removing excessive material from distributor cap terminals and rotor terminals with a file. This action increases the rotor gap and may increase the normal required secondary voltage.

WARNING: Cleaning distributor caps with solvent or compressed air may cause high voltage leaks in these components.

Ignition crossfiring occurs when the spark from one spark plug wire jumps across the distributor cap or spark plug wires and fires another spark plug.

The distributor cap and rotor should be inspected for cracks, corroded terminals, and carbon tracking, indicating high voltage leaks (Figure 9-8). If the distributor cap and rotor have cracks, evidence of leakage, or worn terminals, replacement is necessary. Carefully check the distributor cap terminals for corrosion and excessive wear. Small round wire brushes are available to clean cap terminals. Wipe the cap and rotor with a clean shop towel, but avoid cleaning these components in solvent or blowing them off with compressed air, which may contain moisture. Cleaning these components with solvent or compressed air may result in high voltage leaks.

Testing Secondary Ignition Wires

Spark Plug Wire Inspection

Inspect all the spark plug wires and the secondary coil wire for cracks and worn insulation, which cause high voltage leaks. Inspect all the boots on the ends of the plug wires and coil secondary wire for cracks and hard, brittle conditions. Replace the wires and boots if they show evidence of these conditions. Some vehicle manufacturers recommend spark plug wire replacement only in complete sets.

Spark Plug Wire Testing

The spark plug wires may be left in the distributor cap for test purposes, so the cap terminal connections are tested with the spark plug wires. Calibrate an ohmmeter on the X1,000 scale, and connect the ohmmeter leads from the end of a spark plug wire to the distributor cap terminal inside the cap to which the plug wire is connected (Figure 9-9).

If the ohmmeter reading is more than specified by the vehicle manufacturer, remove the wire from the cap and check the wire alone. If the wire has more resistance than specified, replace the

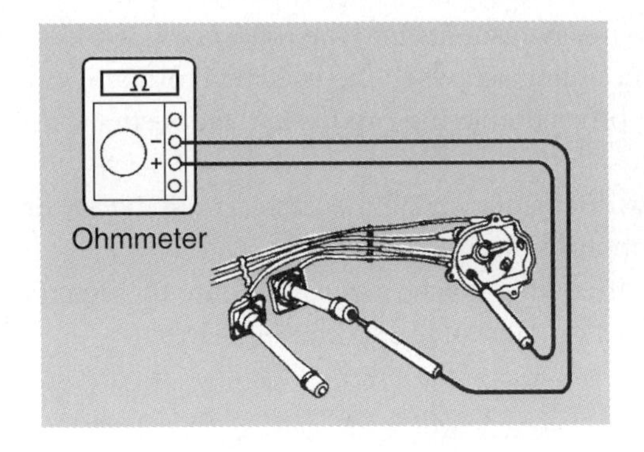

Figure 9-9 Ohmmeter connected to the spark plug wire and the distributor cap terminal to test the plug wire (Courtesy of Toyota Motor Corporation)

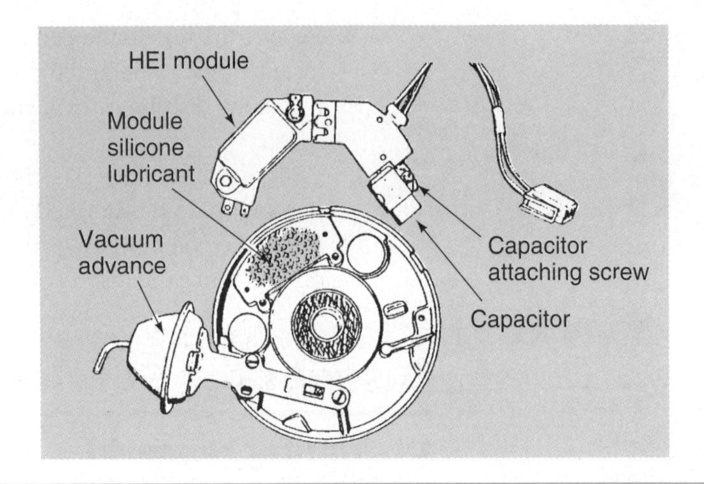

Figure 9-10 Place a light coating of silicone grease on the module mounting surface prior to module installation. (Courtesy of Oldsmobile Division, General Motors Corporation)

wire. When the spark plug wire resistance is satisfactory, check the cap terminal for corrosion. Repeat the ohmmeter tests on each spark plug wire and the coil secondary wire.

Spark Plug Wire Installation

When the spark plug wires are installed, be sure they are routed properly as indicated in the vehicle manufacturer's service manual. Two spark plug wires should not be placed side by side for a long span if these wires fire one after the other in the cylinder firing order. When two spark plug wires that fire one after the other are placed side by side for a long span, the magnetic field from the wire that is firing builds up and collapses across the other wire. This magnetic collapse may induce enough voltage to fire the other spark plug and wire when the piston in this cylinder is approaching TDC on the compression stroke. This action may cause detonation and reduced engine power.

Improperly routed spark plug wires may cause crossfiring and detonation.

Ignition Module Removal and Replacement

The ignition module removal and replacement procedure varies depending on the ignition system. Always follow the procedure in the vehicle manufacturer's service manual. Follow these steps for module removal and replacement on an HEI distributor:

WARNING: Lack of silicone grease on the module mounting surface may cause module overheating and damage.

1. Remove the battery wire from the coil battery terminal, and remove the inner wiring connector on the primary coil terminals. Remove the spark plug wires from the cap.
2. Rotate the distributor latches one-half turn and lift the cap from the distributor.
3. Remove the two rotor retaining bolts and the rotor.
4. Remove the primary leads and the pickup leads from the module.
5. Remove the two module mounting screws, and remove the module from the distributor housing.
6. Wipe the module mounting surface clean, and place a light coating of silicone heat-dissipating grease on the module mounting surface (Figure 9-10).
7. Install the module and tighten the module mounting screws to the specified torque.
8. Install the primary leads and pickup leads on the module.

9. Be sure the lug on the centrifugal advance mechanism fits into the rotor notch while installing the rotor, and tighten the rotor mounting screws to the specified torque.
10. Install the distributor cap, and be sure the projection in the cap fits in the housing notch.
11. Push down on the cap latches with a screwdriver, and rotate the latches until the lower part of the latch is hooked under the distributor housing.
12. Install the coil primary leads and battery wire on the coil terminals. Be sure the notch on the primary leads fits onto the cap projection. Install the spark plug wires.

Distributor Service

Distributor Removal

All distributor service procedures vary depending on the distributor. Always follow the recommended procedure in the vehicle manufacturer's service manual. Following is a typical distributor removal procedure:

1. Disconnect the distributor wiring connector and the vacuum advance hose.
2. Remove the distributor cap and note the position of the rotor. On some vehicles, it may be necessary to remove the spark plug wires from the cap prior to cap removal.
3. Remove the distributor holddown bolt and clamp.
4. Note the position of the vacuum advance, and pull the distributor from the engine.
5. Install a shop towel in the distributor opening to keep foreign material out of the engine block.

Distributor Bushing Check

The distributor bushing checking procedure varies depending on the vehicle manufacturer. Always follow the procedure in the manufacturer's service manual. Some manufacturers recommend clamping the distributor housing lightly in a soft-jaw vise, and clamping a dial indicator on the top of the distributor housing. The dial indicator stem is then positioned against the top of the distributor shaft. When the shaft is pushed horizontally, observe the shaft movement on the dial indicator. If this movement exceeds the manufacturer's specifications, the distributor bushings and/or shaft are worn. Some manufacturers now recommend complete distributor replacement rather than bushing replacement.

Distributor Disassembly

Follow these steps for a typical distributor disassembly procedure:

1. Mark the gear in relation to the distributor shaft so the gear may be installed in the original position.
2. Support the distributor housing on top of a vise and drive the roll pin from the gear and shaft with a pin punch and hammer (Figure 9-11).
3. Pull the gear from the distributor shaft and remove any spacers between the gear and the housing. Note the position of these spacers so they may be installed in their original position.
4. Wipe the lower end of the shaft with a shop towel and inspect this area of the shaft for metal burrs. Remove any burrs with fine emery paper.
5. Pull the distributor shaft from the housing.
6. Remove the pickup coil leads from the module and the pickup retaining clip. Lift the pickup coil from the top of the distributor bushing.
7. Remove the two vacuum advance mounting screws, and remove this advance assembly from the housing (Figure 9-12).

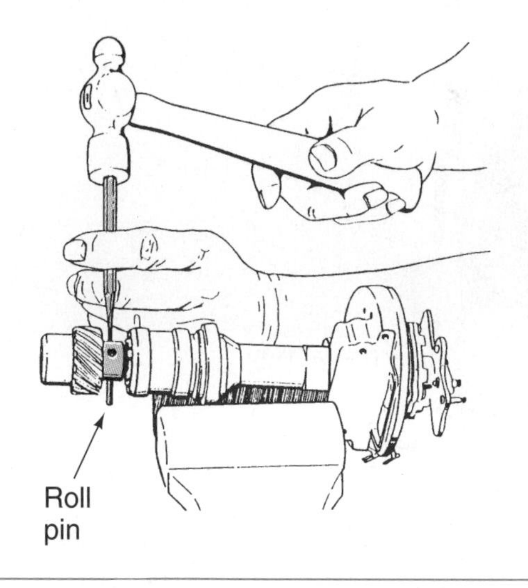

Figure 9-11 Driving the roll pin from the distributor gear (Courtesy of Oldsmobile Division, General Motors Corporation)

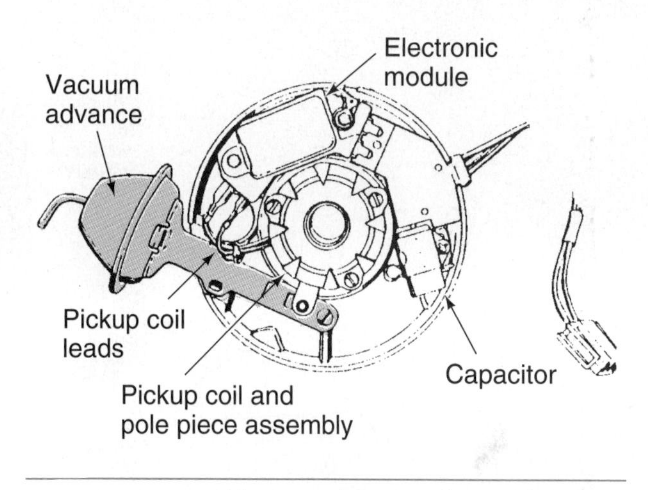

Figure 9-12 Vacuum advance and mounting screws (Courtesy of Oldsmobile Division, General Motors Corporation)

Distributor Inspection

WARNING: Distributor electrical components and the vacuum advance may be damaged by washing them in solvent.

The housing may be washed in solvent, but do not wash electrical components or the vacuum advance. Check these items during a typical distributor inspection:

1. Inspect all lead wires for worn insulation and loose terminals. Replace these wires as necessary.
2. Inspect the centrifugal advance mechanism for wear, particularly checking the weights for wear on the pivot holes. Replace the weights or the complete shaft assembly, if necessary.
3. Inspect the pickup plate for wear and rotation (Figure 9-13). If this plate is loose or seized, replacement is required.
4. Connect a vacuum hand pump to the vacuum advance outlet and apply 20 inches of vacuum. The advance diaphragm should hold this vacuum without leaking.

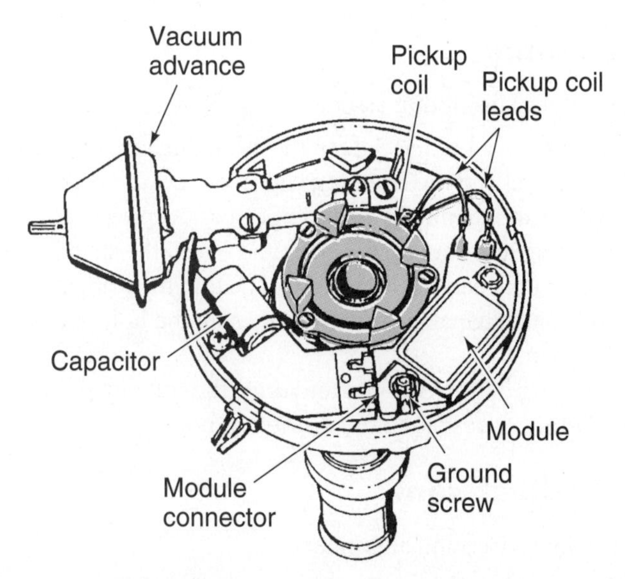

Figure 9-13 Checking pickup assembly for wear and rotation (Courtesy of Oldsmobile Division, General Motors Corporation)

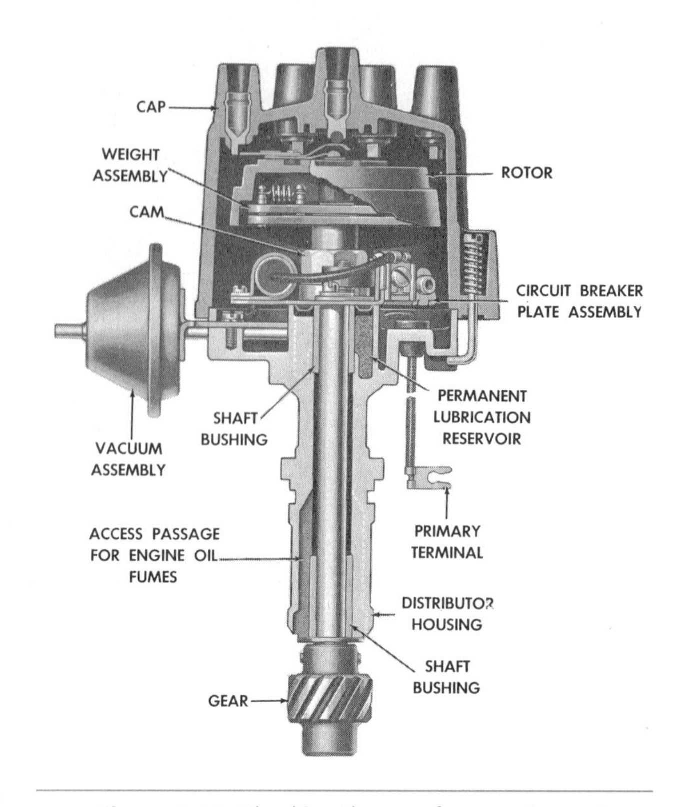

Figure 9-14 Checking the cam for wear in a point-type distributor (Courtesy of Buick Motor Division, General Motors Corporation)

5. Check the distributor gear for worn or chipped teeth.
6. Inspect the reluctor for damage. If the high points are damaged, the distributor bushing probably is worn, allowing the high points to hit the pickup coil. On a point-type distributor, replace the cam assembly if the lobes are worn (Figure 9-14).
7. On a point-type distributor, check the ignition points for burning or pitting. Points with either of these conditions must be replaced.

Distributor Assembly

Follow these steps for a typical distributor assembly procedure:

1. Install the vacuum advance and tighten the mounting screws to the specified torque.
2. Install the pickup coil and the retaining clip. Connect the pickup leads to the module.
3. Install the module and mounting screws as discussed previously.
4. Place some bushing lubricant on the shaft and install the shaft in the distributor.
5. Install the spacers between the housing and gear in their original position.
6. Install the gear in its original position, and be sure the hole in the gear is aligned with the hole in the shaft.
7. Support the housing on top of a vise and drive the roll pin into the gear and shaft.
8. Install a new O-ring or gasket on the distributor housing.

Ignition Point Adjustment

Follow these steps for point service and adjustment:

1. Be sure the breaker plate surface is clean.
2. Install the points on the breaker plate, and be sure the points are flush with the plate.
3. Install the point retaining screws, and leave these screws slightly loose.

4. Install the lead wire and condenser lead on the point terminal, and tighten the nut on this terminal.
5. Clamp the distributor housing in a soft-jaw vise, and rotate the shaft until one of the cam lobe high points is exactly under the point rubbing block.
6. Install a feeler gauge of the specified thickness in the point gap. The feeler gauge should be a light push fit in the gap.
7. If necessary, move the stationary point on the breaker plate until the feeler gauge fits lightly in the point gap (Figure 9-15). Tighten the point retaining screws, and recheck the point gap.
8. Place a small amount of distributor cam lubricant on the cam and rubbing block.

Installing and Timing the Distributor

This procedure may be followed to install the distributor and time it to the engine:

1. Remove the spark plug from the number one cylinder, and place a compression gauge hose fitting in the spark plug hole.
2. Crank the engine a small amount at a time until compression pressure appears on the gauge.
3. Crank the engine a very small amount at a time until the zero degree position on the timing marks is aligned with the timing indicator.
4. Locate the number one spark plug wire position in the distributor cap. The wire terminals in some caps are marked, and the manufacturer's service manual provides this information.
5. Install the distributor in the block with the rotor positioned under the number one spark wire position in the distributor cap and the vacuum advance in the original position (Figure 9-16). The distributor gear goes into mesh easily with the camshaft gear, but many distributors also drive the oil pump with a hex-shaped drive in the lower end of the distributor gear or shaft. It may be necessary to hold down on the distributor housing and crank the engine to get the distributor shaft into mesh with the oil pump drive. When

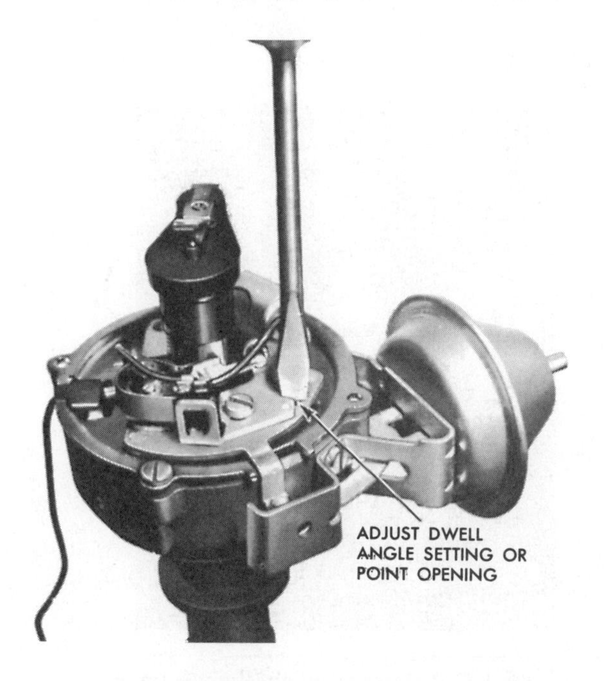

Figure 9-15 Moving stationary point to adjust point gap (Courtesy of Chevrolet Motor Division, General Motors Corporation)

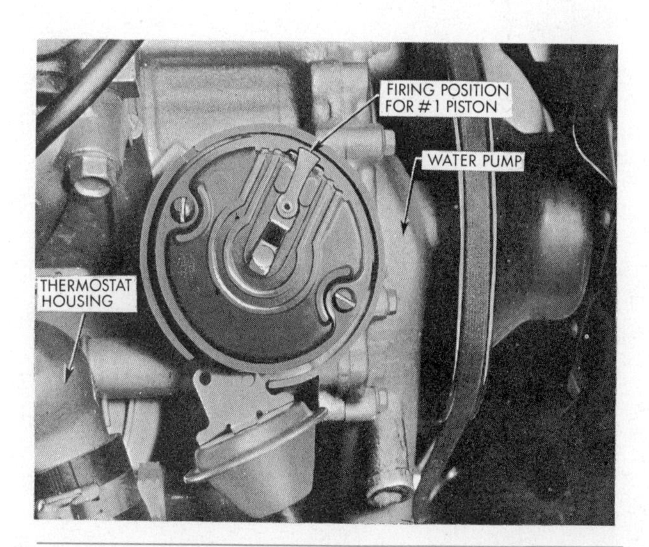

Figure 9-16 Installing the distributor with the rotor under the number one spark plug terminal in the distributor cap (Courtesy of Buick Motor Division, General Motors Corporation)

this action is required, repeat steps 2 and 3 and be sure the rotor is under the number one spark plug wire terminal in the distributor cap with the timing marks aligned.

6. Rotate the distributor a small amount so the timer core teeth and pickup teeth are aligned. With a point-type distributor, rotate the distributor until the points are just beginning to open.
7. Install the distributor holddown clamp and bolt, and leave the bolt slightly loose.

SERVICE TIP: The distributor shaft rotates in the opposite direction to which the vacuum advance pulls the pickup plate.

8. Install the spark plug wires in the direction of distributor shaft rotation and in the cylinder firing order (Figure 9-17).
9. Connect the distributor wiring connectors. The vacuum advance hose is usually left disconnected until the timing is set with the engine running.

Photo Sequence 6 shows a typical procedure for timing the distributor to the engine.

Timing Checking and Adjustment

The ignition timing procedure varies depending on the make and year of the vehicle and the type of ignition system. Ignition timing specifications and instructions are included on the underhood emission label, and more detailed instructions are provided in the vehicle manufacturer's service manual. The ignition timing procedure and specifications recommended by the vehicle manufacturer must be followed. On distributors with advance mechanisms, manufacturers usually recommend disconnecting and plugging the vacuum advance hose while checking ignition timing. On carbureted engines, the manufacturer usually specifies a certain engine rpm while checking the ignition timing. The timing light pickup is connected to the number one spark plug wire, and the power supply wires on the light are connected to the battery terminals with the proper polarity. Follow these steps for ignition timing adjustment:

Special Tools

Advance-type timing light

1. Connect the timing light, and start the engine.
2. The engine must be idling at the manufacturer's recommended rpm, and all other timing procedures must be followed.
3. Aim the timing light marks at the timing indicator, and observe the timing marks (Figure 9-18).

Figure 9-17 The spark plug wires are installed in the direction of distributor shaft rotation and in the cylinder firing order. (Courtesy of Buick Motor Division, General Motors Corporation)

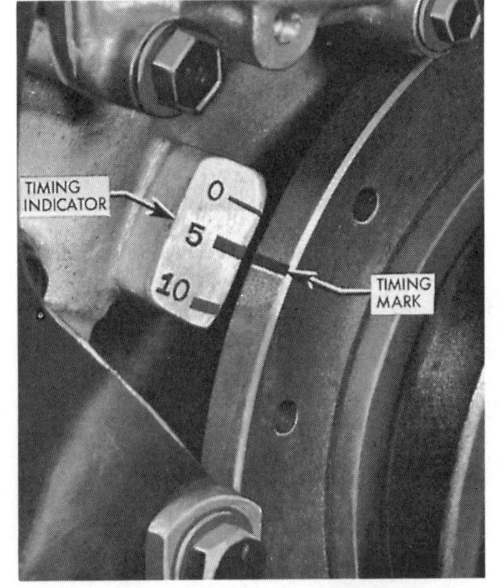

Figure 9-18 When the timing light flashes, the timing mark on the crankshaft pulley must appear at the specified location on the timing indicator above the pulley. (Courtesy of Buick Motor Division, General Motors Corporation)

Photo Sequence 6
Typical Procedure for Timing the Distributor to the Engine

P6-1 Remove the number one spark plug.

P6-2 Place your thumb over the number one spark plug opening and crank the engine until compression is felt.

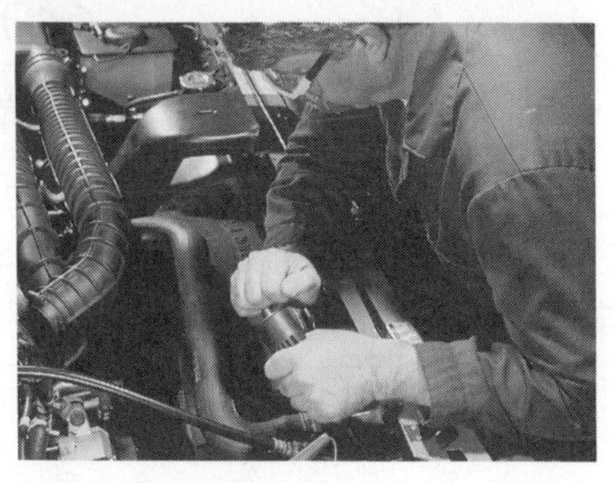

P6-3 Crank the engine a very small amount at a time until the timing marks indicate that the number one piston is at TDC on the compression stroke.

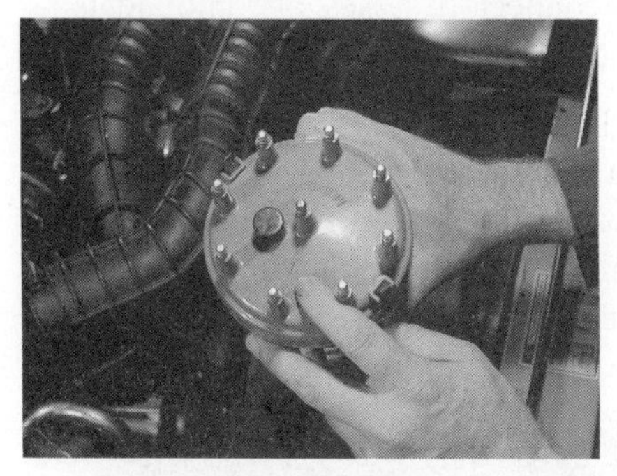

P6-4 Determine the number one spark plug wire position in the distributor cap.

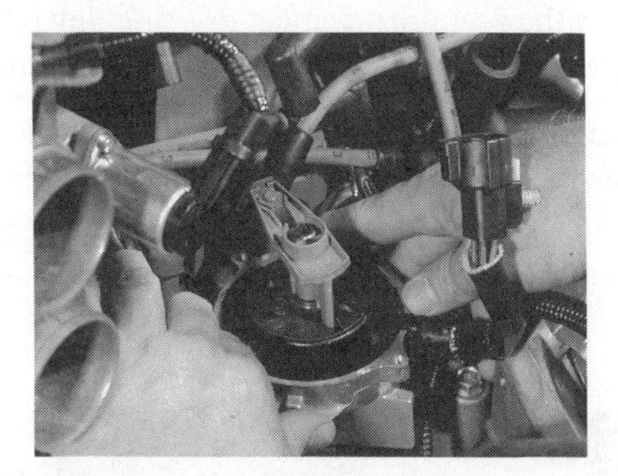

P6-5 Install the distributor with the rotor under the number one spark plug wire terminal in the distributor cap and one of the reluctor high points aligned with the pickup coil.

P6-6 After the distributor is installed in the block, turn the distributor housing slightly so the pickup coil is aligned with the reluctor.

P6-7 Install the distributor clamp bolt, but leave it slightly loose.

P6-8 Connect the pickup leads to the wiring harness.

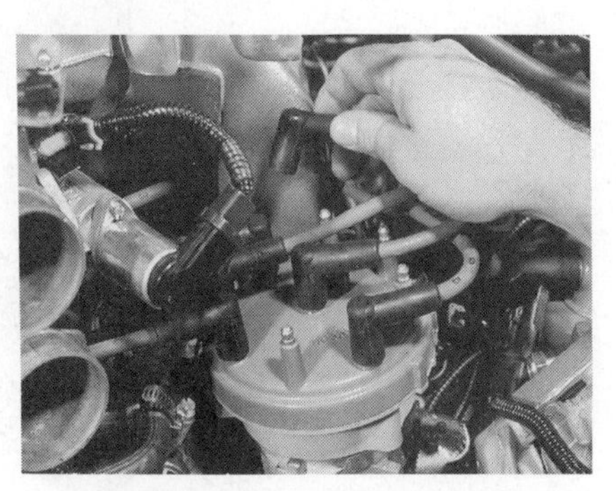

P6-9 Install the spark plug wires in the cylinder firing order and in the direction of distributor shaft rotation.

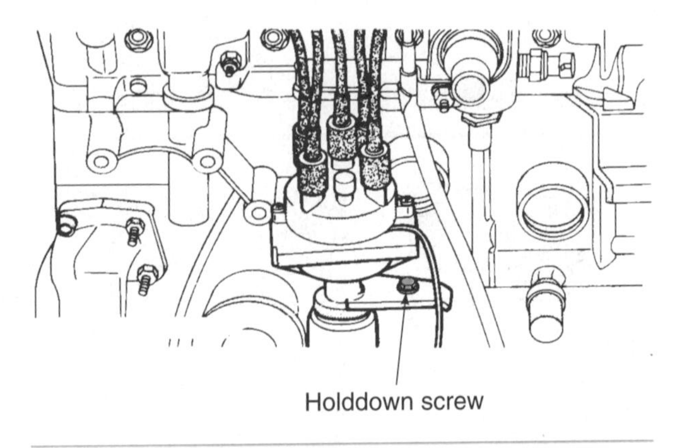

Figure 9-19 Tighten the distributor holddown bolt after the timing is adjusted to specifications. (Courtesy of Chrysler Corporation)

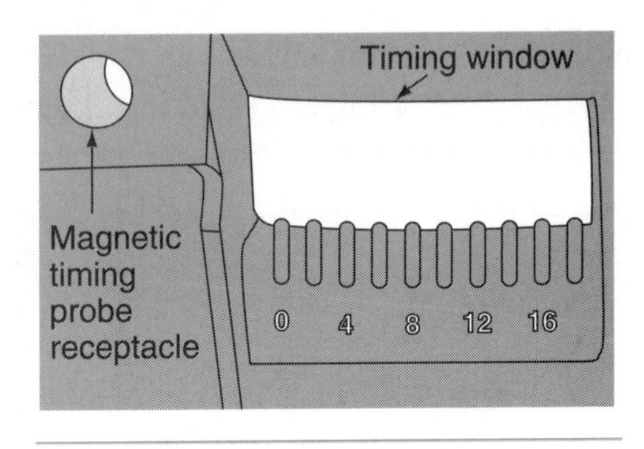

Figure 9-20 Magnetic timing probe receptacle near the timing indicator (Courtesy of Chrysler Corporation)

4. If the timing mark is not at the specified location, rotate the distributor until the mark is at the specified location.
5. Tighten the distributor holddown bolt to the specified torque, and recheck the timing mark position (Figure 9-19).
6. Connect the vacuum advance hose and any other connectors, hoses, or components that were disconnected for the timing procedure.

Special Tools

Magnetic timing meter

The magnetic timing probe receptacle is a small hole near the timing marks in the timing indicator.

Magnetic Timing Procedure. Many later model vehicles have a magnetic timing probe receptacle near the timing indicator (Figure 9-20). Some equipment manufacturers supply a magnetic timing meter with a pickup that fits in the probe hole. The meter pickup must be connected to the number one spark plug wire, and the power supply leads must be connected to the battery terminals (Figure 9-21). Many timing meters have two scales: timing degrees and engine rpm.

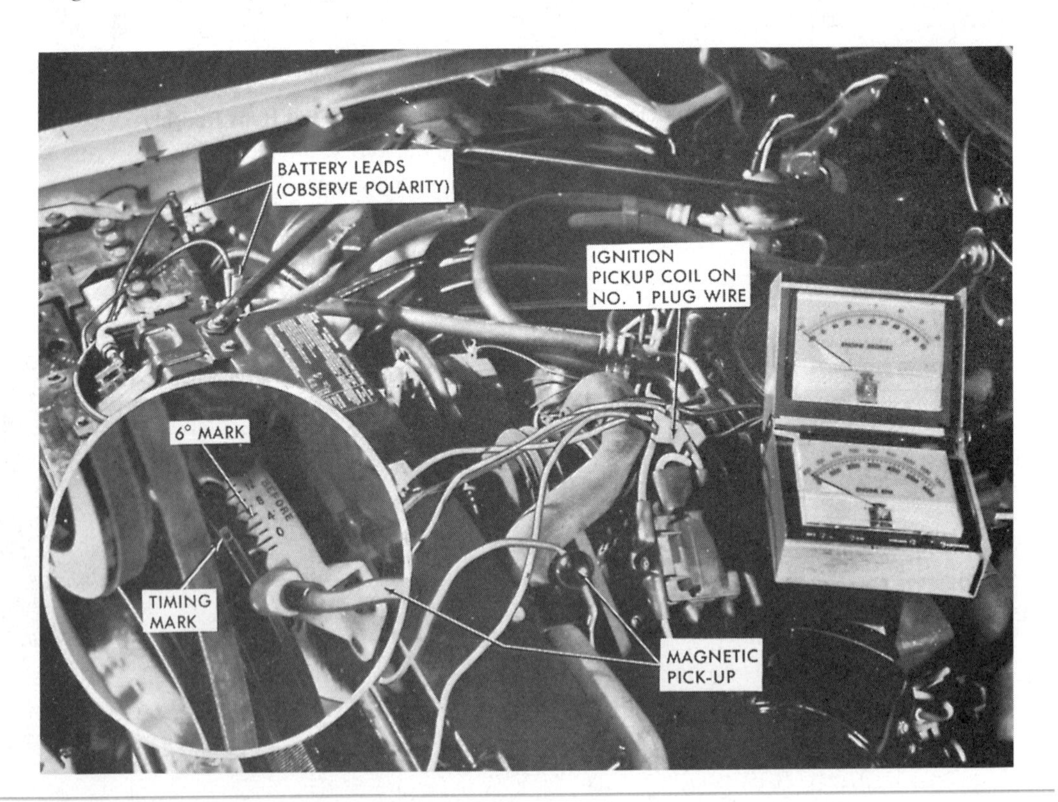

Figure 9-21 Magnetic timing meter and related connections (Courtesy of Cadillac Motor Car Division, General Motors Corporation)

A magnetic timing offset knob on the meter must be adjusted to the vehicle manufacturer's specifications to compensate for the position of the probe receptacle. Once the offset is adjusted, the engine may be started and the timing is indicated on the meter scale. The timing adjustment procedure is the same with the timing meter or light.

Timing Advance Check. Many timing lights have the capability to check the spark advance. An advance control on the light slows down the flashes of the light as the knob is rotated. When the light flashes are slowed with the engine running at higher speed, the timing marks move back to the basic timing setting. Follow these steps for a typical spark advance check:

1. Check the basic timing.
2. Obtain the vehicle manufacturer's spark advance specifications at a specific rpm.
3. Operate the engine at this specific rpm, and rotate the timing light advance control until the timing mark moves back to the basic timing setting (Figure 9-22).
4. Observe the spark advance on the timing light meter, and compare this reading to the specifications.
5. If the spark advance is not equal to the specifications, repeat steps 3 and 4 with the vacuum advance hose disconnected to determine whether the vacuum or centrifugal advance is at fault.

Point Dwell Measurement and Condenser Testing

Ignition point dwell is the number of degrees that the distributor shaft rotates while the points are closed on each cam lobe. In a four-cylinder distribution there are 90° between the cam lobe peaks. In this distributor, the dwell may be 50° while the points are closed on each cam lobe, and the points remain open for 40° on each lobe (Figure 9-23).

Ignition point dwell is the number of degrees that the distributor shaft rotates while the points are closed on each cam lobe.

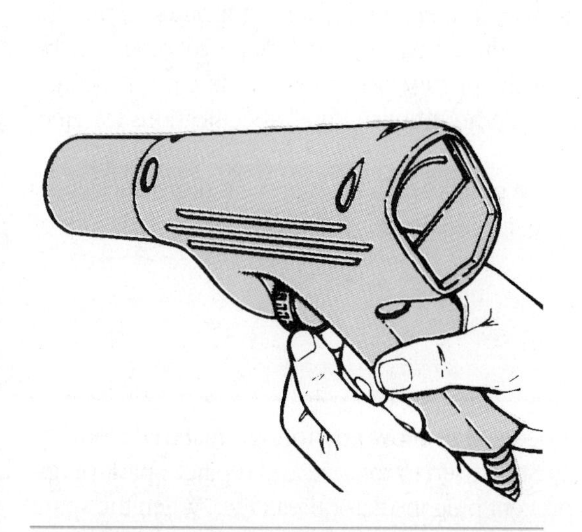

Figure 9-22 Timing light with advance control and spark advance meter (Courtesy of Chrysler Corporation)

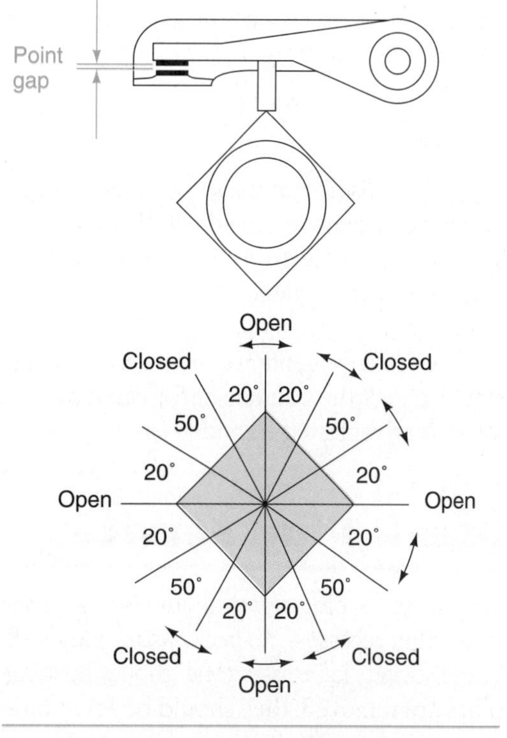

Figure 9-23 Ignition point dwell on a four-cylinder distributor cam

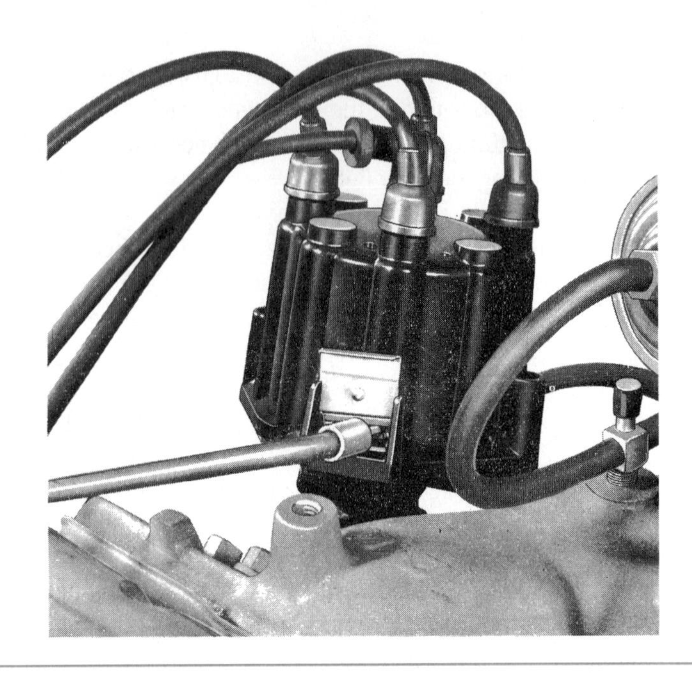

Figure 9-24 Distributor cap door for dwell adjustment (Courtesy of Pontiac Motor Division, General Motors Corporation)

Special Tools

Dwell meter

A dwell meter may be used to measure the point dwell with the engine idling. The dwell meter leads are connected from the negative primary coil terminal to ground with the correct polarity. Always ground the lead with the black clip. Many dwell meters require a calibration procedure, and the meter control must be set for the number of cylinders in the engine being tested. Some vehicle manufacturers recommended that the vacuum hose should be disconnected while checking the dwell. With the engine idling, observe the dwell reading on the meter scale.

If the dwell is not within manufacturer's specifications, the points must be adjusted. Some distributors have a metal door in the side of the distributor cap. When this door is lifted, the point adjustment screw may be turned with an Allen wrench until the proper dwell reading is obtained on the meter scale (Figure 9-24).

SERVICE TIP: Decreasing the ignition point gap increases the dwell reading.

If the distributor does not have an adjustment door, the engine must be stopped and the distributor cap removed to adjust the points and correct the dwell. A dwell meter reading may be obtained on an electronic ignition system. Since dwell is a function of the module on these systems, it is not adjustable. Most vehicle manufacturers do not publish dwell specifications for electronic ignition systems.

A defective ignition condenser may cause burned points and a no-start condition. Condenser testers check the condenser for capacity, resistance, and leakage. If the condenser fails any of these tests, replace the condenser.

Spark Plug Service

Prior to spark plug removal, an air nozzle should be used to blow any foreign material from the spark plug recesses. A special spark plug socket must be used to remove and replace spark plugs. These sockets have an internal rubber bushing to prevent plug insulator breakage. When the spark plugs are removed, they should be set in order so the technician can identify the spark plug from each cylinder. Spark plug carbon conditions are an excellent indicator of cylinder conditions. If the spark plugs all have light brown or gray carbon deposits, the cylinders have been operating normally with a proper air-fuel ratio (Figure 9-25). Abnormal spark plug conditions and their causes are shown in the figure.

Spark plugs may be cleaned in a sand blast-type cleaner (Figure 9-26). Always consult the vehicle manufacturer's recommendations regarding spark plug cleaning. Spark plug electrodes may be filed with a small, fine-toothed file to restore the electrode surfaces. The spark plug gap should be checked with a round feeler gauge and adjusted if necessary (Figure 9-27). Bend the ground electrode to adjust the spark plug gap.

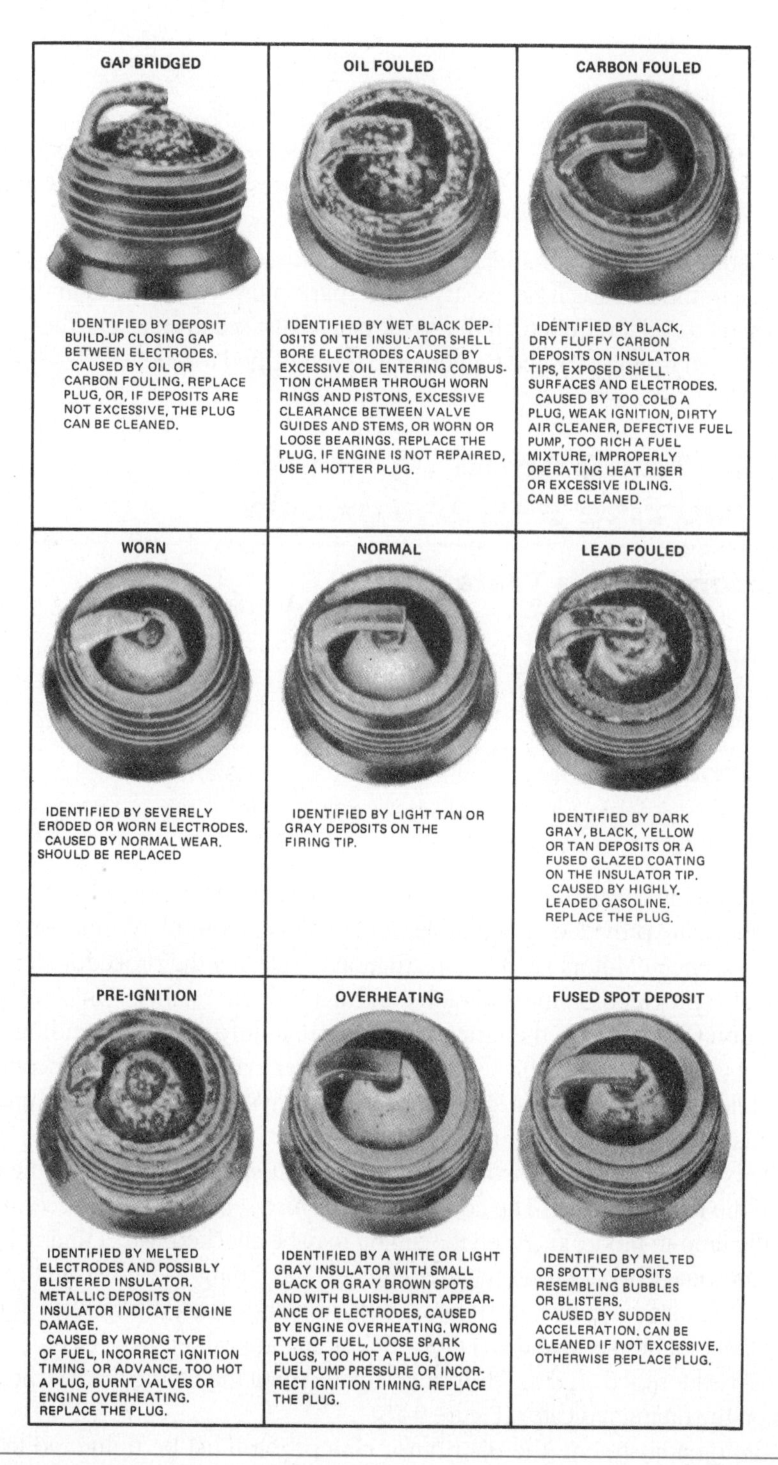

Figure 9-25 Normal and abnormal spark plug conditions (Reprinted with the permission of Ford Motor Company)

Figure 9-26 Spark plug cleaner (Courtesy of Toyota Motor Corporation)

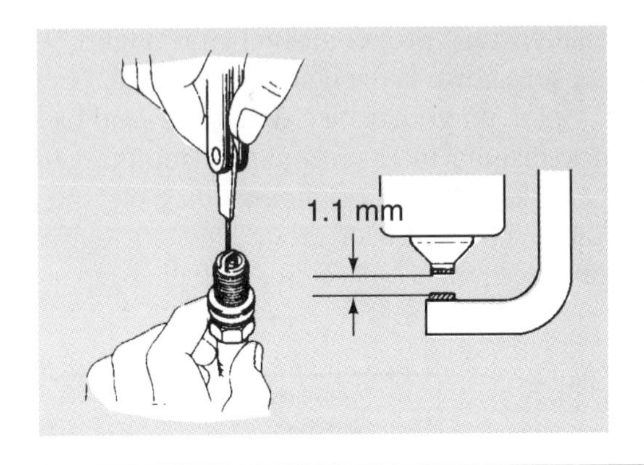

Figure 9-27 Measuring spark plug gap (Courtesy of Toyota Motor Corporation)

Prior to spark plug installation, the spark plug threads and tapered seat must be clean. Use a wire brush to clean the threads if necessary. The spark plug seat in the cylinder head must be cleaned with a shop towel if required. If the spark plugs have sealing gaskets, these gaskets should be replaced. The spark plugs must be tightened to the specified torque.

Computer-Controlled Ignition System Service and Diagnosis

Basic Ignition Timing Tests

CAUTION: To avoid timing light damage and personal injury, keep the timing light and the timing light lead wires away from rotating parts such as fan blades and belts while adjusting basic timing.

CAUTION: To avoid personal injury, keep hands and clothing away from rotating parts such as fan blades and belts while adjusting basic timing. Remember that electric-drive cooling fans may start at any time.

The Society of Automotive Engineers (SAE) J1930 terminology is an attempt to standardize terminology in automotive electronics.

In the SAE J1930 terminology, the term distributor ignition (DI) replaces all previous terms for distributor-type ignition systems that are electronically controlled.

In the SAE J1930 terminology, the term electronic ignition (EI) replaces all previous terms for distributorless ignition systems.

Prior to any basic timing check, the engine must be at normal operating temperature. The ignition timing specifications are provided on the underhood emission control information label. On some vehicles, such as General Motors vehicles, instructions regarding the procedure for checking ignition timing are also provided on the emission control information label. On computer-controlled distributor ignition (DI) systems, the ignition timing procedure varies depending on the vehicle make and model year. On some vehicles, the manufacturer may recommend disconnecting certain components while checking the basic ignition timing. Always follow the ignition timing procedure in the vehicle manufacturer's service manual.

When the basic timing is checked on most Chrysler fuel injected engines, the computer system must be in the limp-in mode. The coolant temperature sensor may be disconnected to place the system in the limp-in mode, and then the timing may be checked with a timing light in the normal manner. On some Chrysler four-cylinder engines, the timing window is in the top of the flywheel housing. On many engines, the timing mark, or marks, is on the crankshaft pulley and the timing indicator is mounted above the pulley. If a timing adjustment is necessary, the distributor clamp bolt must be loosened, and the distributor rotated until the timing mark appears at the specified position on the timing indicator (Figure 9-28).

After a timing adjustment, the distributor clamp bolt must be tightened to the specified torque. Removal of the coolant temperature sensor wires places a fault code in the computer memory. This code should be erased following the timing adjustment.

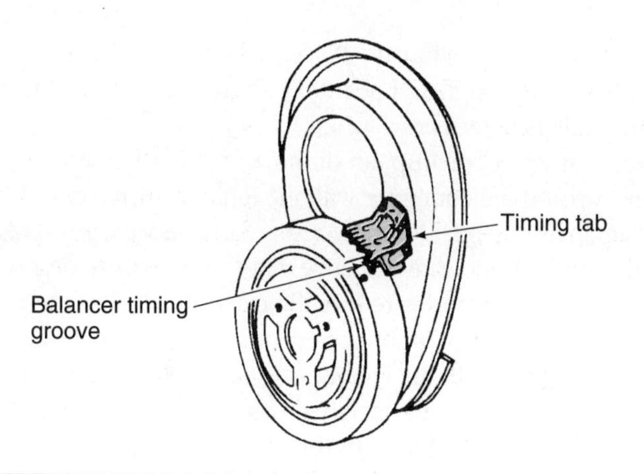

Figure 9-28 Timing mark and timing indicator. (Courtesy of Chevrolet Motor Division, General Motors Corporation)

WARNING: When disconnecting or reconnecting wires on a computer system, always be sure the ignition switch is off. Disconnecting computer system component wires with the ignition switch on may damage system components.

On many DI systems, such as those manufactured by Ford and General Motors, a timing connector located in the engine compartment must be disconnected. The emission control information label usually provides the location of the timing connector on General Motors vehicles. When the timing connector is disconnected, the PCM cannot affect spark advance, and the pickup signal goes directly to the module. The distributor clamp bolt must be loosened and the distributor rotated to adjust the basic timing. After the timing adjustment is completed, the clamp bolt must be tightened and the timing connector reconnected.

A timing connector is a single wire connector sometimes located near the distributor that must be disconnected while checking or adjusting basic ignition timing.

No-Start Ignition Tests

The same no-start tests may be performed on conventional DI systems with centrifugal and vacuum advances and DI systems with computer-controlled spark advance. These tests were explained previously.

Connect a 12-V test light from the coil tachometer (TACH) terminal to ground and crank the engine. If the 12-V test lamp does not flutter while the engine is cranked, the pickup or ignition module is probably defective. Under this condition, always check the voltage supply to the positive primary coil terminal with the ignition switch on before the diagnosis is continued.

The ignition coil tachometer (TACH) terminal is the negative primary coil terminal.

On most Chrysler fuel injected engines, the voltage is supplied through the automatic shutdown (ASD) relay to the coil positive primary terminal and the electric fuel pump. Therefore, a defective ASD relay may cause 0 V at the positive primary coil terminal. This relay is controlled by the PCM. On some Chrysler products, the relay closes when the ignition switch is turned on, whereas on other models it only closes while the engine is cranking or running. If the ASD relay closes with the ignition switch on and the engine not cranking or running, it only remains closed for about one second. This action shuts off the fuel pump and prevents any spark from the ignition system, if the vehicle is involved in a collision with the ignition switch on and the engine stalled. A fault code should be present in the computer memory if the ASD relay is defective.

The automatic shutdown relay supplies voltage to the electric fuel pump, positive primary coil terminal, injectors, and oxygen sensor heater on some Chrysler vehicles.

Pickup Tests

WARNING: Never short across or ground terminals or wires in a computer system unless instructed to do so on the vehicle manufacturer's service manual.

If a magnetic-type pickup is used, the pickup may be checked for open circuits, shorts, and grounds with an ohmmeter. These tests are performed in the same way as the pickup tests on conventional distributors described previously. If the pickup coil tests are satisfactory, and the 12-V test light connected from the coil TACH terminal to ground did not flutter while cranking the engine, the ignition module is defective.

A Hall effect switch contains a Hall element and a permanent magnet. A blade representing each engine cylinder rotates between these components.

Prior to testing a Hall effect pickup, an ohmmeter should be connected across each of the wires between the pickup and the computer with the ignition switch off. A computer terminal and pickup coil wiring diagram is essential for these tests. Satisfactory wires have nearly 0 Ω resistance, while higher or infinite readings indicate defective wires. If the distributor has a Hall effect pickup, the voltage supply wire and the ground wire should be checked before the pickup signal. In the following tests, the distributor connector is connected and this connector may be backprobed to complete the necessary connections. With the ignition switch on, a voltmeter should be connected from the voltage input wire to ground, and the specified voltage must appear on the meter.

The ground wire should be tested with the ignition switch on and a voltmeter connected from the ground wire to a ground connection near the distributor. With this meter connection, the meter indicates the voltage drop across the ground wire, which should not exceed 0.2 V, if the wire has a normal resistance.

Special Tools

Digital volt/ohm-meter

Connect a digital voltmeter from the pickup signal wire to ground. If the voltmeter reading does not fluctuate while cranking the engine, the pickup is defective, whereas a voltmeter reading that fluctuates from nearly 0 V to between 9 V and 12 V indicates a satisfactory pickup. During this test, the voltmeter reading may not be accurate, because of the short duration of the voltage signal. If the Hall effect pickup signal is satisfactory, and the 12-V test lamp did not flutter during the no-start test, the ignition module is probably defective. The ignition module is contained in the PCM on Chrysler products. On Chrysler fuel injected engines, the reference pickup and the SYNC pickup should be tested. If either of these pickups is defective, a fault code may be stored in the computer memory.

The reference pickup is a Hall effect switch located in the distributor on some Chrysler products. This pickup is used for ignition triggering.

The synchronizer pickup is a Hall effect switch used for injector sequencing in some Chrysler distributors.

An optical-type pickup has a slotted plate that rotates between a light emitting diode (LED) and a photo diode.

On Chrysler optical distributors, the pickup voltage supply and the ground wires may be tested at the four-wire connector near the distributor. With the ignition switch on, a voltmeter connected from the orange voltage supply wire to ground should indicate 9.2 V to 9.4 V (Figure 9-29). This voltage reading may vary depending on the model year. Always use the vehicle manufacturer's specifications.

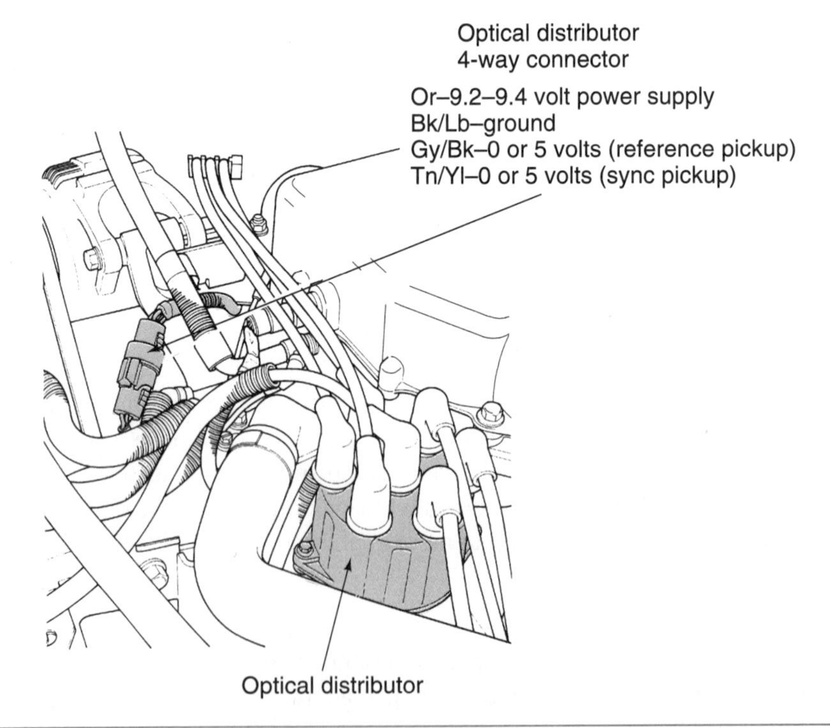

Figure 9-29 Chrysler optical distributor four-wire connector (Courtesy of Chrysler Corporation)

A voltmeter should indicate less than 0.2 V when connected from the black/light blue ground wire to an engine ground connection, if the ground wire is satisfactory. When a digital voltmeter is connected from the gray/black reference pickup wire or the tan/yellow SYNC pickup wire to an engine ground connection, the voltmeter reading should cycle from nearly 0 V to approximately 5 V while the engine is cranking. This is a typical voltage figure. Always use the vehicle manufacturer's specifications for the model year being diagnosed. If the pickup signal is not within specifications, the pickup is defective. A defective SYNC pickup in this distributor should not cause a no-start problem.

Classroom Manual
Chapter 9, page 202

Electronic Ignition (EI) System Diagnosis and Service

No-Start Ignition Diagnosis, Cam and Crank Sensors

The diagnostic procedure for EI systems varies depending on the vehicle make and model year. Always follow the procedure recommended in the vehicle manufacturer's service manual. The following procedure is based on Chrysler EI systems. The crankshaft timing sensor and camshaft reference sensor in these systems are modified Hall effect switches.

When the engine fails to start, follow these steps:

1. Check for fault codes 11 and 43. Code 11, "Ignition Reference Signal," could be caused by a defective camshaft reference signal or crankshaft timing sensor signal. Code 43 is caused by low primary current in coil number 1, 2, or 3.
2. With the engine cranking, check voltage from the orange wire to ground on the crankshaft timing sensor and the camshaft reference sensor (Figure 9-30). Over 7 V is satisfactory. If the voltage is less than specified, repeat the test with the voltmeter connected from PCM terminal 7 to ground. If the voltage is satisfactory at terminal 7, but low at the sensor orange wire, repair the open circuit or high resistance in the orange wire. If the voltage is low at terminal 7, replace the PCM. Be sure 12 V is supplied to PCM terminal 3 with the ignition switch off or on, and 12 V must be supplied to PCM terminal 9 with the ignition switch on. Check the PCM ground connections on terminals 11 and 12 before replacing the PCM.
3. With the ignition switch on, check the voltage drop across the ground circuit (black/light blue wire) on the crankshaft timing sensor and the camshaft reference sensor. A reading below 0.2 V is satisfactory.

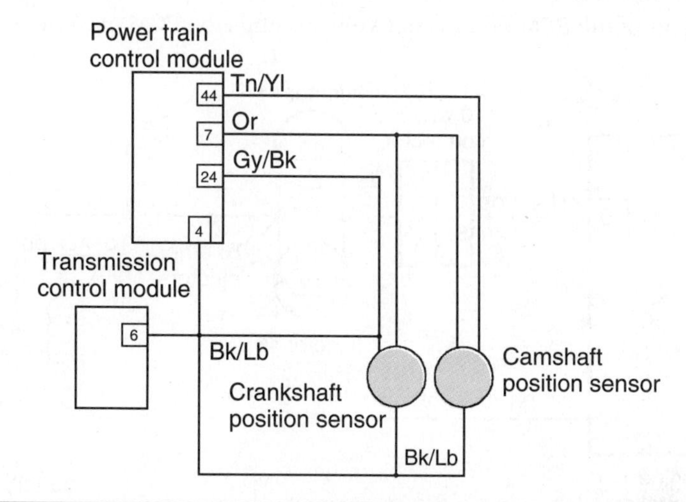

Figure 9-30 Crankshaft timing and camshaft reference sensor terminals (Courtesy of Chrysler Corporation)

SERVICE TIP: When using a digital voltmeter to check a crankshaft or camshaft sensor signal, crank the engine a very small amount at a time and observe the voltmeter. The voltmeter reading should cycle from almost 0 V to a higher voltage of 9 V to12 V. Since digital voltmeters do not react instantly, it is difficult to see the change in voltmeter reading if the engine is cranked continually.

4. If the readings in steps 3 and 4 are satisfactory, connect a 12-V test lamp or a digital voltmeter from the gray/black wire on the crankshaft timing sensor and the tan/yellow wire on the camshaft reference sensor to ground. When the engine is cranking, a flashing 12-V lamp indicates that a sensor signal is present. If the lamp does not flash, sensor replacement is required. Each sensor voltage signal should cycle from low voltage to high voltage as the engine is cranked.

No-Start Ignition Diagnosis Coil and PCM Tests

If the sensor tests are satisfactory, proceed with these coil and PCM tests:

1. Check the spark plug wires with an ohmmeter as explained previously.
2. With the engine cranking, connect a voltmeter from the dark green/black wire on the coil to ground. If this reading is below 12 V, check the automatic shutdown (ASD) relay circuit (Figure 9-31).

SERVICE TIP: Later model Chrysler vehicles have separate fuel pump and ASD relays. Always use the proper wiring diagram for the vehicle being tested.

3. If the reading in step 2 is satisfactory, check the primary and secondary resistance in each coil with the ignition switch off. Primary resistance is 0.52 Ω to 0.62 Ω, and secondary resistance is 11,000 Ω to 15,000 Ω. If these ohm readings are not within specifications, replace the coil assembly.
4. With the ignition switch off, connect an ohmmeter across the three wires from the coil connector to PCM terminals 17, 18, and 19 (Figure 9-32). These terminals are connected from the coil primary terminals to the PCM. If an infinite ohmmeter reading is obtained on any of the wires, repair the open circuits.
5. Connect a 12-V test lamp from the dark blue/black wire, dark blue/gray wire, and black/gray wire on the coil assembly to ground while cranking the engine. If the test lamp does not flutter on any of the three wires, replace the PCM. Since the crankshaft and camshaft sensors, wires from the coils to the PCM, and voltage supply to the coils, have been tested already, the ignition module must be defective. This module is an integral part of the PCM on Chrysler vehicles; thus the PCM must be replaced.

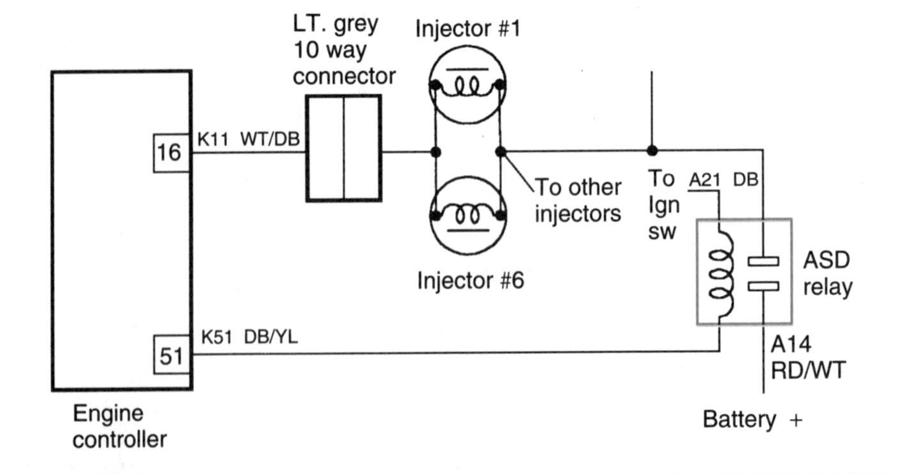

Figure 9-31 Automatic shutdown (ASD) relay circuit (Courtesy of Chrysler Corporation)

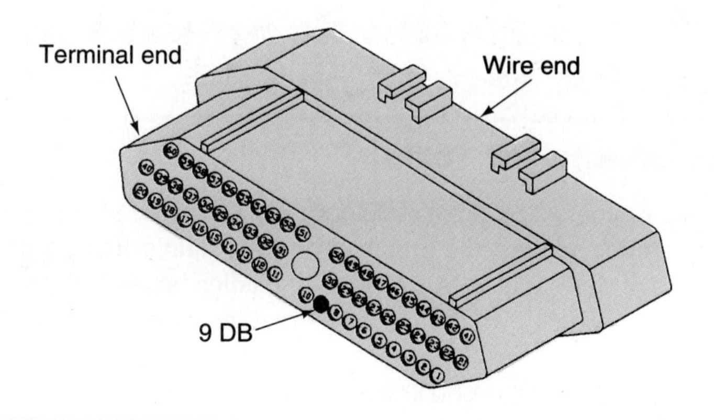

Figure 9-32 PCM terminal identification (Courtesy of Chrysler Corporation)

WARNING: Do not crank or run an EI-equipped engine with a spark plug wire completely removed from a spark plug. This action may cause leakage defects in the coils or spark plug wires.

CAUTION: Since EI systems have more energy in the secondary circuit, electrical shocks from these systems should be avoided. The electrical shock may not injure the human body, but such a shock may cause you to jump and hit your head on the hood or push your hand into contact with a rotating cooling fan.

6. If the tests in steps in 1 to 4 are satisfactory, connect a test spark plug to each spark plug wire and ground and crank the engine. If any of the coils do not fire on the two spark plugs connected to the coil, replace the coil assembly.

Sensor Replacement

If the crankshaft timing sensor or the camshaft reference sensor is removed, follow this procedure when the sensor is replaced:

1. Thoroughly clean the sensor tip and install a new spacer (part number 5252229) on the sensor tip. New sensors should be supplied with the spacer installed (Figure 9-33).
2. Install the sensor until the spacer lightly touches the sensor ring and tighten the sensor mounting bolt to 105 inch pounds.

WARNING: Improper sensor installation may cause sensor, rotating drive plate, or timing gear damage.

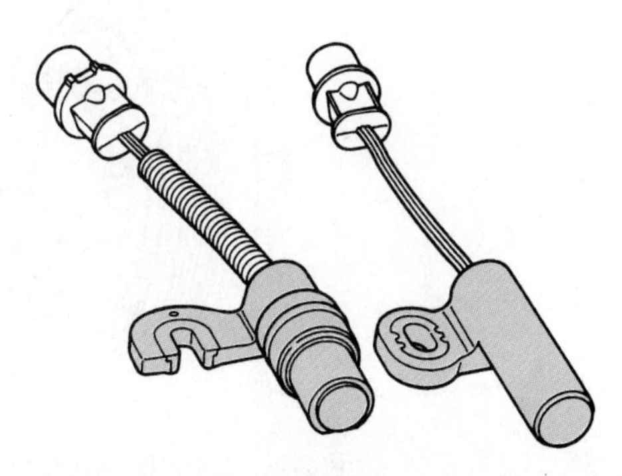

Figure 9-33 Spacer on crankshaft timing sensor and camshaft reference sensor tips (Courtesy of Chrysler Corporation)

General Motors Electronic Ignition (EI) System Service and Diagnosis

Coil Winding Ohmmeter Tests

With the coil terminals disconnected, an ohmmeter calibrated on the X1 scale should be connected to the primary coil terminals to test the primary winding. The primary winding in any EI coil should have 0.35 Ω to 1.50 Ω resistance. An ohmmeter reading below the specified resistance indicates a shorted primary winding. An infinite meter reading proves that the primary winding is open.

An ohmmeter calibrated on the X1,000 scale should be connected to each pair of secondary coil terminals to test the secondary windings. The coil secondary winding in an EI type 1 system should have 10,000 Ω to 14,000 Ω resistance, whereas secondary windings in EI type 2 systems have 5,000 Ω to 7,000 Ω resistance. If the secondary winding is open, the ohmmeter reading is infinite. A shorted secondary winding provides an ohmmeter reading below the specified resistance.

Crankshaft Sensor Adjustment, 3.0 L, 3,300, 3.8 L, and 3,800 Engines

A basic timing adjustment is not possible on any EI system. However, if the gap between the blades and the crankshaft sensor is not correct, the engine may fail to start, stall, misfire, or hesitate on acceleration. Follow these steps during the crankshaft sensor adjustment procedure:

1. Install the sensor loosely on the pedestal.
2. Position the sensor and pedestal on the J37089 adjusting tool.
3. Position the adjusting tool on the crankshaft surface (Figure 9-34).
4. Tighten the pedestal-to-block mounting bolts to 30–35 ft. lb., 20 to 40 Newton meters (N·m).
5. Tighten the pinch bolt to 30–35 in. lb. (3 to 4 N·m).

Special Tools

Crankshaft sensor adjusting tool, dual slot sensor

The interrupter rings on the back of the crankshaft pulley should be checked for a bent condition, and the same crankshaft sensor adjusting tool may be used to check these rings. Place the J37089 tool on the pulley extension surface and rotate the tool around the pulley (Figure 9-35). If any blade touches the tool, replace the pulley.

Crankshaft Sensor Adjustment, Single Slot Sensor

If a single slot crankshaft sensor requires adjustment, follow this procedure:

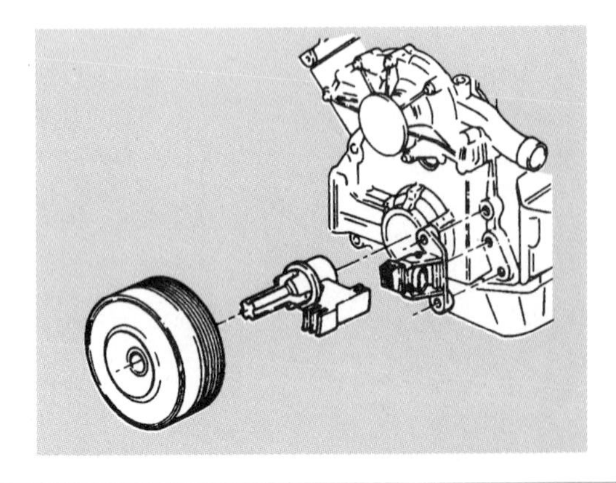

Figure 9-34 Crankshaft sensor adjustment, 30 L, 3,300, 38 L, and 3,800 engines (Courtesy of Oldsmobile Division, General Motors Corporation)

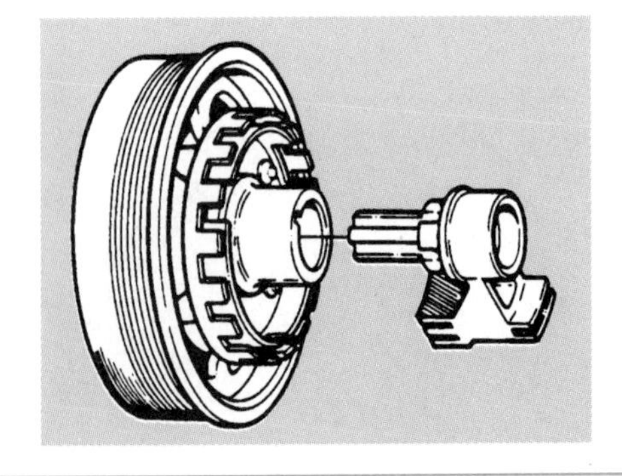

Figure 9-35 Interrupter ring checking procedure (Courtesy of Oldsmobile Division, General Motors Corporation)

CAUTION: Always be sure the ignition switch is off before attempting to rotate the crankshaft with a socket and breaker bar. If the ignition switch is on, the engine may start suddenly and rotate the socket and breaker bar with tremendous force. This action may result in personal injury and vehicle damage.

1. Be sure that the ignition switch is off and then rotate the crankshaft with a pull handle and socket installed on the crankshaft pulley nut. Continue rotating the crankshaft until one of the interrupter blades is in the sensor and the edge of the interrupter window is at the edge of the defector on the pedestal.
2. Insert adjustment tool J-36179 or its equivalent between each side of the blade and the sensor. If the tool does not fit between each side of the blade and the sensor, adjustment is required. The gap measurement should be repeated at all three blades.
3. If a sensor adjustment is necessary, loosen the pinch bolt and insert the adjusting tool between each side of the blade and the sensor. Move the sensor as required to insert the gauge.
4. Tighten the sensor pinch bolt to 30 in. lb. (3 to 4 N·m).
5. Rotate the crankshaft and recheck the gap at each blade.

Special Tools

Crankshaft sensor adjusting tool, single slot sensor

No-Start Ignition Diagnosis, EI Type 1 and Type 2 Systems

If the engine fails to start, follow these steps for a no-start ignition diagnosis:

1. Connect a test spark plug from each spark plug wire to ground and crank the engine while observing the test spark plug.
2. If the test spark plug does not fire on any spark plug, check the 12-V supply wires to the coil module. Some coil modules have two fused 12-V supply wires. Consult the vehicle manufacturer's wiring diagrams for the car being tested to identify the proper coil module terminals.
3. If the test spark plug does not fire on a pair of spark plugs, the coil connected to that pair of spark plugs is probably defective.
4. If the test spark plug did not fire on any of the spark plugs and the 12-V supply circuits to the coil module are satisfactory, disconnect the crankshaft and camshaft sensor connectors and connect short jumper wires between the sensor connector and the wiring harness connector. Be sure the jumper wire terminals fit securely to maintain electrical contact. Each sensor has a voltage supply wire, a ground wire, and a signal wire on 3.8 L engines. On the 3.3 L and 3,300 engines, the dual crankshaft sensor has a voltage supply wire, ground wire, crank signal wire, and SYNC signal wire. Identify each of these wires on the wiring diagram for the system being tested.
5. Connect a digital voltmeter to each of the camshaft and crankshaft sensor black ground wires to an engine ground connection. With the ignition switch on the voltmeter reading should be 0.2 V or less. If the reading is above 0.2 V, the sensor ground wires have excessive resistance.
6. With the ignition switch on, connect a digital voltmeter from the camshaft and crankshaft sensor white/red voltage supply wires to an engine ground (Figure 9-36). The voltmeter readings should be 5 V to 11 V. If the readings are below these values, check the voltage at the coil module terminals that are connected to the camshaft and crankshaft sensor voltage supply wires. When the sensor voltage supply readings are low at the coil module terminals, the coil module should be replaced. If the voltage supply readings are low at either sensor connector but satisfactory at the coil module terminal, the wire from the coil module to the sensor is defective. On the 3.3 L and 3,300 engines, the crankshaft sensor ground wire and voltage supply wire are checked in the same way as explained in steps 5 and 6.

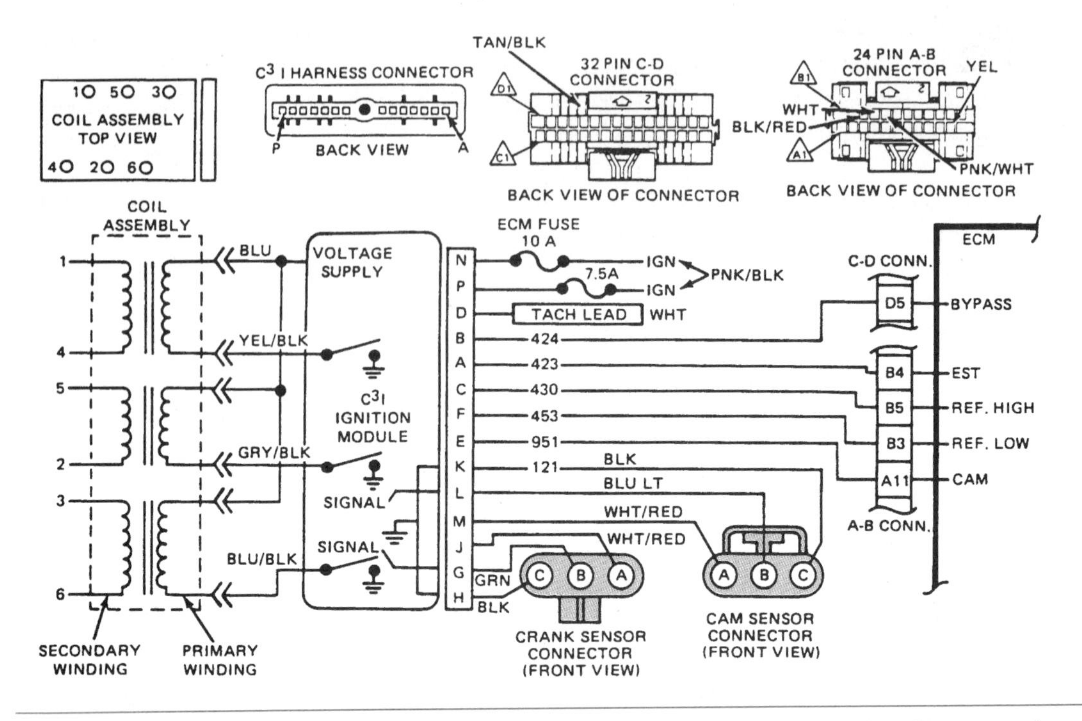

Figure 9-36 Crankshaft and camshaft sensor wiring connections, 38 L engine (Courtesy of Buick Motor Division, General Motors Corporation)

7. If the camshaft and crankshaft sensor ground and voltage supply wires are satisfactory, connect a digital voltmeter to each sensor signal wire and crank the engine. Each sensor should have a 5-V to 7-V fluctuating signal. On the 3.3 L and 3,300 engines, test this voltage signal on the crank and SYNC signal wires at the crankshaft sensor. If the signal is less than specified, replace the sensor with the low signal.
8. When the camshaft and crankshaft sensor signals on 3.8 L engines or crank and SYNC signals on 3.3 L and 3,300 engines are satisfactory and the test spark plug did not fire at any spark plug, the coil module is probably defective.
9. On 3.8 L engines where the coil assembly is easily accessible, the coil assembly screws may be removed and the coil lifted up from the module with the primary coil wires still connected. Connect a 12-V test lamp across each pair of coil primary wires and crank the engine. If the test lamp does not flutter on any of the coils, the coil module is defective, assuming that the crankshaft and camshaft sensor readings are satisfactory.

No-Start Ignition Diagnosis, EI Type 1 Fast-Start Systems

Complete steps 1, 2, and 3 in the No-Start Ignition Diagnosis, EI Type 1 and Type 2 Systems, and then complete these steps:

1. If the 12-V supply circuits to the coil module are satisfactory, disconnect the crankshaft sensor connector and connect four short jumper wires between the sensor connector and the wiring harness connector.
2. Connect a digital voltmeter from the sensor ground wire to an engine ground. With the ignition switch on, the voltmeter should read 0.2 V or less. A reading above this value indicates a defective ground wire.
3. Connect the voltmeter from the sensor voltage supply wire to an engine ground. With the ignition switch on, the voltmeter reading should be 8 V to 10 V. If the reading is lower than specified, check the voltage at coil module terminal N (Figure 9-37). When the voltage at terminal N is satisfactory and the reading at the sensor voltage supply

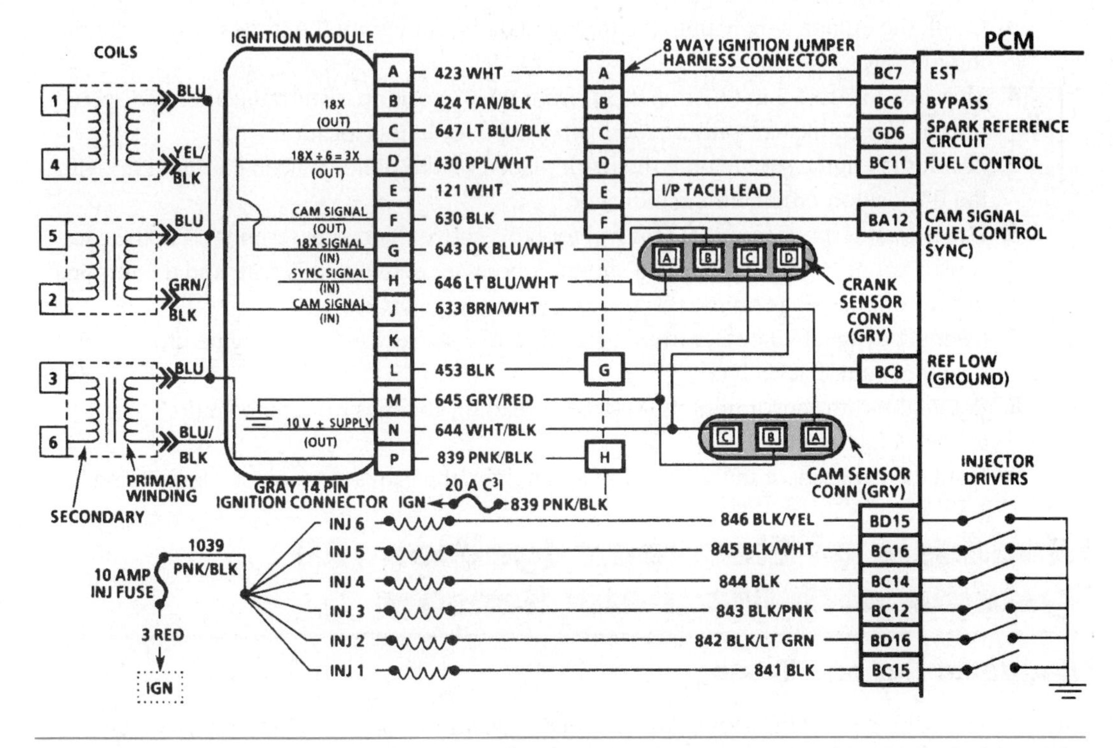

Figure 9-37 Terminal identification, EI fast-start system 3,800 engine (Courtesy of Oldsmobile Division, General Motors Corporation)

wire is low, the wire from terminal N to the sensor is defective. A low voltage reading at terminal N indicates a defective coil module.

4. If the readings in steps 3 and 4 are satisfactory, connect a voltmeter from the 3X and 18X signal wires at the sensor connector to an engine ground and crank the engine. The voltmeter reading should fluctuate from 5 V to 7 V. The exact voltage may be difficult to read especially on the 18X signal, but the reading must fluctuate. If the voltmeter reading is steady on either sensor signal, the sensor is defective.
5. Connect a digital voltmeter from the 18X and 3X signal wires at the coil module to an engine ground and crank the engine. The voltmeter readings should be the same as in step 4. If these voltage signals are satisfactory at the coil module terminals, but low at the sensor, repair the wires between the coil module and the sensor.
6. If the 18X and 3X signals are satisfactory at the coil module terminals, remove the coil assembly-to-module screws and lift the coil assembly up from the module. Connect a 12-V test lamp across each pair of coil primary terminals and crank the engine. If the test lamp does not flash on any pair of terminals, the coil module is defective.

Cam Sensor Timing, 3.8 L Turbocharged Engines

If the cam sensor is removed from the engine on 3.8 L turbocharged engines, the sensor must be timed to the engine when the sensor is installed. The cam sensor gear has a dot that must be positioned opposite the sensor disc window prior to sensor installation. As the cam sensor is installed in the engine, the gear dot must face away from the timing chain toward the passenger's side of the vehicle. When the cam sensor is installed in the engine, the sensor wiring harness must face toward the driver's side of the vehicle. Follow this procedure for cam sensor installation and timing:

1. Remove the spark plug wires from the coil assembly.
2. Remove the number 1 spark plug and crank the engine until compression is felt at the spark plug hole.

3. Crank the engine slowly until the timing mark lines up with the 0° position on the timing indicator.
4. Measure 1.47 to 1.5 in (3.7 to 3.8 cm) from the 0° position toward the after TDC position on the crankshaft pulley, and mark the pulley at this location.
5. Crank the engine slowly until the mark placed on the pulley in step 4 is lined up with the 0° position on the timing indicator.
6. Use a weatherpack terminal removal tool to remove the center terminal B in the cam sensor connector, and connect a short jumper wire between this wire and the terminal in the connector. Terminal B is the cam sensor signal wire.
7. Connect a digital voltmeter from the cam sensor signal wire to an engine ground, and turn the ignition switch on.
8. Rotate the cam sensor until the voltmeter reading changes from high volts (5 V to 12 V) to low volts (0 V to 2 V).
9. Hold the cam sensor in this position and tighten the cam sensor-to-block retaining bolt.

Diagnosis of Electronic Ignition (EI) Systems with Magnetic Sensors

Magnetic Sensor Tests

With the wiring harness connector to the magnetic sensor disconnected and an ohmmeter calibrated on the X10 scale connected across the sensor terminals, the meter should read 900 Ω to 1,200 Ω on 2.0-L, 2.8-L, and 3.1-L engines. The meter should indicate 500 Ω to 900 Ω on a Quad 4 engine, and 800 Ω to 900 Ω on a 2.5-L engine (Figure 9-38).

Meter readings below the specified value indicate a shorted sensor winding, whereas infinite meter readings prove that the sensor winding is open. Since these sensors are mounted in the crankcase, they are continually splashed with engine oil. In some sensor failures, the engine oil enters the sensor and causes a shorted sensor winding. If the magnetic sensor is defective, the engine fails to start.

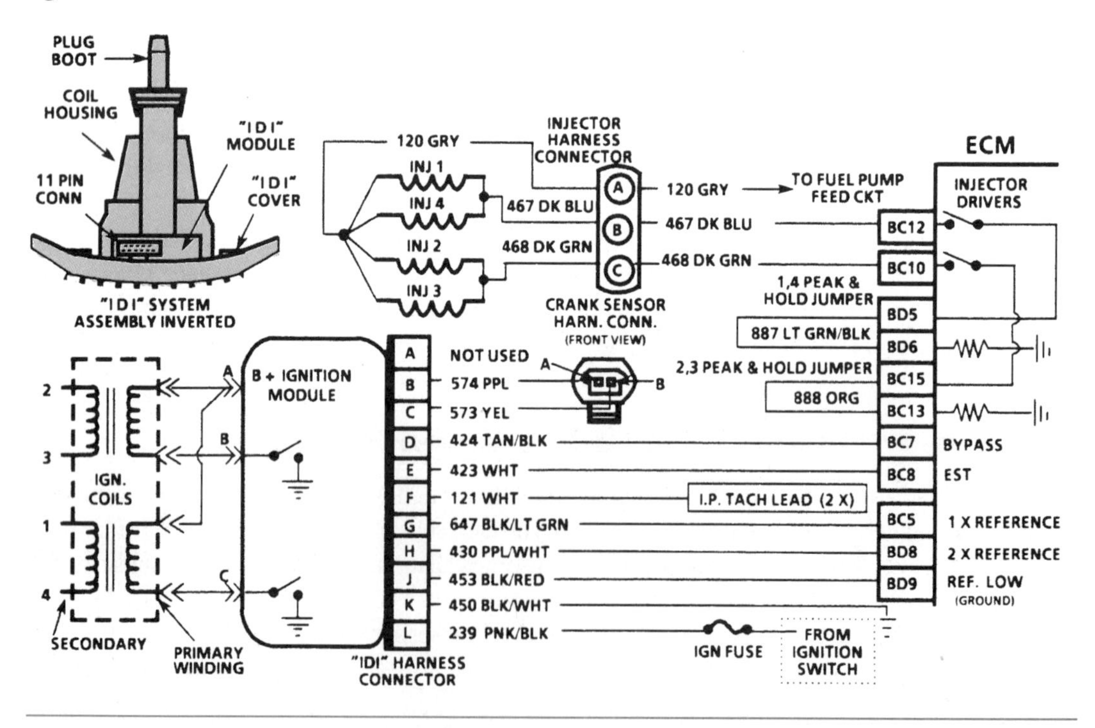

Figure 9-38 Terminal identification, EI system 2.3 L Quad 4 engine (Courtesy of Oldsmobile Division, General Motors Corporation)

With the magnetic sensor wiring connector disconnected, an alternating current (ac) voltmeter may be connected across the sensor terminals to check the sensor signal while the engine is cranking. On 2.0-L, 2.8-L, and 3.1-L engines, the sensor signal should be 100 millivolts (mV) ac. The sensor voltage on a Quad 4 engine should be 200 mV ac. When the sensor is removed from the engine block, a flat steel tool placed near the sensor should be attracted to the sensor if the sensor magnet is satisfactory.

No-Start Diagnosis, EI System with Magnetic Sensor

When an engine with an EI system and a magnetic sensor fails to start, complete steps 1, 2, and 3 of the No-Start Diagnosis, EI Type 1 and Type 2 Systems, and then follow this procedure:

1. If the test spark plug did not fire on any of the spark plugs, check for 12 V at the coil module voltage input terminals. Consult the wiring diagram for the system being tested for terminal identification.
2. If 12 V is supplied to the appropriate coil module terminals, test the magnetic sensor as explained under Magnetic Sensor Tests.
3. When the magnetic sensor tests are satisfactory, the coil module is probably defective.

Some coil module changes were made on 1989 EI systems with magnetic sensors, and 1989 coil modules will operate satisfactorily on 1988 EI systems. However, if a 1988 module is installed on a 1989 EI system, the malfunction indicator light (MIL) is on, and code 41 is stored in the PCM memory.

Engine Misfire Diagnosis

SERVICE TIP: When diagnosing any computer system, never forget the basics. For example, always be sure the engine has satisfactory compression and ignition before attempting to diagnose the computer-controlled fuel injection.

If the engine misfires all the time or on acceleration only, test these components:

1. Engine compression
2. Spark plugs
3. Spark plug wires
4. Ignition coils, test for firing voltage with a test spark plug
5. Crankshaft sensor
6. Fuel injectors on multiport and sequential fuel injection systems

Classroom Manual
Chapter 9, page 206

CUSTOMER CARE: Some intermittent automotive problems are difficult to diagnose unless we can catch the car "in the act." In other words, the problem is hard to diagnose if the symptoms are not present while we are performing diagnostic tests.

One solution to this problem is to have the customer leave the vehicle with the shop, and then drive the car under the conditions when the problem occurs. However, it the problem only appears once every week this solution is not likely to work, because the problem will not occur in the short time we have to drive the car.

A second solution, especially in no-start situations, is to have the customer phone the shop immediately when the problem occurs, and send a technician to diagnose the problem. If this solution is attempted, inform the customer not to attempt starting the car until the technician arrives. This solution may be expensive, but in some cases it may be the only way to diagnose the problem successfully. Always be willing to go the "extra mile" to diagnose and correct the customer's problem. By doing so, you will obtain a lot of satisfied, repeat customers.

Guidelines for Servicing Distributor Ignition (DI) and Electronic Ignition (EI) Systems

1. If a 12-V test light connected from the negative primary coil terminal to ground does not flash when the engine is cranked, the pickup coil or module is defective.
2. When a 12-V test light connected from the negative primary coil terminal to ground flashes while cranking the engine, but a test spark plug connected from the coil secondary wire to ground does not fire while cranking the engine, the coil is defective.
3. When a test spark plug connected from the coil secondary wire to ground fires while cranking the engine, but fails to fire when connected from the spark plug wires to ground, the distributor cap or rotor is defective.
4. Ignition modules may be tested with a module tester to determine if they are capable of triggering the primary ignition circuit on and off.
5. The ohmmeter leads may be connected to the pickup coil leads to test the pickup coil for open and shorted circuits.
6. The ohmmeter leads may be connected from one of the pickup leads to ground to check the pickup for a grounded condition.
7. The gap between the reluctor high points and the pickup coil may be adjusted on some distributors. This gap should be measured with a nonmagnetic feeler gauge.
8. The ohmmeter leads may be connected to the primary coil terminals to check the primary winding for open and shorted circuits.
9. The ohmmeter leads may be connected from one of the primary terminals to the coil tower to check the secondary winding for open and shorted circuits.
10. The ohmmeter tests on the ignition coil windings do not check the coil for insulation leakage.
11. The maximum secondary coil voltage test with a test spark plug is an accurate indication of coil condition.
12. Spark plug wires should be routed so two wires that fire one after the other are not positioned beside each other.
13. Ignition modules mounted on the distributor housing must have silicone grease on the module mounting surface to prevent module overheating.
14. When the distributor is timed to the engine, the number one piston should be at TDC on the compression stroke with the timing marks aligned, and the rotor should be under the number one spark plug wire terminal in the distributor cap with one of the high points aligned with the pickup coil.
15. The ignition spark advance may be checked with the advance control and the spark advance meter on the timing light.
16. A magnetic offset adjustment must be set to the manufacturer's specifications on the magnetic timing meter prior to an ignition timing check.
17. Spark plug carbon conditions are indicators of cylinder and combustion chamber operation.
18. Ignition point dwell is increased when the point gap is decreased.
19. On many vehicles, such as General Motors and Ford vehicles, a timing connector must be disconnected while checking basic timing.
20. The emission control information label provides the location of the timing connector on many General Motors vehicles.
21. On Chrysler vehicles, the computer system must be in the limp-in mode while checking basic timing.
22. A defective automatic shutdown (ASD) relay will cause a no-start problem on Chrysler products.

23. The voltage supply and ground connection should be checked on Hall effect pickups before the voltage signal from the pickup is checked with the engine cranking.
24. After the voltage supply and ground wires have been checked on an optical pickup, the voltage signal from the pickup should fluctuate between 0 V and 5 V while cranking the engine.
25. Prior to checking the voltage signal from a crankshaft or camshaft sensor on an EI system, the sensor voltage supply and ground wires should be checked.
26. The voltage signal from crankshaft or camshaft sensors may be checked with a 12-V test light or a digital voltmeter while cranking the engine.
27. On some crankshaft and camshaft sensors, such as those manufactured by Chrysler, a paper shim must be attached to the sensor tip prior to sensor installation. The sensor must be installed until the paper shim lightly contacts the rotating ring.
28. Primary and secondary coil windings may be checked with an ohmmeter on EI coils.
29. A special tool is used to set the crankshaft sensor position in relation to the crankshaft and crankshaft pulley vanes on some EI systems.
30. The same tool used to adjust the crankshaft sensor is also used to check the crankshaft pulley vanes on some EI systems.
31. If the camshaft sensor is mounted in the previous distributor opening, this sensor must be timed to the engine when it is installed.
32. The winding in magnetic sensors on EI systems may be tested with an ohmmeter.
33. The voltage signal from a magnetic sensor in an EI system may be measured in ac millivolts.

CASE STUDY

A customer complained about a stalling problem on an Oldsmobile 88 with an EI system. When questioned about this problem, the owner said the engine stalled while driving in the city, but it only happened about once a week. Further questioning of the customer indicated the engine would restart after five to ten minutes, and the owner said the engine seemed to be flooded.

The technician performed voltmeter and ohmmeter tests, and an oscilloscope diagnosis on the EI system. There were no defects in the system and the engine operation was satisfactory. The fuel pump pressure was tested and the filter checked for contamination, but no problems were discovered in these components. The customer was informed that it was difficult to diagnose this problem when the symptoms were not present. The service writer asked the customer to phone the shop immediately the next time the engine stalled, without attempting to restart the car. The service writer told the customer that a technician would be sent out to check the problem when she phoned the shop.

Approximately ten days later, the customer phoned and said her car had stalled. A technician was dispatched immediately, and she connected a test spark plug to several of the spark plug wires. The ignition system was not firing any of the spark plugs. The car was towed to the shop without any further attempts to start the engine.

The technician discovered that the voltage supply and the ground wires on the crankshaft sensor were normal, but there was no voltage signal from this sensor while cranking the engine. The crankshaft sensor was replaced and adjusted properly. When the customer returned later for other service, she reported the stalling problem had been eliminated.

Terms to Know

Test spark plug
Ignition module tester
Ignition crossfiring
Silicone grease
Magnetic timing probe receptacle
Magnetic timing meter
Magnetic timing offset
Timing connector
Tachometer (TACH) terminal
Automatic shutdown (ASD) relay
Reference pickup
Synchronizer (SYNC) pickup
Hall effect pickup
Optical-type pickup
Distributor ignition (DI) system
Electronic ignition (EI) system
Magnetic sensor

ASE Style Review Questions

1. While discussing no-start diagnosis with a test spark plug:
Technician A says if the test light flutters at the coil tach terminal, but the test spark plug does not fire when connected from the coil secondary wire to ground with the engine cranking, the ignition coil is defective.
Technician B says if the test spark plug fires when connected from the coil secondary wire to ground with the engine cranking, but the test spark plug does not fire when connected from the spark plug wires to ground, the cap or rotor is defective.
Who is correct?
A. A only **C.** Both A and B
B. B only **D.** Neither A nor B

2. While discussing a pickup coil test with an ohmmeter connected to the pickup leads:
Technician A says an ohmmeter reading below the specified resistance indicates the pickup coil is grounded.
Technician B says an ohmmeter reading below the specified resistance indicates the pickup coil is open.
Who is correct?
A. A only **C.** Both A and B
B. B only **D.** Neither A nor B

3. While discussing ignition coil ohmmeter tests:
Technician A says the ohmmeter should be placed on the X1,000 scale to test the secondary winding.
Technician B says the ohmmeter tests on the coil check the condition of the winding insulation.
Who is correct?
A. A only **C.** Both A and B
B. B only **D.** Neither A nor B

4. While discussing timing of the distributor to the engine with the number one piston at TDC compression and the timing marks aligned:
Technician A says the distributor must be installed with the rotor under the number one spark plug terminal in the distributor cap and one of the reluctor high points aligned with the pickup coil.
Technician B says the distributor must be installed with the rotor under the number one spark plug terminal in the distributor cap and the reluctor high points out of alignment with the pickup coil.
Who is correct?
A. A only **C.** Both A and B
B. B only **D.** Neither A nor B

5. While discussing ignition point gap and dwell adjustment:
Technician A says if the point gap is reduced, the dwell reading decreases.
Technician B says if the point gap is reduced, the dwell reading increases.
Who is correct?
A. A only **C.** Both A and B
B. B only **D.** Neither A nor B

6. While discussing basic ignition timing adjustment on vehicles with computer-controlled distributor ignition (DI):
Technician A says on some DI systems a timing connector must be disconnected.
Technician B says the distributor must be rotated until the timing mark appears at the specified location on the timing indicator.
Who is correct?
A. A only **C.** Both A and B
B. B only **D.** Neither A nor B

7. While discussing the diagnosis of an electronic ignition (EI) system in which the crankshaft and camshaft sensor tests are satisfactory, but a test spark plug connected from the spark plug wires to ground does not fire:
Technician A says the coil assembly may be defective.
Technician B says the voltage supply wire to the coil assembly may be open.
Who is correct?
A. A only **C.** Both A and B
B. B only **D.** Neither A nor B

8. While discussing EI service and diagnosis:
Technician A says the crankshaft sensor may be rotated to adjust the basic ignition timing.
Technician B says the crankshaft sensor may be moved to adjust the clearance between the sensor and the rotating blades on some EI systems.
Who is correct?
A. A only **C.** Both A and B
B. B only **D.** Neither A nor B

9. While discussing the EI system crankshaft and camshaft sensors that require a paper spacer on the sensor tip prior to installation:
Technician A says the sensor should be installed so the paper spacer lightly touches the rotating sensor ring.
Technician B says the sensor should be installed so the paper spacer lightly touches the rotating sensor ring and then pulled outward 0.125 in.
Who is correct?
A. A only **C.** Both A and B
B. B only **D.** Neither A nor B

10. While discussing engine misfire diagnosis:
Technician A says a defective EI coil may cause cylinder misfiring.
Technician B says the engine compression should be verified first if the engine is misfiring continually.
Who is correct?
A. A only **C.** Both A and B
B. B only **D.** Neither A nor B

TABLE 9-1 ASE TASK

Diagnose no-starting, hard starting, engine misfire, poor driveability, spark knock, power loss, and poor mileage problems on vehicles with point-type ignition systems; determine needed repairs.

Problem Area	Symptoms	Possible Causes	Classroom Manual	Shop Manual
ENGINE PERFORMANCE	No-starting or hard starting	1. Defective coil	184	161
		2. Defective cap and rotor	187	161
		3. Open secondary coil wire	184	161
		4. Low or zero primary voltage at the coil	190	161
		5. Fouled spark plugs	188	161
		6. Burned, pitted, or improperly adjusted points, defective condenser	199	161
	Engine misfire	1. Low cylinder compression	184	161
		2. Intake manifold vacuum leaks	184	161
		3. High resistance in spark plug wires, coil secondary wire, or cap terminals	189	161

TABLE 9-1 ASE TASK (continued)

Problem Area	Symptoms	Possible Causes	Classroom Manual	Shop Manual
	Engine misfire	4. Electrical leakage in the cap, rotor, plug wires, coil wire, or coil tower	187	161
		5. Defective coil	184	161
		6. Defective spark plugs`	188	161
		7. Low primary voltage and current	190	162
		8. Improperly routed spark plug wires	189	162
		9. Improper dwell	199	162
		10. Worn distributor bushings	186	162
	Poor driveability, power loss	1. Low engine compression		
		2. Restricted exhaust or air intake	184	162
		3. Late ignition timing	191	162
		4. Insufficient centrifugal or vacuum advance	192	162
		5. Cylinder misfiring	191	162
	Spark knock	1. Higher than normal compression	193	162
		2. Ignition timing too far advanced	194	162
		3. Excessive centrifugal or vacuum advance	194	162
		4. Spark plug heat range too hot	188	162
		5. Improperly routed spark plug wires	189	162
FUEL ECONOMY	Low fuel mileage	1. Low engine compression	193	162
		2. Late ignition timing	193	162
		3. Lack of spark advance	194	162
		4. Cylinder misfiring	192	162

Table 9-2 ASE TASK

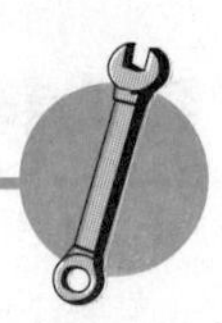

Diagnose no-starting, hard starting, engine misfire, poor driveability, spark knock, power loss, and poor mileage problems on vehicles with electronic ignition systems; determine needed repairs.

Problem Area	Symptoms	Possible Causes	Classroom Manual	Shop Manual
ENGINE PERFORMANCE	No-starting or hard starting	1. Defective coil	184	161
		2. Defective cap and rotor	187	161
		3. Open secondary coil wire	189	161
		4. Low or zero coil primary voltage	190	161
		5. Fouled spark plugs	188	161
		6. Defective pickup	185	161
	Engine misfire	1. Low cylinder compression	192	161
		2. Intake manifold vacuum leaks	192	161
		3. High resistance in spark plug wires, coil secondary wire, or cap terminals	187	161
		4. Electrical leakage in the cap, rotor, plug wires, coil wire, or coil tower	187	161
		5. Defective coil	184	161
		6. Defective spark plugs	188	161
		7. Low primary voltage and current	190	162
		8. Improperly routed spark plug wires	189	162
		9. Worn distributor bushings	186	162
	Poor driveability, power loss	1. Low engine compression	193	162
		2. Restricted exhaust or air intake	194	162
		3. Late ignition timing	193	162
		4. Insufficient centrifugal or vacuum advance	193	162
		5. Cylinder misfiring	192	162
	Spark knock	1. Higher than normal compression	194	162
		2. Ignition timing too far advanced	193	162
		3. Excessive centrifugal or vacuum advance	193	162
		4. Spark plug heat range too hot	188	162
		5. Improperly routed spark plug wires	189	162
FUEL ECONOMY	Low fuel mileage	1. Low engine compression	193	162
		2. Late ignition timing	193	162
		3. Lack of spark advance	193	162
		4. Cylinder misfiring	192	162

Table 9-3 ASE TASK

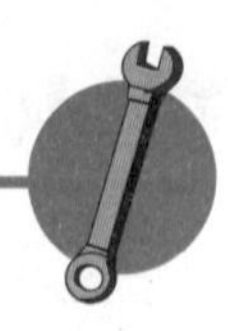

Inspect, test, repair, or replace ignition primary circuit wiring and components.

Problem Area	Symptoms	Possible Causes	Classroom Manual	Shop Manual
ENGINE PERFORMANCE	No-start	1. Defective pickup coil	185	164
		2. Defective ignition coil	184	165
		3. Defective module	186	164
		4. Defective primary circuit wiring	190	169
		5. Defective ballast resistor	190	162
	Engine misfiring	1. Defective coil	184	165
		2. Defective pickup coil or improper gap	185	164
		3. High resistance in primary circuit wiring	190	162

Table 9-4 ASE TASK

Inspect, test, and service distributor, including drives, shaft, bushings, cam, breaker plate, and advance/retard (vacuum, mechanical, or electric) units.

Problem Area	Symptoms	Possible Causes	Classroom Manual	Shop Manual
ENGINE PERFORMANCE	Engine misfiring	1. Worn distributor bushings	186	169
		2. Worn, loose, pickup, or breaker plate	185	169
	Power loss, poor driveability	1. Insufficient centrifugal advance	192	175
		2. Insufficient vacuum advance	193	175
NOISE	Spark knock	1. Excessive centrifugal advance	192	175
		2. Excessive vacuum advance	193	175

Table 9-5 ASE TASK

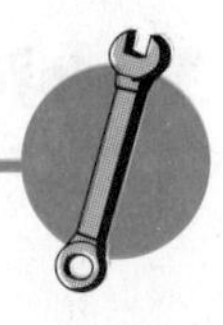

Inspect, test, adjust, service, repair, or replace ignition points and condenser.

Problem Area	Symptoms	Possible Causes	Classroom Manual	Shop Manual
ENGINE PERFORMANCE	No-start	1. Burned, pitted, or improperly adjusted points	199	170
		2. Defective condenser	199	170
	Engine misfiring	1. Burned, pitted, or improperly adjusted points	199	170
		2. Weak point spring tension	199	170

Table 9-6 ASE TASK

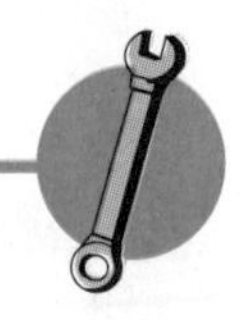

Inspect, test, service, repair, or replace ignition system secondary circuit wiring and components.

Problem Area	Symptoms	Possible Causes	Classroom Manual	Shop Manual
ENGINE PERFORMANCE	No-start	1. Defective coil	184	165
		2. Defective cap and rotor	187	166
		3. Fouled spark plugs	188	177
		4. Defective secondary coil wire	189	166
	Engine misfiring	1. Defective coil	184	165
		2. Defective cap and rotor	187	166
		3. Defective spark plug wires or secondary coil wire	189	166

Table 9-7 ASE TASK

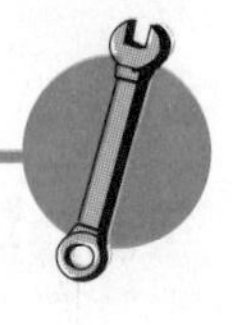

Inspect, test, and replace ignition coils.

Problem Area	Symptoms	Possible Causes	Classroom Manual	Shop Manual
ENGINE PERFORMANCE	No-start	Defective coil	184	165
	Engine misfiring	Defective coil	184	165

Table 9-8 ASE TASK

Check and adjust ignition system timing and timing advance/retard.

Problem Area	Symptoms	Possible Causes	Classroom Manual	Shop Manual
NOISE	Spark knock	1. Ignition timing too far advanced	194	172
		2. Excessive centrifugal or vacuum advance	193	175
ENGINE PERFORMANCE	Power loss, poor driveability	1. Late ignition timing	183	172
		2. Insufficient centrifugal or vacuum advance	193	175
FUEL ECONOMY	Low fuel mileage	1. Late ignition timing	183	172
		2. Insufficient centrifugal or vacuum advance	193	175

Table 9-9 ASE TASK

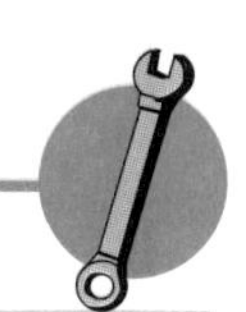

Inspect, test, and replace electronic ignition wiring harness and connectors.

Problem Area	Symptoms	Possible Causes	Classroom Manual	Shop Manual
ENGINE PERFORMANCE	No-start	Open circuit in primary ignition wiring	190	182

Table 9-10 ASE TASK

Inspect, test, and replace electronic ignition system pickup sensor or triggering devices.

Problem Area	Symptoms	Possible Causes	Classroom Manual	Shop Manual
ENGINE PERFORMANCE	No-start	Defective or improperly adjusted pickup coil	185	164

Table 9-11 ASE TASK

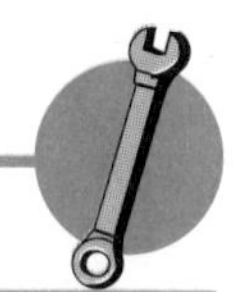

Inspect, test, and replace electronic ignition system control unit (module).

Problem Area	Symptoms	Possible Causes	Classroom Manual	Shop Manual
ENGINE PERFORMANCE	No-start	Defective ignition module	186	164

Fuel Tank, Line, Filter, and Pump Service

Upon completion and review of this chapter, you should be able to:

- ❑ Test alcohol content in the fuel.
- ❑ Relieve fuel system pressure.
- ❑ Inspect fuel tanks.
- ❑ Drain fuel tanks.
- ❑ Remove and replace fuel tanks.
- ❑ Remove, inspect, service, and replace electric fuel pumps and gauge sending units.
- ❑ Flush fuel tanks.
- ❑ Purge fuel tanks.
- ❑ Inspect and service nylon fuel lines.
- ❑ Inspect and service steel fuel tubing.
- ❑ Remove and replace fuel filters.
- ❑ Service and test mechanical fuel pumps.
- ❑ Remove and replace mechanical fuel pumps.
- ❑ Test electric fuel pumps, carbureted engines.

Alcohol in Fuel Test

Basic Tools

Basic technician's tool set

Service manual

100-milliliter graduated cylinder

Approved gasoline containers

Some gasoline may contain a small quantity of alcohol. The percentage of alcohol mixed with the fuel usually does not exceed 10%. An excessive quantity of alcohol mixed with gasoline may result in fuel system corrosion, fuel filter plugging, deterioration of rubber fuel system components, and a lean air-fuel ratio. These fuel system problems caused by excessive alcohol in the fuel may cause driveability complaints such as lack of power, acceleration stumbles, engine stalling, and no-start.

Some vehicle manufacturers supply test equipment to check the level of alcohol in the gasoline. The following alcohol-in-fuel test procedure requires only the use of a calibrated cylinder:

1. Obtain a 100-milliliter (mL) cylinder graduated in 1-mL divisions.
2. Fill the cylinder to the 90-mL mark with gasoline.
3. Add 10 mL of water to the cylinder so it is filled to the 100-mL mark.
4. Install a stopper in the cylinder, and shake it vigorously for 10 to 15 seconds.
5. Carefully loosen the stopper to relieve any pressure.
6. Install the stopper and shake vigorously for another 10 to 15 seconds.
7. Carefully loosen the stopper to relieve any pressure.
8. Place the cylinder on a level surface for 5 minutes to allow liquid separation.
9. Any alcohol in the fuel is absorbed by the water and settles to the bottom. If the water content in the bottom of the cylinder exceeds 10 mL, there is alcohol in the fuel. For example, if the water content is now 15 mL, there was 5% alcohol in the fuel.

Since this procedure does not extract 100% of the alcohol from the fuel, the percentage of alcohol in the fuel may be higher than indicated.

Fuel System Pressure Relief

CAUTION: Failure to relieve the fuel pressure on electronic fuel injection (EFI) systems prior to fuel system service may result in gasoline spills, serious personal injury, and expensive property damage.

CAUTION: Make sure no one smokes near a vehicle while servicing gasoline fuel system components. Such action may result in serious personal injury and property damage.

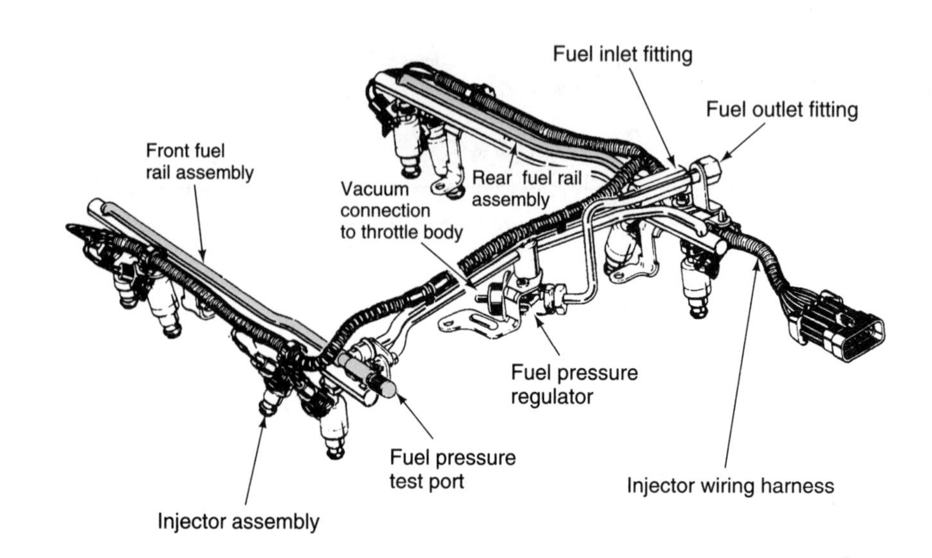

Figure 10-1 Fuel rail with pressure test port on a port fuel injected (PFI) engine (Courtesy of Cadillac Motor Car Division, General Motors Corporation)

Electronic fuel injection (EFI) is a generic term that may be applied to any computer-controlled fuel injection system.

Port fuel injection (PFI) systems have a fuel injector located in each intake port. These systems usually have a fuel pressure test port on the fuel rail to which a pressure gauge may be connected for test purposes.

The fuel pressure test port may be called a Schrader valve.

Special Tools

Fuel pressure gauge, electronic fuel injection (EFI)

CAUTION: When servicing gasoline fuel system components, flames, sparks, or sources of ignition in the area may result in an explosion, causing serious personal injury and property damage.

CAUTION: Always wear eye protection and observe all other safety rules to avoid personal injury when servicing fuel system components.

Since electronic fuel injection (EFI) systems have a residual fuel pressure, this pressure must be relieved before disconnecting any fuel system component. Most port fuel injection (PFI) systems have a fuel pressure test port on the fuel rail (Figure 10-1). Follow this procedure for fuel system pressure relief:

1. Disconnect the negative battery cable to avoid fuel discharge if an accidental attempt is made to start the engine.
2. Loosen the fuel tank filler cap to relief any fuel tank vapor pressure.
3. Wrap a shop towel around the fuel pressure test port on the fuel rail and remove the dust cap from this valve.
4. Connect the fuel pressure gauge to the fuel pressure test port on the fuel rail (Figure 10-2).
5. Install the bleed hose on the gauge in an approved gasoline container and open the gauge bleed valve to relieve fuel pressure from the system into the gasoline container. Be sure all the fuel in the bleed hose is drained into the gasoline container.

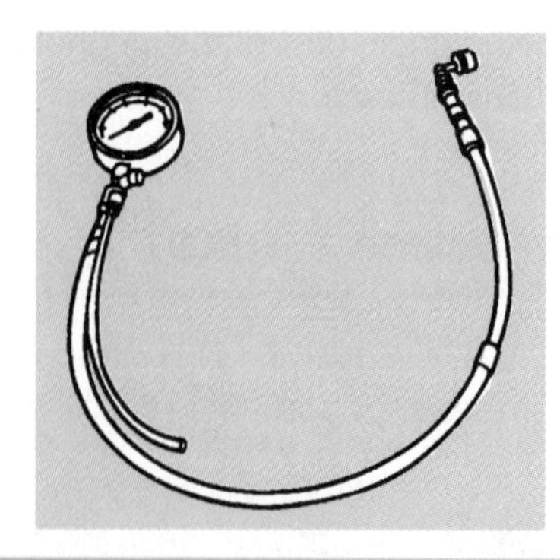

Figure 10-2 Fuel pressure gauge for engines with port fuel injection (PFI) (Courtesy of Cadillac Motor Car Division, General Motors Corporation)

On EFI systems that do not have a fuel pressure test port, such as most throttle body injection (TBI) systems, follow these steps for fuel system pressure relief:

1. Loosen the fuel tank filler cap to relieve any tank vapor pressure.
2. Remove the fuel pump fuse.
3. Start and run the engine until the fuel is used up in the fuel system and the engine stops.
4. Engage the starter for 3 seconds to relieve any remaining fuel pressure.
5. Disconnect the negative battery terminal to avoid possible fuel discharge if an accidental attempt is made to start the engine.

Throttle body injection (TBI) systems have one or two fuel injectors positioned above the throttle, or throttles, in the throttle body assembly.

Fuel Tank Service

Fuel Tank Inspection

The fuel tank should be inspected for leaks, road damage, corrosion, and rust on metal tanks, loose, damaged, or defective seams, loose mounting bolts, and damaged mounting straps. Leaks in the fuel tank, lines, or filter may cause gasoline odor in and around the vehicle, especially during low-speed driving and idling. In most cases, the fuel tank must be removed for servicing.

Fuel Tank Draining

SERVICE TIP: When a fuel tank must be removed, if possible inform the customer to bring the vehicle to the shop with a minimal amount of fuel in the tank.

WARNING: Always drain gasoline into an approved container, and use a funnel to avoid gasoline spills.

WARNING: When servicing fuel system components, always place a Class B fire extinguisher near the work area.

The fuel tank must be drained prior to tank removal. If the tank has a drain bolt, this bolt may be removed and the fuel drained into an approved container (Figure 10-3).

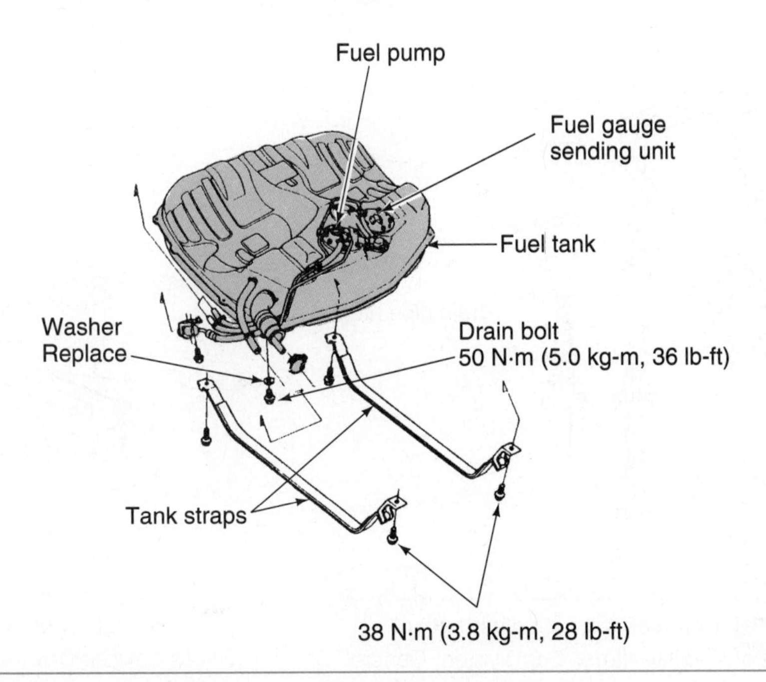

Figure 10-3 Fuel tank with drain bolt (Courtesy of Honda Motor Co., Ltd.)

If the fuel tank does not have a drain bolt, follow these steps to drain the fuel tank:

1. Remove the negative battery cable.
2. Raise the vehicle on a hoist.
3. Locate the fuel tank drain pipe, and remove the drain pipe plug.
4. Install the appropriate adaptor in the fuel tank drain pipe, and connect the intake hose from a hand-operated or air-operated pump to this adaptor (Figure 10-4). If the fuel tank does not have a drain pipe, install the pump hose through the filler pipe into the fuel tank.
5. Install the discharge hose from the hand-operated or air-operated pump into an approved gasoline container, and operate the pump until all the fuel is removed from the tank.

Special Tools

Hand- or air-operated pump

Fuel Tank Removal

The fuel tank removal procedure varies depending on the vehicle make and year. Always follow the procedure in the vehicle manufacturer's service manual. Following is a typical fuel tank removal procedure:

1. Relieve the fuel system pressure, and drain the fuel tank.
2. Raise the vehicle on a hoist or lift the vehicle with a floor jack and lower the chassis onto jack stands.
3. Use compressed air to blow dirt from the fuel line fittings and wiring connectors.
4. Remove the fuel tank wiring harness connector from the body harness connector.
5. Remove the ground wire retaining screw from the chassis if used.
6. Disconnect the fuel lines from the fuel tank. If these lines have quick-disconnect fittings, follow the manufacturer's recommended removal procedure in the service manual. Some quick-disconnect fittings are hand releasable, and others require the use of a special tool (Figure 10-5).
7. Wipe the filler pipe and vent pipe hose connections with a shop towel, and then disconnect the hoses from the filler pipe and vent pipe to the fuel tank.
8. Support the fuel tank with a transmission jack, and remove the front and rear tank strap attaching bolts (Figure 10-6).
9. Remove the tank straps, and then lower the transmission jack to remove the fuel tank.

Special Tools

Quick-disconnect fuel line fitting removal tools

Classroom Manual
Chapter 10, page 228

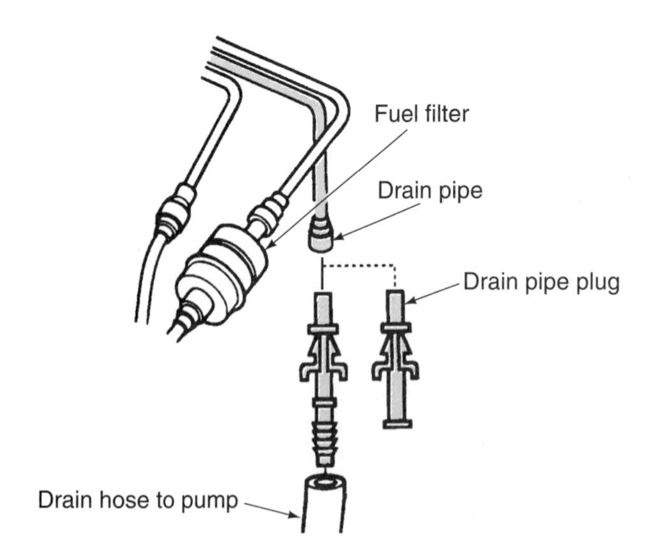

Figure 10-4 Fuel tank drain pipe with adaptor (Courtesy of Cadillac Motor Car Division, General Motors Corporation)

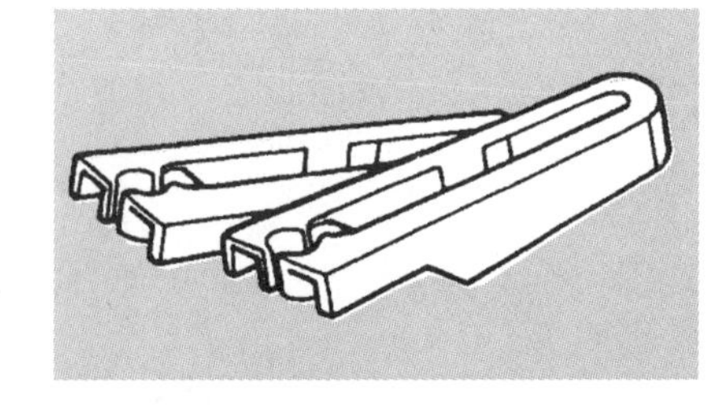

Figure 10-5 Tools required for quick-disconnect fuel line fittings (Courtesy of Cadillac Motor Car Division, General Motors Corporation)

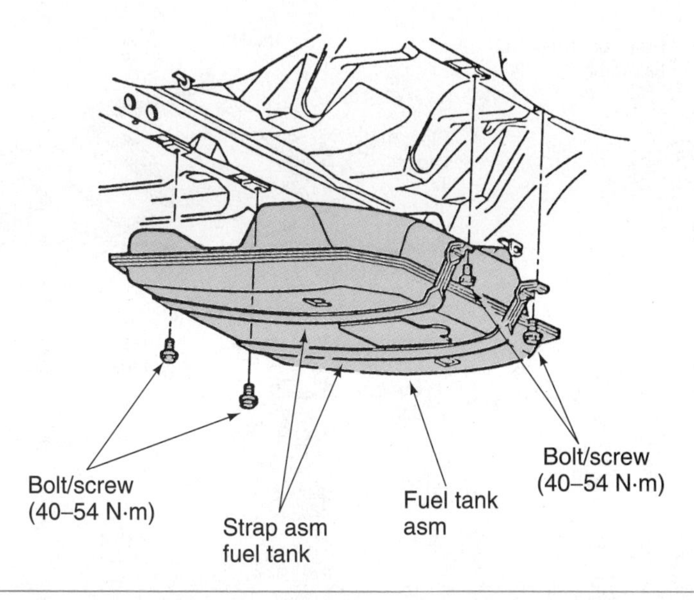

Figure 10-6 Front and rear tank strap retaining bolts (Courtesy of Oldsmobile Division, General Motors Corporation)

Electric Fuel Pump Removal and Replacement and Fuel Tank Cleaning

1. Remove the fuel tank from the vehicle.
2. Follow the vehicle manufacturer's recommended procedure to remove the fuel pump and gauge sending unit from the fuel tank. In many cases, a special tool must be used to remove this assembly (Figure 10-7).
3. Check the filter on the fuel pump inlet. If the filter is contaminated or damaged, replace the filter.
4. Inspect the fuel pump inlet for dirt and debris. Replace the fuel pump if these foreign particles are found in the pump inlet.
5. If the pump inlet filter is contaminated, flush the tank with hot water for at least 5 minutes.
6. Dump all the water from the tank through the pump opening in the tank. Shake the tank to be sure all the water is removed.
7. Check all fuel hoses and tubing on the fuel pump assembly. Replace fuel hoses that are cracked, deteriorated, or kinked. When fuel tubing on the pump assembly is damaged, replace the tubing or the pump.

Special Tools

Electric fuel pump removal and replacement tools

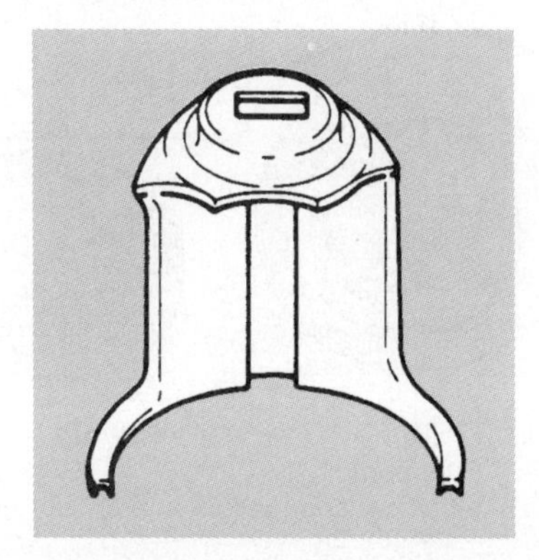

Figure 10-7 Special tool for removing fuel pump and gauge sending unit from the fuel tank (Courtesy of Oldsmobile Division, General Motors Corporation)

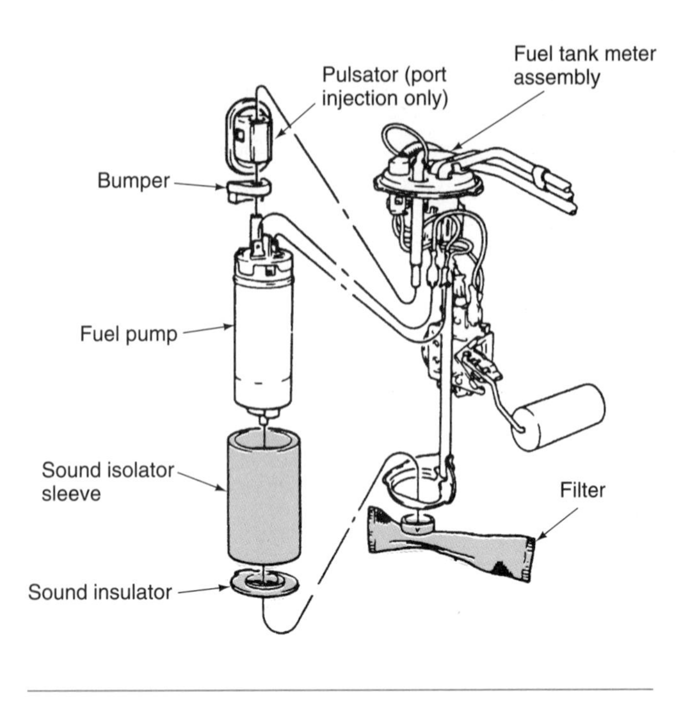

Figure 10-8 Electric fuel pump and gauge sending unit with filter, sound insulator sleeve, and sound insulator (Courtesy of Oldsmobile Division, General Motors Corporation)

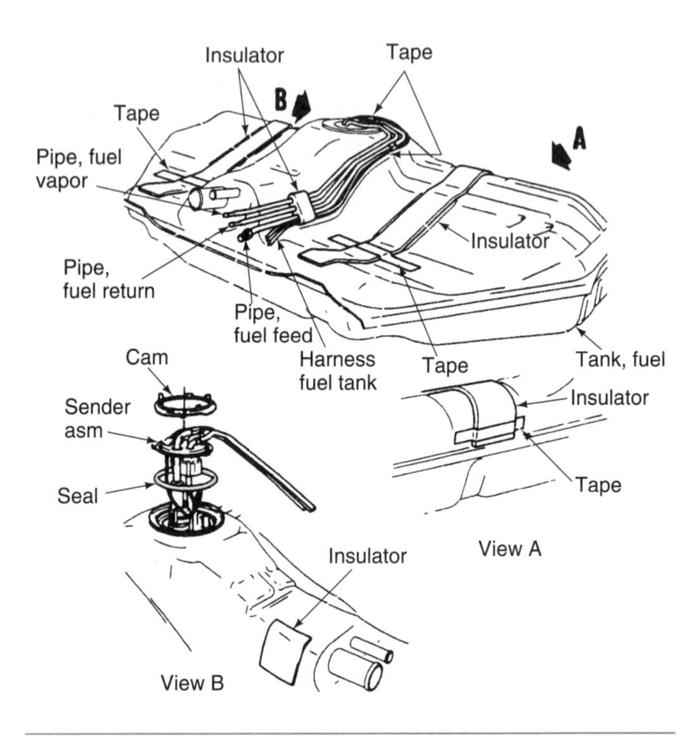

Figure 10-9 New gasket installed on fuel pump and gauge sending unit assembly (Courtesy of Oldsmobile Division, General Motors Corporation)

Classroom Manual
Chapter 10, page 236

8. Be sure the sound insulator sleeve is in place on the electric fuel pump, and check the position of the sound insulator on the bottom of the pump (Figure 10-8).
9. Clean the pump and sending unit mounting area in the fuel tank with a shop towel, and install a new gasket or O-ring on the pump and sending unit (Figure 10-9). On some tanks, the gauge sending unit and fuel pump are mounted separately (Figure 10-10).
10. Install the fuel pump and gauge sending unit assembly in the fuel tank and secure this assembly in the tank using the vehicle manufacturer's recommended procedure. On some vehicles, this procedure involves the use of a special tool. On some vehicles with a separate fuel pump and gauge sending unit, a lock ring must be rotated into place with a brass drift and a hammer to secure each of these units (Figure 10-11).

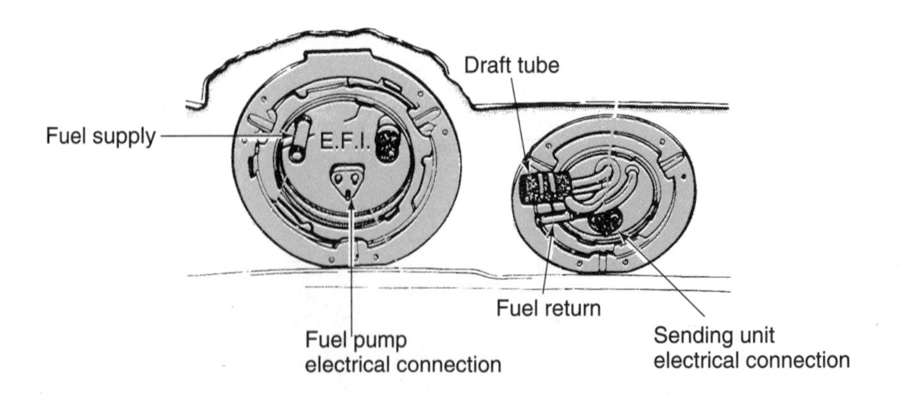

Figure 10-10 Electric fuel pump and gauge sending unit mounted separately in the fuel tank (Courtesy of Chrysler Corporation)

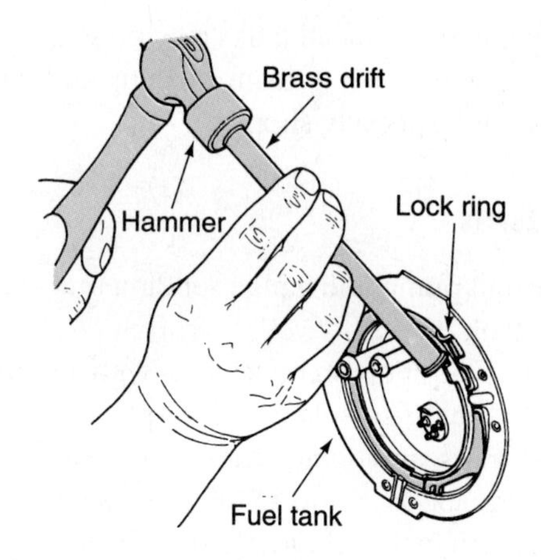

Figure 10-11 Installing the fuel pump lock ring with a brass drift and a hammer (Courtesy of Chrysler Corporation)

Fuel Tank Purging

CAUTION: Always wear eye protection and protective gloves when purging a fuel tank.

CAUTION: When handling emulsifying agents, always follow the precautions recommended by the agent manufacturer. Failure to follow these precautions may result in personal injury.

WARNING: When disposing of contaminated fuel or emulsifying agents, obey all local environmental regulations.

The fuel tank purging procedure provides additional cleaning and removal of gasoline vapors. Following is a typical fuel tank purging procedure:

1. Remove the fuel tank from the vehicle.
2. Follow the vehicle manufacturer's recommended procedure to remove the fuel pump and gauge sending unit from the fuel tank.
3. Be sure all the gasoline is removed from the tank, and fill the tank with tap water.
4. Agitate the tank vigorously and drain the tank.
5. Mix an emulsifying agent such as Product-Sol No. 913 or its equivalent with the amount of water recommended by the emulsifying agent manufacturer, and pour this mixture into the fuel tank. Then fill the tank with water.
6. Agitate the tank for 10 minutes, and then drain the tank.
7. Refill the tank completely with water, and then completely empty the tank.

After the purging procedure, an explosion meter should be used to determine if gasoline vapors remain in the fuel tank.

Fuel Tank Steam Cleaning and Repairing

WARNING: Do not steam clean plastic fuel tanks. This procedure may damage these tanks.

WARNING: Empty fuel tanks may contain gasoline vapors, making them extremely dangerous! Do not allow any flames, sparks, or other sources of ignition near an empty gasoline fuel tank.

Some manufacturers of plastic or metal fuel tanks recommend tank replacement if the tank is leaking. Repair kits are available for plastic fuel tanks; if these kits are used, the kit manufacturer's instructions must be carefully followed. Metal tanks may be steam cleaned prior to tank repairs to remove all gasoline residue and vapors. The steam cleaning and repair of metal fuel tanks should be done by a radiator and fuel tank specialty shop.

Fuel Tank Installation

1. Be sure the electric fuel pump and gauge sending unit are securely and properly installed in the fuel tank.
2. Raise the fuel tank into position on the chassis with a transmission jack if the vehicle is raised on a hoist. Be sure the insulators are in place on top of the fuel tank.
3. Install the fuel tank straps, and tighten the strap mounting bolts to the specified torque.
4. If the fuel lines have quick-disconnect fittings, be sure the large collar on these fittings is rotated back to the original position. Be sure the springs are visible on the inside diameter of the quick connector.
5. Place one or two drops of clean engine oil on the male tube ends where the quick connectors will be installed. Install all the fuel lines and electrical connections to the fuel pump and gauge sending unit. Be sure these lines and wires are properly secured, and check to be sure they do not interfere with other components.
6. Install and tighten the filler pipe hose and vent hose connections to the fuel tank.
7. Check the filler cap for damage. If the cap has pressure and vacuum valves, be sure these valves are working freely and are not damaged.
8. Install some fuel in the tank and cycle the ignition switch several times on an EFI system to pressurize the fuel system.
9. Start the engine and check the fuel tank and all line connections for leaks.

Fuel Line Service

Nylon Fuel Pipe Inspection and Service

WARNING: Always cover a nylon fuel pipe with a wet shop towel before using a torch or other source of heat near the line. Failure to observe this precaution may result in fuel leaks, personal injury, and property damage.

WARNING: If a vehicle has nylon fuel pipes, do not expose the vehicle to temperatures above 239°F (115°C) for more than one hour to avoid damage to the fuel pipes.

WARNING: If a vehicle has nylon fuel pipes, do not expose the vehicle to temperatures above 194°F (90°C) for any extended period to avoid damage to the nylon fuel pipes.

WARNING: Do not nick or scratch nylon fuel pipes. If damaged, these fuel pipes must be replaced.

Nylon fuel pipes should be inspected for leaks, nicks, scratches and cuts, kinks, melting, and loose fittings. If these fuel pipes are kinked or damaged in any way, they must be replaced. Nylon fuel pipes must be secured to the chassis at regular intervals to prevent fuel pipe wear and vibration.

Nylon fuel pipes provide a certain amount of flexibility and can be formed around gradual curves under the vehicle. Do not force a nylon fuel pipe into a sharp bend, because this action may kink the pipe and restrict the flow of fuel. When nylon fuel pipes are exposed to gasoline, they may become stiffer, making them more susceptible to kinking. Be careful not to nick or scratch nylon fuel pipes.

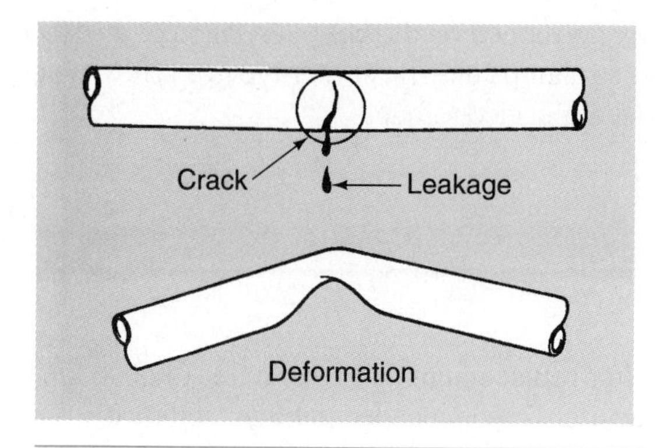

Figure 10-12 Steel tubing should be inspected for leaks, kinks, and deformation. (Courtesy of Toyota Motor Corporation)

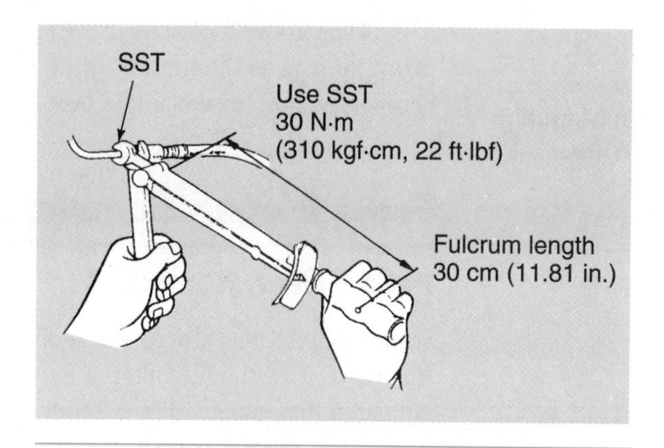

Figure 10-13 If fittings on fuel tubing are loose, they must be tightened to the specified torque. (Courtesy of Toyota Motor Corporation)

Steel Fuel Tubing Inspection and Service

Steel fuel tubing should be inspected for leaks, kinks, and deformation (Figure 10-12). This tubing should also be checked for loose connections and proper clamping to the chassis. If the fuel tubing threaded connections are loose, they must be tightened to the specified torque (Figure 10-13). Damaged fuel tubing should be replaced.

WARNING: O-rings in fuel line fittings are usually made from fuel-resistant Viton. Other types of O-rings must not be substituted for fuel fitting O-rings.

Some threaded fuel line fittings contain an O-ring. If the fitting is removed, the O-ring should be replaced. Flared ends are used on some steel fuel tubing. If a new flare is required, the old flare may be cut from the tubing with a pipe cutter, and a new flare made on the end of the tubing with a double flaring tool (Figure 10-14).

Special Tools

Double flaring tool

Rubber Fuel Hose Inspection and Service

Rubber fuel hose should be inspected for leaks, cracks, cuts, kinks, oil soaking, and soft spots or deterioration. If any of these conditions are found, the fuel hose should be replaced. When rubber fuel hose is installed, the hose should be installed to the proper depth on the metal fitting or line (Figure 10-15).

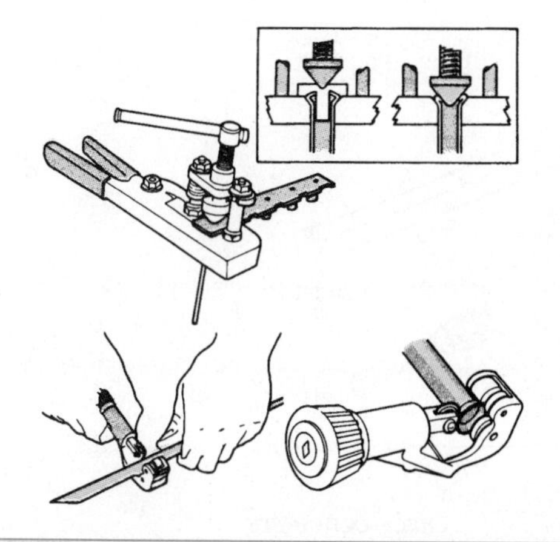

Figure 10-14 Cutting old flare from steel fuel tubing and making a new double flare (Courtesy of Chrysler Corporation)

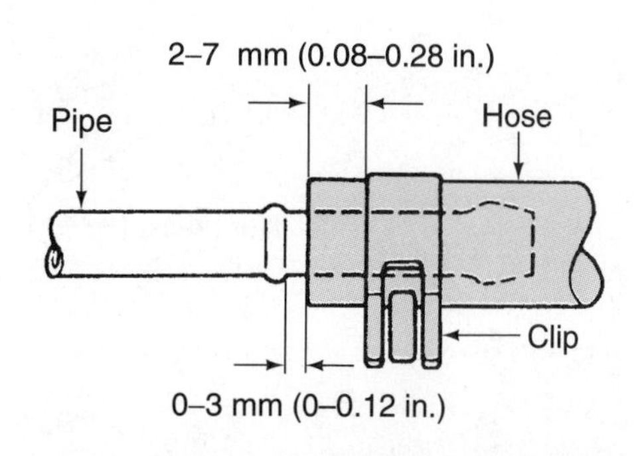

Figure 10-15 Rubber fuel hose installation on steel fitting or line (Courtesy of Toyota Motor Corporation)

Classroom Manual
Chapter 10, page 230

The rubber fuel hose clamp must be properly positioned on the hose in relation to the steel fitting or line as illustrated in the figure. Fuel hose clamps may be spring-type or screw-type. Screw-type fuel hose clamps must be tightened to the specified torque.

Fuel Filter Service

Fuel Filter Removal

Some vehicle manufacturers recommend fuel filter replacement at 30,000 miles (48,000 km). Always replace the fuel filter at the vehicle manufacturer's recommended mileage. If dirty or contaminated fuel is placed in the fuel tank, the filter may require replacing before the recommended mileage. A plugged fuel filter may cause the engine to surge and cut-out at high speed, or hesitate on acceleration. If a plastic fuel filter is used on a carbureted engine, contaminants may be seen in the filter. A restricted fuel filter causes low fuel pump pressure.

The fuel filter replacement procedure varies depending on the make and year of the vehicle, and the type of fuel system. Always follow the filter replacement procedure in the vehicle manufacturer's service manual. Following is a typical filter replacement procedure on a vehicle with EFI:

1. Relieve fuel system pressure as mentioned previously.
2. Raise the vehicle on a hoist.
3. Flush the quick connectors on the filter with water and use compressed air to blow debris from the connectors.
4. Disconnect the inlet connector first. Grasp the large connector collar, twist in both directions and pull the connector off the filter.
5. Disconnect the outlet connector using the same procedure used on the inlet connector (Figure 10-16).
6. Loosen and remove the filter mounting bolts, and remove the filter from the vehicle.

Photo Sequence 7 shows a typical procedure for relieving fuel pressure and removing a fuel filter.

Fuel Filter Installation

Following is a typical filter installation procedure on a vehicle with EFI:

1. Use a clean shop towel to wipe the male tube ends of the new filter.
2. Apply a few drops of clean engine oil to the male tube ends on the filter.

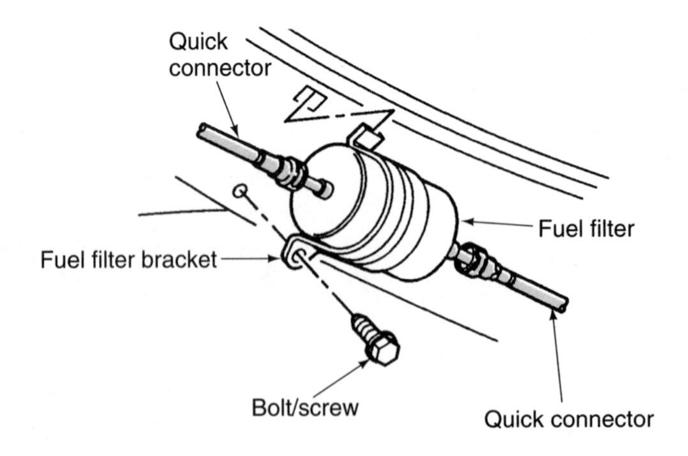

Figure 10-16 Removing fuel filter quick connectors (Courtesy of Oldsmobile Division, General Motors Corporation)

Photo Sequence 7
Typical Procedure for Relieving Fuel Pressure and Removing Fuel Filter

P7-1 Disconnect the negative battery cable.

P7-2 Loosen the fuel tank filler cap to relieve any fuel tank vapor pressure.

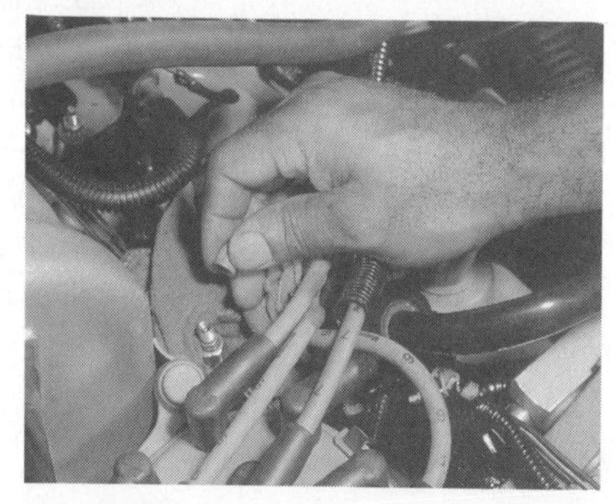

P7-3 Wrap a shop towel around the Schrader valve on the fuel rail, and remove the dust cap from this valve.

P7-4 Connect a fuel pressure gauge to the Schrader valve.

P7-5 Install the gauge bleed hose into an approved gasoline container, and open the gauge bleed valve to relieve the fuel pressure.

P7-6 Place the lift arms under the manufacturer's specified lift points on the vehicle, and lift the vehicle.

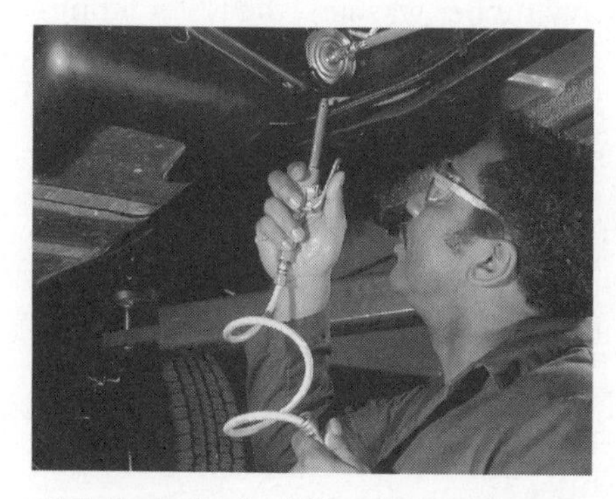

P7-7 Flush the fuel filter line connectors with water, and use compressed air to blow debris from the connectors.

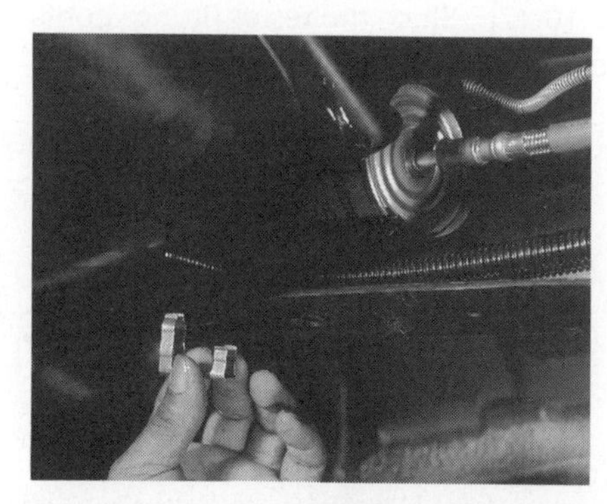

P7-8 Follow the vehicle manufacturer's recommended procedure to remove the inlet connector.

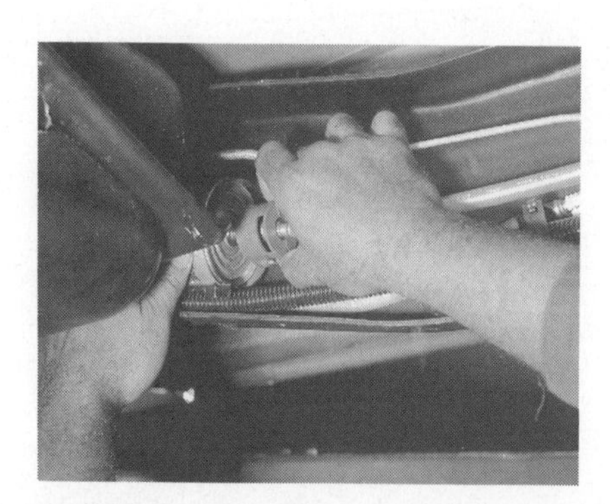

P7-9 Follow the vehicle manufacturer's recommended procedure to remove the outlet connector and the fuel filter.

3. Check the quick connectors to be sure the large collar on each connector has rotated back to the original position. The springs must be visible on the inside diameter of each quick connector.
4. Install the filter on the vehicle in the proper direction, and leave the mounting bolt slightly loose.
5. Install the outlet connector onto the filter outlet tube and press the connector firmly in place until the spring snaps into position.
6. Grasp the fuel line and try to pull this line from the filter to be sure the quick connector is locked in place.
7. Repeat steps 5 and 6 on the inlet connector.
8. Tighten the filter retaining bolt to the specified torque.
9. Lower the vehicle, start the engine, and check for leaks at the filter.

Classroom Manual
Chapter 10, page 234

Mechanical Fuel Pump Service and Diagnosis

Mechanical Fuel Pump Inspection

A mechanical fuel pump should be inspected for fuel leaks. If gasoline is leaking at the line fittings, these fittings should be tightened to the specified torque. If the fuel leak is still present, replace the fittings and/or fuel line. When gasoline is leaking from the vent opening in the pump housing, the pump diaphragm is leaking, and fuel pump replacement is necessary.

If engine oil is leaking from the vent opening in the pump housing, the pull rod seal is worn, and pump replacement is required. A loose pivot pin may cause oil leaks between the pin and the pump housing. When oil is leaking between the pump housing and the engine block, the pump mounting bolts should be tightened to the specified torque. If the oil leak continues, the gasket between the pump housing and the engine block must be replaced.

The mechanical fuel pump should be checked for excessive noise. If the pump makes a clicking noise when the engine is operating at a fast idle, the small spring between the rocker arm and the pump housing is broken or weak.

Mechanical Fuel Pump Testing

The mechanical fuel pump should be tested for pressure and volume. Mechanical fuel pump testers vary depending on the tester manufacturer. Always use the fuel pump tester according to the tester manufacturer's recommended procedure. Following is a mechanical fuel pump test procedure with a typical fuel pump tester.

1. Connect the fuel pump tester between the carburetor inlet fuel line and the inlet nut (Figure 10-17). Since the tester fittings contain heavy rubber washers, the tester fittings only require hand tightening.

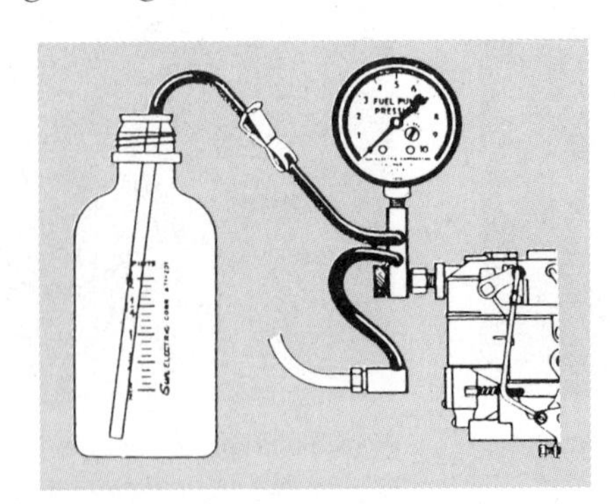

Figure 10-17 Tester connected to test mechanical fuel pump pressure and volume (Courtesy of Sun Electric Corporation)

2. Place the clipped fuel hose from the tester into the calibrated container supplied with the tester. Be sure the clip is closing the fuel hose. Have a co-worker hold the calibrated container in an upright position away from rotating or hot components.
3. Start the engine and immediately check for fuel leaks at the gauge connections. If fuel leaks exist, shut off the engine immediately and repair the leaks.
4. If no fuel leaks exist, record the fuel pressure on the tester gauge with the engine idling.
5. Release the clip on the fuel discharge hose, and allow fuel to discharge into the calibrated container for the time specified in the vehicle manufacturer's fuel pump volume specifications. Usually this time is 30 to 45 seconds. At the end of the specified time, immediately close the clip on the fuel discharge hose, and check the amount of fuel in the calibrated container.
6. Pour the fuel in the calibrated container back into the fuel tank. Use a funnel to avoid gasoline spills.
7. Compare the fuel pump pressure recorded in step 4 and the volume obtained in step 5 to the vehicle manufacturer's specifications.

Fuel pump volume is the amount of fuel the pump will deliver in a specific length of time.

If the pressure or volume is less than specified, check the fuel filter for restrictions, and check the fuel lines for restrictions and leaks before replacing the fuel pump. When air leaks into the fuel line between the fuel pump and the fuel tank, the fuel pump discharges some air with the fuel, which reduces pump volume. When the fuel pump is removed, always check the camshaft lobe or eccentric for wear.

Remove and Replace Mechanical Fuel Pump

Follow these steps for a typical mechanical fuel pump removal and replacement procedure:

1. Place shop towels under the inlet and and outlet fittings and loosen these fittings. Remove the inlet and outlet lines from the pump.
2. Remove the fuel pump mounting bolts, and remove the fuel pump from the engine.
3. Use a scraper to clean the pump mounting surface on the engine block. Remove all the old gasket material.
4. Check the fuel pump cam lobe for wear.
5. If the fuel pump is to be reused, clean the pump mounting surface with a scraper.
6. Install a new gasket on the fuel pump mounting surface.
7. Install the fuel pump in the engine block, and tighten the mounting bolts alternately to the specified torque.
8. Connect the inlet and outlet fuel lines to the fuel pump, and tighten these fittings to the specified torque.

Electric Fuel Pump Testing, Carbureted Engines

CAUTION: Never turn on the ignition switch or crank the engine with a fuel line disconnected. This action will result in gasoline discharge from the disconnected line, which may result in a fire, causing personal injury and/or property damage.

The electric fuel pump test procedure on carbureted engines varies depending on the vehicle make and year. Always follow the electric fuel pump test procedure in the vehicle manufacturer's service manual. Following is a typical electric fuel pump test procedure on a carbureted engine:

1. Disconnect the starting motor cable from the starter solenoid (Figure 10-18).
2. Leave the fuel pump power supply wire connected to this terminal, and tighten the nut on the solenoid terminal.

Special Tools

Fuel pressure and volume tester, carbureted engines

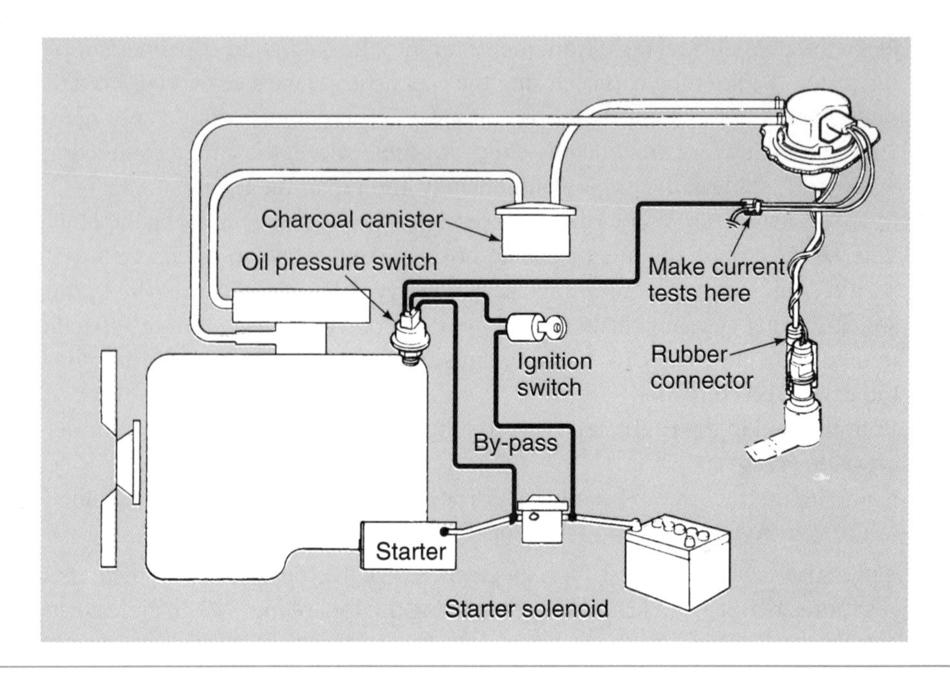

Figure 10-18 Electric fuel pump wiring diagram on a carbureted engine (Reprinted with the permission of Ford Motor Company)

3. Follow the same procedure used in the mechanical fuel pump tests to connect the fuel pump pressure and volume tester at the carburetor inlet nut.
4. Connect a jumper wire from the positive battery cable at the starter solenoid to the starting motor terminal on the solenoid. This action supplies full voltage to the fuel pump and operates the pump at high speed.
5. Follow the same procedure used in the mechanical fuel pump tests to test the electric fuel pump pressure and volume. Remove the jumper wire immediately when the specified volume test time is over.
6. Pour the gasoline in the graduated container into the fuel tank, and reconnect the starting motor cable to the solenoid terminal. Tighten the nut on this terminal to the specified torque.
7. Compare the fuel pump pressure and volume test results to the vehicle manufacturer's specifications. If the fuel pump pressure or volume is less than specified, check the fuel filter and fuel lines before replacing the pump.

Since electric fuel pumps in EFI systems are computer-controlled, the testing of these pumps is discussed later in the appropriate chapter.

CUSTOMER CARE: When performing any undercar service, always perform a quick visual inspection of the fuel tank, lines, filter, and pump for fuel leaks and damaged components. If there is any evidence of a fuel leak or damaged components, advise the customer regarding the potential danger and the necessary repairs. In most cases, the customer will approve the necessary repairs. The customer usually will be impressed with your thorough inspection and probably will return to the shop for other service.

Guidelines for Fuel Tank, Line, Filter, and Pump Service and Diagnosis

1. Excessive amounts of alcohol mixed with gasoline may cause fuel system corrosion, fuel filter plugging, deterioration of rubber fuel system components, and a lean air-fuel ratio.
2. Excessive amounts of alcohol in the fuel may cause lack of power, acceleration stumbles, engine stalling, or no-start.
3. On electronic fuel injection (EFI) systems, fuel system pressure relief is necessary before disconnecting fuel system components.
4. Fuel tanks should be inspected for leaks, road damage, corrosion or rust on metal tanks, loose, damaged, or defective seams, loose mounting bolts, and damaged mounting straps.
5. Fuel tanks may be drained through the drain bolt opening, or the hose on a hand-operated pump may be connected to the tank drain pipe. If the tank does not have a drain pipe, the pump hose may be installed through the filler pipe.
6. Electric in-tank fuel pumps should be inspected for a contaminated filter, deteriorated rubber hoses, dirt in the pump inlet, and damaged steel fuel tubing.
7. Fuel tanks containing dirt or contaminants may be flushed with hot water.
8. Fuel tanks may be purged with an emulsifying agent mixed with water.
9. Some vehicle manufacturers recommend fuel tank replacement if the tank is leaking.
10. Fuel tank repairs should be done by a specialty radiator and fuel tank repair shop.
11. Nylon fuel pipes must not be subjected to excessive heat above 239°F (115°C) for one hour or above 194°F (90°C) for an extended time period.
12. When a torch or heat source must be used near a nylon fuel line, cover the fuel line with a wet shop towel.
13. Steel fuel tubing should be inspected for leaks, kinks, and deformation.
14. Rubber fuel hose should be inspected for leaks, cracks, cuts, kinks, oil soaking, soft spots, and deterioration.
15. Fuel filters must be installed in the proper direction.
16. Mechanical fuel pumps should be inspected for gasoline leaks, oil leaks, and excessive clicking noise.
17. Mechanical fuel pumps should be tested for pressure and volume.
18. Electric fuel pumps on carbureted engines may be tested with a test procedure similar to that for mechanical fuel pumps.

CASE STUDY

A customer complained about severe engine surging on a Dodge station wagon with a 5.2-L carbureted engine. When the technician lifted the hood, he noticed many fuel system and ignition system components had been replaced recently. The customer was asked about previous work done on the vehicle, and he indicated this problem had existed for some time. Several shops had worked on the vehicle, but the problem still persisted. The carburetor had been overhauled, and the fuel pump replaced. Many ignition components such as the coil, distributor cap and rotor, spark plugs, and spark plug wires had also been replaced.

The technician road tested the vehicle and found it did have a severe surging problem at freeway cruising speeds. From past experience, the technician thought this severe surging problem was caused by lack of fuel supply. The technician decided to connect a fuel pressure gauge at the carburetor inlet nut. The gauge was securely taped to one of the windshield wiper blades so the gauge could be observed from the passenger compartment. The technician drove the vehicle on a second road test and found when the surging problem occurred, the fuel pump pressure dropped well below the vehicle manufacturer's specifications. Since the fuel pump and filter had been replaced, the technician concluded the problem must be in the fuel line or tank.

The technician returned to the shop and raised the vehicle on a hoist. The steel fuel tubing appeared to be in satisfactory condition. However, a short piece of rubber fuel hose between the steel fuel tubing and the fuel line entering the fuel tank was flattened and soft in the center. This fuel hose was replaced and routed to avoid kinking. Another road test proved the surging problem was eliminated.

The flattened fuel hose restricted fuel flow, and at higher speeds the increased vacuum from the fuel pump made the flattened condition worse, which restricted the fuel flow and caused the severe surging problem.

Terms to Know

Electronic fuel injection (EFI)
Port fuel injection (PFI)
Throttle body injection (TBI)
Fuel pressure test port
Schrader valve
Quick-disconnect fuel line fittings
Fuel tank purging
Fuel pump volume

ASE Style Review Questions

1. While discussing alcohol content in gasoline:
Technician A says excessive quantities of alcohol in gasoline may cause fuel filter plugging.
Technician B says excessive quantities of alcohol in gasoline may cause lack of engine power.
Who is correct?
A. A only **C.** Both A and B
B. B only **D.** Neither A nor B

2. While discussing an alcohol-in-fuel test with a 100 milliliter (mL) cylinder filled with 90 mL of gasoline and 10 mL of water:
Technician A says water and alcohol in the gasoline remain separate during the test.
Technician B says if the water content is 20 mL at the end of the test, the gasoline contains 10% alcohol.
Who is correct?
A. A only **C.** Both A and B
B. B only **D.** Neither A nor B

3. While discussing fuel system service on electronic fuel injection (EFI) systems:
Technician A says the fuel system pressure must be relieved before fuel system components are removed.
Technician B says the fuel system pressure may be relieved by connecting a pressure gauge to the fuel pressure test port and opening the bleed valve on the gauge with the bleed hose installed in an approved container.
Who is correct?
A. A only **C.** Both A and B
B. B only **D.** Neither A nor B

4. While discussing fuel tank draining:
Technician A says the fuel tank may be drained with a hand-operated pump.
Technician B says some fuel tanks have a drain pipe for draining the tank.
Who is correct?
A. A only **C.** Both A and B
B. B only **D.** Neither A nor B

5. While discussing quick-disconnect fuel line fittings:
Technician A says some quick-disconnect fittings may be disconnected with a pair of snap ring pliers.
Technician B says some quick-disconnect fittings are hand releasable.
Who is correct?
A. A only **C.** Both A and B
B. B only **D.** Neither A nor B

6. While discussing fuel tank and electric pump service:
Technician A says if the filter on the pump inlet is contaminated, the fuel tank should be flushed with hot water.
Technician B says if there is dirt in the pump inlet, the inlet may be cleaned and the pump reused.
Who is correct?
A. A only **C.** Both A and B
B. B only **D.** Neither A nor B

7. While discussing nylon fuel pipes:
Technician A says nylon fuel pipes may be subjected to temperatures up to 300°F (149°C).
Technician B says nylon fuel pipes may be bent at a 90° angle.
Who is correct?
A. A only **C.** Both A and B
B. B only **D.** Neither A nor B

8. While discussing quick-disconnect fuel line fittings:
Technician A says on some hand-releasable, quick-disconnect fittings, the fitting may be removed by pulling on the fuel line.
Technician B says some hand-releasable, quick-disconnect fittings may be disconnected by twisting the large connector collar in both directions and pulling on the connector.
Who is correct?
A. A only **C.** Both A and B
B. B only **D.** Neither A nor B

9. While discussing mechanical fuel pump diagnosis:
Technician A says if gasoline is leaking from the vent opening in the pump housing, the pump diaphragm is leaking.
Technician B says if oil is leaking from the vent opening in the pump housing, the pump diaphragm is leaking.
Who is correct?
A. A only **C.** Both A and B
B. B only **D.** Neither A nor B

10. While discussing mechanical fuel pump testing:
Technician A says if the fuel pump pressure is lower than specified, the fuel pump camshaft lobe may be worn.
Technician B says if the fuel pump volume is lower than specified, there may be an air leak in the fuel line between the tank and the pump.
Who is correct?
A. A only **C.** Both A and B
B. B only **D.** Neither A nor B

Table 10-1 ASE TASK

Inspect fuel tank, tank filter, and gas cap; inspect and replace fuel lines, fittings, and hoses; check fuel for contaminants and quality.

Problem Area	Symptoms	Possible Causes	Classroom Manual	Shop Manual
ENGINE PERFORMANCE	No-start, stalling	Excessive alcohol, contaminants in the fuel	228	199
	Lack of power	Excessive alcohol contaminants in the fuel	228	199
	Acceleration stumbles	Excessive alcohol contaminants in the fuel	228	199
	Surging, cutting out at high speed	1. Restricted fuel lines	230	206
		2. Contaminated fuel pump inlet filter	236	204
ODOR	Gasoline odor, low-speed driving, idling	Fuel leaks in tank, filter lines	230	201

Table 10-2 ASE TASK

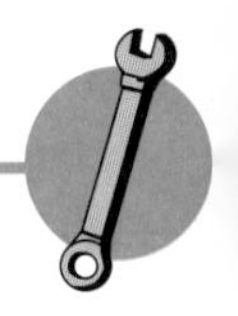

Inspect, test, and replace mechanical and electrical fuel pumps and pump control systems; inspect, service, and replace fuel filters.

Problem Area	Symptoms	Possible Causes	Classroom Manual	Shop Manual
ENGINE PERFORMANCE	Surging, cutting out at high speed	1. Restricted fuel filter	233	208
		2. Defective fuel pump	236	210

Conventional and Computer-Controlled Carburetor Diagnosis and Service

Upon completion and review of this chapter, you should be able to:

- ❑ Remove and replace carburetors.
- ❑ Disassemble carburetors.
- ❑ Inspect carburetor components.
- ❑ Clean carburetor components.
- ❑ Assemble carburetors.
- ❑ Adjust carburetor floats.
- ❑ Adjust secondary throttle linkage.
- ❑ Adjust secondary air valve.
- ❑ Perform secondary air valve opening adjustment.
- ❑ Adjust secondary air valve spring.
- ❑ Adjust choke control lever.
- ❑ Perform choke diaphragm connector rod adjustment.
- ❑ Adjust choke vacuum kick.
- ❑ Adjust choke unloader.
- ❑ Perform secondary throttle lockout adjustment.
- ❑ Adjust fast idle cam position.
- ❑ Adjust accelerator pump stroke.
- ❑ Perform carburetor adjustments with a choke valve angle gauge.
- ❑ Adjust idle speed.
- ❑ Adjust fast idle speed.
- ❑ Adjust idle mixture.
- ❑ Adjust A/C idle speed.
- ❑ Adjust engine idle speed.
- ❑ Adjust idle air-fuel mixture.
- ❑ Perform an antidieseling adjustment.
- ❑ Perform a computer-controlled carburetor system performance test.
- ❑ Diagnose the results of a computer-controlled carburetor system performance test.
- ❑ Perform a flash code diagnosis.
- ❑ Obtain fault codes with an analog voltmeter.
- ❑ Perform a computed timing test.
- ❑ Perform an output state test.
- ❑ Erase fault codes.
- ❑ Perform a continuous self-test.
- ❑ Diagnose a computer-controlled carburetor system with a scan tester.

Carburetor Service

Basic Tools

Basic technician's tool set

Service manual

Jumper wires

Carburetor Removal

All carburetor service procedures vary depending on the vehicle make and year. Always follow the recommended procedure in the vehicle manufacturer's service manual. Following is a typical removal procedure:

1. Disconnect the negative battery cable.
2. Remove the air cleaner vacuum hose from the intake manifold and any air cleaner flex hoses. Remove the air cleaner.
3. Disconnect the throttle linkage from the carburetor.
4. Disconnect the fuel line from the carburetor.
5. Disconnect all vacuum hoses from the carburetor. Note the position of each hose so they may be installed in their original positions.
6. Disconnect all electrical wires from the carburetor.
7. Remove the carburetor mounting nuts or bolts.
8. Remove the carburetor from the engine, and discard the carburetor-to-intake mounting gasket.

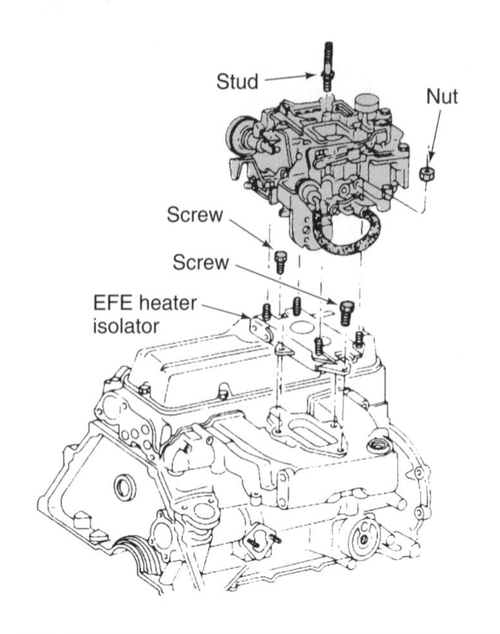

Figure 11-1 Carburetor removal (Courtesy of Chevrolet Motor Division, General Motors Corporation)

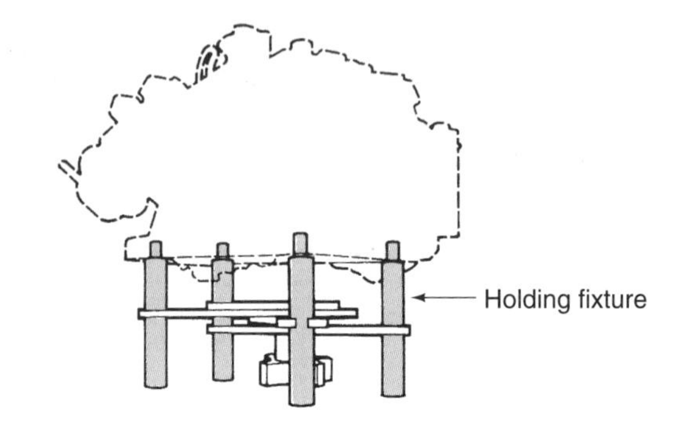

Figure 11-2 A carburetor holding fixture protects throttles from striking the work bench. (Courtesy of Chevrolet Motor Division, General Motors Corporation)

9. If a spacer is positioned between the carburetor and the intake, remove this spacer. Discard the gasket between the spacer and the intake manifold. On some engines, two spacer retaining screws must be removed prior to spacer removal (Figure 11-1). If coolant hoses are connected to the spacer or carburetor, the coolant must be drained before the carburetor and spacer are removed.

Carburetor Disassembly

Following is a typical carburetor disassembly procedure:

1. Place a funnel in an approved gasoline container and invert the carburetor over the funnel to drain fuel out of the float bowl. Since this gasoline may be contaminated, discard it following hazardous waste disposal regulations.
2. Place the carburetor on a holding fixture to protect the throttles from striking the work bench (Figure 11-2).
3. Remove the clip from the choke operating rod, and disconnect this rod (Figure 11-3).

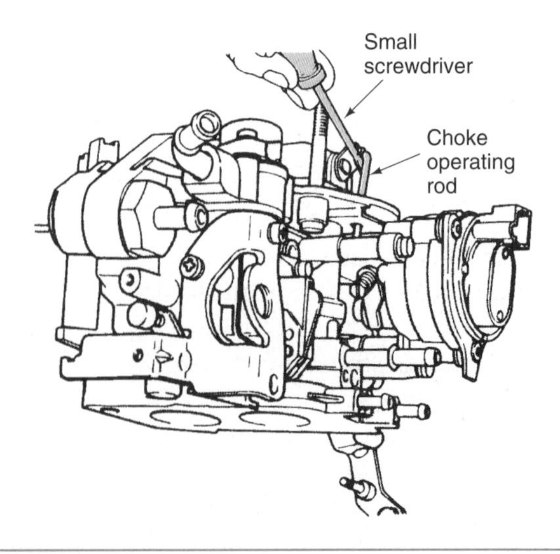

Figure 11-3 Removing choke operating rod (Courtesy of Chrysler Corporation)

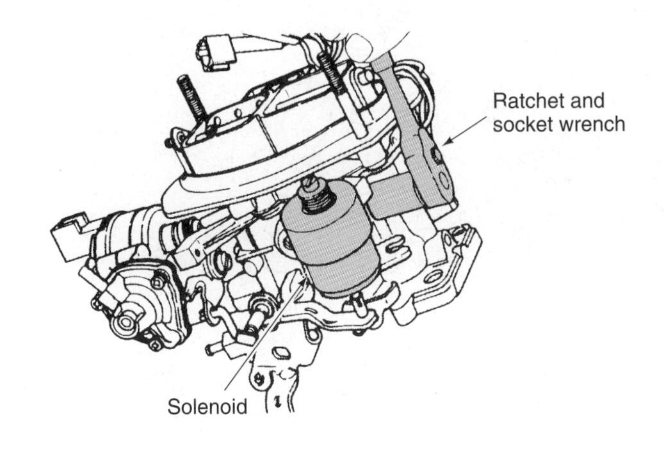

Figure 11-4 Removing carburetor solenoid (Courtesy of Chrysler Corporation)

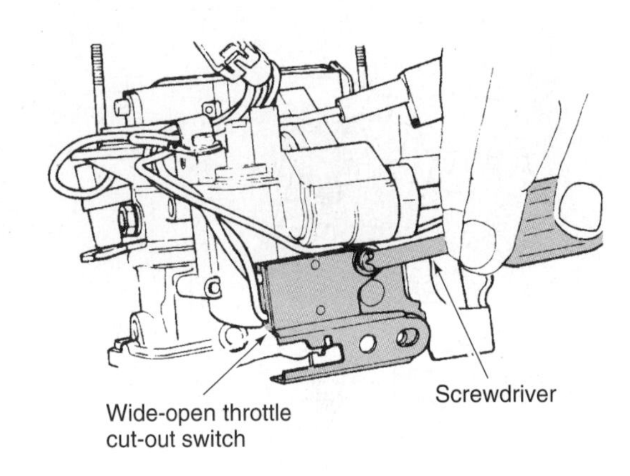

Figure 11-5 Removing the wide-open throttle cut-out A/C switch (Courtesy of Chrysler Corporation)

4. Remove any solenoids from the carburetor (Figure 11-4).
5. Remove any electric switches, such as the wide-open throttle cut-out A/C switch, from the carburetor (Figure 11-5).
6. Remove any diaphragms, such as a secondary throttle diaphragm (Figure 11-6). Discard the gasket between the diaphragm and the mounting surface.
7. Remove all the air horn mounting screws (Figure 11-7), and remove the air horn from the float bowl. Discard the air horn gasket.
8. Remove the float lever pin and the float (Figure 11-8).
9. Remove the needle valve and seat (Figure 11-9), and discard the gasket on the seat.
10. Remove the main metering jets, and check the size of each jet as it is removed (Figure 11-10). The primary and secondary main jet sizes are different in many two-stage, two-barrel carburetors.
11. Remove the secondary high speed bleed (Figure 11-11) and the high speed primary bleed (Figure 11-12). Identify each bleed size as it is removed, because they are usually different in size.
12. Remove the accelerator pump discharge nozzle (Figure 11-13). Invert the carburetor and drop the weight ball and check ball out of the pump discharge passage. These balls are the same size.

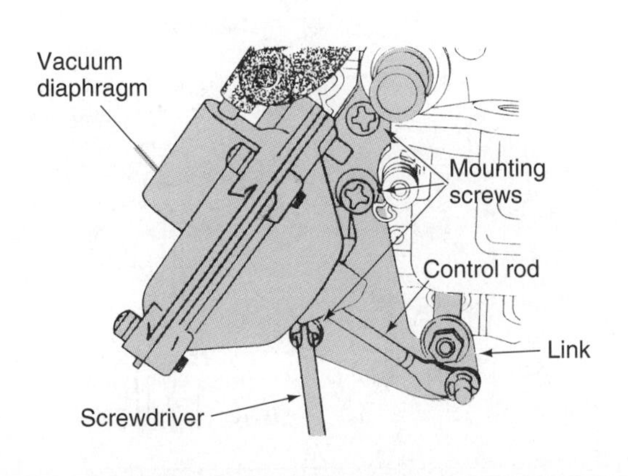

Figure 11-6 Removing the secondary throttle diaphragm (Courtesy of Chrysler Corporation)

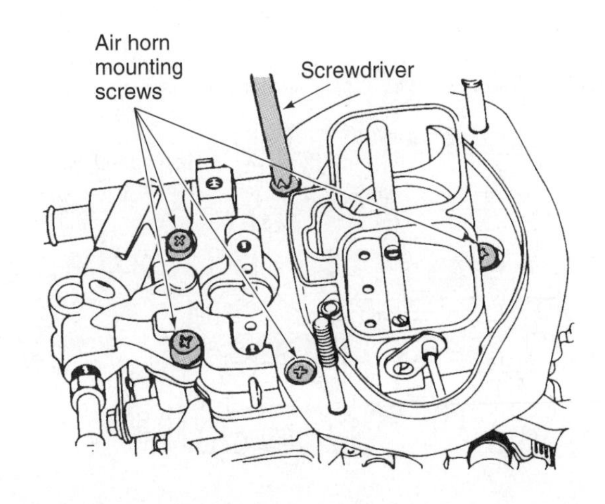

Figure 11-7 Removing the air horn mounting screws (Courtesy of Chrysler Corporation)

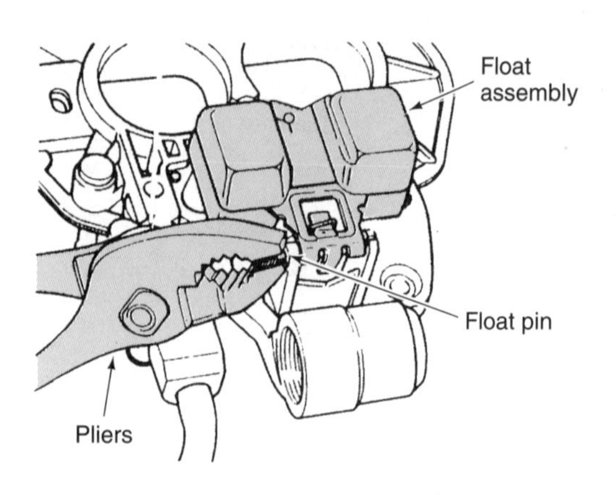

Figure 11-8 Removing the float lever pin and float (Courtesy of Chrysler Corporation)

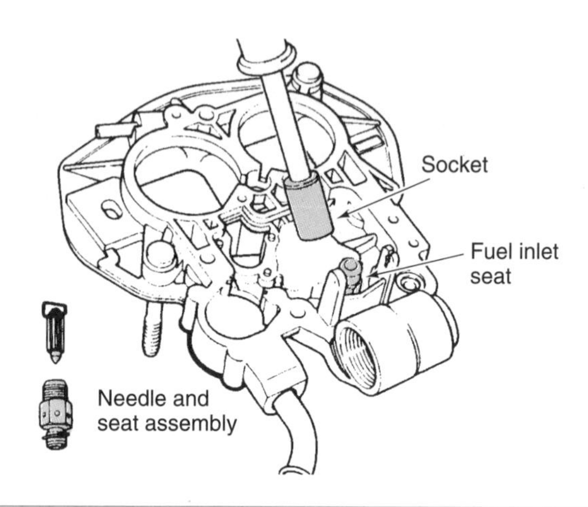

Figure 11-9 Removing needle valve and seat (Courtesy of Chrysler Corporation)

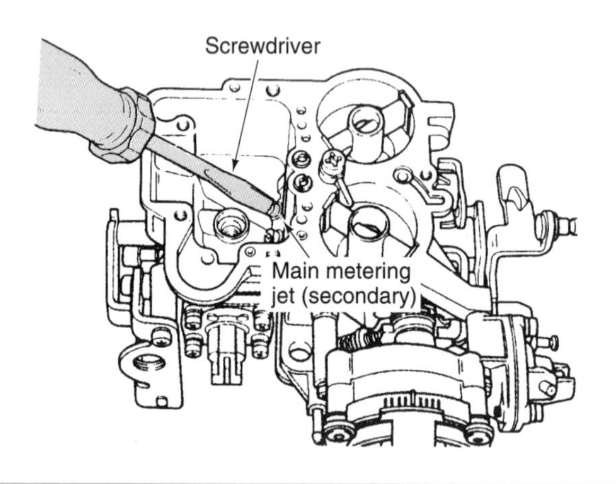

Figure 11-10 Removing main metering jets (Courtesy of Chrysler Corporation)

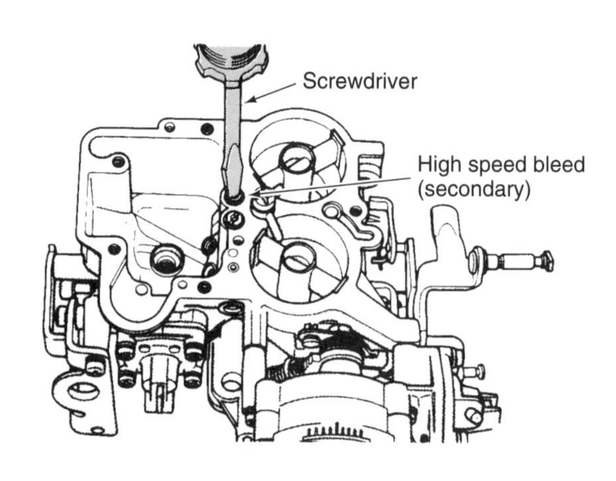

Figure 11-11 Removing secondary high-speed bleed (Courtesy of Chrysler Corporation)

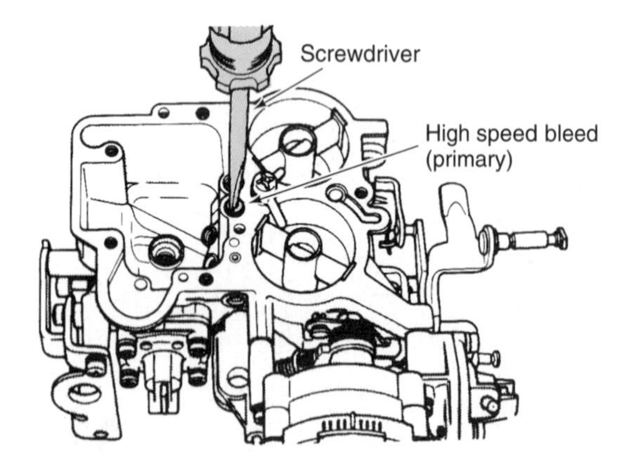

Figure 11-12 Removing primary high speed bleed (Courtesy of Chrysler Corporation)

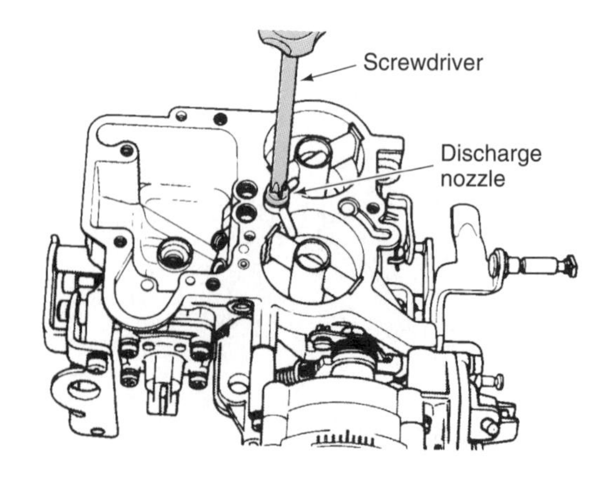

Figure 11-13 Removing accelerator pump nozzle (Courtesy of Chrysler Corporation)

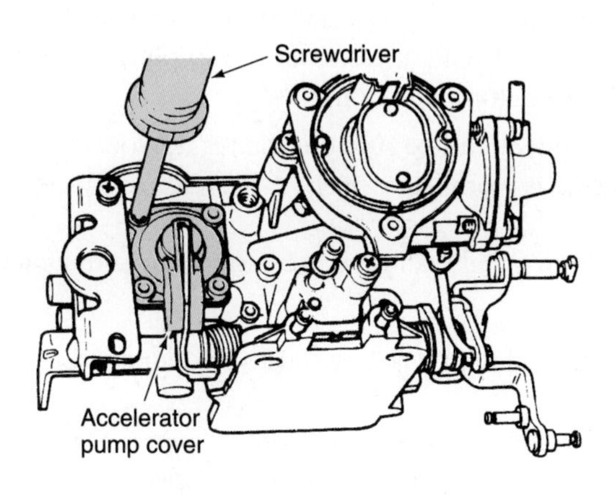

Figure 11-14 Removing the accelerator pump cover screws (Courtesy of Chrysler Corporation)

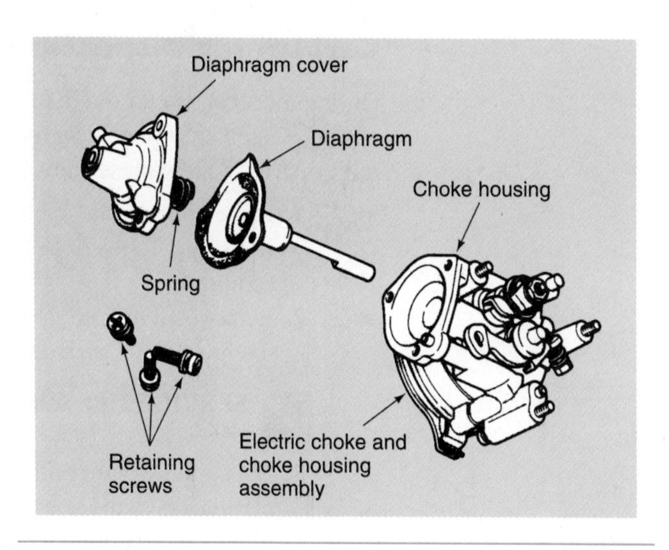

Figure 11-15 Removing choke cover and diaphragm (Courtesy of Chrysler Corporation)

13. Remove the screws from the accelerator pump cover and diaphragm, and remove these components (Figure 11-14).
14. Remove the screws from the choke diaphragm cover, and remove the cover and diaphragm (Figure 11-15).
15. Remove the choke cover retaining screws or rivets. If the choke cover is retained with rivets, the heads must be drilled from the rivets with the drill size specified in the service manual (Figure 11-16). After the heads are drilled from the rivets, use a pin punch to drive the rivets out of the housing. Then remove the choke cover.
16. If the carburetor base is detachable, remove the base screws and remove the base. Discard the base-to-bowl gasket.
17. Center punch the carburetor casting 0.250 inch from the end of the idle mixture screw housing. Drill through the outer housing at the center-punched location with a 0.187-inch drill bit. Insert a small pin punch through this drilled opening and pry out the idle mixture screw concealment plug (Figure 11-17).
18. Use the proper size Allen wrench to remove the idle mixture screw.

Special Tools

Electric drill

Drill bit set

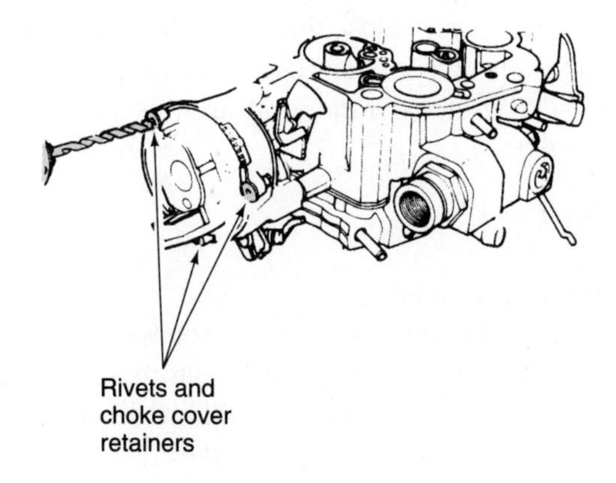

Figure 11-16 Drilling the choke cover retaining rivets (Courtesy of Chrysler Corporation)

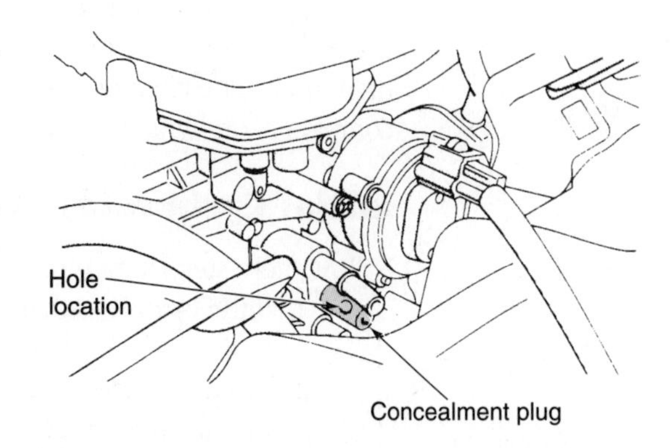

Figure 11-17 Removing idle mixture screw concealment plug (Courtesy of Chrysler Corporation)

Carburetor Inspection

During most carburetor overhaul procedures, a carburetor kit is installed. This kit usually includes all the gaskets and fiber washers, needle valve and seat, accelerator pump plunger, and other minor parts. Follow these steps for a typical carburetor inspection:

1. Inspect the choke valve for free movement and be sure the shaft is not loose in the housing. If the choke shaft or housing is worn excessively, carburetor replacement is required.
2. Remove all the old gasket material from metal mating surfaces, and use a straightedge to check these surfaces for warping.

SERVICE TIP: A float that is saturated with gasoline causes flooding problems.

3. Check the weight of the float to determine if it is saturated with gasoline. Some service manuals give the specified float weight. If the float is heavier than specified, it must be replaced. If a hollow copper float is used, shake the float beside your ear. Fuel sloshing will be heard if the float contains gasoline. Check the pivot pin opening in the float arm and the float pin for wear, and replace these components if necessary.
4. Check all carburetor linkages for wear.
5. Check the throttle shaft for wear and looseness in the base casting. If the throttle shaft is excessively loose, the carburetor base or the complete carburetor must be replaced.
6. Check the power piston and power valve for free movement and wear. Replace these components as required. If the carburetor had metering rods, check these rods for wear and bending.
7. Check the choke cover for cracks and warping, and inspect the choke spring for distortion.

Carburetor Cleaning

CAUTION: Always wear eye protection in the shop. Carburetor cleaners contain strong acids that are harmful to the eyes.

CAUTION: Carburetor cleaner will irritate human flesh. Use protective gloves when using this solution.

Place all the metal carburetor parts in an approved carburetor cleaner. Do not place solenoids, floats, choke covers and springs, gaskets, filters, O-rings, diaphragms, plastic parts, or fiber washers in carburetor cleaner. Some nylon bushings such as throttle shaft bushings may be placed in carburetor cleaner for a short time. The carburetor castings should be completely immersed in the cleaning solution and left in the solution until they are clean. Rinse the carburetor components in hot water or solvent after they are removed from the cleaner. Always use the rinsing agent recommended by the cleaner manufacturer. After the carburetor components are rinsed, blow them completely dry with compressed air.

Initial Carburetor Assembly

Follow these steps for a typical carburetor assembly procedure:

1. If the carburetor has a detachable base, install the base gasket and screws. Be sure the gasket is properly positioned and tighten the screws to the specified torque.
2. Install the idle mixture screw. Bottom the screw lightly and back it out about 1.5 turns.
3. Install a new gasket on the seat and tighten the seat to the specified torque.
4. Install a new air horn gasket and install the needle valve, float, and float pin.
5. Install the choke diaphragm and cover, and tighten the cover retaining screws alternately.

6. Install the accelerator pump diaphragm and cover, and tighten the cover retaining screws alternately.
7. Install the accelerator pump discharge ball and weight ball. Place a new gasket under the pump nozzle and install the nozzle. Tighten the nozzle screw.
8. Install the primary and secondary high speed bleeds in their original locations.
9. Install the primary and secondary main metering jets in their original locations.

Float Level Adjustment

Float adjusting procedures vary depending on the vehicle make and year. Always follow the vehicle manufacturer's recommended procedure in the service manual.

All the carburetor adjustments are very important to providing proper engine performance and fuel economy. When the float is adjusted to specifications, the desired float level is obtained in the float bowl. If the float level in the float bowl is too high, fuel consumption is excessive and engine stalling may occur, especially on deceleration. A lower-than-normal fuel level in the float bowl causes a lean air-fuel ratio, which may result in acceleration stumbles and surging at high speed.

A high fuel level in the float bowl increases fuel consumption and may cause engine stalling.

A low fuel level in the float bowl results in a lean air-fuel ratio and may result in acceleration stumbles.

WARNING: Do not exert excessive pressure on the needle valve when performing the float adjustment. This action results in an inaccurate float setting and may damage the needle valve.

With the float resting on the needle and the needle valve seated, measure the distance from the air horn gasket surface to the lower edge of the float with a T-scale (Figure 11-18). If the float level must be adjusted, bend the float arm near the needle valve. Repeat the same procedure on both floats.

Special Tools

T-scale

On some carburetors, such as a Holley 6520, the float setting is measured between the air horn surface and the top of the float with the air horn gasket removed (Figure 11-19). Always check the vehicle manufacturer's specifications to determine if the air horn gasket should be installed or removed while measuring the float setting. If this float must be adjusted, bend the tang on the float arm that rests against the needle valve (Figure 11-20).

Float Drop Adjustment

On some carburetors, a float drop adjustment is specified by the vehicle manufacturer. If the float does not drop down far enough in the float bowl, the engine may starve for fuel and surge at high speed. When the float drops down too far, the needle may stick in the seat, resulting in engine flooding.

Excessive float drop may cause the needle to stick, resulting in engine flooding.

Insufficient float drop may result in fuel starvation and engine surging at high speed.

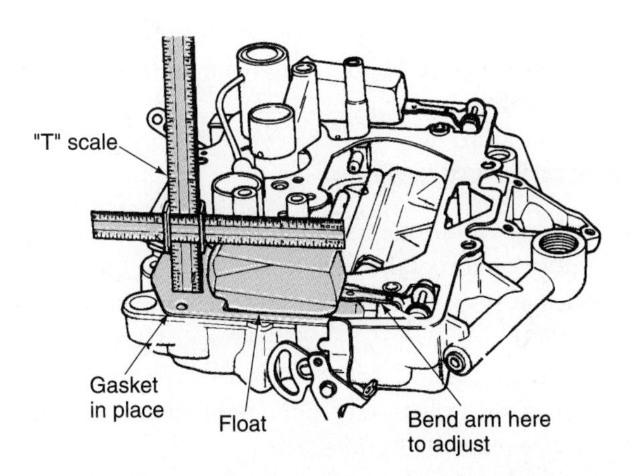

Figure 11-18 Float level adjustment (Courtesy of Chrysler Corporation)

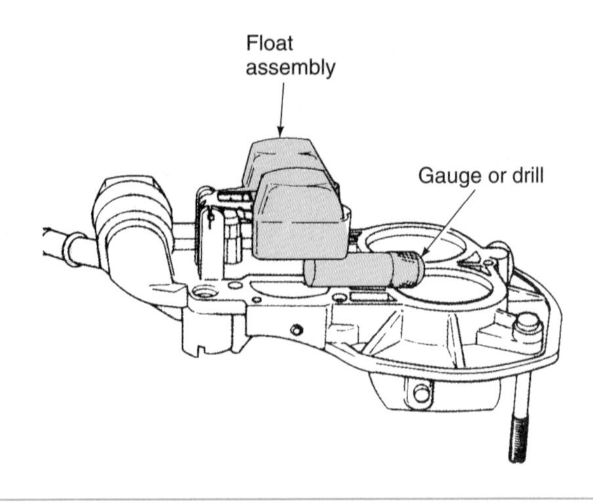

Figure 11-19 Measuring float setting, Holley 6520 carburetor (Courtesy of Chrysler Corporation)

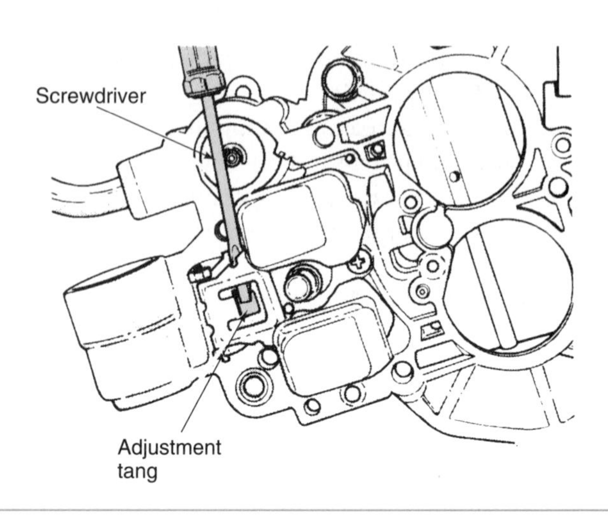

Figure 11-20 Bending float tang to adjust float setting (Courtesy of Chrysler Corporation)

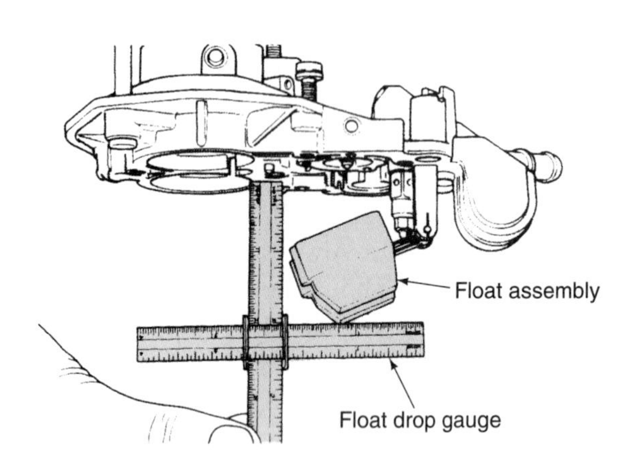

Figure 11-21 The specified float drop is measured from the air horn surface to the bottom of the float with the air horn in the upright position. (Courtesy of Chrysler Corporation)

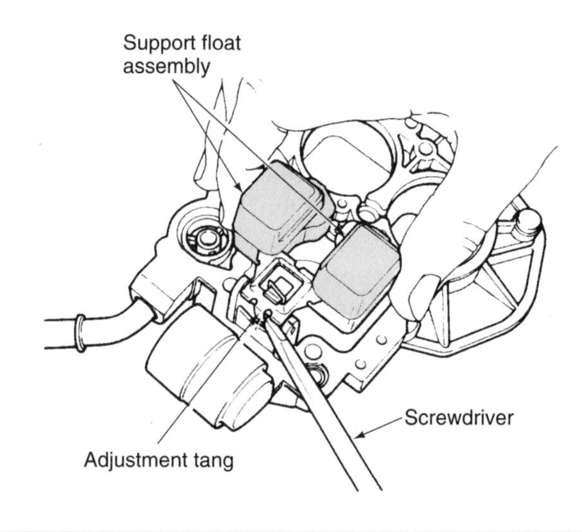

Figure 11-22 Bend the stop tang on the back of the float arm to adjust the float drop. (Courtesy of Chrysler Corporation)

Hold the air horn in the normal upright position, and measure the distance from the air horn surface to the bottom of the float (Figure 11-21). When a float drop adjustment is required, bend the stop tang on the back of the float arm (Figure 11-22).

Complete Carburetor Assembly

Follow these steps for a complete carburetor assembly after the float adjustment is completed:

1. Assemble the air horn on top of the float bowl. Be sure the air horn gasket and the float pin are in place. Check for float interference with the sides of the float bowl as the air horn is installed.
2. Install all the air horn screws and start the screws into their threaded openings. Tighten the air horn screws alternately in the tightening sequence shown in the vehicle manufacturer's service manual (Figure 11-23).
3. Install all carburetor solenoids, and tighten the solenoid retaining screws to the specified torque.

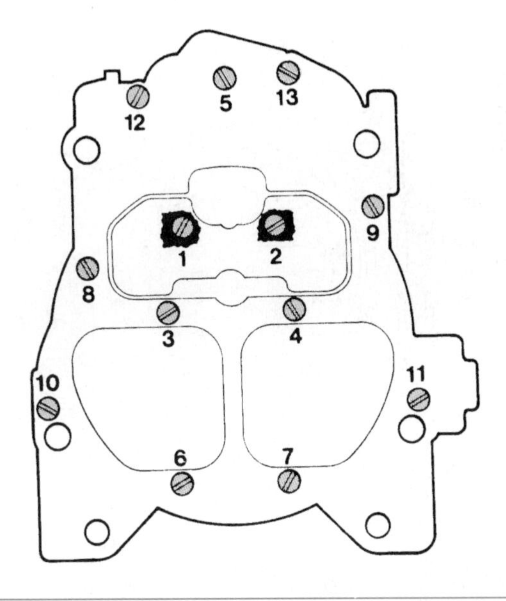

Figure 11-23 Air horn screw tightening sequence (Courtesy of Chrysler Corporation)

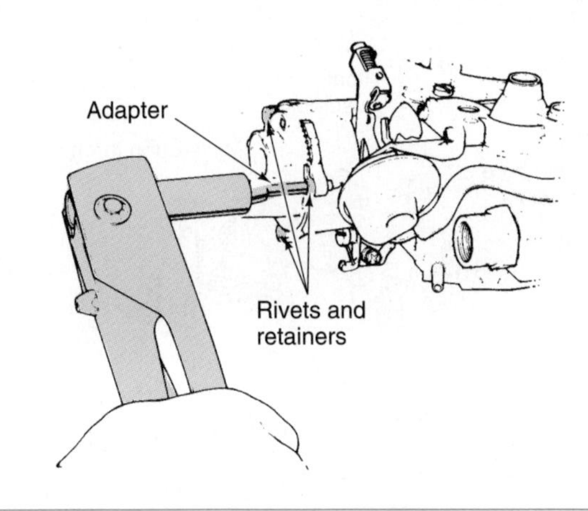

Figure 11-24 Installing new rivets to retain the choke cover to the housing (Courtesy of Chrysler Corporation)

4. Install the choke operating rod and clip.
5. Install all choke linkages and retaining clips in their original locations.
6. Install the choke cover. On hot air chokes, a gasket is positioned between the cover and the housing. This gasket is not used on electric chokes since the cover must ground on the housing.

SERVICE TIP: When the choke thermostatic spring is properly set, the choke valve should be very lightly closed with the spring at room temperature.

7. Rotate the housing until the mark on the choke cover is aligned with the specified mark on the housing. Hold the cover in this position and tighten the cover retaining screws. If the choke cover is riveted to the housing, install new rivets with a riveting tool (Figure 11-24).
8. Install all electric switches on the carburetor and tighten the retaining screws.
9. Install the secondary throttle diaphragm with a new gasket between the diaphragm housing and the carburetor casting. Tighten the secondary diaphragm retaining screws to the specified torque.

Special Tools

Riveting tool

Carburetor Linkage Adjustments

Secondary Throttle Linkage Adjustment

SERVICE TIP: Carburetor adjustments must be done in the order they are described in the vehicle manufacturer's service manual. If this adjustment order is not followed, one adjustment may change another adjustment already completed.

The types of carburetor adjustments and the adjustment procedures vary depending on the make and year of the carburetor. Always follow the vehicle manufacturer's carburetor adjustment procedures in the service manual. Since a four-barrel carburetor usually has more adjustments than a single-barrel or two-barrel carburetor, the carburetor adjustments in our discussion are based on a Thermoquad four-barrel carburetor unless otherwise indicated.

If the secondary throttle linkage is not adjusted properly, a loss of power may result on acceleration or at wide-open throttle.

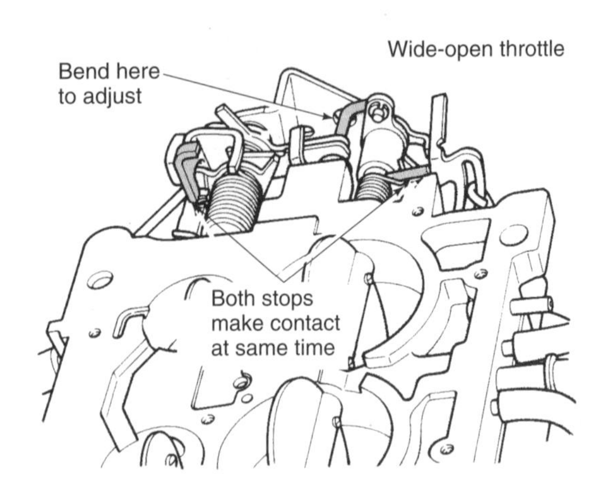

Figure 11-25 Secondary throttle linkage adjustment (Courtesy of Chrysler Corporation)

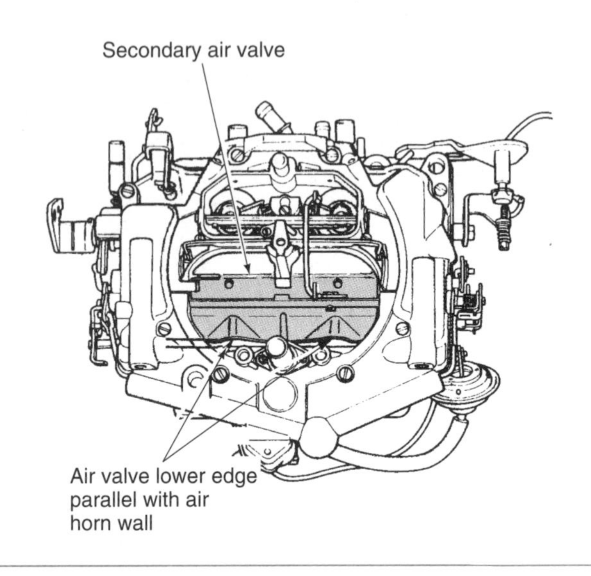

Figure 11-26 Secondary air valve alignment (Courtesy of Chrysler Corporation)

When the secondary throttle linkage adjustment is checked, hold the fast idle lever in the curb idle position and invert the carburetor. Open the primary throttle to the wide-open position, and observe the primary and secondary throttle stops. These stops should contact the carburetor casting at the same time (Figure 11-25). When the throttle stops do not contact the castings at the same time, bend the linkage between the primary and secondary throttles to provide the proper stop setting.

If the secondary throttles open too soon in relation to the primary throttles, a lose of power may occur when accelerating at medium speeds. If the secondary throttles do not open soon enough in relation to the primary throttles, the secondary throttles will not be wide open when the primary throttles are wide open. This problem may cause a loss of power at wide-open throttle.

If the secondary air valve is not adjusted properly, the air valve may stick closed, causing a loss of power at high speed and during hard acceleration.

Secondary Air Valve Alignment

To check the secondary air valve adjustment, visually check to be sure the lower edge of the air valve is parallel with the air horn wall (Figure 11-26). If the air valve is not parallel to the air horn wall at this location, loosen the air valve mounting screws on the air valve shaft and reposition the air valve. After this adjustment, retighten the air valve screws.

If the secondary air valve does not open far enough, a loss of power may be experienced at high speed and on hard acceleration.

Secondary Air Valve Opening Adjustment

Push the secondary air valves wide open and measure the distance from the upper edge of the air valve to the air horn wall. If this measurement does not equal the vehicle manufacturer's specifications, bend the corner of the air valve that contacts the air valve stop until the specified distance is obtained (Figure 11-27).

Secondary Air Valve Spring Adjustment

SERVICE TIP: Rotating the secondary air valve spring adjusting screw clockwise may cause this spring to be disconnected inside the carburetor. Disassembly of the carburetor is necessary to correct this problem.

Special Tools

Secondary air valve spring adjusting tool

Loosen the secondary air valve lock screw with the T-handle on the secondary air valve spring adjusting tool. Engage the screwdriver in this adjusting tool in the secondary air valve spring adjusting screw, and rotate the screwdriver handle clockwise until the secondary air valve is lightly

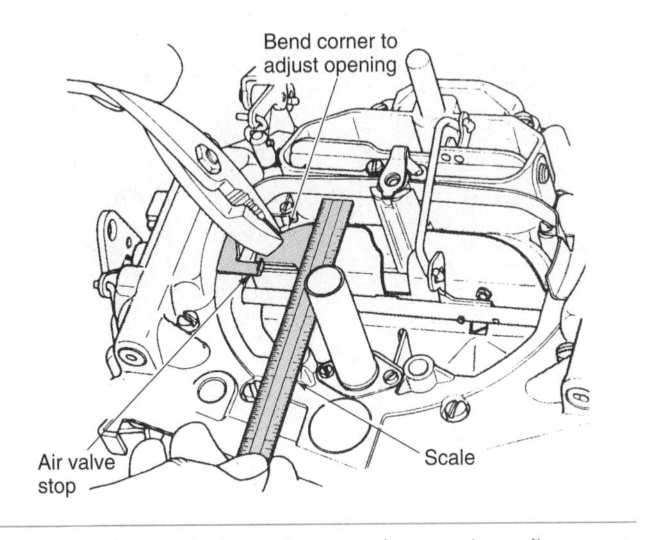

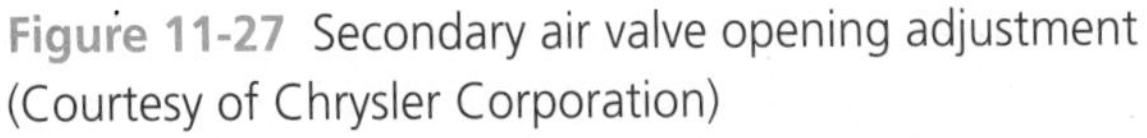
Figure 11-27 Secondary air valve opening adjustment (Courtesy of Chrysler Corporation)

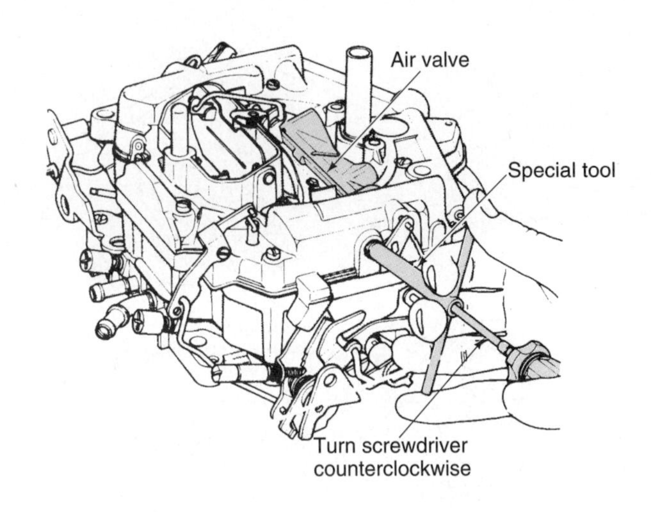

Figure 11-28 Secondary air valve spring adjustment (Courtesy of Chrysler Corporation)

closed. Turn the screwdriver the specified number of turns clockwise from this point, and tighten the lock ring with the T-handle (Figure 11-28).

If the secondary air valve spring is weaker than specified, a stumble may occur on hard acceleration.

Choke Control Lever Adjustment

SERVICE TIP: The screw in the thermostatic spring linkage has a left-hand thread.

The carburetor should be positioned on top of a smooth bench surface to perform the choke control lever adjustment. The bottom of the carburetor base must be flat on the bench surface. Hold the choke closed by pushing on the linkage where the choke thermostatic spring rod is connected. Use a ruler to measure from the bench surface to the top of the thermostatic spring rod hole (Figure 11-29). If the specified distance is not obtained on the ruler, loosen the screw in the thermostatic spring rod linkage and rotate this linkage to obtain the specified measurement.

When the secondary air valve spring has more than the specified tension, a power loss may result at high speed.

Choke Diaphragm Connector Rod Adjustment

A restriction is located in the vacuum break diaphragm outlet. When the primary throttles are open far enough to begin opening the secondary throttles, the intake manifold vacuum decreases and the stem on the vacuum break diaphragm slowly extends, because of the restriction in the vacuum passage. This vacuum break action provides a cushioning action on the secondary air valves and slows their opening to prevent an acceleration stumble.

The vacuum break, fast idle cam, and unloader adjustments are affected by an improper choke control lever adjustment.

If the choke diaphragm connector rod measurement is more than specified, an acceleration stumble may occur on hard acceleration.

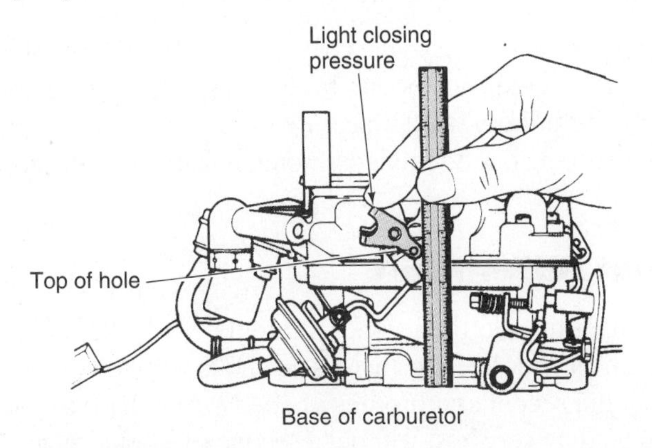

Figure 11-29 Choke control lever adjustment (Courtesy of Chrysler Corporation)

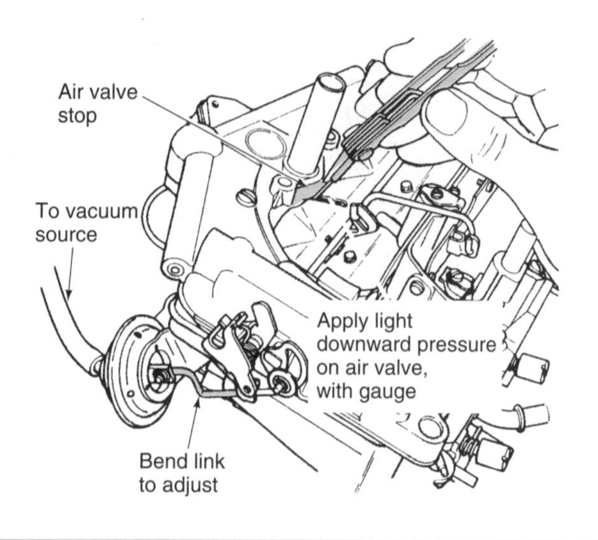

Figure 11-30 Choke diaphragm connector rod adjustment (Courtesy of Chrysler Corporation)

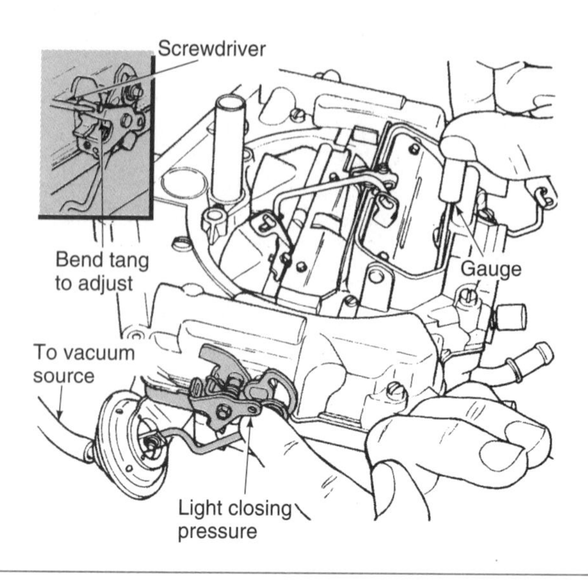

Figure 11-31 Choke vacuum kick adjustment (Courtesy of Chrysler Corporation)

Special Tools

Round carburetor gauges

If the choke vacuum kick clearance is more than specified, a hesitation, or stalling, may occur on acceleration with a cold engine.

When the choke vacuum break clearance is less than specified, the air-fuel ratio may be excessively rich after a cold engine is started.

Apply 20 inches of vacuum to the vacuum break diaphragm with a hand pump. Measure the clearance between the lower edge of the air valve and the air valve stop tang (Figure 11-30). If this clearance is not correct, bend the vacuum break linkage to obtain the proper clearance.

Choke Vacuum Kick Adjustment

The choke vacuum kick diaphragm pulls the choke open a specific amount when a cold engine is started to provide the proper air-fuel ratio under this condition.

Apply 20 inches of vacuum to the vacuum break diaphragm with a hand pump, and rotate the choke toward the closed position by pushing on the thermostatic spring rod linkage. Push the choke valve lightly toward the open position, and place the specified round gauge between the lower edge of the choke valve and the air horn wall. When the gauge is held flat against the air horn wall, the gauge should be a light push fit. If the specified clearance is not obtained, place a screwdriver in the choke linkage tang, and bend the tang until the specified measurement is available (Figure 11-31).

Choke Unloader Adjustment

The unloader mechanism forces the choke open a specific amount when the primary throttle is wide open. This action provides a method of opening the choke if a cold engine becomes flooded. If the unloader mechanism does not force the choke open far enough, it may be difficult to restart a cold flooded engine.

Apply a light closing pressure to the fast idle control lever above the choke valve to hold the choke closed and the primary throttle wide open. Insert the specified gauge between the lower edge of the choke valve and the air horn wall to measure the unloader clearance. If an adjustment is required, bend the unloader tang on the primary throttle linkage until the specified opening is obtained (Figure 11-32).

Secondary Lockout Adjustment

The secondary lockout mechanism locks the secondary throttles while the choke valve is closed or partly closed. This action is necessary because there is no choke valve on the secondary barrels of a four-barrel carburetor. If the secondary throttles opened during hard acceleration on a cold engine, the air-fuel ratio would be very lean without a secondary choke valve. This lean air-fuel ratio would cause a power loss and acceleration stumbles.

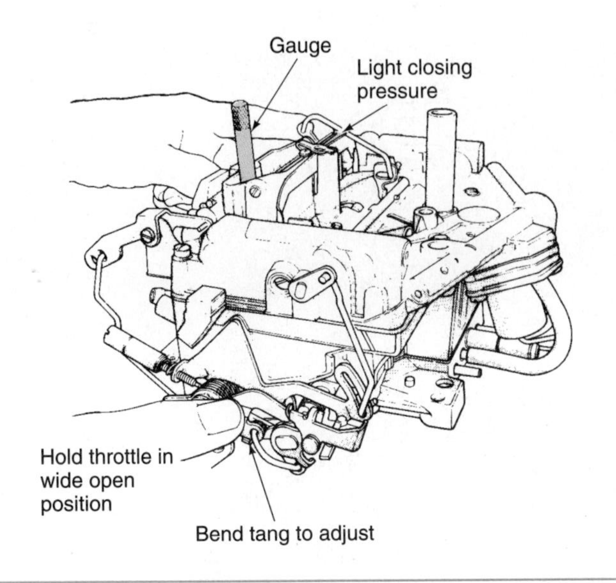

Figure 11-32 Unloader adjustment (Courtesy of Chrysler Corporation)

Light closing pressure
Bend tang to adjust
Gauge

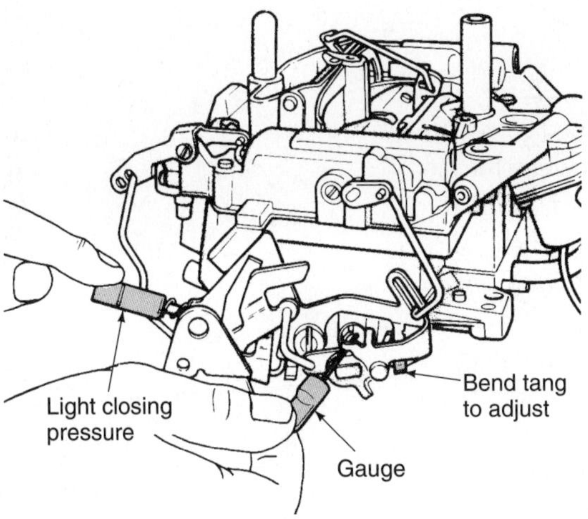

Figure 11-33 Secondary throttle lockout adjustment (Courtesy of Chrysler Corporation)

With the primary throttle in the closed position, hold downward lightly on the fast idle screw to keep the choke valve wide open. Measure the specified clearance between the lockout lever and the stop. Bend the tang on the fast idle control lever under the lockout lever to obtain the specified clearance (Figure 11-33).

Fast Idle Cam Position Adjustment

The fast idle cam must be lifted the proper amount to provide the correct fast idle rpm while the engine is warming up and to position the throttle properly while starting a cold engine. If the fast idle cam is lifted up more than specified, the cam remains upward for a longer time period, maintaining the fast idle speed too long as the engine warms up. When the fast idle cam is not lifted upward as much as specified, the fast idle speed may be excessively slow on a cold engine, and the cam drops to the off position too soon in relation to engine temperature.

Place the fast idle screw on the second highest step of the fast idle cam, against the shoulder of the highest step. Hold downward lightly on the fast idle control lever, and measure the specified clearance between the lower edge of the choke valve and the air horn wall. If an adjustment is necessary, bend the fast idle connector rod at the lower bend (Figure 11-34).

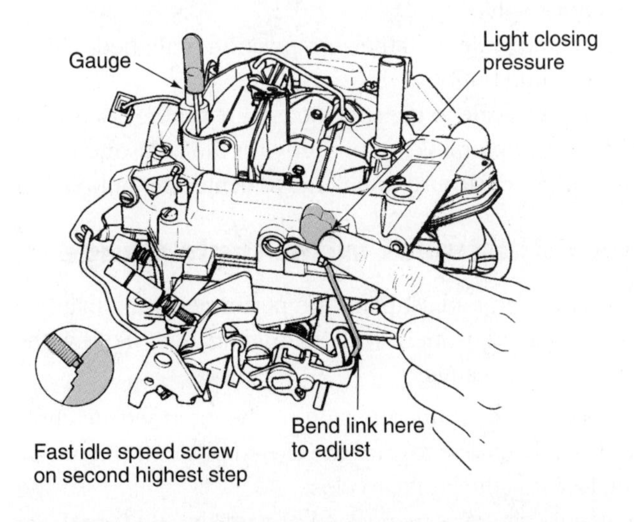

Figure 11-34 Fast idle cam position adjustment (Courtesy of Chrysler Corporation)

A flooded condition occurs when the air-fuel ratio becomes extremely rich. This problem may be caused by a carburetor defect such as a needle valve or a float loaded with gasoline.

A driver may cause a flooded condition by excessive pumping of the gas pedal on a carbureted engine. With each gas pedal pump, the accelerator pump squirts a stream of fuel into the venturi.

An improper accelerator pump stroke adjustment may cause a hesitation on low-speed acceleration.

Special Tools

Choke valve angle gauge

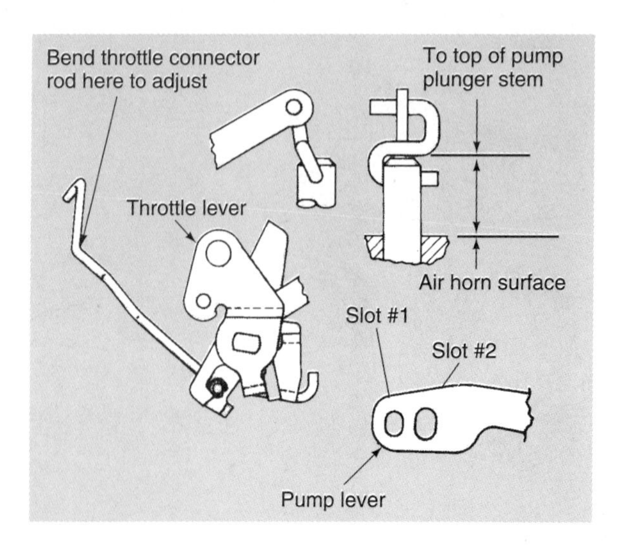

Figure 11-35 Accelerator pump stroke adjustment (Courtesy of Chrysler Corporation)

Accelerator Pump Stroke Adjustment

When the accelerator pump is adjusted properly, the correct pump stroke length is provided. If the pump stroke is too short, the pump does not discharge enough fuel, and an acceleration stumble may occur. When the pump stroke is longer than specified, the pump discharges excessive fuel, which is wasted.

Be sure the accelerator pump linkage is in the specified slot in the pump lever. With the throttle in the curb idle position, measure the distance between the top of the air horn casting and the top of the pump plunger. If this distance requires adjusting, bend the accelerator pump linkage at the upper bend (Figure 11-35).

Using a Choke Valve Angle Gauge Procedure

Some vehicle manufacturers specify the use of a choke valve angle gauge while performing carburetor adjustments. This gauge contains a degree scale, pointer, and a center leveling bubble. A magnet on the lower end of the gauge holds the gauge on the choke valve. This procedure must be followed while using the choke valve angle gauge:

1. Allow the magnet to hold the gauge on the choke valve.
2. Close the choke valve.
3. Rotate the degree scale until the zero position is aligned with the pointer.
4. Center the leveling bubble.
5. Rotate the degree scale to the specified angle for the adjustment being performed.
6. Position the carburetor linkage properly for the adjustment being performed.
7. Adjust the specified carburetor linkage to center the bubble if necessary (Figure 11-36).

Fast Idle Cam Adjustment with Angle Gauge

Since the same basic carburetor adjustments are performed with an angle gauge or a round feeler gauge, we will discuss one adjustment with an angle gauge. Follow these steps for the fast idle cam adjustment with an angle gauge:

1. Install a rubber band around the tang on the intermediate choke shaft and the air horn casting to hold the choke toward the closed position (Figure 11-37). Be sure the throttle is open to allow the choke to close.
2. Set up the angle gauge as mentioned previously. Be sure the angle gauge is set to the specified fast idle cam angle.

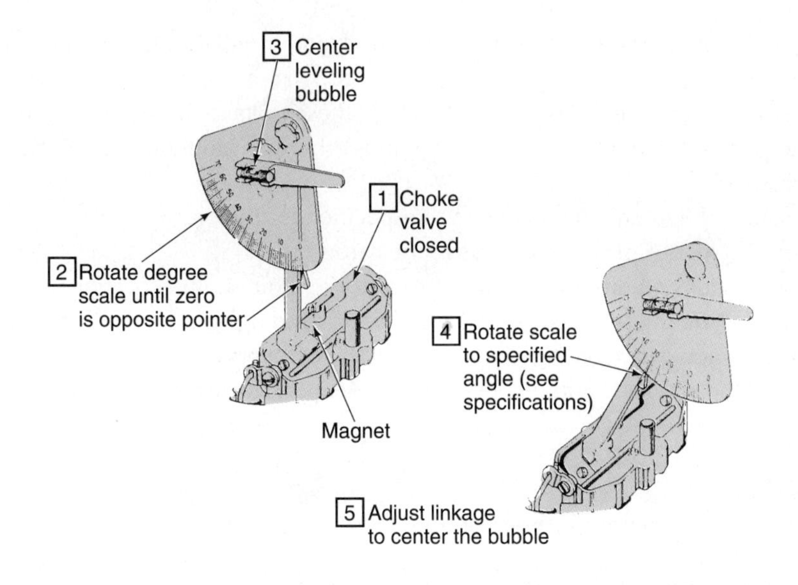

Figure 11-36 Choke valve angle gauge (Courtesy of Chrysler Corporation)

3. Open the choke valve until the cam follower is on the second highest step of the fast idle cam against the shoulder of the high step.
4. If an adjustment is necessary, bend the tang under the fast idle cam until the bubble is centered.

Carburetor Installation

Follow these steps for a typical carburetor installation:

1. Scrape all gasket material from the carburetor mounting surface on the intake manifold.
2. If a spacer is positioned between the carburetor and the intake manifold, scrape all gasket material from both sides of the spacer. Use a straightedge to check the spacer for warping.

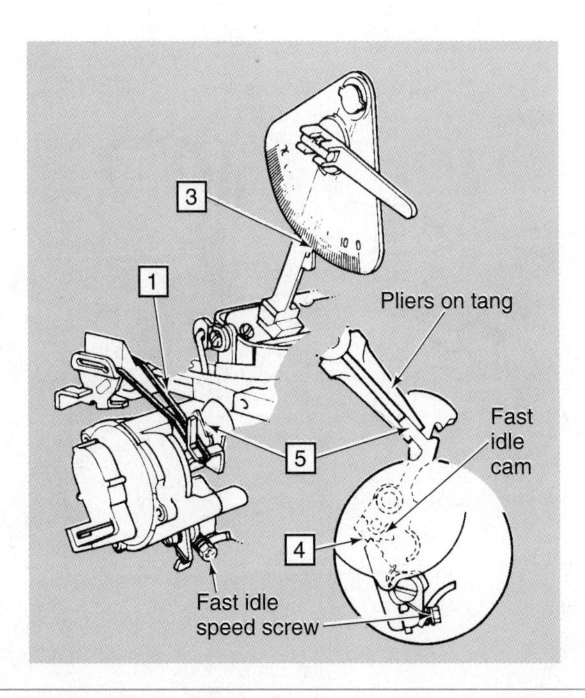

Figure 11-37 Fast idle cam adjustment with an angle gauge (Courtesy of Chrysler Corporation)

3. Install a new gasket on the intake manifold, and install the spacer with a new gasket on top of the spacer.
4. If spacer mounting bolts are used, tighten these bolts to the specified torque. Install the carburetor and alternately tighten the carburetor mounting nuts to the specified torque.
5. Install and tighten the fuel line to the specified torque.
6. Connect all electrical connectors to the carburetor.
7. Connect all vacuum hoses to the carburetor in their original locations.
8. Connect the throttle linkage to the carburetor and tighten the retaining nut.
9. Install a new air cleaner gasket on top of the carburetor.
10. Install the air cleaner, and connect all air cleaner hoses in their original locations.

Idle Mixture and Speed Adjustments

Idle Speed Adjustment

The engine must be at normal operating temperature before any idle speed and mixture adjustments are performed. The idle speed, fast idle speed, and idle mixture adjustment procedures vary depending on the vehicle make and year. Always follow the exact procedure in the vehicle manufacturer's service manual. Following is a typical idle speed adjustment procedure:

1. Connect a tachometer from the negative primary coil terminal to ground. Connect the black clip to ground.
2. Disconnect all items recommended in the vehicle manufacturer's service manual. In some cases, this may include disconnecting and plugging the vacuum hoses at the exhaust gas recirculation (EGR) valve and air cleaner heated air temperature sensor. Other manufacturers recommend removing the positive crankcase ventilation (PCV) valve, and allowing the valve to pull in underhood air.
3. Start the engine and wait one minute for the speed to stabilize.
4. Adjust the idle speed to specifications with the idle speed screw (Figure 11-38).

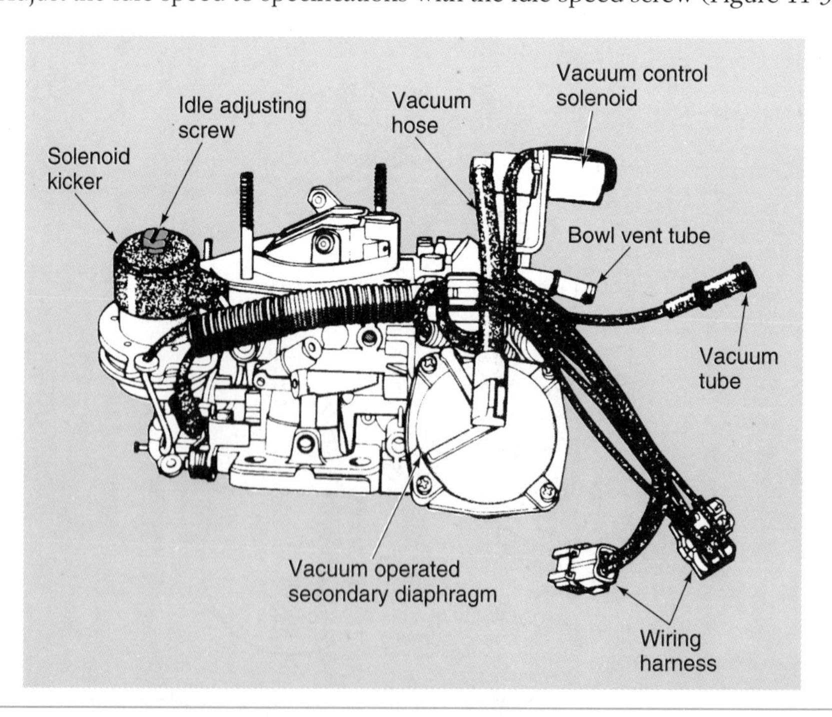

Figure 11-38 Idle speed adjusting screw (Courtesy of Chrysler Corporation)

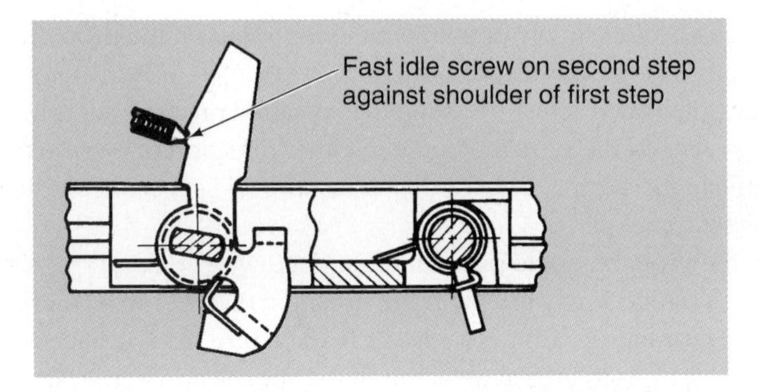

Figure 11-39 Adjusting fast idle speed (Courtesy of Chrysler Corporation)

Fast Idle Speed Adjustment

The fast idle speed adjustment may be completed after the idle speed adjustment. Following is a typical fast idle speed adjustment procedure:

1. Disconnect all vacuum hoses and components recommended in the vehicle manufacturer's service manual. Usually these are the same components disconnected for the idle speed adjustment.
2. Leave the tachometer connected as it was in the idle speed adjustment procedure.
3. Place the fast idle screw on the second highest step of the fast idle cam against the shoulder of the highest step, and be sure the choke is wide open (Figure 11-39).
4. Adjust the fast idle screw to obtain the specified fast idle rpm.
5. Touch the throttle once to allow the fast idle cam to drop downward. Repeat steps 3 and 4 to verify the adjustment.

Idle Mixture Adjustment

WARNING: Tampering with a carburetor is a violation of federal law in the United States.

Tampering with a carburetor may be defined as any carburetor adjustment that is not recommended by the vehicle manufacturer and causes higher-than-specified exhaust emissions levels for that year of vehicle.

The idle mixture adjustment procedure in the vehicle manufacturer's service manual must be followed, because this procedure varies widely depending on the year and make of vehicle. Before the idle mixture adjustment is performed, the idle speed and basic timing must be set to specifications. All vacuum hose connections and underhood wiring connectors should be checked. The idle mixture adjustment should be performed after a carburetor overhaul. When the idle mixture is adjusted using the vehicle manufacturer's recommended procedure, the vehicle should meet emission level standards, assuming the engine compression and ignition system are in satisfactory condition. A significant number of vehicle manufacturers recommend the propane enrichment method of idle mixture adjustment. Follow these steps for a typical propane enrichment idle mixture adjustment:

1. Be sure the engine is at normal operating temperature.
2. Disconnect the vacuum hoses at the EGR valve, distributor vacuum advance, and air cleaner heated air temperature sensor.
3. Connect the hose from the propane bottle to the vacuum outlet on the intake manifold to which the air cleaner heated air temperature sensor was connected.
4. Place the propane bottle in a safe, upright position.
5. Remove the PCV valve from the rocker arm cover, and allow this valve to pull in underhood air.
6. Disconnect and plug the 0.187-inch hose from the vapor canister.

Special Tools

Propane bottle with metering valve

7. Start the engine and slowly open the metering valve on the propane cylinder until the highest rpm is obtained (Figure 11-40).
8. Adjust the idle speed screw to obtain the specified propane set rpm. Slightly turn the metering valve on the propane bottle in either direction to be sure the engine is running at the highest rpm. If necessary, adjust the idle speed screw slightly to obtain the propane set rpm.
9. Turn off the propane bottle main valve, and allow the engine speed to stabilize. Turn the mixture screw slowly to obtain the specified idle set rpm (Figure 11-41). Since two-barrel and four-barrel carburetors have dual mixture screws, alternately turn these screws inward 1/16th inch at a time until the specified idle set rpm is obtained.

Special Tools

Tachometer

Computer-Controlled Carburetor Service and Diagnosis

Preliminary Diagnostic Procedure

When a problem occurs in an automotive computer system, the tendency is to blame the computer system. However, before any computer system diagnosis is performed, these items must be checked first and proven to be in satisfactory condition:

1. Engine compression
2. Intake manifold vacuum leaks
3. Emission devices such as the EGR valve
4. Ignition system components

Prior to a computer system diagnosis, all wiring harness connections to computer system components should be checked for damage or loose connections. The engine must be warmed up to normal operating temperature before fault codes are obtained on any computer system.

Idle Speed Adjustment

A/C Idle Speed Check. The idle speed adjustment procedure varies with each vehicle, engine, or model year. Always follow the vehicle manufacturer's recommended procedure in the service manual. The following A/C idle rpm check applies to a carburetor with a combined vacuum and

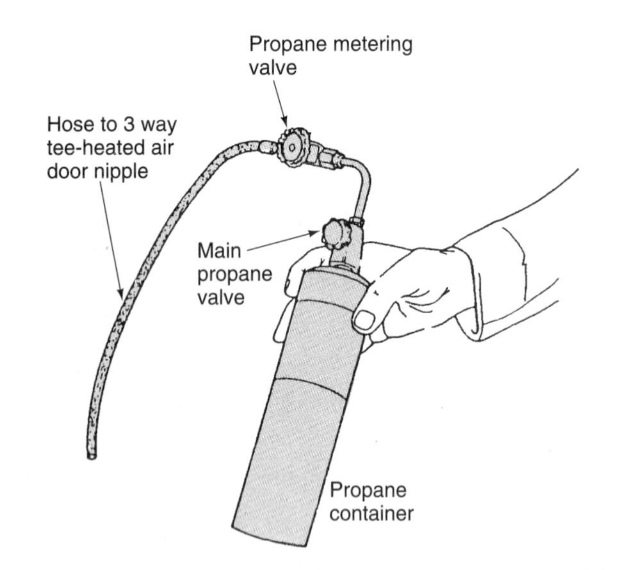

Figure 11-40 Propane bottle with hose and metering valve (Courtesy of Chrysler Corporation)

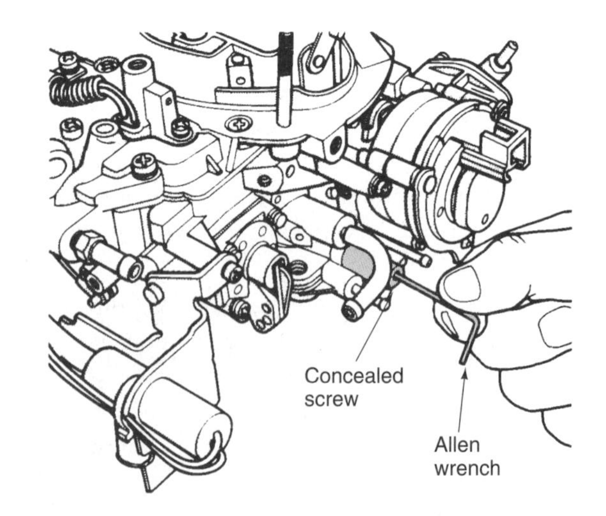

Figure 11-41 Idle mixture screw adjustment (Courtesy of Chrysler Corporation)

electric throttle kicker. The vacuum throttle kicker is activated when the driver selects the A/C mode, and the electric kicker solenoid is energized when the ignition switch is turned on. Prior to any idle speed check, the engine must be at normal operating temperature. Follow these steps for the A/C idle speed check:

1. Select the A/C mode and set the temperature control to the coldest position.
2. Each time the A/C compressor cycles off and on, the kicker solenoid should be energized and the vacuum kicker plunger should move in and out. If the plunger reacts properly, the system is satisfactory. There is no adjustment on the vacuum kicker stem. When the vacuum kicker plunger does not react properly, check all the vacuum hoses and the solenoid, and test the vacuum kicker diaphragm for leaks.

Engine Idle Speed Check. The ignition timing should be checked and adjusted as necessary prior to the idle speed check. Follow these steps for the idle rpm check:

1. With the transaxle in neutral and the parking brake applied, turn all the accessories and lights off. Be sure the engine is at normal operating temperature and connect a tachometer from the coil negative primary terminal to ground.
2. Disconnect the cooling fan motor connector and connect 12 V to the motor terminal so the fan runs continually.
3. Remove the PCV valve from the crankcase vent module and allow this valve to draw in underhood air.
4. Disconnect the O_2 feedback test connector on the left fender shield.
5. Disconnect the wiring connector from the kicker solenoid on the left fender shield.
6. If the idle rpm is not as specified on the underhood emission label, adjust the idle rpm with the screw on the kicker solenoid (Figure 11-42).
7. Reconnect the O_2 connector, PCV valve, and kicker solenoid connector on the left fender shield.
8. Increase the engine rpm to 2,500 for 15 seconds and then allow the engine to idle. If the idle speed changes slightly, this is normal and a readjustment is not required.
9. Disconnect the jumper wire and reconnect the fan motor connector.

Idle Mixture Adjustment Propane-Assisted Method

The concealment plug is a metal plug placed in the outer end of the idle mixture screw bore to try to prevent improper idle air-fuel mixture adjustments.

Concealment Plug Removal. The idle mixture adjustment procedure varies depending on the vehicle model and year. Always follow the procedure in the vehicle manufacturer's service manual. The carburetor concealment plug must be removed prior to an idle mixture adjustment. Following is a typical concealment plug removal procedure:

1. Center punch the carburetor casting 1/4 in (6.35 mm) from the end of the mixture screw housing (Figure 11-43).

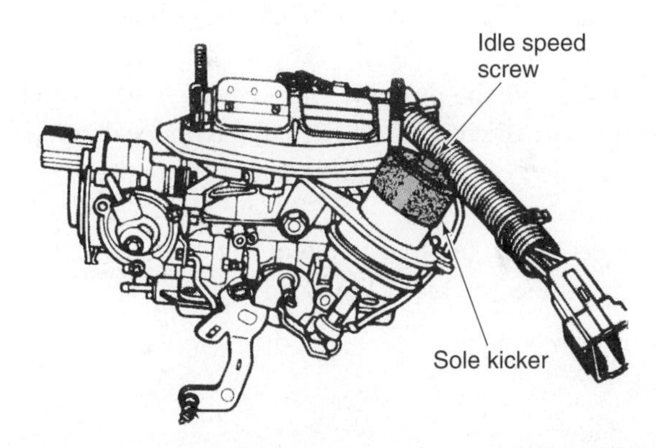

Figure 11-42 Idle speed screw on kicker solenoid (Courtesy of Chrysler Corporation)

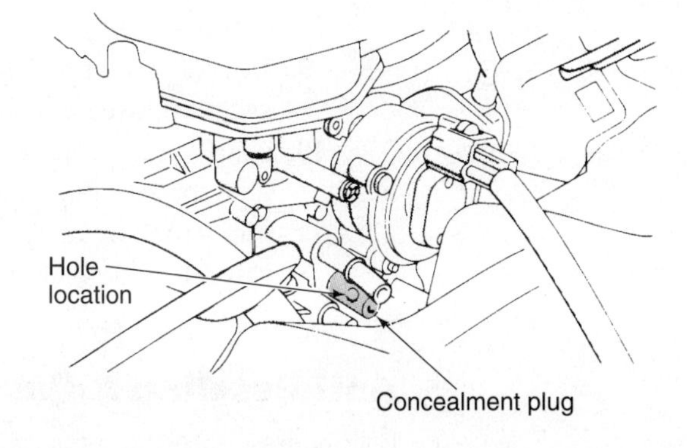

Figure 11-43 Concealment plug removal (Courtesy of Chrysler Corporation)

2. Drill the outer housing at the center-punched location with a 3/16 in (4.762 mm) drill bit.
3. Pry out the concealment plug and save the plug for reinstallation.

Propane-Assisted Idle Mixture Adjustment. The propane-assisted idle mixture adjustment may vary depending on the vehicle make and year. Always follow the idle mixture adjustment procedure in the vehicle manufacturer's service manual. After the concealment plug is removed, follow these steps for a typical propane-assisted idle mixture adjustment:

1. With the transaxle in neutral and the parking brake on, turn off all the lights and accessories. Be sure the engine is at normal operating temperature, and connect a tachometer from the negative primary coil terminal to ground.
2. Disconnect the cooling fan motor connector, and connect a jumper wire from a 12-V source to the fan motor terminal so the fan motor runs all the time. Remove the PCV valve from the crankcase vent module and allow this valve to draw in underhood air. Disconnect the O_2 feedback system test connector on the left fender shield.
3. Disconnect the vacuum hoses from the coolant vacuum switch cold closed (CVSCC) valve on the driver's side of the thermostat housing and plug both hoses. This valve controls the vacuum to the EGR valve on some engines. On 2.2-L engines, disconnect the wiring harness connector from the throttle kicker solenoid on the left fender shield.
4. Locate the vacuum hose from the air cleaner heated air sensor to the intake manifold and disconnect this hose from the intake manifold. Connect the supply hose from the propane bottle to the intake manifold connection where the air cleaner hose was removed. Be sure the propane bottle valves are closed and the bottle is in an upright position. Position the propane bottle in a safe place where it will not come in contact with rotating parts.

After an idle air-fuel mixture adjustment is completed, the vehicle must conform to all federal and state emission standards.

CAUTION: During the propane-assisted idle mixture adjustment, the propane bottle must be kept in an upright position and located in a safe place where it will not contact rotating parts.

5. With the air cleaner in place, slowly open the main valve on the propane bottle. Slowly open the propane metering valve until the highest engine rpm is reached. When excessive propane is added, the engine starts to slow down. Slowly rotate the propane metering valve until the highest possible rpm is reached.
6. With the propane still flowing, adjust the idle speed screw on the carburetor electric kicker solenoid to obtain the specified propane idle speed. Increase the engine speed to 2,500 rpm for 15 seconds, and allow the engine speed to return to idle. Slowly rotate the propane metering valve in either direction, recheck the propane rpm, and adjust as necessary.
7. Close the main valve on the propane bottle and allow the engine to idle. Adjust the mixture screw to obtain the specified idle set rpm. Increase the engine rpm to 2,500 for 15 seconds and allow the engine to return to idle speed. Recheck the idle set rpm and adjust the idle mixture screw to correct the idle set rpm if necessary.
8. Turn on the main propane bottle valve and slowly rotate the metering valve to obtain the highest rpm. If the rpm is more than 25 rpm above or below the propane set rpm, repeat steps 5 through 8.
9. Turn off both propane bottle valves, remove the propane supply hose, and install the vacuum hose from the air cleaner to the intake manifold. Install the concealment plug.
10. Place the fast idle screw on the lowest step of the fast idle cam and rotate this screw until the specified fast idle speed is obtained.
11. Allow the engine to idle and reconnect the O_2 feedback system test connector, PCV valve, throttle kicker solenoid wiring connector, and CVSCC valve hoses.

Antidieseling Adjustment

Follow these steps for a typical antidieseling adjustment:

1. Be sure the engine is at normal operating temperature and place the transaxle in neutral with the parking brake applied. Turn off all the lights and accessories.

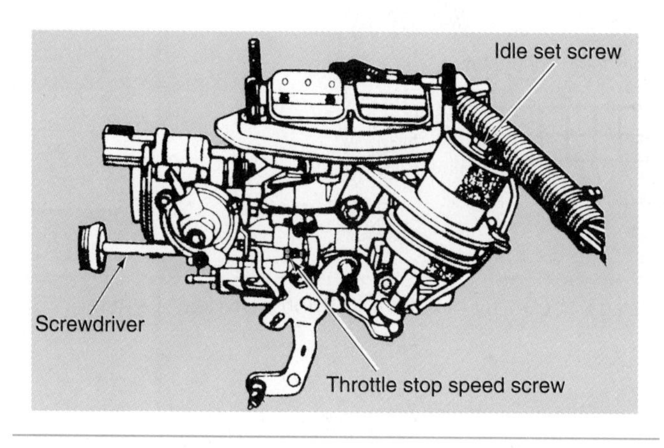

Figure 11-44 Throttle stop screw (Courtesy of Chrysler Corporation)

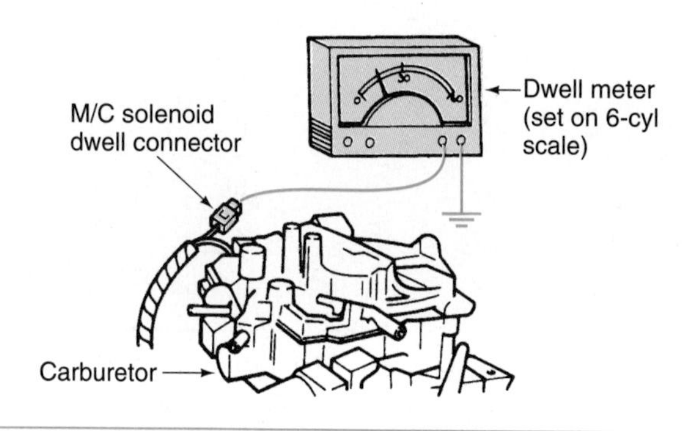

Figure 11-45 Dwell meter connection to the mixture control (MC) solenoid dwell connector (Courtesy of Chevrolet Motor Division, General Motors Corporation)

2. Remove the red wire from the six-way carburetor connector on the carburetor side of the connector, and disconnect the O_2 feedback test connector on the left fender shield.
3. Adjust the throttle stop screw to 700 rpm (Figure 11-44).
4. Reconnect the red wire in the six-way carburetor connector and the O_2 feedback test connector.

Classroom Manual
Chapter 11, page 257

Computer-Controlled Carburetor System Performance Test

During a computer-controlled carburetor performance test, the engine rpm is checked under specific conditions with the mixture control solenoid connected and disconnected to determine if this solenoid is functioning properly. The following performance test procedure applies to General Motors vehicles. However, a similar performance test may be recommended on other vehicles.

Prior to any computer command control (3C) diagnosis, complete the preliminary diagnostic procedure mentioned previously in this chapter. Connect a dwell meter to the mixture control (MC) solenoid dwell connector and ground to begin the performance test (Figure 11-45). The dwell connector is a green plastic connector on a blue wire located near the carburetor on many models. Since the MC solenoid operates on a duty cycle of ten times per second regardless of the engine type, the dwell meter must be used on the six-cylinder scale on all engines.

Special Tools
Dwellmeter

WARNING: Never ground or short across any terminals in a computer system unless instructed to do so in the vehicle manufacturer's service manual. This action may damage computer system components.

WARNING: Never disconnect any computer system electrical connectors with the ignition switch on because this action may damage system components.

Be sure that the engine is at normal operating temperature, and then complete these steps in the performance test:

1. Start the engine.
2. Connect a jumper wire between terminals A and B in the data link connector (DLC) under the dash (Figure 11-46).
3. Connect a tachometer from the ignition coil tach terminal to ground.
4. Disconnect the MC solenoid wiring connector and ground the MC solenoid dwell lead. This action moves the MC solenoid plunger upward and provides a rich air-fuel ratio.
5. Run the engine at a constant 3,000 rpm.
6. Reconnect the MC solenoid connector and note the engine rpm. This action moves the MC solenoid plunger downward and creates a lean air-fuel ratio because the MC solenoid

The SAE J1930 terminology is an attempt to standardize terminology in automotive electronics.

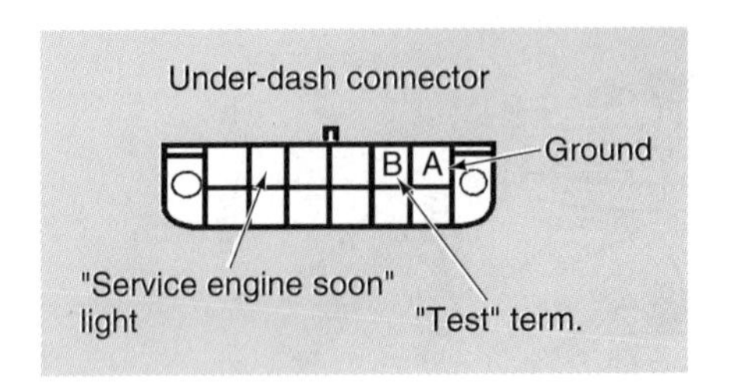

Figure 11-46 Data link connector (DLC) (Courtesy of Chevrolet Motor Division, General Motors Corporation)

In the SAE J1930 terminology, the term power train control module (PCM) replaces all other terms for engine computers.

In the SAE J1930 terminology, the term malfunction indicator light (MIL) replaces all other terms for check engine light.

In the SAE J1930 terminology, the term data link connector (DLC) replaces the terms self-test connector and assembly line diagnostic link (ALDL).

dwell connector is grounded. If the rpm decreases 300 or more, the MC solenoid plunger is reacting properly. When the rpm drop is less than 300, check the MC solenoid wires, and inspect the evaporative canister for excessive fuel loading. If these items are satisfactory, the MC solenoid and carburetor require servicing. Remove the ground wire from the MC solenoid dwell connector.

7. Be sure that the engine is at normal operating temperature, and observe the dwell meter at idle speed and 3,000 rpm. If the 3C system is operating correctly, the dwell should be varying between 10° and 54°.

Photo Sequence 8 shows a typical procedure for testing computer command control (3C) performance.

Diagnostic Procedure if the Dwell is Fixed Below 10°

1. If the dwell reading is fixed below 10° in the performance test, the MC solenoid is upward most of the time and the system is trying to provide a richer air-fuel ratio. Therefore, the power train control module (PCM) may be receiving a continually lean signal from the O_2 sensor. Run the engine at 2,000 rpm and momentarily choke the engine. Wait one minute while observing the dwell meter. If the dwell meter reading increases above 50°, the PCM and O_2 sensor are responding to the rich mixture provided by the choking action. When the dwell meter responds properly, check for vacuum leaks, an air management system that is pumping air continually upstream to the exhaust ports, vacuum hose routing, EGR operation, and deceleration valve, if so equipped. When these items and components are satisfactory, the carburetor is providing a continually lean mixture.
2. If the dwell reading did not reach 50° when the engine was choked, the PCM and O_2 sensor are not responding properly. Select the 20-V scale on a digital voltmeter. Disconnect the O_2 sensor wire and connect the digital voltmeter from the battery positive terminal to the purple O_2 sensor wire from the PCM. Operate the engine at part throttle and observe the dwell meter. If the meter reading increases, the PCM is responding to a rich signal. Check the O_2 sensor ground wire for an open circuit. If the ground wire is satisfactory, the O_2 sensor is defective. When the dwell does not increase with the digital voltmeter connected to the purple wire from the PCM to the O_2 sensor, check the coolant temperature sensor and sensor wires, and theTPS and TPS wires. If these items are satisfactory, replace the PCM.

Diagnostic Procedure if the Dwell is Fixed Above 50°

When the dwell is fixed above 50° in the performance test, the MC solenoid plunger is downward most of the time, and the system is trying to provide a leaner air-fuel ratio. Therefore, the O_2 sensor signal must be continually rich. If the dwell meter reading is fixed above 50° in the performance test, follow these steps to locate the problem:

Photo Sequence 8
Typical Procedure for Testing Computer Command Control (3C) Performance

P8-1 Start the engine.

P8-2 Connect terminals A and B in the data link connector (DLC).

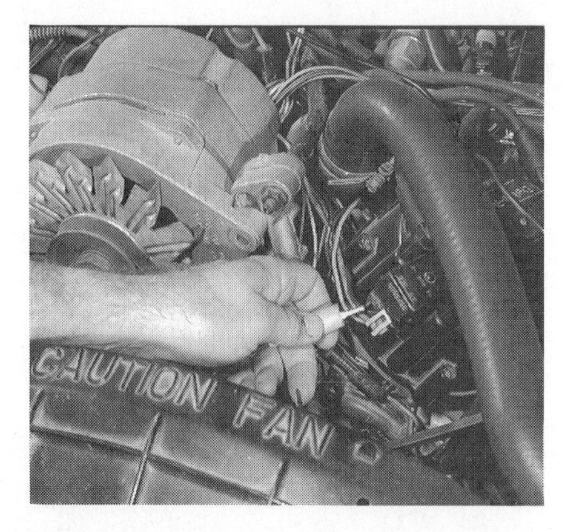

P8-3 Connect the tachometer leads from the coil tach terminal to ground.

P8-4 Disconnect the mixture control (MC) solenoid and ground the MC solenoid dwell lead.

P8-5 Operate the engine at 3,000 rpm.

P8-6 Reconnect the MC solenoid connector and note the engine rpm on the tachometer.

P8-7 Compare the rpm in step 6 to the vehicle manufacturer's specifications in the service manual.

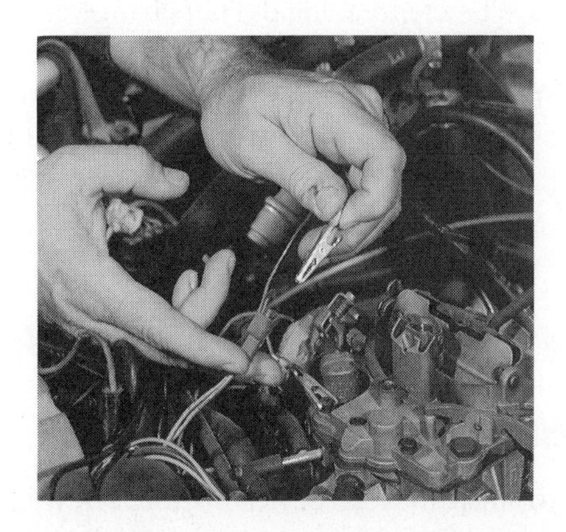

P8-8 Return the engine to idle speed and disconnect the ground wire from the MC solenoid dwell lead.

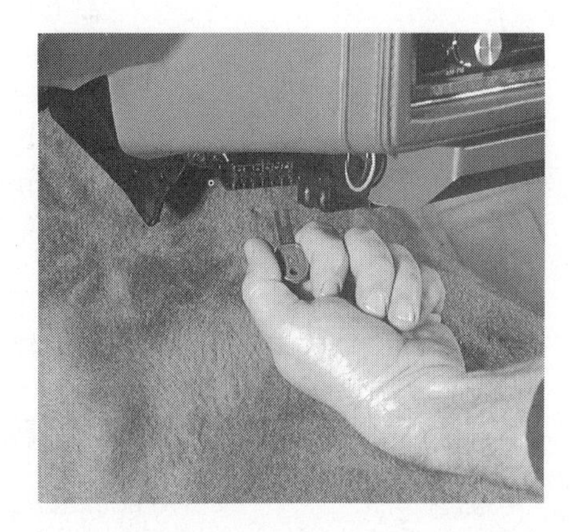

P8-9 Remove the connection between the A and B terminals in the DLC.

1. Start the engine and connect a jumper wire between terminals A and B in the DLC. Connect a dwell meter to the MC solenoid dwell connector and operate the engine at part throttle for two minutes. Return the throttle to idle speed.
2. Remove the PCV valve from the PCV hose and cover the hose with your thumb. Slowly uncover the PCV hose and create a large intake manifold vacuum leak. Run the engine for two minutes and observe the dwell meter. If the dwell reading drops 20° or more, the O_2 sensor and the PCM are responding to the lean air-fuel ratio. Check the evaporative fuel canister for fuel loading, and check the carburetor bowl venting system. If these components and systems are satisfactory, the carburetor is providing a continually rich air-fuel ratio, and carburetor service is required.
3. If the dwell meter reading did not drop 20° in step 2, the PCM and the O_2 sensor are not responding to a lean air-fuel ratio. Disconnect the O_2 sensor connector and ground the O_2 sensor signal wire from the PCM to the sensor. Operate the engine at part throttle and observe the dwell meter. If the dwell reading does not change, the PCM is not responding to a simulated lean O_2 sensor signal, and the PCM is faulty. When the dwell drops to 10° or less, the PCM is responding to a simulated lean O_2 sensor signal, and this sensor or the sensor signal wire is defective. Remove the O_2 sensor wire from ground and connect the digital voltmeter to this wire and ground. With the ignition switch on and the engine stopped, the voltmeter reading should be under 0.55 V. If this voltage reading is satisfactory, replace the O_2 sensor. When the voltage reading is above 0.55 V, the signal wire to the O_2 sensor may be shorted to battery voltage. If this wire is satisfactory, the PCM is faulty.

Diagnostic Procedure if the Dwell is Fixed Between 10° and 50°

Special Tools

Digital voltmeter

1. Start the engine and connect a jumper wire between terminals A and B in the DLC. Run the engine at 2,000 rpm for two minutes and then allow the engine to idle. Disconnect the O_2 sensor wire and ground the purple wire from the PCM to the sensor. Increase the engine rpm to 2,000 for two minutes. Then allow the engine to idle, and observe the MC solenoid dwell reading. If the dwell reading decreases, the PCM is responding to a simulated lean O_2 sensor signal, and this sensor or the sensor ground wire must be defective. If the sensor ground wire is not open, leave the purple wire grounded, and connect a digital voltmeter to the O_2 sensor wire and ground. Run the engine at part throttle; the sensor voltage should be over 0.8 V. If the voltmeter reading is satisfactory, check for a loose sensor connection. When the voltmeter reading is below 0.8 V, replace the O_2 sensor.
2. If the dwell meter reading did not change in step 1, check the coolant temperature sensor and sensor wires. A defective coolant temperature sensor or sensor wires causes the system to remain in open loop with a fixed dwell. Check for an open circuit in the purple wire from the PCM to the O_2 sensor, and be sure that the coolant sensor and O_2 sensor wiring connections at the PCM are satisfactory. If these items are satisfactory, the PCM is faulty.

Mixture Control Solenoid Adjustments

Lean Mixture Screw Adjustment

Do not attempt to adjust the lean or rich mixture screws unless a system performance test indicates a carburetor problem. The mixture control solenoid winding should be tested with an ohmmeter. If this winding has less than 10 Ω, the winding is shorted, and solenoid replacement is necessary. Prior to adjusting the mixture control solenoid, the type of carburetor must be identified. A 4-point carburetor has a rich mixture screw and no letters stamped on the idle air bleed screw, whereas a 3-point carburetor has a rich mixture screw and a letter stamped on the idle air bleed screw. A 2-point carburetor does not have a rich mixture screw, but it does have a letter stamped on the idle air bleed screw. On 3-point and 4-point carburetors follow these steps to adjust the lean mixture screw:

1. With the air horn and the mixture control solenoid plunger removed, place the proper bullet gauge on top of a main metering jet.
2. Install the mixture control solenoid plunger, and hold the plunger fully downward.
3. Rotate the lean mixture screw until the plunger plate just touches the top of the bullet tool (Figure 11-47).
4. Install the bullet tool on the opposite main metering jet and repeat steps 1 through 3. If the adjustment is not the same on both main metering jets, the plunger is bent, or one of the main jets is not fully seated.

On 2-point carburetors a rich limit stop is attached to the lean mixture screw in place of the rich mixture screw. The rich limit stop automatically adjusts the mixture control solenoid to the proper travel. On this type of carburetor follow these steps to adjust the lean mixture screw:

1. Remove the air horn, solenoid lean mixture screw, rich limit stop, plunger, metering rods and springs, and the float bowl insert.
2. Install the bullet gauge on one of the main metering jets.
3. Install the solenoid plunger, solenoid screw, and rich limit stop.
4. Hold the plunger downward against the bullet gauge, and turn the lean mixture screw downward until the solenoid is bottomed in the float bowl. Back the lean mixture screw outward carefully counting the number of turns, until the plunger plate just begins to move away from the bullet gauge. Record the number of turns.
5. Remove the lean mixture screw, rich limit stop, plunger, and bullet gauge.
6. Assemble all the mixture control solenoid and float components in the float bowl.

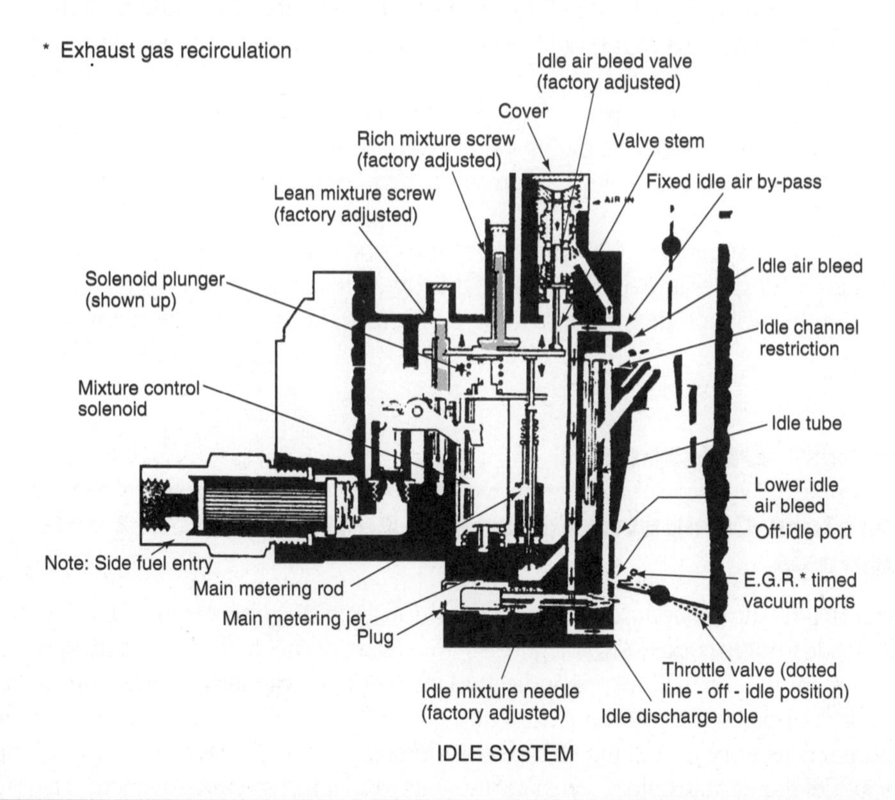

Figure 11-47 Mixture control solenoid adjustments (Courtesy of Chevrolet Motor Division, General Motors Corporation)

7. Turn the lean mixture screw all the way downward until the solenoid bottoms in the float bowl.
8. Back out the lean mixture screw exactly the same number of turns recorded in step 4.

Rich Mixture Screw Adjustment

After the air horn is assembled on the carburetor follow these steps to adjust the rich mixture screw on 3-point and 4-point carburetors:

1. Insert a float gauge tool into the D-shaped hole in the air horn tower to the lower end of the tool rests on the mixture control solenoid plunger plate.
2. Push down on the tool until the mixture control solenoid plungers bottoms, and measure the plunger travel.
3. Repeat steps 1 and 2 to be sure the plunger travel reading is accurate.
4. The specified plunger travel is usually 2/32 to 6/32 with a preferred setting of 4/32. Always use the vehicle manufacturer's specifications in the service manual.
5. If necessary, adjust the rich mixture screw to obtain the specified plunger travel.

After the lean and rich mixture screw adjustments, plugs must be installed on top of the lean and rich mixture screws.

Idle Mixture Screw and Dwell Adjustments

Be sure the engine is at normal operating temperature, and connect the dwell meter leads from the mixture control solenoid dwell connector to ground. Be sure to ground the black meter lead, and place the dwell meter control on the 6 cylinder scale. On 4-point carburetors rotate the air bleed screw until the dwell reading is 25 degrees to 35 degrees. Clockwise rotation increases the dwell and turning this screw counterclockwise reduces the dwell. If the proper dwell reading canot be obtained the mixture screws may require adjusting. For example, if the idle air bleed screw is turned all the way clockwise, and the dwell reading is only 18 degrees, the idle mixture screws should be adjusted. Turn each idle mixture screw outward one turn, and then adjust the dwell with the idle air bleed screw.

Never adjust the idle air bleed screw on a 2-point or 3-point carburetor. On these carburetors adjust the idle mixture screws to obtain the proper mixture control solenoid dwell. Outward idle mixture screw rotation increases the dwell, and turning the idle mixture screws inward lowers the dwell.

Increase the engine speed to 3,000 rpm and observe the dwell reading. The dwell reading should be between 30 degrees and 45 degrees. If this dwell reading is not correct, turn the rich mixture screw slightly to obtain the proper dwell. Never turn the rich mixture screw more than one full turn.

Flash Code Diagnosis

General Motors Computer Command Control System Fault Code Diagnosis

Hard fault DTCs represent faults that are present at the time of testing.

The preliminary diagnostic procedure mentioned previously in this chapter must be completed before the diagnostic trouble code (DTC) diagnosis. If a defect occurs in the 3C system and a DTC is set in the PCM memory, the malfunction indicator lamp (MIL) on the instrument panel is illuminated. If the fault disappears, the MIL lamp goes out, but a DTC is likely set in the computer memory. If a 3C system defect occurs and the MIL lamp is illuminated, the system is usually in a limp-in mode. In this mode, the PCM provides a rich air-fuel ratio and a fixed spark advance. Therefore, engine performance and economy decrease, and emission levels increase, but the vehicle may be driven to a service center.

Connect a jumper wire from terminals A to B in the DLC and turn the ignition switch on to obtain the DTCs. The DTCs are flashed out by the MIL lamp. Two lamp flashes followed by a brief pause and two more flashes indicate code 22. Each code is repeated three times. When more than one code is present, the codes are given in numerical order. Code 12, which indicates that the PCM is capable of diagnosis, is given first. The DTC sequence continues to repeat until the ignition switch is turned off. A DTC indicates a problem in a specific area. Some voltmeter or ohmmeter tests may be necessary to locate the exact defect. Additional information on flash code diagnosis of electronic fuel injection systems is provided in the following chapter.

Memory DTCs represent intermittent faults that occurred sometime in the past and were placed in the computer memory.

Erasing Fault Codes

A quick-disconnect wire connected to the battery positive cable may be disconnected for 10 seconds with the ignition switch off to erase DTCs. If the vehicle does not have a quick-disconnect wire, remove the PCM battery (B) fuse to erase codes. The DTCs vary depending on the vehicle model and year, and the technician must be sure the DTC list is correct for the vehicle being tested.

Memory DTCs may be referred to as history codes.

Voltmeter Diagnosis

Key On, Engine Off Test

A DLC is located under the hood on Ford vehicles with electronic engine control III (EEC III) or EEC IV systems with computer-controlled carburetors. A separate self-test input wire is located near the DLC on many of these systems (Figure 11-48). However, on some Ford systems, the self-test input wire is an integral part of the DLC (Figure 11-49).

Many Ford products with computer-controlled carburetors do not have a MIL light in the instrument panel. On these systems, an analog voltmeter may be connected to the DLC to obtain fault codes. The positive voltmeter lead must be connected to the positive battery terminal, and the negative voltmeter lead must be connected to the proper DLC terminal. With the ignition switch off, connect a jumper wire from the separate self-test input wire to the proper DLC terminal (Figure 11-50). If the vehicle does not have a separate self-test input wire, connect the jumper wire to the proper self-test connector terminals (Figure 11-51).

Special Tools

Analog voltmeter

During the Ford diagnosis, the voltmeter pointer sweeps out the DTCs. For example, if the voltmeter pointer sweeps upward twice, pauses, and then sweeps upward three times, code 23 is displayed (Figure 11-52).

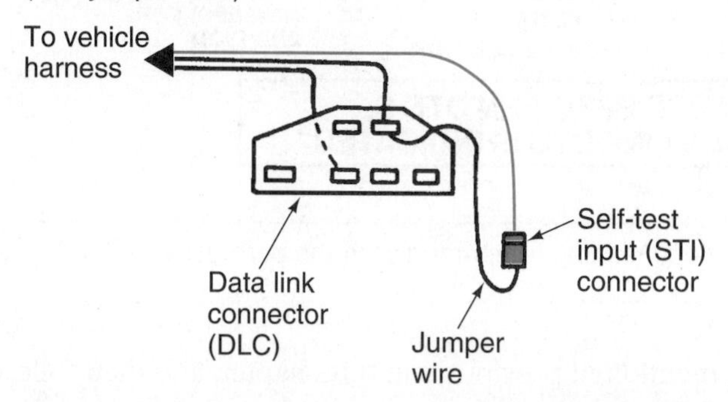

Figure 11-48 DLC with separate self-test input wire (Reprinted with the permission of Ford Motor Company)

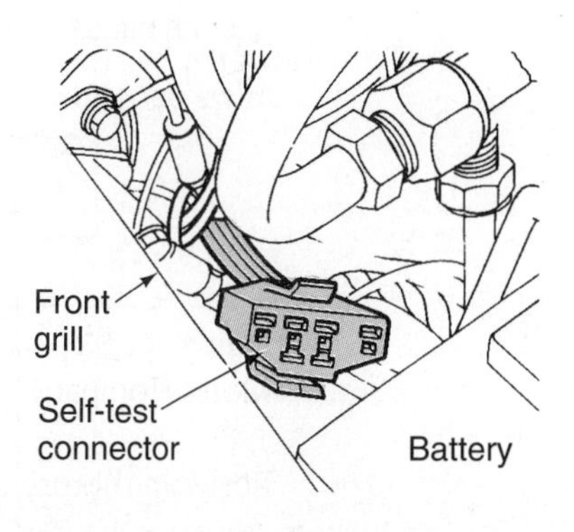

Figure 11-49 DLC with integral self-test input wire (Reprinted with the permission of Ford Motor Company)

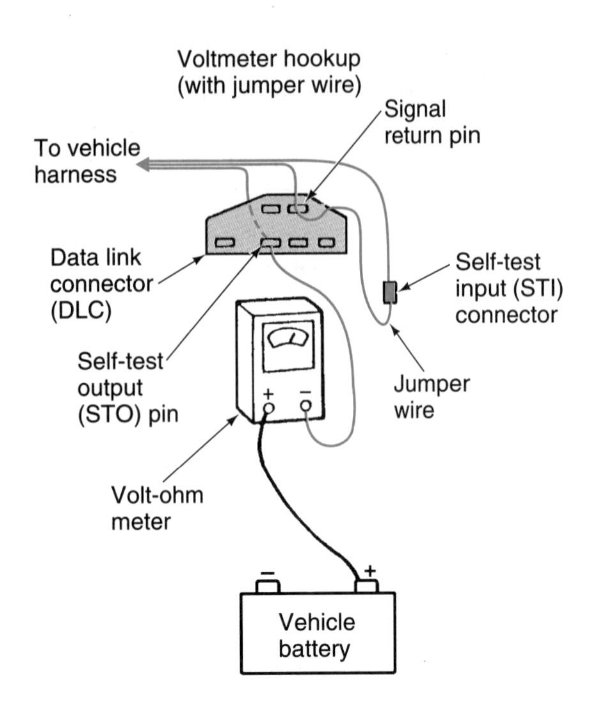

Figure 11-50 Voltmeter and jumper wire connections to the DLC and self-test input wire (Reprinted with the permission of Ford Motor Company)

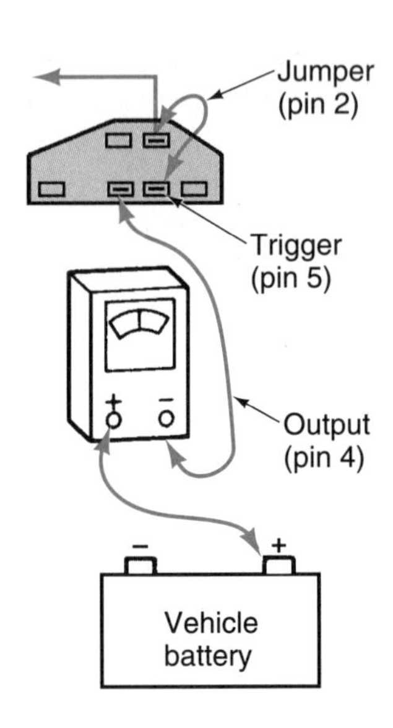

Figure 11-51 Voltmeter and jumper wire connections to the DLC (Reprinted with the permission of Ford Motor Company)

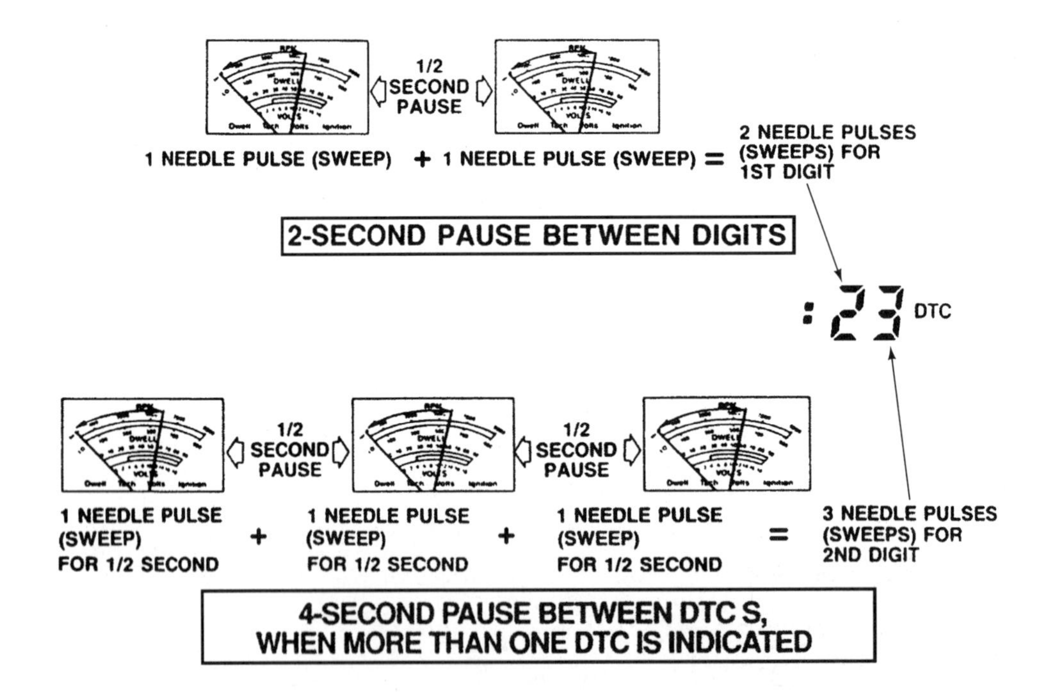

Figure 11-52 Analog voltmeter fault code reading (Reprinted with the permission of Ford Motor Company)

Perform the preliminary diagnosis mentioned previously in this chapter, and then follow these steps for the key on, engine off (KOEO) test:

1. Be sure the ignition switch is off and then connect the jumper wire and voltmeter to the DLC and the self-test input wire.

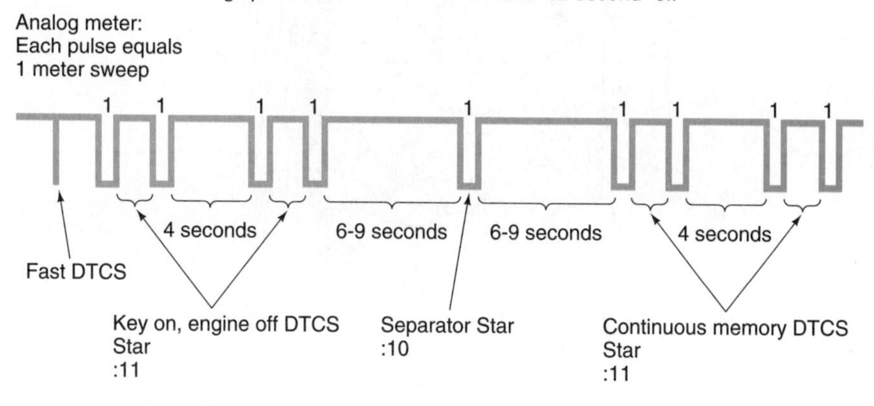

Figure 11-53 Key on, engine off (KOEO) test sequence (Reprinted with the permission of Ford Motor Company)

2. Turn on the ignition switch, and observe the voltmeter. Hard fault DTCs are displayed, followed by a separator code 10, and then memory fault codes. Each DTC is displayed twice. If no DTCs are present, code 11, system pass code, is shown (Figure 11-53).

Key On, Engine Running Test

Before any test is repeated or another test initiated, the ignition switch must be turned off for 10 seconds on these computer-controlled carburetor systems. When the key on, engine running (KOER) test is performed, leave the voltmeter and jumper wire connected to the DLC. Turn the ignition switch off, wait 10 seconds, and then start the engine.

An engine identification (ID) code is displayed first. This code is equal to half the engine cylinders. For example, on a V6 engine, the engine ID code is 30. This code is displayed on a voltmeter as 3 upward pointer sweeps. After the engine ID code, a separator code 10 is provided on some Ford products. When this code is received, the technician has 20 seconds to momentarily move the throttle to the wide-open position. During this time, the PCM monitors the input sensors for defects. After the separator code 10, engine running DTCs are displayed representing faults that are present at the time of testing (Figure 11-54).

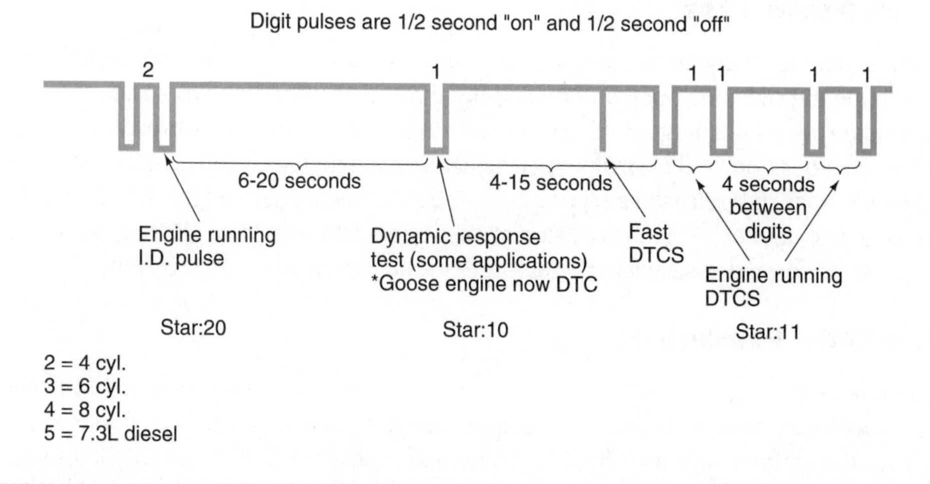

Figure 11-54 Key on, engine running (KOER) test sequence (Reprinted with the permission of Ford Motor Company)

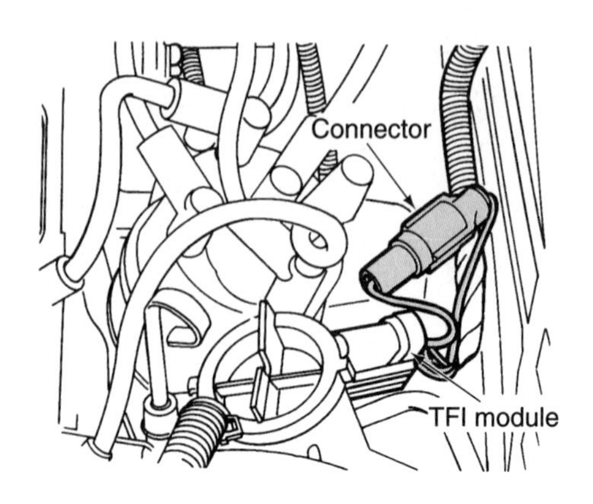

Figure 11-55 In-line timing connector (Reprinted with the permission of Ford Motor Company)

DTCs vary depending on the model year and vehicle. The technician must use the correct fault code list for the vehicle being tested.

Computed Timing Test

The computed timing test checks the spark advance provided by the PCM. Complete these steps for the computed timing check:

Special Tools

Advance-type timing light

1. Disconnect the in-line timing connector near the distributor and check the basic timing with a timing light. Leave the timing light connected, and reconnect the in-line timing connector (Figure 11-55).
2. Leave the analog voltmeter connected, and disconnect the self-test input wire. Be sure the ignition switch has been off for 10 seconds.
3. Start the engine and connect the jumper wire to the same terminals used for the KOEO and KOER tests.
4. The spark advance should be 20° more than the basic timing setting if the PCM is providing the correct spark advance. The timing marks may be observed with the timing light to check the advance. If an advance-type timing light is used, the timing knob on the light may be rotated until the timing marks return to the basic timing setting. The scale on the light then indicates the degrees of advance.

Output State Test

The output state test may be referred to as an output cycling test. Complete the KOEO test, and then push the throttle wide open and release the throttle to enter the output state test. When this action is taken, the PCM energizes all the actuators in the EEC IV system. Normally closed actuators go to the open position, and normally open actuators move to the closed position. If the throttle is pushed wide open and released a second time, the PCM deenergizes all the actuators. During the output state test, the technician may listen to suspected actuators for a clicking noise. Voltmeter tests may be performed at suspected actuators to locate the cause of the problem.

Erase Code Procedure

Do not erase DTCs until the defects represented by the codes have been corrected. To erase DTCs, perform a KOEO test and disconnect the jumper wire at the self-test connector during the code display. Turn the ignition switch off for 10 seconds and repeat the KOEO and KOER tests to be sure the codes are erased. If DTCs are left in the PCM memory, they will be erased after the problem is corrected and the engine is stopped and started 50 times. This action applies to many computer systems.

Continuous Self-Test

The continuous self-test may be referred to as a wiggle test. Test procedures may vary depending on the vehicle year and model. Always use the procedure recommended in the vehicle manufacturer's service manual.

This test allows the technician to wiggle suspected wiring harness connectors while the EEC IV system is monitored. The continuous monitor test may be performed at the end of the KOER test. Leave the jumper wire connected to the self-test connector. Approximately two minutes after the last code is displayed, the continuous monitor test is started. If suspected wiring connectors are wiggled, the voltmeter pointer will deflect when a loose connection is present.

Scan Tester Diagnosis

Chrysler Oxygen (O_2) Feedback System Fault Code Diagnosis

A scan tester may be used to diagnose different types of computer-controlled carburetor systems. Scan testers are supplied by the vehicle manufacturer and several independent suppliers. Many scan testers have a removable module plugged into the tester. This module is designed for a specific vehicle make and model year. Always be sure the proper module is installed in the tester for the vehicle being diagnosed.

Special Tools

Scan tester

Some Chrysler O_2 feedback systems have self-diagnostic capabilities, whereas older models of these systems had to be diagnosed with voltmeter and ohmmeter tests. Always use the diagnostic procedure recommended in the vehicle manufacturer's appropriate service manual. Some Chrysler products with O_2 feedback systems and self-diagnostic capabilities do not have a check engine light. These systems have to be diagnosed with a scan tester (Figure 11-56). The scan tester must be connected to the diagnostic connector in the engine compartment (Figure 11-57).

After the scan tester is connected to the diagnostic connector, follow the scan tester manufacturer's recommended procedure to obtain the DTCs. These fault codes are displayed once in numerical order. If the technician wants to repeat the DTC display, the ignition switch must be turned off and the fault code procedure repeated. DTCs in any computer system indicate a fault in a specific area, not necessarily in a specific component. For example, if a code representing a throttle position sensor (TPS) is present, the wires from the sensor to the computer may be defective, the sensor may require replacement, or the computer may not be able to receive this sensor signal. The technician may have to perform some voltmeter or ohmmeter tests on the sensor wiring to locate the exact cause of the code.

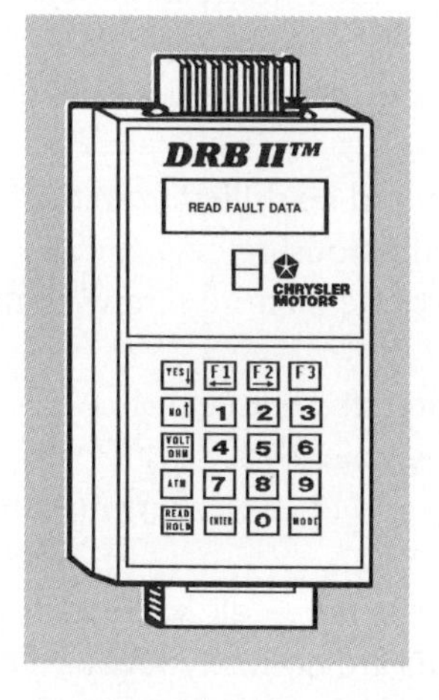

Figure 11-56 Chrysler digital readout box II (DRB II) (Courtesy of Chrysler Corporation)

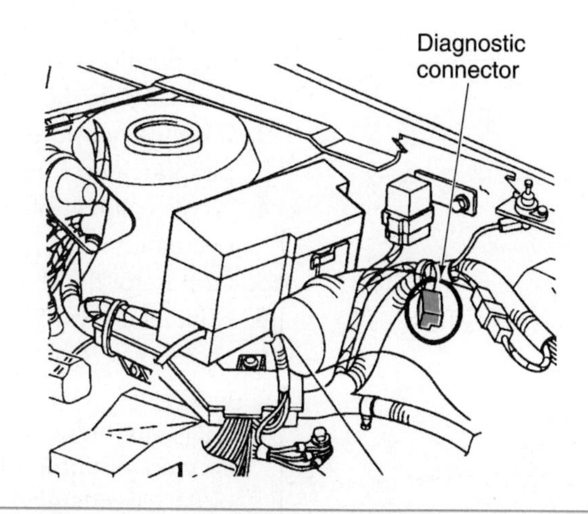

Figure 11-57 DLC in the engine compartment (Courtesy of Chrysler Corporation)

Switch Test Mode

If a defective input switch is suspected, the switch test mode may be used to check the switch inputs. Switch inputs include the brake switch, neutral park switch, and A/C switch. The switch inputs vary depending on the vehicle and model year.

When the fault code display is completed, be sure all the input switches are off, and then follow the scan tester manufacturer's recommended procedure to enter the switch test mode. During the switch test mode, each switch is turned on and off. The reading on the scan tester should change, indicating this switch input signal was received by the computer. For example, the brake switch is turned on and off by depressing and releasing the brake pedal. If the scan tester display does not change when any input switch is activated, the switch or connecting wires are defective.

Actuation Test Mode (ATM)

If a fault code is obtained representing a solenoid or relay, the ATM may be used to cycle the component on and off to locate the exact cause of the problem. When code 55 is displayed at the end of the fault code diagnosis, follow the scan tester manufacturer's recommended procedure to enter the ATM mode.

Typical components that may be cycled during the ATM are the canister purge solenoid, cooling fan relay, and throttle kicker solenoid. The actual components that may be cycled in the ATM vary depending on the vehicle and model year. Each relay or solenoid may be cycled for five minutes, or until the ignition switch is turned off. When the appropriate scan tester button is pressed, the tester begins cycling the next relay or solenoid. The following chapter provides more information on scan tester diagnosis of electronic fuel injection systems.

Classroom Manual
Chapter 11, page 268

CUSTOMER CARE: Always concentrate on quality workmanship and customer satisfaction. Most customers do not mind paying for vehicle repairs if the work is done properly and their vehicle problem is corrected. A follow-up phone call to determine customer satisfaction a few days after repairing a car indicates that you are interested in the car and that you consider quality work and satisfied customers a priority.

Guidelines for Conventional and Computer-Controlled Carburetor System Diagnosis and Service

1. When a carburetor is serviced on the work bench, the carburetor should be placed in a holding fixture.
2. The position of carburetor components such as jets should be noted so they may be reinstalled in their original positions.
3. If a choke cover is riveted in place, the heads should be drilled from the rivets, and a pin punch may be used to drive the rivets from the housing.
4. A concealment plug must be removed from most idle mixture screws during carburetor overhaul.
5. Carburetor flooding may be caused by a float saturated with gasoline.
6. Nonmetallic carburetor parts such as diaphragms, solenoids, plastic parts, gaskets, choke covers and springs, O-rings, filters, floats, and fiber washers must not be placed in carburetor cleaner.
7. A carburetor kit contains such components as gaskets, needle valve and seat, and the accelerator pump plunger. Therefore, these parts are normally replaced during a carburetor overhaul.
8. A high fuel level in the float bowl results in excessive fuel consumption and possible engine stalling.

9. A low fuel level in the float bowl causes a lean air-fuel ratio with possible loss of power and engine surging at high speed.
10. Less-than-specified float drop may cause fuel starvation and engine surging at high speed.
11. Excessive float drop may result in needle valve sticking and flooding.
12. Carburetor adjustments must be done in the order recommended in the service manual to prevent one adjustment from changing an adjustment performed previously.
13. The secondary throttle linkage adjustment sets the secondary throttle opening in relation to the primary throttle opening.
14. The secondary air valve opening adjustment adjusts the wide-open air valve position.
15. If the secondary air valve spring tension is weaker than specified, a hesitation may occur on hard acceleration.
16. The choke diaphragm connector rod adjustment sets the cushioning action of the vacuum kick diaphragm on the air valve opening.
17. If the vacuum kick diaphragm pulls the choke valve open further than specified, a hesitation may occur when a cold engine is accelerated.
18. When the vacuum kick diaphragm does not pull the choke open as far as specified, the air-fuel ratio will be excessively rich during engine warm-up, and stalling may occur.
19. The choke unloader forces the choke open when the primary throttles are wide open. This action provides a means of clearing a cold, flooded engine.
20. The secondary lockout prevents the secondary throttles from opening while the choke is closed or partially closed.
21. The fast idle cam adjustment positions this cam properly on a cold engine and during engine warm-up.
22. If the accelerator pump stroke is shorter than specified, a hesitation may occur on low speed acceleration.
23. Some vehicle manufacturers specify the use of a choke valve angle gauge in place of round gauges while performing carburetor adjustments.
24. The engine must be at normal operating temperature before any idle speed, fast idle speed, and idle mixture adjustment.
25. After any carburetor idle speed and mixture adjustment, the vehicle must meet emission level standards for that vehicle year.
26. An A/C idle speed check or adjustment may be required on vehicles equipped with A/C.
27. The ignition timing must be checked and adjusted as necessary prior to the engine idle speed check.
28. The concealment plug must be removed to gain access to the idle mixture screw.
29. After the idle mixture adjustment is completed, the vehicle must conform to federal and state emission standards.
30. During the propane-assisted idle mixture adjustment, the propane set rpm is adjusted with the idle speed screw, and the idle set rpm is adjusted with the mixture screw.
31. The throttle stop screw is adjusted to complete the antidieseling adjustment.
32. During a computer-controlled carburetor system performance test, the engine rpm is checked with the mixture control solenoid disconnected and connected to determine the operation of the mixture control solenoid and carburetor.
33. If the computer-controlled carburetor system is working properly with the engine at normal operating temperature, the mixture control solenoid dwell should be fluctuating between 10° and 54°.
34. If the dwell on the mixture control solenoid is fixed below 10°, the system is trying to provide a rich air-fuel ratio, and the O_2 sensor signal must be continually lean.
35. When the mixture control solenoid dwell is fixed above 50°, the system is attempting to provide a lean air-fuel ratio, and the O_2 sensor signal must be continually rich.

36. If the mixture control solenoid dwell is fixed between 10° and 54°, the system may be in open loop, or there may be a defect in one of the input sensors or the computer.
37. On General Motors vehicles, terminals A and B may be connected in the DLC to read DTCs on the MIL light.
38. Some vehicles with computer-controlled carburetors do not have a MIL light. On some of these vehicles, an analog voltmeter may be connected to the self-test connector to read the DTCs.
39. A scan tester may be connected to the DLC on many vehicles to obtain DTCs and perform other system tests.
40. A computed timing test allows the technician to check the spark advance provided by the computer.
41. An output state test allows the technician to cycle the solenoids and relays on and off in a computer-controlled carburetor system.
42. A continuous self-test, or wiggle test, allows the technician to check for loose wiring connections on system components.
43. A switch test checks the switch inputs such as the brake switch and park/neutral switch to the computer.
44. An actuation test mode (ATM) allows the technician to cycle the solenoids and relays in the system individually.

CASE STUDY

A customer complained about the malfunction indicator light (MIL) coming on on a Buick Riviera with a computer-controlled carburetor system. When questioned further about the driving conditions when this light came on, the customer indicated the light came on when the car was driven at lower speeds. However, when driving on the freeway, the light did not come on. The technician asked the customer if the car had any other performance and economy problems. The customer replied that the fuel consumption had increased recently, and the engine performance seemed erratic at low speed.

The technician checked the computer for fault codes and found a code 45, indicating a rich exhaust condition. Next, a performance test was completed on the computer-controlled carburetor system. During this test, the system did not provide the specified rpm change when the mixture control solenoid was disconnected and connected. This test result indicated a possible internal carburetor problem.

The technician remembered the diagnostic procedure learned during her Automotive Technology training. She recalled this procedure said something about always performing the easiest, quickest tests first when eliminating the causes of a problem. The technician thought about other causes of this problem, and it occurred to her that one cause might be the vapor control system. This car had a vacuum-operated purge control valve. When the technician checked this valve with the engine idling, she found the valve was wide open and pulling large quantities of fuel vapors from the carburetor float bowl once the engine was at normal operating temperature.

The purge control valve was replaced and the car driven at lower speeds during a road test. The check engine light remained off and the engine performance was normal. From this experience, the technician proved the diagnostic procedure she had learned years before was very helpful. Part of this procedure stated that a technician should always think about the causes of a problem and test to locate the defect beginning with the easiest, quickest tests first. If the technician had overhauled the carburetor, the customer's money would have been wasted and the problem would have been unsolved.

Terms to Know

Float level adjustment
Float drop adjustment
Secondary throttle linkage adjustment
Secondary air valve alignment
Secondary air valve opening adjustment
Secondary air valve spring adjustment
Choke control lever adjustment
Choke diaphragm connector rod adjustment
Choke vacuum kick adjustment
Choke unloader adjustment
Secondary lockout adjustment
Fast idle cam position adjustment
Accelerator pump adjustment
Choke valve angle gauge
Concealment plug
Propane-assisted idle mixture adjustment
Antidieseling adjustment
Computer-controlled carburetor performance test
Data link connector (DLC)
Flash code diagnosis
Diagnostic trouble code (DTC)
Self-test input wire
Key on, engine off (KOEO) test
Key on, engine running (KOER) test
Computed timing test
Output state test
Continuous self-test
Switch test
Actuation test mode

ASE Style Review Questions

1. While discussing carburetor adjustments:
Technician A says if the fuel level in the float bowl is higher than specified, the engine may detonate.
Technician B says if the fuel level in the float bowl is higher than specified, the engine may stall on deceleration.
Who is correct?
A. A only **C.** Both A and B
B. B only **D.** Neither A nor B

2. While discussing the vacuum kick adjustment:
Technician A says if the vacuum kick diaphragm pulls the choke valve open further than specified, the air-fuel ratio will be too rich during engine warm-up.
Technician B says if the vacuum kick diaphragm pulls the choke valve open less than specified, a hesitation may occur on acceleration with a cold engine.
Who is correct?
A. A only **C.** Both A and B
B. B only **D.** Neither A nor B

3. While discussing the accelerator pump adjustment:
Technician A says if the accelerator pump stroke is less than specified, a hesitation may occur on low speed acceleration.
Technician B says when the accelerator pump adjustment is performed, the throttle should be wide open.
Who is correct?
A. A only **C.** Both A and B
B. B only **D.** Neither A nor B

4. While discussing idle mixture adjustment:
Technician A says the vehicle must meet emission standards for that year of vehicle after the idle mixture adjustment.
Technician B says during the propane enrichment method of idle mixture adjustment, the idle set rpm is adjusted with the idle mixture screws.
Who is correct?
A. A only **C.** Both A and B
B. B only **D.** Neither A nor B

5. While discussing propane-assisted idle mixture adjustment:
Technician A says the proper propane idle speed is adjusted with the idle speed screw.
Technician B says the propane metering valve must be set to obtain the highest idle speed prior to adjusting the idle speed screw.
Who is correct?
A. A only **C.** Both A and B
B. B only **D.** Neither A nor B

6. While discussing a computer-controlled carburetor system performance test:
Technician A says if the system is operating normally, the mixture control solenoid dwell should be varying between 10° and 20°.
Technician B says if the system is operating normally, the mixture control solenoid dwell should be varying between 10° and 54°.
Who is correct?
A. A only **C.** Both A and B
B. B only **D.** Neither A nor B

7. While discussing a computer-controlled carburetor system performance test:
   Technician A says if the mixture control solenoid dwell is fixed below 10°, the system is trying to provide a rich air-fuel ratio.
   Technician B says if the mixture control solenoid dwell is fixed below 10°, the O_2 sensor voltage signal is continually high.
   Who is correct?
   A. A only **C.** Both A and B
   B. B only **D.** Neither A nor B

8. While discussing flash code diagnosis:
   Technician A says the check engine light flashes fault codes in numerical order.
   Technician B says if a fault code is set in the computer memory, the system may be in the limp-in mode.
   Who is correct?
   A. A only **C.** Both A and B
   B. B only **D.** Neither A nor B

9. While discussing erase code procedures:
   Technician A says on many computer-controlled carburetor systems, fault codes are erased by disconnecting battery power from the computer for 10 seconds.
   Technician B says if fault codes are left in a computer after the fault is corrected, the code is erased after the vehicle is started 50 times.
   Who is correct?
   A. A only **C.** Both A and B
   B. B only **D.** Neither A nor B

10. While discussing scan tester diagnosis:
   Technician A says some computer-controlled carburetor systems do not have a malfunction indicator light (MIL) in the instrument panel.
   Technician B says an actuation test mode (ATM) allows the technician to obtain fault codes.
   Who is correct?
   A. A only **C.** Both A and B
   B. B only **D.** Neither A nor B

Table 11-1 ASE TASK

Perform analytic/diagnostic procedures on vehicles with on-board or self-diagnostic type computer systems; determine needed repairs.

Problem Area	Symptoms	Possible Causes	Classroom Manual	Shop Manual
ENGINE PERFORMANCE	Low fuel economy, performance problems, MIL light on	Engine compression, computer system faults, ignition system defects	257	234

CHAPTER 12

Electronic Fuel Injection Diagnosis and Service

Upon completion and review of this chapter, you should be able to:

- ❑ Perform a preliminary diagnostic procedure on a throttle body injection (TBI), multiport fuel injection (MFI), or sequential fuel injection (SFI) system.
- ❑ Test fuel pump pressure.
- ❑ Explain the results of high fuel pressure in a TBI, MFI, or SFI system.
- ❑ Describe the results of low fuel pump pressure.
- ❑ Perform an injector balance test and determine the injector condition.
- ❑ Clean injectors on an MFI or SFI system.
- ❑ Perform an injector sound test.
- ❑ Perform an injector ohmmeter test.
- ❑ Perform an injector noid light test.
- ❑ Perform an injector flow test and determine injector condition.
- ❑ Perform an injector leakage test.
- ❑ Check for leakage in the fuel pump check valve and the pressure regulator valve.
- ❑ Remove and replace the fuel rail, injectors, and pressure regulator.
- ❑ Remove, test, and replace cold start injectors.
- ❑ Perform a minimum idle speed adjustment.
- ❑ Remove, clean, inspect, and install throttle body assemblies.
- ❑ Check the fuel cut system.
- ❑ Perform a flash code diagnosis on various vehicles.
- ❑ Erase fault codes.
- ❑ Perform a scan tester diagnosis on various vehicles.
- ❑ Perform a cylinder output test.
- ❑ Remove and replace computer chips.
- ❑ Diagnose problems in TBI, MFI, and SFI systems.

Throttle Body, Multiport, and Sequential Fuel Injection Service and Diagnosis

Preliminary Diagnostic Procedure

When engine performance or economy complaints occur on fuel injected vehicles, the tendency of many technicians is to think that the problem is in the fuel injection and computer system. However, many other defects can affect the engine and fuel injection system operation. For example, an intake manifold vacuum leak causes a rough idle condition and engine surging at low speed. If the engine has a MAF sensor, an intake manifold vacuum leak allows additional air into the intake, and this air does not flow through the MAF sensor. Therefore, the MAF sensor signal indicates to the power train control module (PCM) that less air flow is entering the engine in relation to the throttle opening. Under this condition, the PCM supplies less fuel through the injectors. This creates a lean air-fuel ratio, which results in engine surging and acceleration stumbles. Expensive hours of diagnostic time may be saved if these items are proven to be satisfactory before the fuel injection system is diagnosed:

1. Intake manifold vacuum leaks
2. Emission devices such as the EGR valve and related controls
3. Ignition system condition
4. Engine compression
5. Battery fully charged
6. Engine at normal operating temperature
7. All accessories turned off

Basic Tools

Basic technician's tool set

Service manual

Special jumper wires for connection to various DLCs and other components

In a sequential fuel injection (SFI) system, each injector has an individual ground wire connected into the computer.

In a multiport fuel injection (MFI) system, the injector ground wires are connected to the computer in pairs or in groups of three or four, depending on the engine.

The SAE J1930 terminology is an attempt to provide a universal terminology for automotive electronics.

In the SAE J1930 terminology, the terms MFI and SFI replace the previous EFI and PFI terminology.

In the SAE J1930 terminology, the term power train control module (PCM) replaces all previous terms for engine computers.

Service Precautions

These precautions must be observed when TBI, MFI, and SFI systems are diagnosed and serviced:

1. Always relieve the fuel pressure before disconnecting any component in the fuel system.
2. Never turn on the ignition switch when any fuel system component is disconnected.
3. Use only the test equipment recommended by the vehicle manufacturer.
4. Always turn off the ignition switch before connecting or disconnecting any system component or test equipment.
5. When arc welding is necessary on a computer-equipped vehicle, disconnect both battery cables before welding is started. Always disconnect the negative cable first.
6. Never allow electrical system voltage to exceed 16 V. This could be done by disconnecting the circuit between the alternator and the battery with the engine running.
7. Avoid static electric discharges when handling computers, modules, and computer chips.

Disconnecting Battery Cables

During the diagnosis of TBI, MFI, and SFI systems, many procedures indicate the removal of the negative battery cable or both battery cables. The negative battery cable may be disconnected during diagnostic and service procedures, but disconnecting the battery has these effects:

1. Deprograms the radio
2. Deprograms other convenience items such as memory seats or mirrors
3. Erases the trip odometer if the vehicle has digital instrumentation
4. Erases the adaptive strategy in the computer

Disconnecting the battery has the same effect on any vehicle with TBI, MFI, or SFI and adaptive strategy in the PCM. If the adaptive strategy in the computer is erased, the engine operation may be rough at low speeds when the engine is restarted, because the computer must relearn the computer system defects. Under this condition, the vehicle should be driven for 5 minutes on the road with the engine at normal operating temperature. Some manufacturers recommend that a 12-V dry cell battery be connected from the positive battery cable to ground if the battery is disconnected. The 12 V supplied by the dry cell prevents deprogramming and memory erasing. Some 12-V sources for this purpose are designed to plug into the cigarette lighter socket.

Fuel Pressure Testing

SERVICE TIP: Remember that Ford products have an inertia switch in the fuel pump circuit. If there is no fuel pump pressure, always push the inertia switch reset button first and determine if this is a problem.

SERVICE TIP: Many fuel pump circuits are connected through a fuse in the fuse panel or a fusible link. If there is no fuel pump pressure, always check the fuel pump fuse or fusible link first.

When tests are performed to diagnose any automotive problem, always start with the tests that are completed quickly and easily. The fuel pressure test is usually one of the first tests to consider when TBI, MFI, and SFI systems are diagnosed. Remember that low fuel pressure may cause lack of power, acceleration stumbles, engine surging, and limited top speed, whereas high fuel pressure results in excessive fuel consumption, rough idle, engine stalling, and excessive sulphur smell from the catalytic converter.

In some cases, in-tank fuel pumps have the specified pressure when the ignition switch is turned on or when the engine is idling, but the fuel pump cannot meet the engine demand for fuel at or near wide-open throttle. Therefore, if the customer complains about the engine quitting momentarily or completely at higher speeds, the fuel pump pressure should be tested at higher speeds during a road test. The hose is long enough on some fuel pressure gauges to allow the gauge hose to be connected under the hood, and the gauge placed in the passenger compartment.

Relieving Fuel Pressure. Prior to pressure gauge connection on TBI, MFI, or SFI systems, the fuel pressure should be relieved. This is accomplished by momentarily supplying 12 V to one injector terminal and grounding the other injector terminal. This action lifts the injector plunger and the fuel discharges from the injector to relieve the fuel pressure. Do not supply 12 V and a ground to an injector for more than 5 seconds unless a vehicle manufacturer's recommended procedure specifies a longer time period.

WARNING: Never energize an injector with a 12-V source for more than 5 seconds, unless a vehicle manufacturer's recommended procedure specifies a longer time period. This action may damage the injector winding.

Connecting Fuel Pressure Gauge. In a TBI system, the inlet fuel line at the throttle body assembly must be removed and the pressure gauge hose installed in series between the inlet line and the inlet fitting (Figure 12-1). On other TBI systems, the vehicle manufacturer recommends connecting the fuel pressure gauge at the fuel filter inlet (Figure 12-2). Use new gaskets on the union bolt when the pressure gauge is connected at this location.

CAUTION: Never turn on the ignition switch or crank the engine with a fuel line disconnected. This action causes the fuel pump to discharge fuel from the disconnected line, which may result in a fire, causing personal injury and/or property damage.

In an MFI or SFI system, the pressure gauge must be connected to the Schrader valve on the fuel rail (Figure 12-3). On some SFI systems, the vehicle manufacturer recommends connecting the fuel pressure gauge to the cold start injector fuel line (Figure 12-4). Install new gaskets on the union bolt when the pressure gauge is installed at this location.

Special Tools

Fuel pressure gauge

Operating the Fuel Pump to Test Pressure. The technician must have pressure specifications for the make and model year of the vehicle being tested. Once the pressure gauge is connected, the ignition switch may be cycled several times to read the fuel pressure, or the pressure may be read with the engine idling. In cases where the engine will not start or when further diagnosis of the fuel pump circuit is required, it may be helpful to operate the fuel pump continually. Many fuel pump circuits have a provision for operating the fuel pump continually to test fuel pump pressure if the engine will not run. Some manufacturers of import vehicles, such as Toyota, recommend

In the SAE J1930 universal terminology, the term for self-test connector is data link connector (DLC).

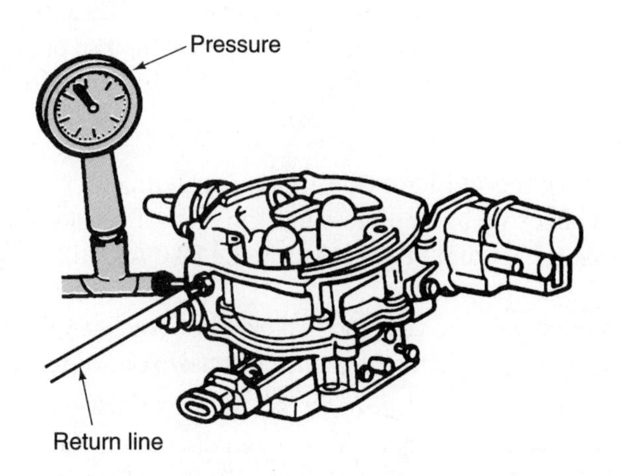

Figure 12-1 Connecting fuel pressure gauge in series in the fuel inlet line at the throttle body assembly (Courtesy of Chrysler Corporation)

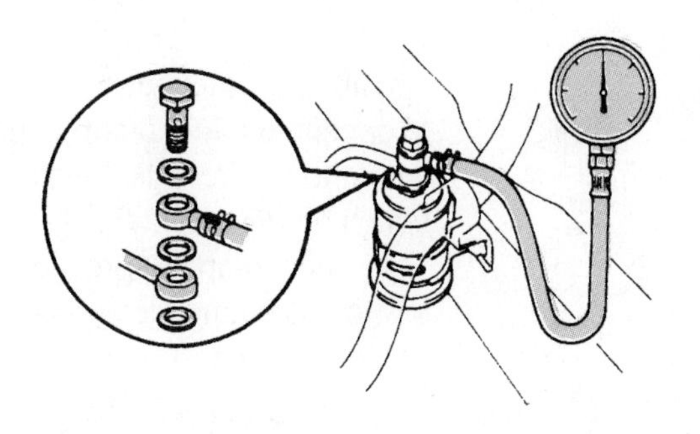

Figure 12-2 Connecting fuel pressure gauge in series in the fuel filter inlet line (Courtesy of Toyota Motor Corporation)

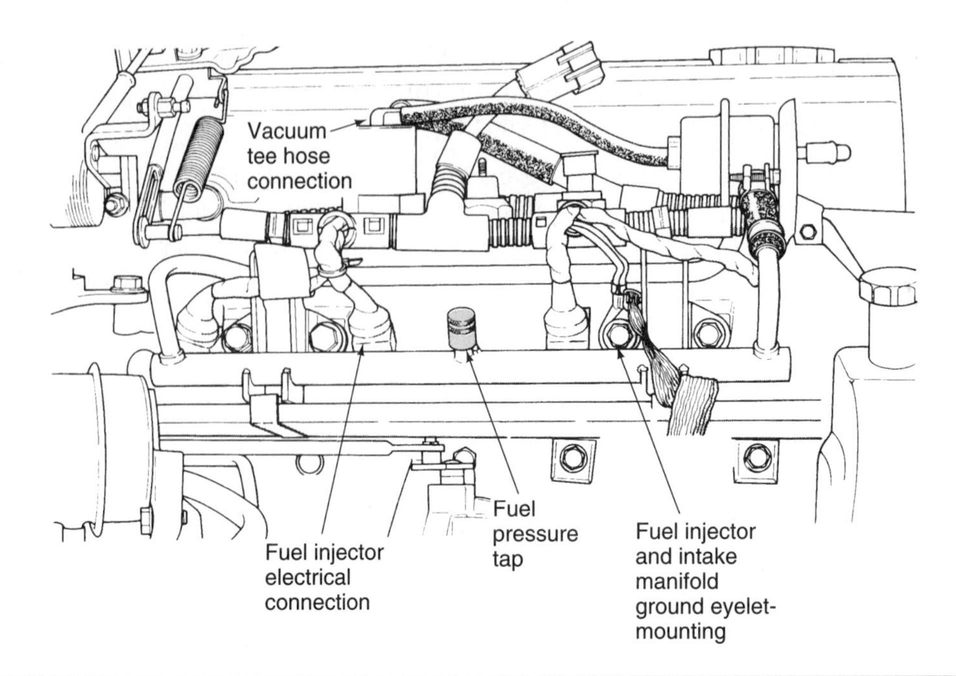

Figure 12-3 Connecting fuel pressure gauge to the Schrader valve on the fuel rail (Courtesy of Chrysler Corporation)

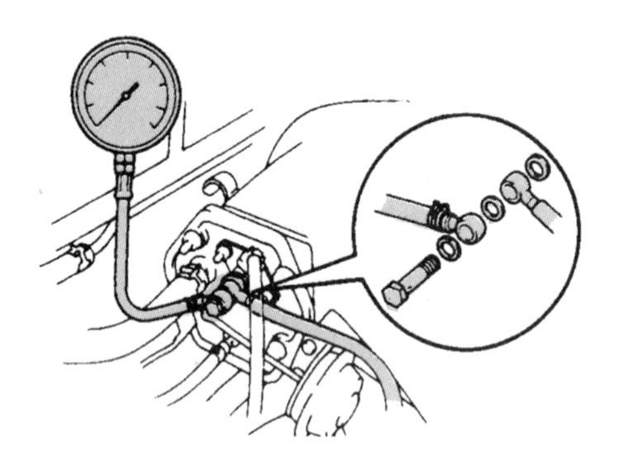

Figure 12-4 Connecting fuel pressure gauge to the cold start injector fuel line (Courtesy of Toyota Motor Corporation)

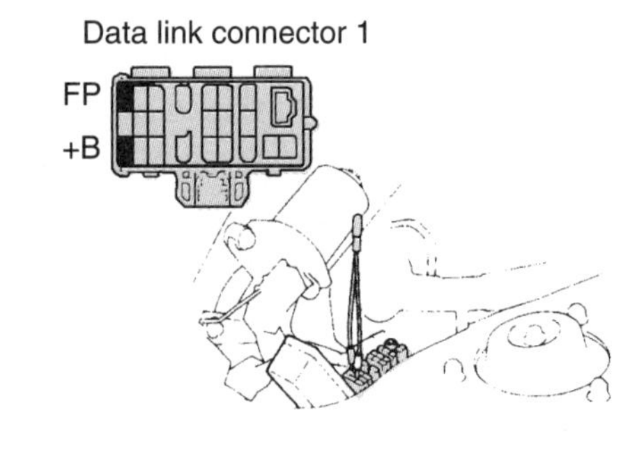

Figure 12-5 A jumper wire may be connected across terminals B+ and FP in the DLC with the ignition switch on to operate the fuel pump while testing fuel pressure. (Courtesy of Toyota Motor Corporation)

operating the fuel pump with a jumper wire connected across the appropriate terminals in the DLC to operate the fuel pump continually and check fuel pump pressure. On many Toyota products, the jumper wire must be connected across the B+ and FP terminals in the DLC with the ignition switch on (Figure 12-5).

Fuel pump test procedures vary depending on the year and make of the vehicle. Always follow the recommended procedure in the vehicle manufacturer's service manual. Following is a typical fuel pump test procedure on a Toyota vehicle:

1. Connect a 12-V power supply to the cigarette lighter socket and disconnect the negative battery cable. If the vehicle is equipped with an air bag, wait one minute.
2. Bleed pressure from the fuel system as mentioned previously.
3. Connect the fuel pressure gauge as outlined previously. Use a shop towel to wipe up any spilled gasoline.

4. Connect the jumper wire across the B+ and FP terminals in the DLC.
5. Reconnect the battery negative cable, and turn the ignition switch on.
6. Observe the fuel pressure on the gauge.
7. Disconnect the jumper wire from the DLC terminals.
8. Disconnect and plug the vacuum hose from the pressure regulator, and start the engine.
9. Observe the fuel pressure on the gauge with the engine idling.
10. Reconnect the vacuum hose to the pressure regulator and observe the fuel pressure on the gauge.

The fuel pump pressure must equal the manufacturer's specifications under all conditions. This pressure is usually about 10 psi (70 kPa) higher with the vacuum hose removed from the pressure regulator than when this vacuum hose is connected. If the pressure is higher than specified, check the return fuel line and pressure regulator.

When there is no fuel pump pressure, check the fusible link, fuses, SFI main relay, fuel pump, PCM, and wiring connections. If the fuel pump pressure is lower than specified, check the fuel lines and hoses, fuel pump, fuel filter, pressure regulator, and cold start injector.

On many Ford products, a self-test connector is located in the engine compartment. This connector is tapered on both ends, but one tapered end is longer than the other end. A wire is connected from the fuel pump relay to the outer terminal in the short tapered end of the DLC (Figure 12-6).

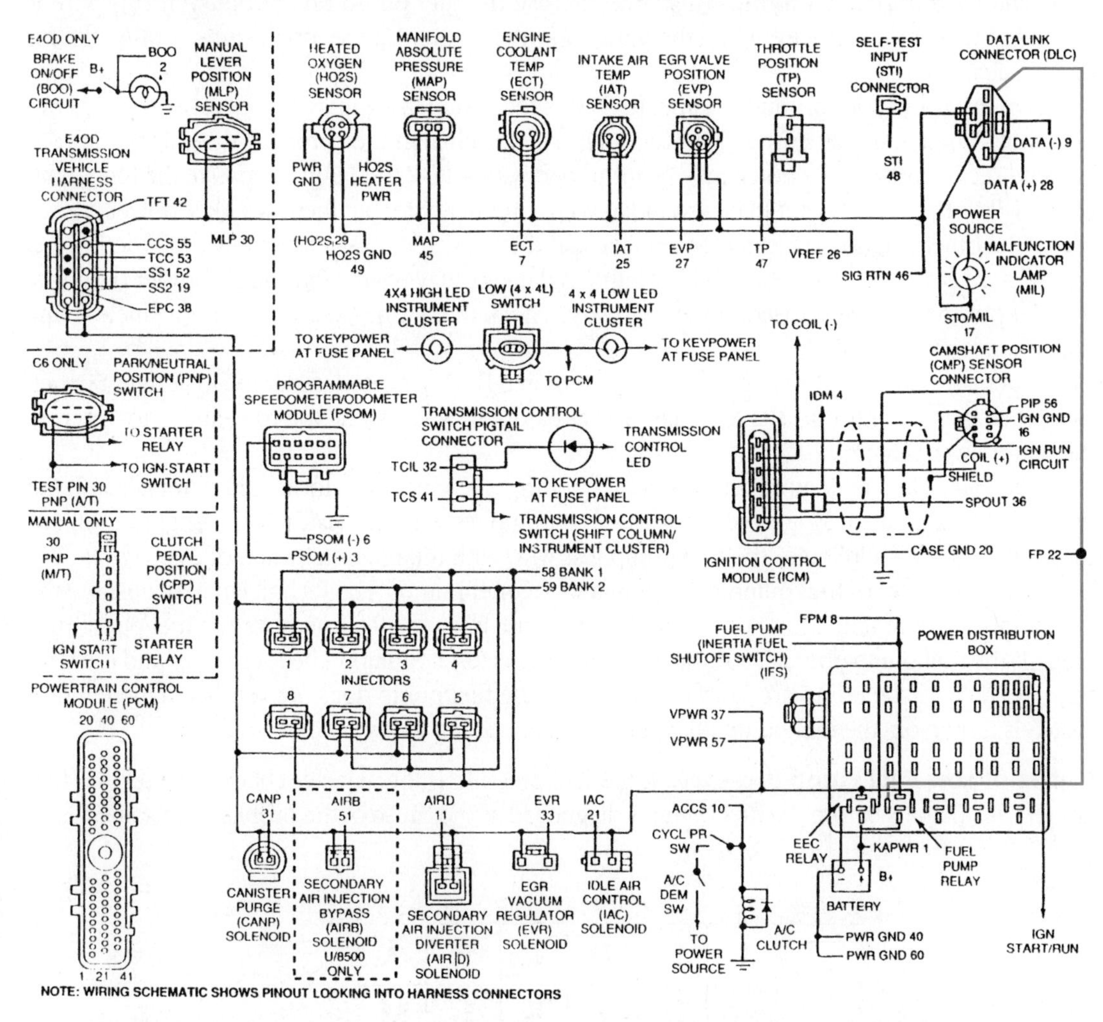

Figure 12-6 The wire from the fuel pump relay to the outer terminal in the short tapered end of the DLC may be grounded to operate the fuel pump continually. (Reprinted with the permission of Ford Motor Company)

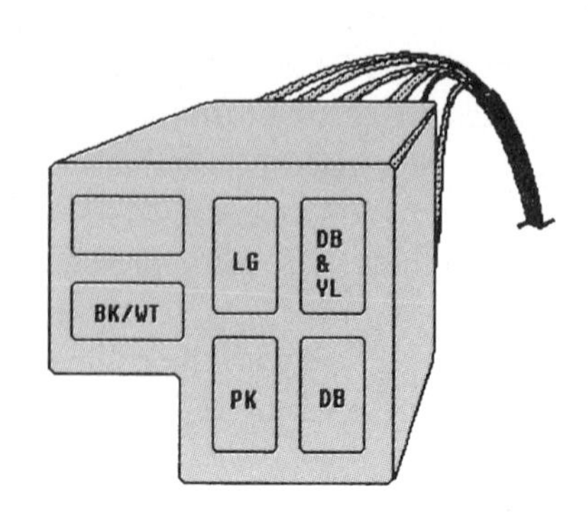

Figure 12-7 On many Chrysler products, the terminal directly opposite the notch in the corner of the DLC may be grounded with a 12-V test lamp to operate the fuel pump continually. (Courtesy of Chrysler Corporation)

WARNING: When instructed to ground a wire for diagnostic purposes, always be sure you are grounding the proper wire under the specified conditions. Improper grounding of computer system terminals may damage computer system components.

The PCM normally grounds this wire to close the fuel pump relay points. If this wire is grounded with a jumper wire when the ignition switch is on, the fuel pump runs continually for diagnostic purposes.

On many Chrysler products, a square DLC is located in the engine compartment. This connector has a notch in one corner (Figure 12-7). The terminal in the corner of the diagnostic connector directly opposite the notch may be grounded with a 12-V test lamp to operate the fuel pump continually. This terminal could be grounded with a jumper wire, but there is a 12-V power wire in one of the other diagnostic connector terminals. If this power wire is accidentally grounded with a jumper wire, severe computer and wiring harness damage may result. On some Chrysler products, the fuel pump test wire is discontinued. Always check the wiring diagram for the vehicle being diagnosed.

On many General Motors products, a 12-terminal DLC is located under the instrument panel. In most of these connectors, the terminals are lettered A to F across the top row, and G to M across the bottom row (Figure 12-8).

On some General Motors vehicles, a fuel pump test connector is located in terminal G on the DLC. On other General Motors vehicles, this fuel pump test wire is located in the engine compartment. If 12 V are supplied to the fuel pump test wire with the ignition switch off, voltage is supplied through a pair of fuel pump relay points to the fuel pump. The technician may observe the fuel pump pressure under this condition or listen at the fuel tank filler neck for the fuel pump running. If the fuel pump operates satisfactorily under this test condition, the fuel pump and the wire from the relay to the pump are satisfactory. When the fuel pump does not run when the ignition switch is turned on, the fuel pump relay, PCM, or connecting wires are defective.

Causes of Low Fuel Pump Pressure. If the fuel pressure is low, always check the filter and fuel lines for restrictions before the fuel pump is diagnosed as the cause of the problem. In some cases,

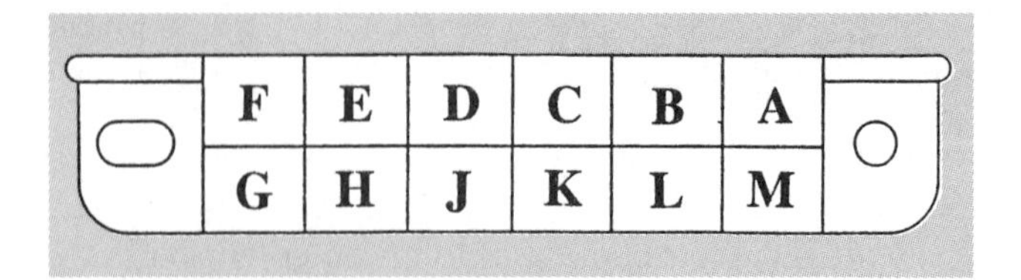

Figure 12-8 Data link connector (DLC)

water or dirt in the fuel tank covers and plugs the pickup sock on the in-tank fuel pump. This action shuts off the fuel supply to the pump and the engine stops. Usually this problem occurs at highway speeds. Technicians must keep this problem in mind when fuel pump pressure is tested.

Classroom Manual
Chapter 12, page 277

Injector Testing

Since injectors on MFI and SFI systems are subject to more heat than TBI injectors, port injectors have more problems with tip deposits. The symptoms of restricted injectors are:

1. Lean surge at low speeds
2. Acceleration stumbles
3. Hard starting
4. Acceleration sag, cold engine
5. Engine misfiring
6. Rough engine idle
7. Lack of engine power
8. Slow starting when cold
9. Stalling after a cold start

An injector balance test may be performed to diagnose restricted injectors on MFI and SFI systems. A fuel pressure gauge and an injector balance tester are required for this test, and the fuel pressure should be checked before the injector balance test is performed. The injector balance tester contains a timer circuit which energizes each injector for an exact time period when the timer button is pressed. When the injector balance test is performed, follow these steps:

1. Connect the fuel pressure gauge to the Schrader valve on the fuel rail.
2. Connect the injector tester leads to the battery terminals with the correct polarity. Remove one of the injector wiring connectors and install the tester lead to the injector terminals (Figure 12-9).
3. Cycle the ignition switch on and off until the specified fuel pressure appears on the fuel gauge. Many fuel pressure gauges have an air bleed button that must be pressed to bleed air from the gauge. Cycle the ignition switch or start the engine to obtain the specified pressure on the fuel gauge, and then leave the ignition switch off.
4. Push the timer button on the tester and record the gauge reading. When the timer energizes the injector, fuel is discharged from the injector into the intake port, and the fuel pressure drops in the fuel rail.
5. Repeat steps 2, 3, and 4 on each injector, and record the fuel pressure after each injector is energized by the timer.

Special Tools

Injector balance tester

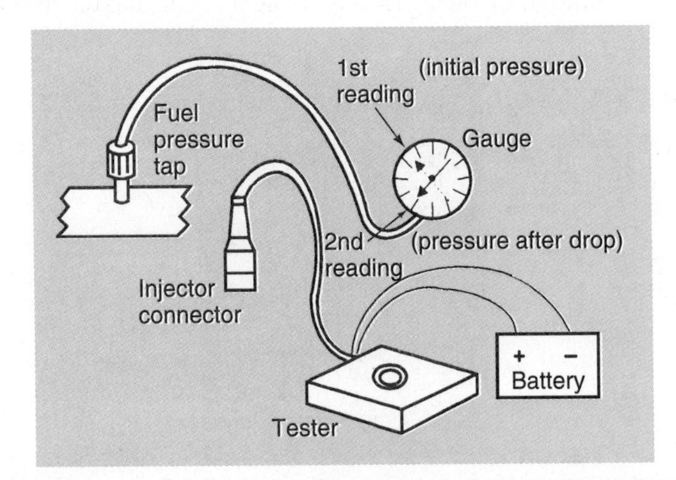

Figure 12-9 Injector balance tester and pressure gauge connections for injector balance test (Courtesy of Oldsmobile Division, General Motors Corporation)

CYLINDER	1	2	3	4	5	6
HIGH READING	225	225	225	225	225	225
LOW READING	100	100	100	90	100	115
AMOUNT OF DROP	125	125	125	135	125	110
	OK	OK	OK	FAULTY, RICH (TOO MUCH) (FUEL DROP)	OK	FAULTY, LEAN (TOO, LITTLE) (FUEL DROP)

Figure 12-10 Pressure readings from injector balance test indicating defective injectors (Courtesy of Oldsmobile Division, General Motors Corporation)

6. Compare the gauge readings on each injector. When the injectors are in satisfactory condition, the fuel pressure will be the same after each injector is energized by the timer. If an injector orifice or tip is restricted, the fuel pressure does not drop as much when the injector is energized by the timer. When an injector plunger is sticking in the open position, the fuel pressure drop is excessive. If the fuel pressure on an injector is 1.4 psi (10 kPa) below or above the average pressure when the injectors are energized by the timer, the injector is defective (Figure 12-10).

Photo Sequence 9 shows a typical procedure for testing injector balance.

Injector Service and Diagnosis

Injector Cleaning

If the injector balance test indicates that some of the injectors are restricted, the injectors may be cleaned. Tool manufacturers market a variety of injector cleaning equipment. The injector cleaning solution is poured into a canister on some injector cleaners, and the shop air supply is used to pressurize the canister to the specified pressure. The injector cleaning solution contains unleaded fuel mixed with injector cleaner. The container hose is connected to the Schrader valve on the fuel rail (Figure 12-11).

Automotive parts stores usually sell a sealed pressurized container of injector cleaner with a hose for Schrader valve attachment. During the cleaning process, the engine is operated on the pressurized container of unleaded fuel and injector cleaner. The fuel pump operation must be stopped to prevent the pump from forcing fuel up to the fuel rail, and the fuel return line must be

Figure 12-11 Injector cleaner connected to Schrader valve on fuel rail and pressurized by the shop air supply (Courtesy of OTC Division, SPX Corp.)

Photo Sequence 9
Typical Procedure for Testing Injector Balance

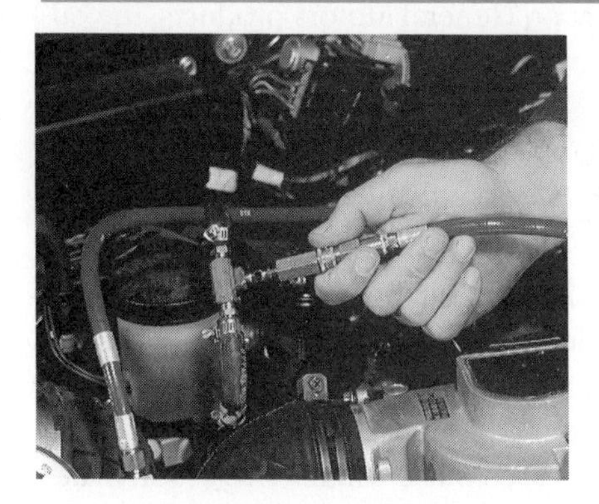

P9-1 Connect the fuel pressure gauge to the Schrader valve on the fuel rail.

P9-2 Disconnect the number 1 injector, and connect the injector tester lead to the injector terminals.

P9-3 Connect the injector tester power supply leads to the battery terminals with the proper polarity.

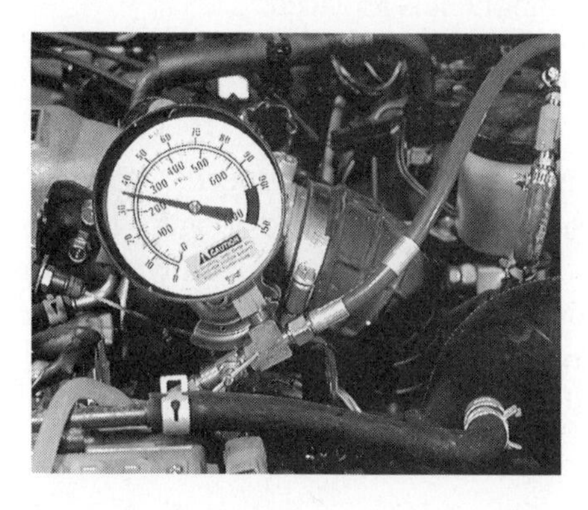

P9-4 Cycle the ignition switch several times until the specified pressure appears on the pressure gauge.

P9-5 Push the injector tester switch, and record the pressure on the pressure gauge.

P9-6 Move the injector tester lead to the number 2 injector, and cycle the ignition switch several times to restore the fuel pressure.

P9-7 Touch the injector tester switch and observe the fuel pressure.

P9-8 Move the injector tester lead to the number 3 injector, and cycle the ignition switch several times to restore the fuel pressure.

P9-9 Touch the injector tester switch and observe the fuel pressure. Follow the same procedure to test the number 4 injector. Compare the pressure readings to the vehicle manufacturer's specifications.

plugged to prevent the solution in the cleaning container from flowing through the return line into the fuel tank. Follow these steps for the injector cleaning procedure:

1. Disconnect the wires from the in-tank fuel pump or the fuel pump relay to disable the fuel pump. If you disconnect the fuel pump relay on General Motors products, the oil pressure switch in the fuel pump circuit must also be disconnected to prevent current flow through this switch to the fuel pump.
2. Plug the fuel return line from the fuel rail to the tank.
3. Connect a can of injector cleaner to the Schrader valve on the fuel rail and run the engine for about 20 minutes on the injector solution.

Special Tools

Injector cleaner

After the injectors are cleaned or replaced, rough engine idle may still be present. This problem occurs because the adaptive memory in the computer has learned previously about the restricted injectors. If the injectors were supplying a lean air-fuel ratio, the computer increased the pulse width to try to bring the air-fuel ratio back to stoichiometric. With the cleaned or replaced injectors, the adaptive computer memory is still supplying the increased pulse width. This action makes the air-fuel ratio too rich now that the restricted injector problem does not exist. With the engine at normal operating temperature, drive the vehicle for at least 5 minutes to allow the adaptive computer memory to learn about the cleaned or replaced injectors. After this time, the computer should supply the correct injector pulse width and the engine should run smoothly. This same problem may occur when any defective computer system component is replaced.

Injector Sound Test

Special Tools

Stethoscope

A port injector that is not functioning may cause a cylinder misfire at low engine speeds. With the engine idling, a stethoscope pickup may be placed on the side of the injector body (Figure 12-12). Each injector should produce the same clicking noise. If an injector does not produce any clicking noise, the injector, connecting wires, or PCM may be defective. When the injector clicking noise is erratic, the injector plunger may be sticking. If there is no injector clicking noise, proceed with the injector ohms test and noid light test to locate the cause of the problem.

Injector Ohmmeter Test

An ohmmeter may be connected across the injector terminals to check the injector winding (Figure 12-13) after the injector wires are disconnected. If the ohmmeter reading is infinite, the injector winding is open. An ohmmeter reading below the specified value indicates the injector winding is shorted. A satisfactory injector winding has the amount of resistance specified by the manufacturer. Injector replacement is necessary if the injector winding does not have the specified resistance.

Special Tools

Digital volt-ohmmeter

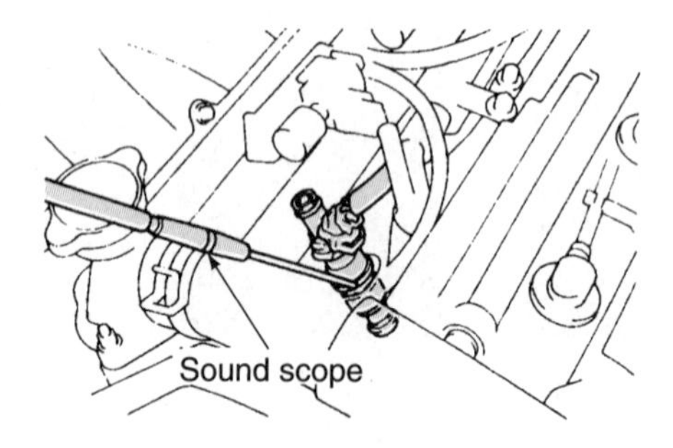

Figure 12-12 Checking for a clicking noise at each injector with a stethoscope (Courtesy of Toyota Motor Corporation)

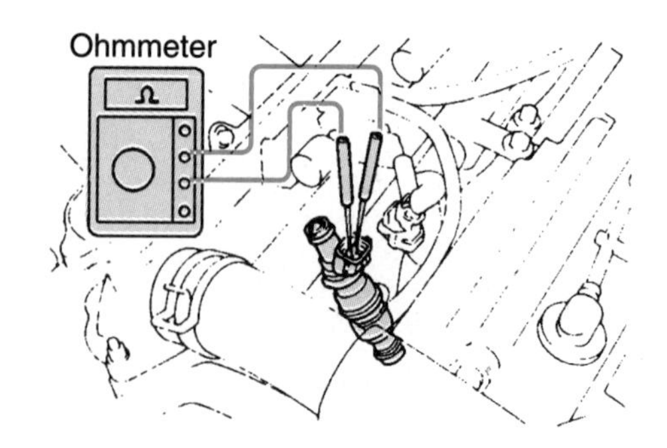

Figure 12-13 An ohmmeter may be connected across the injector terminals to test the injector winding. (Courtesy of Toyota Motor Corporation)

Noid Light Test

Special Tools
Noid light

Some manufacturers of automotive test equipment market noid lights, which have terminals designed to plug into most injector wiring connectors after these connectors are disconnected from the injector. When the engine is cranked, the noid light flashes if the computer is cycling the injector on and off. If the light is not flashing, the computer or connecting wires are defective.

Injector Flow Testing

Some vehicle manufacturers recommend an injector flow test rather than the balance test. Follow these steps to perform an injector flow test:

1. Connect a 12-V power supply to the cigarette lighter socket and disconnect the negative battery cable. If the vehicle is equipped with an air bag, wait one minute.
2. Remove the injectors and fuel rail and place the tip of the injector to be tested in a calibrated container. Leave the injectors in the fuel rail.
3. Connect a jumper wire between the B+ and FP terminals in the DLC as in the fuel pump pressure test.
4. Turn on the ignition switch.
5. Connect a special jumper wire from the terminals of the injector being tested to the battery terminals (Figure 12-14).
6. Disconnect the jumper wire from the negative battery cable after 15 seconds.
7. Record the amount of fuel in the calibrated container.
8. Repeat the procedure on each injector. If the volume of fuel discharged from any injector varies more than 0.3 cu in (5 cc) from the specifications, the injector should be replaced.
9. Connect the negative battery cable and disconnect the 12-V power supply.

Special Tools
Graduated plastic container

Injector, Fuel Pump, and Pressure Regulator Leakage Test

Connect the fuel pressure gauge to the fuel system as explained previously in this chapter. While the fuel system is pressurized with the injectors removed from the fuel rail after the flow test, observe each injector for leakage from the injector tip (Figure 12-15). Injector leakage must not exceed the manufacturer's specifications. Injector leakage may cause slow starting when hot or cold.

If the injectors leak into the intake ports on a hot engine, the air-fuel ratio may be too rich when a restart is attempted a short time after the engine is shut off. When the injectors leak, they drain all the fuel out of the rail after the engine is shut off for several hours. This may result in slow starting after the engine has been shut off for a longer period of time.

While checking leakage at the injector tips, observe the fuel pressure in the pressure gauge. If the fuel pressure drops off and the injectors are not leaking, the fuel may be leaking back

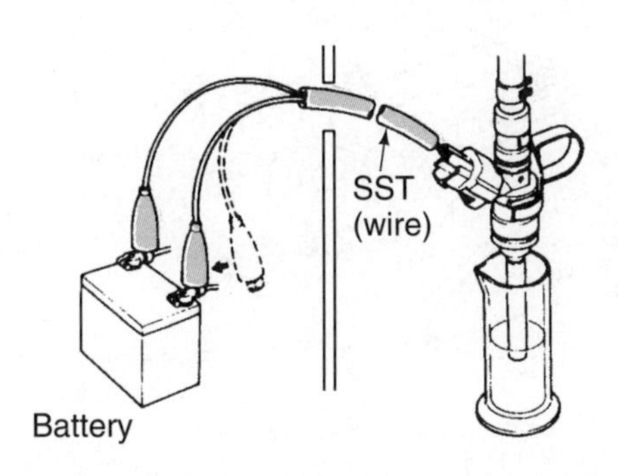

Figure 12-14 Special jumper wire connected from the injector terminals to the battery terminals (Courtesy of Toyota Motor Corporation)

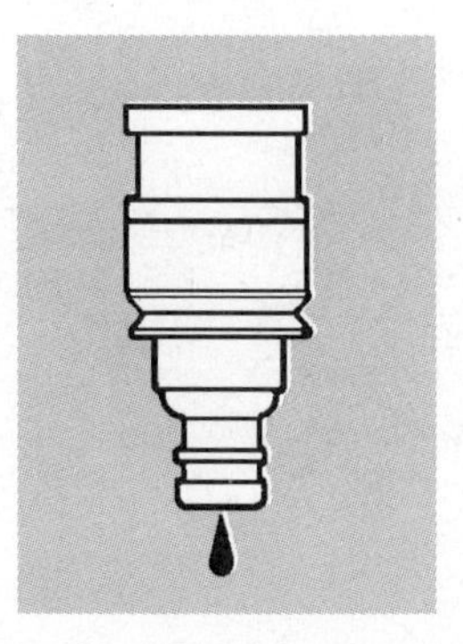

Figure 12-15 Checking injector leakage with the fuel system pressurized and the injectors in the fuel rail (Courtesy of Toyota Motor Corporation)

through the check valve in the fuel pump. Repeat the test with the fuel line plugged. If the fuel pressure no longer drops, the fuel pump check valve is leaking. When the fuel pressure drops off and the injectors are not leaking, the fuel pressure may be leaking through the pressure regulator and the return fuel line. Repeat the test with the return line plugged. If the fuel pressure no longer drops off, the pressure regulator valve is leaking.

Removing and Replacing Fuel Rail, Injectors, and Pressure Regulator

Fuel Rail, Injector, and Pressure Regulator Removal

WARNING: Cap injector openings in the intake manifold to prevent the entry of dirt and other particles.

WARNING: After the injectors and pressure regulator are removed from the fuel rail, cap all fuel rail openings to keep dirt out of the fuel rail.

WARNING: Do not use compressed air to flush or clean the fuel rail. Compressed air contains water, which may contaminate the fuel rail.

WARNING: Do not immerse the fuel rail, injectors, or pressure regulator in any type of cleaning solvent. This action may damage and contaminate these components.

The procedure for removing and replacing the fuel rail, injectors, and pressure regulator varies depending on the vehicle. On some applications, certain components must be removed to gain access to these components. Always follow the procedure recommended in the vehicle manufacturer's service manual. Following is a typical removal and replacement procedure for the fuel rail, injectors, and pressure regulator on a General Motors 3,800 engine:

1. Connect a 12-V power supply to the cigarette lighter and disconnect the battery negative cable. If the vehicle is equipped with an air bag, wait one minute.
2. Bleed the pressure from the fuel system.
3. Wipe excess dirt from the fuel rail with a shop towel.
4. Loosen fuel line clamps on the fuel rail if clamps are present on these lines. If these lines have quick-disconnect fittings, grasp the larger collar on the connector and twist in either direction while pulling on the line to remove the fuel supply and return lines (Figures 12-16 and 12-17).

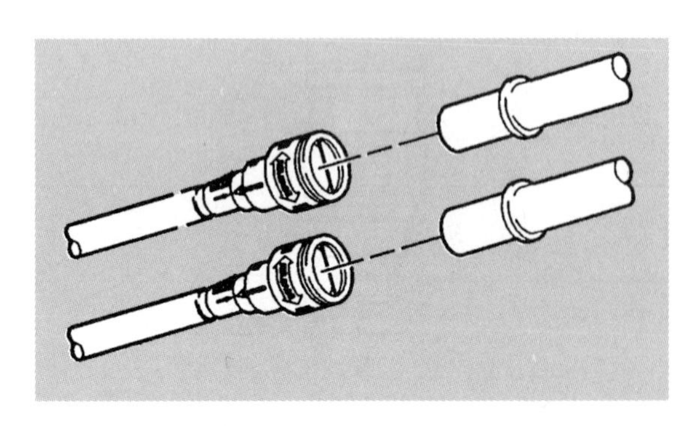

Figure 12-16 Quick-disconnect fuel line fittings (Courtesy of Oldsmobile Division, General Motors Corporation)

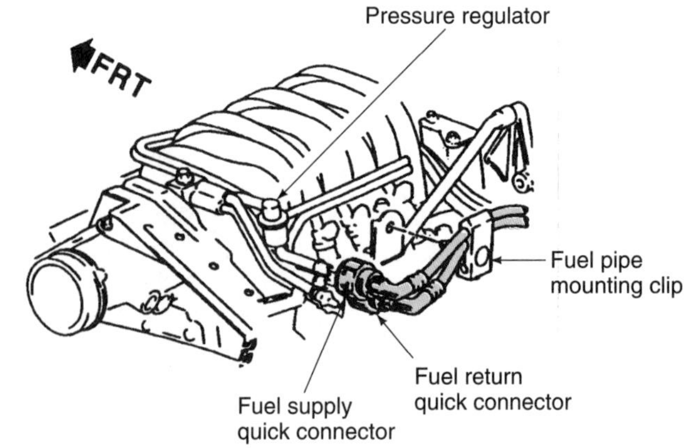

Figure 12-17 Fuel supply and return lines on fuel rail (Courtesy of Oldsmobile Division, General Motors Corporation)

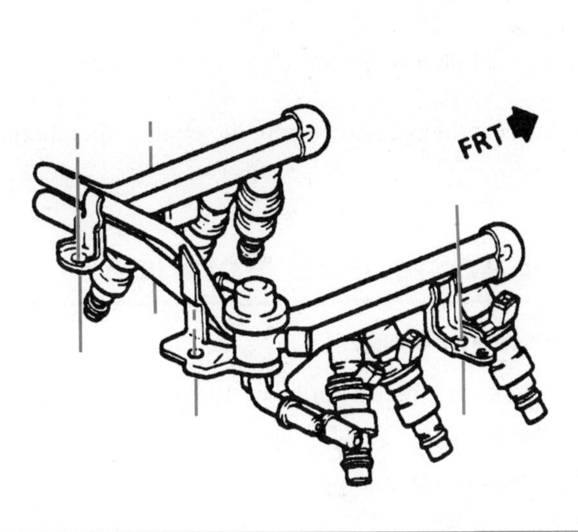

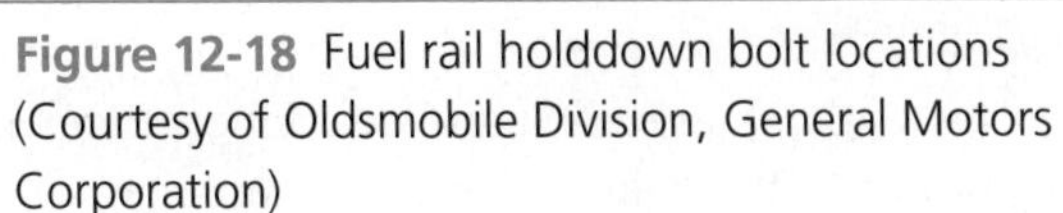

Figure 12-18 Fuel rail holddown bolt locations (Courtesy of Oldsmobile Division, General Motors Corporation)

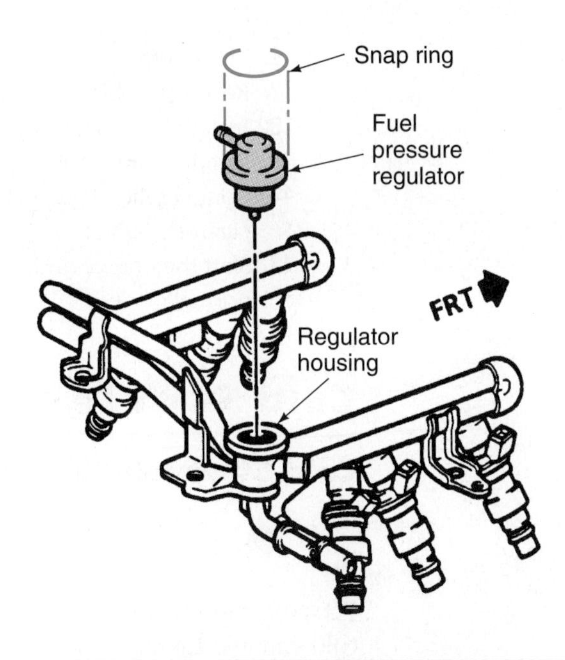

Figure 12-19 Removing snap ring and pressure regulator from the fuel rail (Courtesy of Oldsmobile Division, General Motors Corporation)

5. Remove the vacuum line from the pressure regulator.
6. Disconnect the electrical connectors from the injectors.
7. Remove the fuel rail holddown bolts (Figure 12-18).
8. Pull with equal force on each side of the fuel rail to remove the rail and injectors.

Fuel Rail, Injector, and Pressure Regulator Cleaning and Inspection

1. Prior to injector and pressure regulator removal, the fuel rail may be cleaned with a spray-type engine cleaner such as AC Delco X-30A or its equivalent.
2. Pull the injectors from the fuel rail.
3. Use snap ring pliers to remove the snap ring from the pressure regulator cavity. Note the original direction of the vacuum fitting on the pressure regulator and pull the pressure regulator from the fuel rail (Figure 12-19).
4. Clean all components with a clean shop towel. Be careful not to damage fuel rail openings and injector tips.
5. Check all injector and pressure regulator openings in the fuel rail for metal burrs and damage.

Installation of Fuel Rail, Injectors, and Pressure Regulator

1. If the same injectors and pressure regulator are reinstalled, replace all O-rings and lightly coat each O-ring with engine oil.
2. Install the pressure regulator in the fuel rail and position the vacuum fitting on the regulator in the original direction.
3. Install the snap ring above the pressure regulator.
4. Install the injectors in the fuel rail.
5. Install the fuel rail while guiding each injector into the proper intake manifold opening. Be sure the injector terminals are positioned so they are accessible to the electrical connectors.

6. Tighten the fuel rail holddown bolts alternately, and torque them to specifications.
7. Reconnect the vacuum hose on the pressure regulator.
8. Install the fuel supply and fuel return lines on the fuel rail.
9. Install the injector electrical connectors.
10. Connect the negative battery terminal, and disconnect the 12-V power supply from the cigarette lighter.
11. Start the engine and check for fuel leaks at the rail and be sure the engine operation is normal.

Cold Start Injector Diagnosis and Service

Cold Start Injector Removal and Testing

WARNING: Energizing the cold start injector for more than five seconds may damage the injector winding.

The cold start injector service and diagnosis procedure varies depending on the vehicle. A typical cold start injector removal and testing procedure follows:

1. Bleed the pressure from the fuel system.
2. Connect a 12-V power supply to the cigarette lighter socket and disconnect the negative battery cable. If the vehicle is equipped with an air bag, wait one minute.
3. Wipe excess dirt from the cold start injector with a shop towel.
4. Remove the electrical connector from the cold start injector.
5. Remove the union bolt and the cold start injector fuel line (Figure 12-20).
6. Remove the cold start injector retaining bolts and remove the cold start injector.
7. Connect an ohmmeter across the cold start injector terminals (Figure 12-21). If the resistance is more or less than specified, replace the injector.
8. Connect the fuel line and union bolt to the cold start injector and place the injector tip in a container.

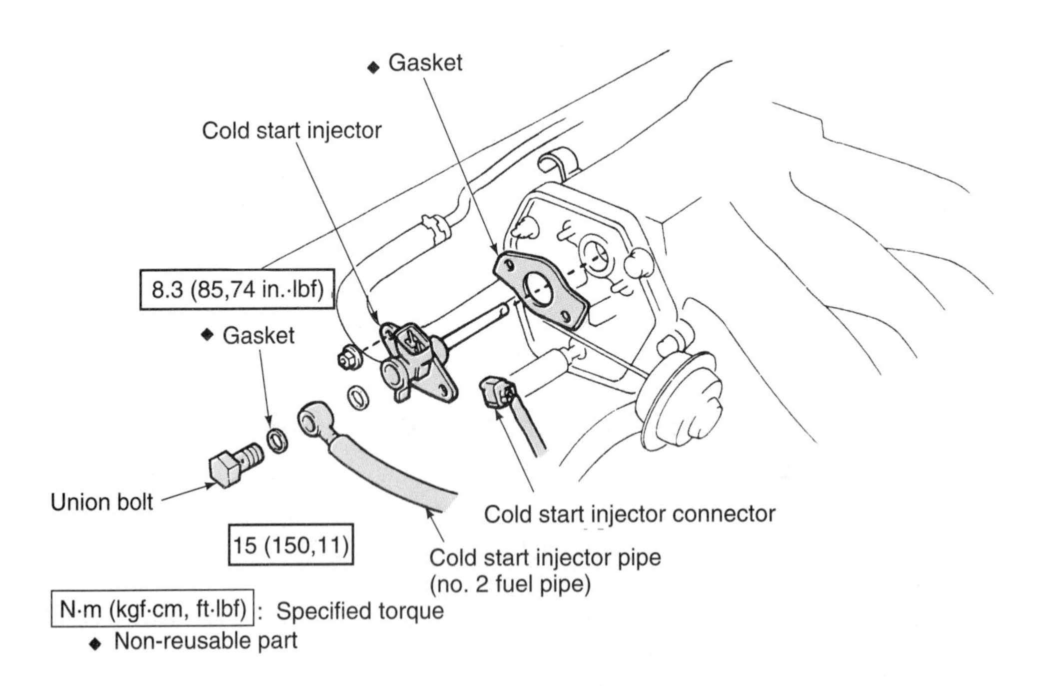

Figure 12-20 Removing cold start injector union bolt and fuel line (Courtesy of Toyota Motor Corporation)

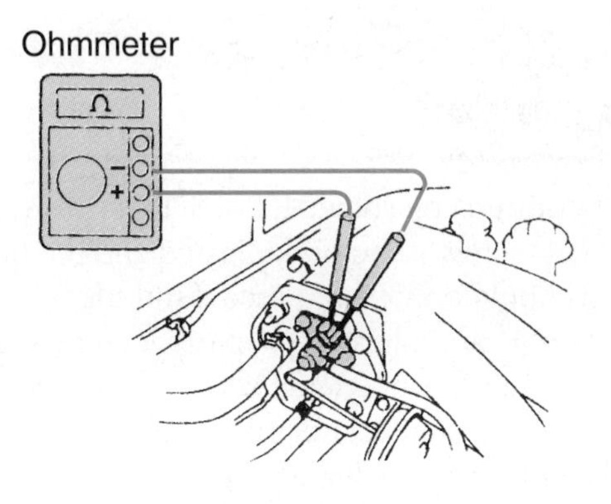

Figure 12-21 Testing cold start injector resistance with an ohmmeter (Courtesy of Toyota Motor Corporation)

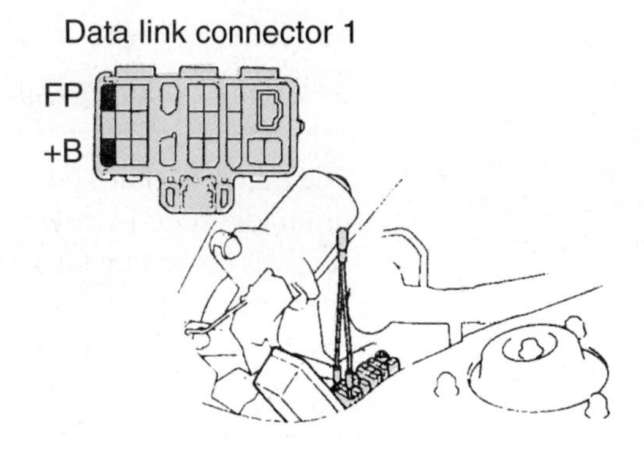

Figure 12-22 A jumper wire is connected from the B+ to the FP terminal in the DLC to operate the fuel pump and check the cold start injector. (Courtesy of Toyota Motor Corporation)

9. Connect a jumper wire to the B+ and FP terminals in the data link connector (DLC) and turn on the ignition switch (Figure 12-22).
10. Connect a special jumper wire from the cold start injector terminals to the battery terminals (Figure 12-23).
11. Check the fuel spray pattern from the injector. This pattern should be as illustrated in Figure 12-23. If the pattern is not as shown, replace the injector. Do not energize the cold start injector for more than five seconds.

Cold Start Injector Installation

1. Replace all cold start injector gaskets, and check all mounting surfaces for metal burrs, scratches, and warping.
2. Install the cold start injector gasket and injector, and tighten the injector mounting bolts to the specified torque.
3. Install the cold start injector fuel line, gaskets, and union bolt, and tighten this bolt to the specified torque.
4. Connect the cold start injector electrical connector.
5. Connect the negative battery cable and disconnect the 12-V power supply.
6. Start the engine and check for fuel leaks at the cold start injector.

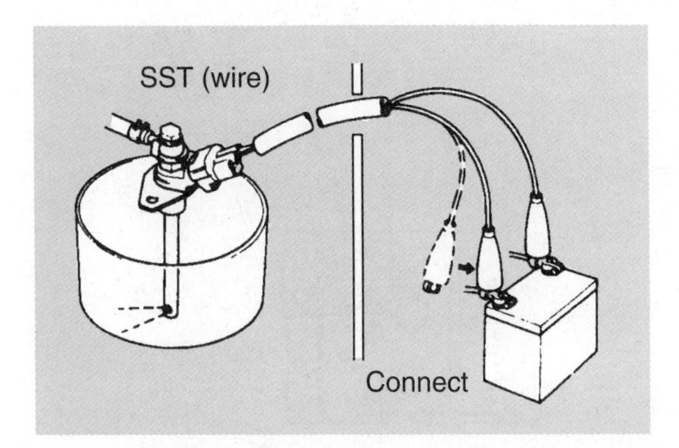

Figure 12-23 Connecting a jumper wire from the cold start injector terminals to the battery terminals (Courtesy of Toyota Motor Corporation)

Minimum Idle Speed Adjustment and Throttle Position Sensor Adjustment

The minimum idle speed adjustment may be referred to as a minimum air rate or air flow adjustment.

The minimum idle speed adjustment may be performed on some MFI or SFI systems with a minimum idle speed screw in the throttle body. This screw is factory-adjusted and the head of the screw is covered with a plug. This adjustment should only be required if throttle body parts are replaced. If the minimum idle speed adjustment is not adjusted properly, engine stalling may result. The procedure for performing a minimum idle speed adjustment varies considerably depending on the vehicle. Always follow the vehicle manufacturer's recommended procedure in the service manual. Following is a typical minimum idle speed adjustment procedure for a General Motors vehicle:

1. Be sure the engine is at normal operating temperature, and turn off the ignition switch.
2. Connect terminals A and B in the DLC, and connect a tachometer from the ignition tach terminal to ground.
3. Turn on the ignition switch and wait 30 seconds. Under this condition, the idle air control (IAC) motor is driven completely inward by the PCM.
4. Disconnect the idle air control (IAC) motor connector.
5. Remove the connection between terminals A and B in the DLC and start the engine.
6. Place the transmission selector in drive with an automatic transmission or neutral with a manual transmission.
7. Adjust the idle stop screw if necessary to obtain 500 to 600 rpm with an automatic transmission, or 550 to 650 rpm with a manual transmission. A plug must be removed to access the idle stop screw (Figure 12-24).
8. Turn off the ignition switch and reconnect the IAC motor connector.
9. Turn on the ignition switch and connect a digital voltmeter from the TPS signal wire to ground. If the voltmeter does not indicate the specified voltage of 0.55 V, loosen the TPS mounting screws and rotate the sensor until this voltage reading is obtained. Hold the TPS in this position and tighten the mounting screws.

Before an IAC motor is installed in a General Motors throttle body on TBI, MFI, or SFI systems, the distance from the end of the valve to the shoulder on the motor body must not exceed 1.125 in (28 mm). If the IAC motor is installed with the plunger extended beyond this measurement, the motor may be damaged.

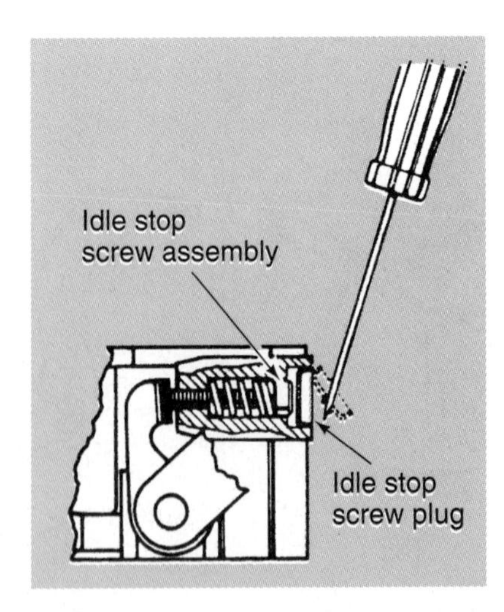

Figure 12-24 Adjusting the idle stop screw for minimum air adjustment (Courtesy of Chevrolet Motor Division, General Motors Corporation)

Minimum Idle Speed Adjustment, Throttle Body Injection

The minimum idle speed adjustment procedure varies depending on the year and make of vehicle. Always follow the adjustment procedure in the vehicle manufacturer's service manual. The minimum idle speed adjustment is only required if the TBI assembly or TBI assembly components are replaced. If the minimum idle speed adjustment is not adjusted properly, engine stalling may result. Proceed as follows for a typical minimum idle speed adjustment on a General Motors TBI system:

1. Be sure that the engine is at normal operating temperature and remove the air cleaner and TBI-to-air cleaner gasket. Plug the air cleaner vacuum hose inlet to the intake manifold.
2. Disconnect the throttle valve (TV) cable to gain access to the minimum air adjustment screw. A tamper-resistant plug in the TBI assembly must be removed to access this screw.
3. Connect a tachometer from the ignition tach terminal to ground, and disconnect the IAC motor connector.
4. Start the engine and place the transmission in park with an automatic transmission, or neutral with a manual transmission.
5. Plug the air intake passage to the IAC motor. Tool J-33047 is available for this purpose (Figure 12-25).
6. On 2.5-L four-cylinder engines, use the appropriate torx bit to rotate the minimum air adjustment screw until the idle speed on the tachometer is 475 to 525 rpm with an automatic transaxle, or 750 to 800 rpm with a manual transaxle.
7. Stop the engine and remove the plug from the idle air passage. Cover the minimum air adjustment screw opening with silicone sealant, reconnect the TV cable, and install the TBI gasket and air cleaner.

Special Tools

Tachometer

Throttle Body Service

Throttle Body On-Vehicle Cleaning

WARNING: Use only approved throttle body cleaners. Other cleaners may damage throttle body components.

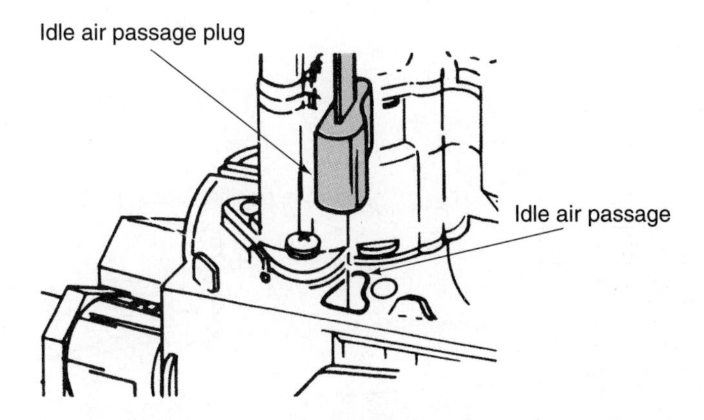

Figure 12-25 Special tool to plug the air passage to the IAC motor while checking minimum idle speed adjustment on a TBI system (Courtesy of Chevrolet Motor Division, General Motors Corporation)

WARNING: When cleaning a throttle body on the vehicle, be careful not to get the cleaner into the TPS or idle air control (IAC) valve. Throttle body cleaner will damage these components.

After many miles, or kilometers, of operation, an accumulation of gum and carbon deposits may occur around the throttle area in TBI, MFI, and SFI systems. This condition may cause rough idle operation. A pressurized can of throttle body cleaner may be used to spray around the throttle area without removing and disassembling the throttle body. If this cleaning method does not remove the deposits, the throttle body will have to be removed, disassembled, and placed in an approved cleaning solution. Never place the IAC motor or the TPS in cleaning solution, or damage to these components will result! Always remove the TPS, MAF, IAC motor, injectors, seals, gaskets, and pressure regulator before the throttle body is placed in a cleaning solution. Since MFI and SFI systems do not have injectors and a pressure regulator in the throttle body, removal of these components is not required.

Throttle Body On-Vehicle Inspection

Throttle body inspection and service procedures vary widely depending on the year and make of the vehicle. However, some components such as the TPS are found on nearly all throttle bodies. Since throttle bodies have some common components, inspection procedures often involve checking common components with the procedure recommended in the vehicle manufacturer's service manual. The following throttle body service procedures are based on a Toyota MFI system:

1. Check for smooth movement of the throttle linkage from the idle position to the wide-open position. Check the throttle linkage and cable for wear and looseness.
2. With the engine idling and operating at higher speed, check for vacuum with your finger at each vacuum port in the throttle body (Figure 12-26).
3. Apply vacuum from a hand vacuum pump to the throttle opener, and disconnect the TPS connector. Test the TPS with an ohmmeter connected across the appropriate terminals (Figure 12-27), and the specified thickness gauge inserted between the throttle stop screw and the stop lever (Figure 12-28).
4. Check the ohmmeter reading when the ohmmeter is connected to each of the specified terminals on the TPS (Figure 12-29).

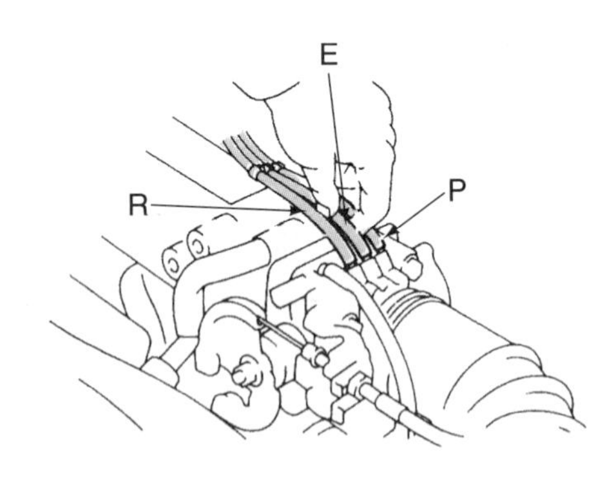

Port name	At idle	Other than idle
P	No vacuum	Vacuum
E	No vacuum	Vacuum
R	No vacuum	No vacuum

Figure 12-26 Throttle body vacuum ports and appropriate vacuum in relation to throttle position (Courtesy of Toyota Motor Corporation)

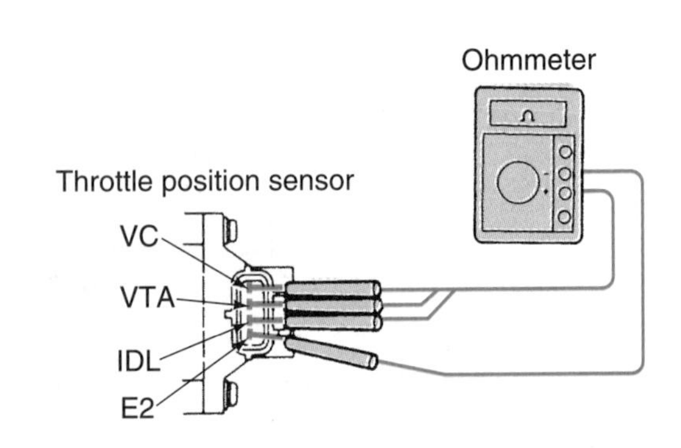

Figure 12-27 Ohmmeter connected to various TPS terminals to test TPS condition (Courtesy of Toyota Motor Corporation)

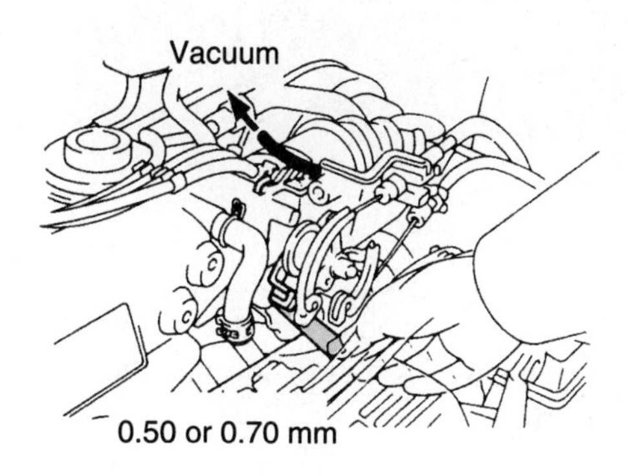

Figure 12-28 Thickness gauge inserted between the throttle stop screw and the stop lever while testing the TPS (Courtesy of Toyota Motor Corporation)

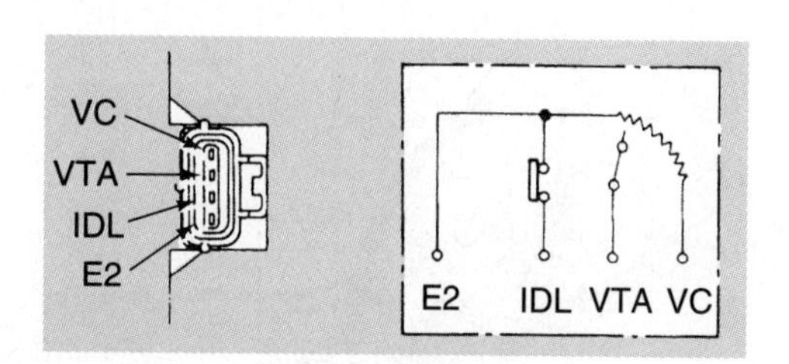

Clearance between lever and stop screw	Between terminals	Resistance
0 mm (0 in.)	VTA–E2	0.2–5.7 kΩ
0.50 mm (0.20 in.)	IDL–E2	2.3 kΩ or less
0.70 mm (0.028 in.)	IDL–E2	Infinity
Throttle valve fully open	VTA–E2	2.0–10.2 kΩ
——	VC–E2	2.5–5.9 kΩ

Figure 12-29 Specified ohmmeter reading at the TPS terminals (Courtesy of Toyota Motor Corporation)

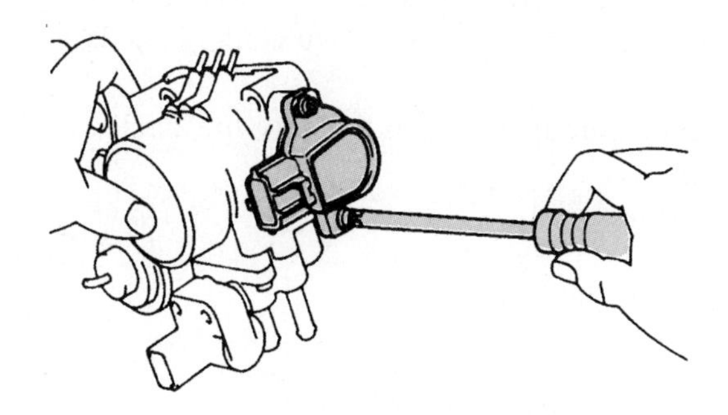

Figure 12-30 Loosening the TPS mounting screws to adjust the TPS until the specified ohmmeter readings are obtained (Courtesy of Toyota Motor Corporation)

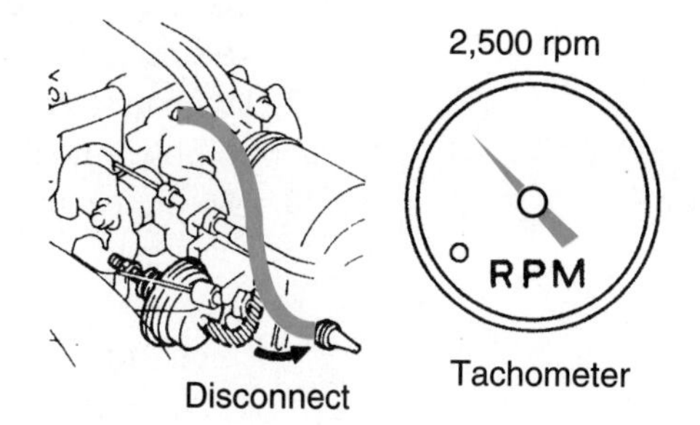

Figure 12-31 Throttle opener hose disconnected and plugged, and engine running at 2,500 rpm prior to throttle opener test (Courtesy of Toyota Motor Corporation)

5. Loosen the two TPS mounting screws and rotate the TPS as required to obtain the specified ohmmeter readings (Figure 12-30), and retighten the mounting screws. If the TPS cannot be adjusted to obtain the proper ohmmeter readings, replace the TPS.
6. Operate the engine until it reaches normal operating temperature, and check the idle speed on a tachometer. The idle speed should be 700 to 800 rpm.
7. Disconnect and plug the vacuum hose from the throttle opener, and maintain 2,500 engine rpm (Figure 12-31).
8. Be sure the cooling fan is off. Release the throttle valve, and observe the tachometer reading. When the throttle linkage strikes the throttle opener stem, the engine rpm should be 1,300 to 1,500.
9. Adjust the throttle opener as necessary (Figure 12-32), and reconnect the throttle opener vacuum hose.

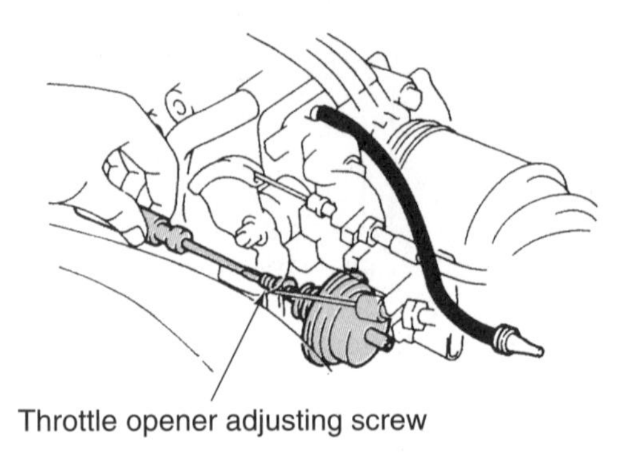

Figure 12-32 Throttle opener adjustment (Courtesy of Toyota Motor Corporation)

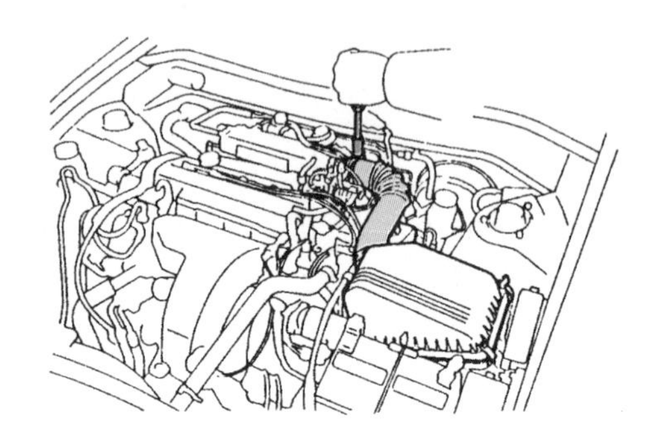

Figure 12-33 Loosening the air cleaner hose clamp at the throttle body (Courtesy of Toyota Motor Corporation)

Throttle Body Removal and Cleaning

Follow these steps for throttle body removal:

1. Connect a 12-V power supply to the cigarette lighter socket and disconnect the negative battery cable. If the vehicle is equipped with an air bag, wait one minute.
2. Drain the engine coolant from the radiator.
3. Disconnect the accelerator cable from the throttle linkage. If the vehicle has an automatic transmission, disconnect the throttle cable from the throttle linkage.
4. Disconnect the air intake temperature sensor connector.
5. Remove the cruise control cable from the clamp on the air cleaner resonator.
6. Loosen the air cleaner hose clamp bolt at the throttle body (Figure 12-33), and disconnect the four air cleaner cap clips.
7. Disconnect the air cleaner hose from the throttle body, and remove the air cleaner cap, air hose, and resonator (Figure 12-34).
8. Disconnect the TPS wiring connector.

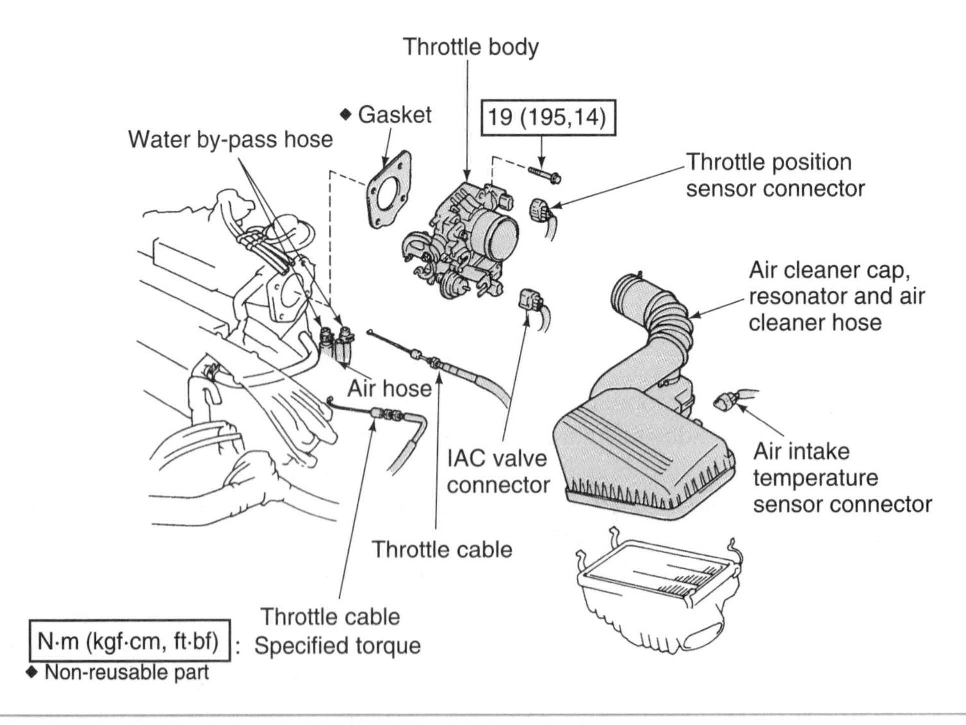

Figure 12-34 Throttle body and related components including air cleaner, hose, and resonator (Courtesy of Toyota Motor Corporation)

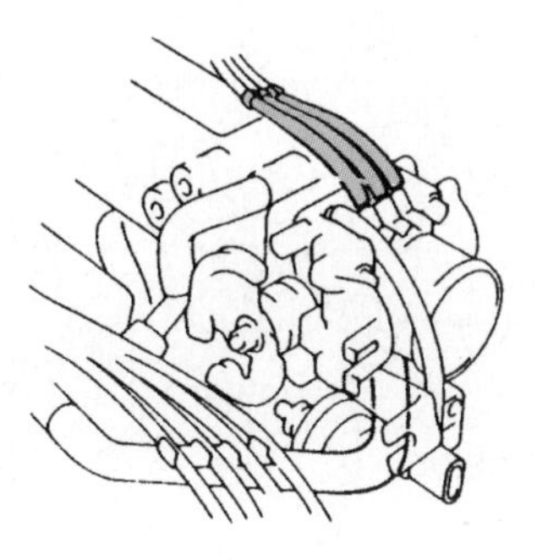

Figure 12-35 Removing vacuum hoses from the throttle body (Courtesy of Toyota Motor Corporation)

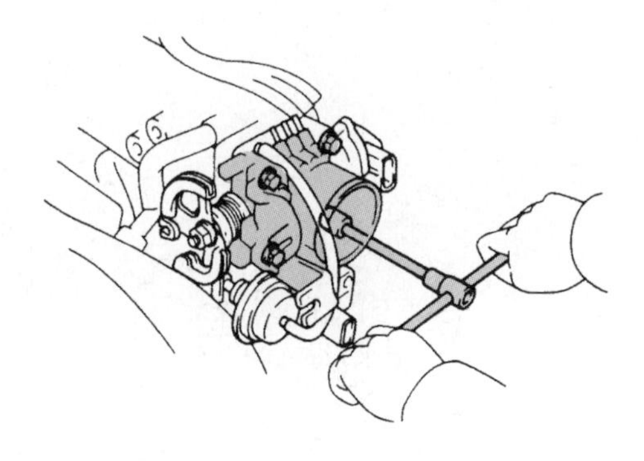

Figure 12-36 Removing four throttle body attaching bolts (Courtesy of Toyota Motor Corporation)

9. Disconnect the idle air control (IAC) connector.
10. Remove the vacuum hoses from the throttle body, and note the position of each hose so they may be installed in the same location (Figure 12-35).
    (a) Positive crankcase ventilation (PCV) hose
    (b) Exhaust gas recirculation (EGR) modulator hoses
    (c) Vacuum hoses from fuel evaporation vacuum switching valve (VSV)
11. Remove the four throttle body mounting bolts, and remove the throttle body and gasket from the intake manifold (Figure 12-36).
12. Disconnect two water by-pass hoses and the air hose from the throttle body, and note the position of each hose so they can be reconnected properly (Figure 12-37).
    (a) Water by-pass hose from water outlet
    (b) Water by-pass hose from water by-pass pipe
    (c) Air hose from air tube
13. Remove all nonmetallic parts such as the TPS, IAC valve, throttle opener, and the throttle body gasket from the throttle body.
14. Clean the throttle body assembly in the recommended throttle body cleaner and blow dry with compressed air. Blow out all passages in the throttle body assembly.

Special Tools

Throttle body cleaner

Throttle Body Assembly and Installation

1. Be sure all metal mating surfaces are clean and free from metal burrs and scratches. Install a new IAC valve gasket and install the IAC valve (Figure 12-38). Tighten the four valve mounting screws to the proper torque (Figure 12-39).
2. With the TPS screws loose, connect the ohmmeter leads to the IDL and E2 terminals on the TPS. Apply vacuum to the throttle opener with a hand vacuum pump, and place a

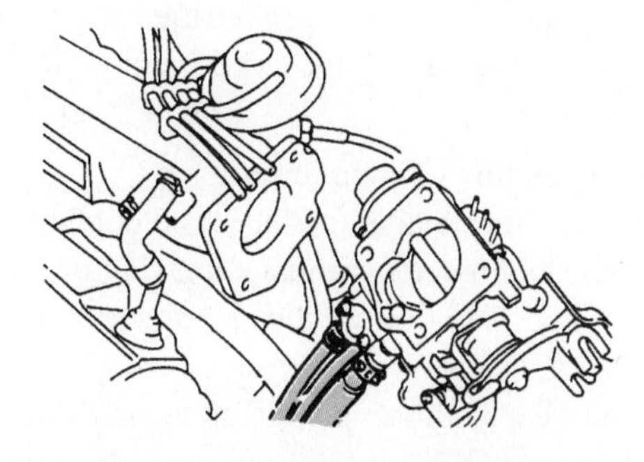

Figure 12-37 Disconnecting hoses from the throttle body (Courtesy of Toyota Motor Corporation)

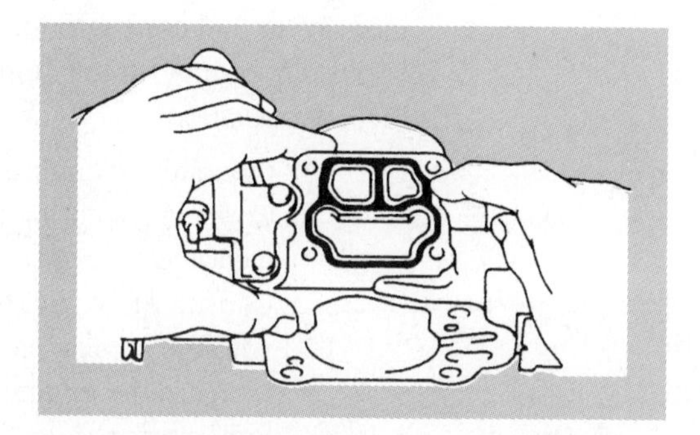

Figure 12-38 Installing new IAC valve gasket (Courtesy of Toyota Motor Corporation)

Figure 12-39 Tightening IAC valve retaining screws (Courtesy of Toyota Motor Corporation)

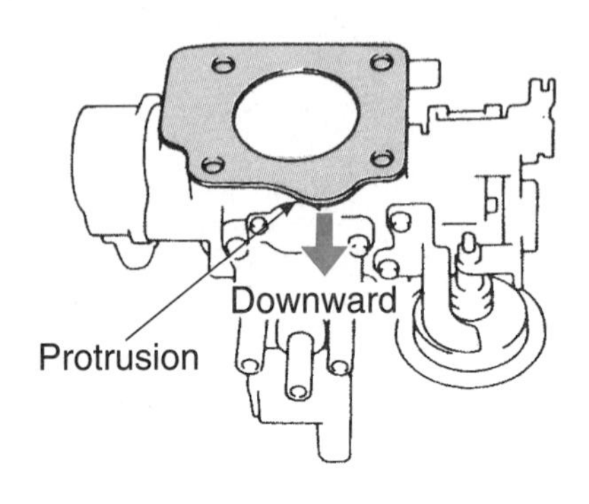

Figure 12-40 Install the new throttle body gasket with the protrusion facing downward (Courtesy of Toyota Motor Corporation)

0.024-in (0.60-mm) gauge between the throttle stop screw and the stop lever. Slowly rotate the TPS clockwise until the ohmmeter deflects, and tighten the TPS screws.

3. Perform the TPS measurements as outlined previously in the Throttle Body Inspection.
4. Install the water by-pass hoses and the air hose in their original locations on the throttle body. Be sure the hose clamps are tight.
5. Install a new throttle body gasket with the gasket protrusion facing downward (Figure 12-40).
6. Install the four throttle body attaching bolts and tighten these bolts to the specified torque.
7. Install the vacuum hoses on the throttle body in their original location.
8. Connect the IAC valve and TPS wiring connectors.
9. Check the air cleaner element, and replace this element if necessary. Inspect the air cleaner box, cap, hose, and resonator for cracks and distortion. Remove any debris from the air cleaner box. Connect the air cleaner hose and tighten the hose clamp. Install the air cleaner cap and the four retaining clamps.
10. Connect the air intake temperature sensor connector, and install the cruise control cable in the clamp on the air cleaner resonator.
11. Connect the throttle cable and accelerator cable, and replace the engine coolant.
12. Connect the negative battery cable, and disconnect the 12-V power supply.

Fuel Cut RPM Check

If the fuel cut mode is not operating properly, emission levels are high during deceleration and an increase in fuel consumption is experienced. The checking procedure for the fuel cut operation varies depending on the vehicle year and model. Following is a typical procedure for checking fuel cut operation:

1. Operate the engine until it is at normal operating temperature.
2. Connect a tachometer pickup lead to the IG terminal in the DLC, and connect the other tachometer leads as recommended by the tachometer manufacturer (Figure 12-41). Consult the vehicle manufacturer's information to be sure the tachometer is compatible with the vehicle electrical system.
3. Increase the engine rpm to 2,500, and place a stethoscope pickup against the body of a fuel injector.
4. Allow the engine speed to return to idle, and listen to the injector operation with the stethoscope (Figure 12-42). The injector should stop clicking momentarily and then

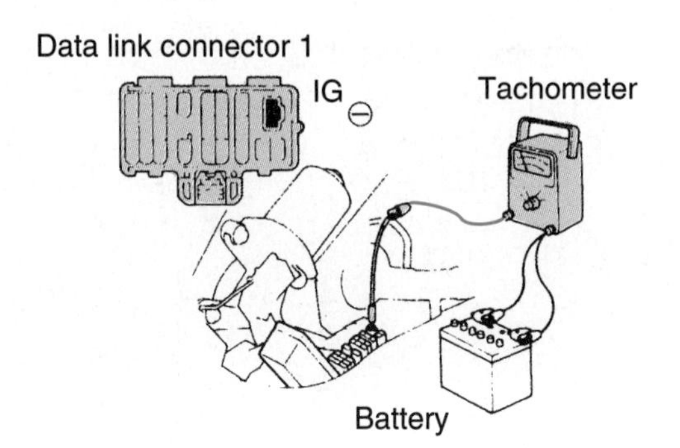

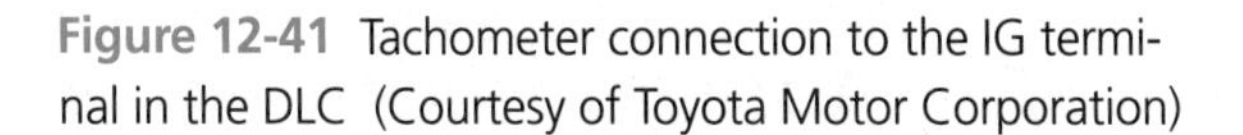
Figure 12-41 Tachometer connection to the IG terminal in the DLC (Courtesy of Toyota Motor Corporation)

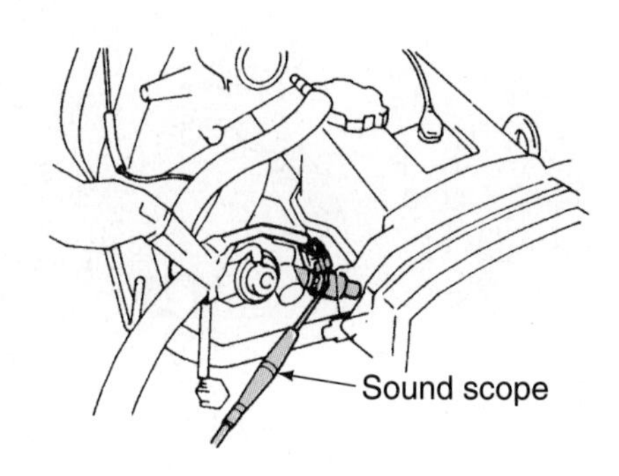

Figure 12-42 Stethoscope pickup placed on the injector body (Courtesy of Toyota Motor Corporation)

resume clicking as the engine speed decreases. The injector should begin clicking again at 1,500 rpm as the engine decelerates.

5. Shut the engine off and disconnect the tachometer.

Flash Code Diagnosis of TBI, MFI, and SFI Systems

Chrysler Flash Code Diagnosis

If a TBI, MFI, or SFI system is working normally, the malfunction indicator light (MIL) is illuminated when the ignition switch is turned on, and this light goes out a few seconds after the engine is started. The MIL light should remain off while the engine is running.

In the SAE J1930 terminology, the term malfunction indicator light (MIL) replaces other terms such as check engine light or service engine soon light.

If a defect occurs in a sensor and a diagnostic trouble code (DTC) is set in the computer memory, the computer may enter a limp-in mode. In this mode, the malfunction indicator light (MIL) or check engine light is on, the air-fuel ratio is rich, and the spark advance is fixed, but the vehicle can be driven to an automotive service center. When a vehicle is operating in the limp-in mode, fuel consumption and emission levels increase, and engine performance may decrease.

Prior to any DTC diagnosis, the Preliminary Diagnostic Procedure mentioned previously in this chapter must be completed, and the engine must be at normal operating temperature. If the engine is not at normal operating temperature, the computer may provide erroneous DTCs. The battery in the vehicle must be fully charged prior to DTC diagnosis.

Follow these steps to read the DTCs from the flashes of the MIL light on most Chrysler products:

1. Cycle the ignition switch on and off, on and off, and on in a five-second interval.
2. Observe the MIL lamp flashes to read the DTCs. Two quick flashes followed by a brief pause and three quick flashes indicates code 23. The DTCs are flashed once in numerical order.
3. When code 55 is flashed, the DTC sequence is completed. The ignition switch must be turned off, and steps 1 and 2 repeated to read the DTCs a second time.

On any TBI or PFI system, a DTC indicates a defect in a specific area. For example, a TPS code indicates a defective TPS, defective wires between the TPS and the computer, or the computer may be unable to receive the TPS signal. Specific ohmmeter or voltmeter tests may be necessary to locate the exact cause of the fault code. On logic module and power module systems, disconnect the quick-disconnect connector at the positive battery cable for 10 seconds with the ignition switch off to erase DTCs. On later module PCMs, this connector must be disconnected for 30 minutes to erase fault codes.

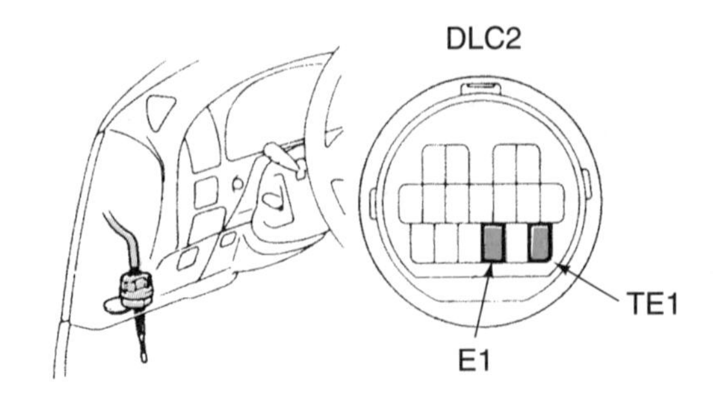

Figure 12-43 E1 and TE1 terminals in round data link connector (DLC) located under the instrument panel (Courtesy of Toyota Motor Corporation)

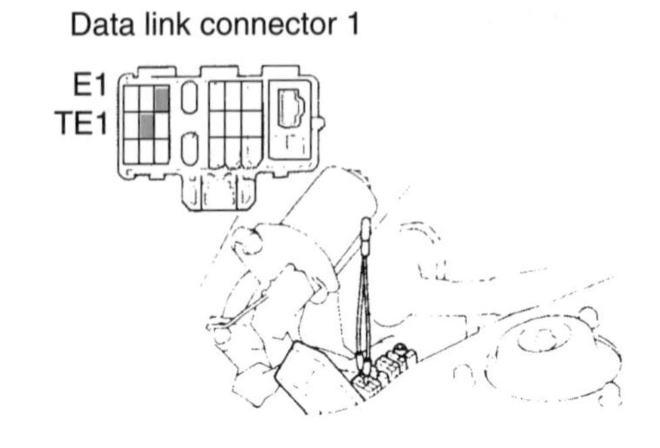

Figure 12-44 E1 and TE1 terminals in rectangular DLC positioned in the engine compartment (Courtesy of Toyota Motor Corporation)

Toyota Flash Code Diagnosis

Flash Code Output. Prior to the flash code output, the Preliminary Diagnostic Procedure must be performed as mentioned at the beginning of this chapter. Follow these steps for DTC diagnosis:

1. Turn on the ignition switch and connect a jumper wire between terminals E1 and TE1 in the data link connector (DLC). Some round DLCs are located under the instrument panel (Figure 12-43), while other rectangular-shaped DLCs are positioned in the engine compartment (Figure 12-44).
2. Observe the MIL light flashes. If the light flashes on and off at 0.26-second intervals, there are no DTCs in the computer memory (Figure 12-45).
3. If there are DTCs in the computer memory, the MIL light flashes out the DTCs in numerical order. For example, one flash followed by a pause and three flashes is code 13, and three flashes followed by a pause and one flash represents code 31 (Figure 12-46). The codes will be repeated as long as terminals E1 and TE1 are connected and the ignition switch is on.
4. Remove the jumper wire from the DLC.

Driving Test Mode. Follow this procedure to obtain fault codes during a driving test mode:

1. Turn on the ignition switch and connect terminals E1 and TE2 in the DLC (Figure 12-47).
2. Start the engine and drive the vehicle at speeds above 6 mph (10 km/h). Simulate the conditions when the problem occurs.
3. Connect a jumper wire between terminals E1 and TE1 on the DLC.
4. Observe the flashes of the MIL light to read the DTCs, and remove the jumper wire from the DLC.

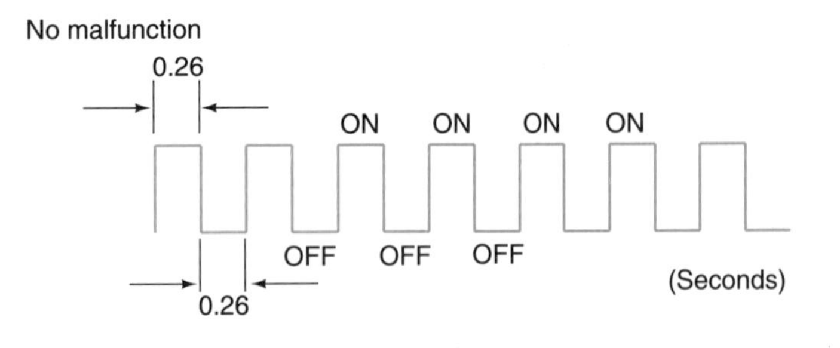

Figure 12-45 If the MIL light flashes at 0.26-second intervals, there are no DTCs in the computer memory. (Courtesy of Toyota Motor Corporation)

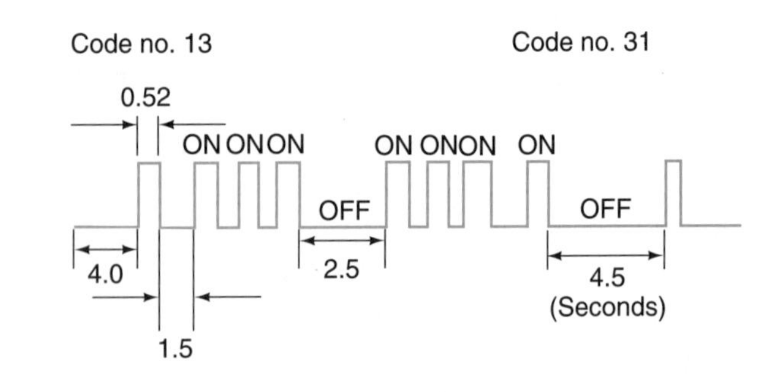

Figure 12-46 DTCs 13 and 31 (Courtesy of Toyota Motor Corporation)

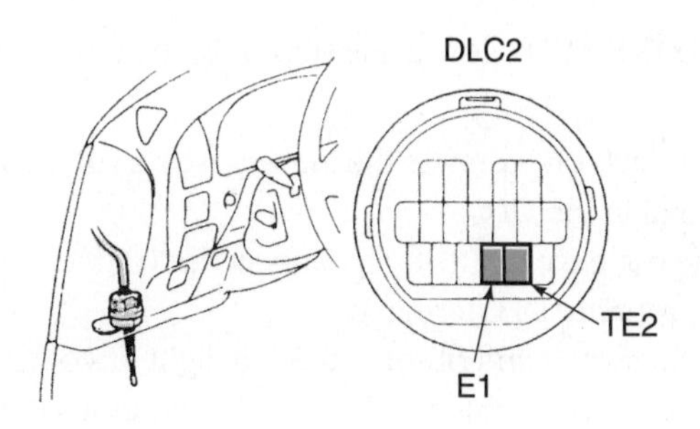

Figure 12-47 Terminals E1 and TE2 in the DLC (Courtesy of Toyota Motor Corporation)

Ford Flash Code Diagnosis

Most Ford vehicles have a MIL light on the instrument panel, and this light flashes the DTCs in the diagnostic mode. When a defect occurs in a major sensor, the PCM illuminates the MIL light and enters the limp-in mode in which the air-fuel ratio is rich and the spark advance is fixed. In this mode, engine performance decreases, and fuel consumption and emission levels increase.

Jumper Wire Connection. Prior to any fault code diagnosis the engine must be at normal operating temperature and the Preliminary Diagnostic Procedure mentioned previously in the chapter must be completed. A jumper wire must be connected from the self-test input wire to the appropriate DLC terminal to enter the self-test mode. When the ignition switch is turned on after this jumper wire connection, the MIL light begins to flash any DTCs in the PCM memory.

Optional Voltmeter Connection. If the vehicle does not have a check engine light, a voltmeter may be connected from the positive battery terminal to the proper DLC terminal (Figure 12-48). The voltmeter must be connected with the correct polarity as indicated in the figure.

When the ignition switch is turned on after the jumper wire and voltmeter connections are completed, the DTCs may be read from the sweeps of the voltmeter pointer or the flashes of the check engine light. For example, if three upward sweeps of the voltmeter pointer are followed by a pause and then four upward sweeps, code 34 is displayed.

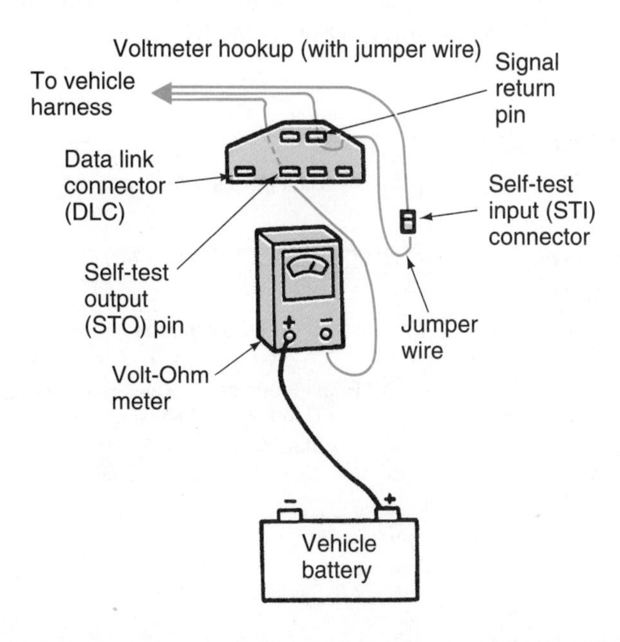

Figure 12-48 Jumper wire and voltmeter connection to Ford DLC (Reprinted with the permission of Ford Motor Company)

Key On, Engine Off (KOEO) Test. Follow these steps for the key on engine off (KOEO) fault code diagnostic procedure:

1. With the ignition switch off, connect the jumper wire to the self-test input wire and the appropriate terminal in the DLC.
2. If the vehicle does not have a MIL light, connect the voltmeter to the positive battery terminal and the appropriate DLC terminal.
3. Turn on the ignition switch and observe the MIL light or voltmeter. Hard fault DTCs are displayed, followed by a separator code 10 and continuous memory DTCs (Figure 12-49).

Hard fault diagnostic trouble codes (DTCs) are present in the computer memory at the time of testing.

Memory DTCs represent intermittent faults that occurred previously and were set in the computer memory at that time.

Memory DTCs may be called continuous memory codes or history codes.

Hard fault DTCs are present at the time of testing, whereas memory DTCs represent intermittent faults that occurred sometime ago and are set in the computer memory. Separator code 10 is displayed as one flash of the MIL light or one sweep of the voltmeter pointer. Each fault DTC is displayed twice and provided in numerical order. If there are no DTCs, system pass code 11 is displayed. If the technician wants to repeat the test or proceed to another test, the ignition switch must be turned off for 10 seconds.

Key On, Engine Running (KOER) Test. Follow these steps to obtain the fault codes in the Key On, Engine Running (KOER) test sequence:

1. Connect the jumper wire and the voltmeter as explained in steps 1 and 2 of the KOEO test.
2. Start the engine and observe the MIL lamp or the voltmeter. The engine identification code is followed by the separator code 10 and hard fault codes (Figure 12-50).

The engine identification (ID) code represents half of the engine cylinders. On a V8 engine, the MIL light flashes four times or the voltmeter pointer sweeps upward four times during the engine ID display.

On some Ford products, the brake on/off (BOO) switch and the power steering pressure switch (PSPS) must be activated after the engine ID code or DTCs 52 and 74, representing these switches, are present. Step on the brake pedal and turn the steering wheel to activate these switches immediately after the engine ID display.

Separator code 10 is presented during the KOER test on many Ford products. When this code is displayed, the throttle must be pushed momentarily to the wide-open position. The best way to provide a wide-open throttle is to push the gas pedal to the floor momentarily. On some Ford products, the separator code 10 is not displayed during the KOER test, and this throttle action is not required.

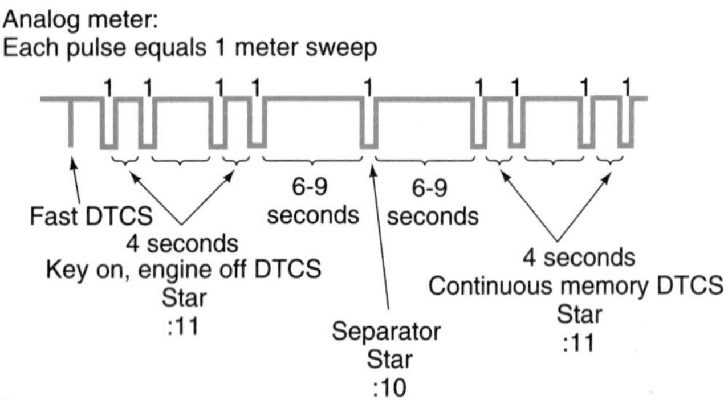

Figure 12-49 Key on, engine off (KOEO) test procedure (Reprinted with the permission of Ford Motor Company)

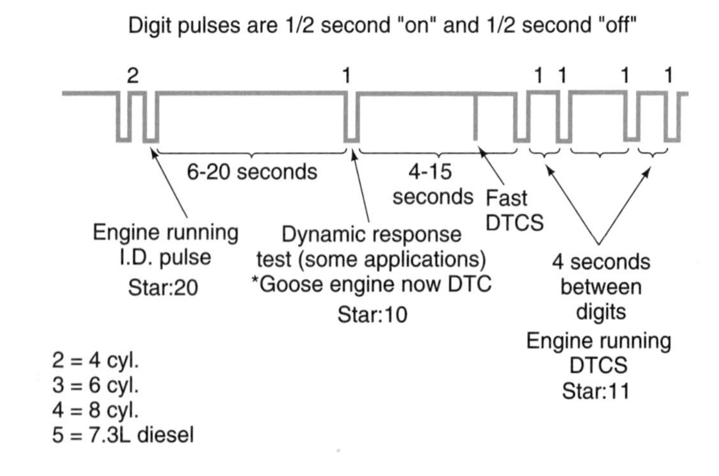

Figure 12-50 Key on, engine running (KOER) test (Reprinted with the permission of Ford Motor Company)

Hard fault DTCs are displayed twice in numerical order, and if no faults are present, system pass code 11 is given.

Fault Code Erasing Procedure. DTCs may be erased by entering the KOEO test procedure and disconnecting the jumper wire between the self-test input wire and the DLC during the code display.

General Motors Flash Code Testing

When a fault occurs in a major sensor, the PCM illuminates the MIL light, and a fault code is set in the PCM memory. Once this action takes place, the PCM is usually operating in a limp-in mode, and the air-fuel ratio is rich with a fixed spark advance. In this mode, driveability is adversely affected and fuel consumption and emission levels increase. Prior to any fault code diagnosis, the engine must be at normal operating temperature and the Preliminary Diagnostic Procedure mentioned previously must be completed. Follow these steps to obtain the DTCs with the MIL light flashes:

1. With the ignition switch off, connect a jumper wire between terminals A and B in the DLC under the instrument panel (Figure 12-51). A special tool that has two lugs that fit between these terminals is available. Usually terminals A and B are located at the top right corner of the DLC, but some DLCs are mounted upside down or vertically. Always consult the vehicle manufacturer's service manual for exact terminal location.
2. Turn on the ignition switch, and observe the MIL lamp.
3. One lamp flash followed by a brief pause and two more flashes indicates code 12. This code indicates that the PCM is capable of diagnosis. Each code is flashed three times, and codes are given in numerical order. If there are no DTCs in the PCM, only code 12 is provided. The code sequence keeps repeating until the ignition switch is turned off.

Complete DLC Terminal Explanation. The actual number of wires in the DLC varies depending on the vehicle make and year. The purpose of each DLC terminal may be explained as follows:

- **A** Ground terminal.
- **B** Diagnostic request.
- **C** Air injection reactor. When this terminal is grounded, the air injection reactor (AIR) pump air is directed upstream to the exhaust ports continually because this connection grounds the AIR system port solenoid. This applies to AIR systems with a converter solenoid and a port solenoid; however, this action does not apply to newer AIR systems with an electric diverter valve (EDV) solenoid.
- **D** MIL light. When this terminal is grounded on some systems, the check engine light is illuminated continually.
- **E** Serial data slow speed 160-baud PCM. This terminal supplies input sensor data to the scan tester on 160-baud PCM systems.
- **F** Torque converter clutch (TCC). If the vehicle is lifted and the engine accelerated until the transmission shifts through all the gears, a 12-V test light may be connected from this terminal to ground to diagnose the TCC system. The light is on when the TCC is not locked up, and the light goes out when TCC lockup occurs.

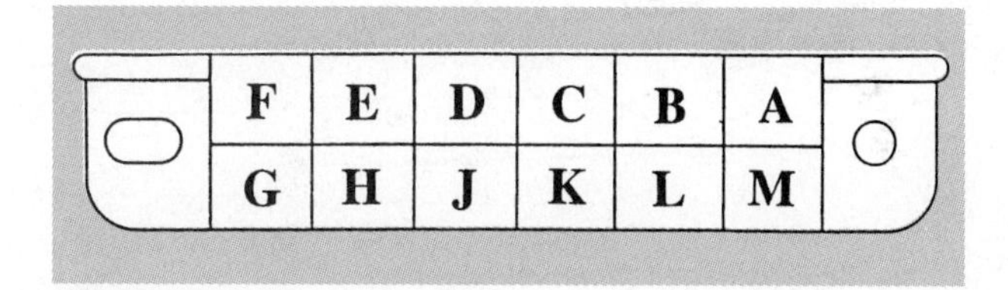

Figure 12-51 Terminals in the DLC

G Fuel pump test. On some models, when 12 V are supplied to this terminal with the ignition off, current flows through the top fuel pump relay contacts to the fuel pump, and the pump should run. Other models have a separate fuel pump test lead located under the hood.

H Antilock brake system (ABS), cars and trucks. When a jumper wire is connected from this terminal to terminal A, the ABS computer flashes the ABS warning light to provide fault codes.

J Not used.

K Air bag, supplemental inflatable restraint (SIR) system. When this terminal is connected to terminal A, the SIR computer flashes fault codes on the SIR warning light.

L Not used.

M High-speed serial data P4 PCM. This terminal supplies sensor data to the scan tester on P4 PCM systems.

DTC Erasing Procedure. The fault codes may be erased by disconnecting the quick-disconnect connector at the positive battery terminal for 10 seconds with the ignition switch off. If the vehicle does not have a quick-disconnect connector, the PCM B fuse may be disconnected to erase fault codes. On later model General Motors vehicles with P4 PCMs, the quick-disconnect, or PCM B, fuse may have to be disconnected for a longer time to erase codes.

If DTCs are left in a computer after the defect is corrected, the codes are erased automatically when the engine is stopped and started 30 to 50 times. This applies to most computer-equipped vehicles.

Field Service Mode. If the A and B terminals are connected in the DLC and the engine is started, the PCM enters the field service mode. In this mode, the speed of the MIL lamp flashes indicate whether the system is in open loop or closed loop. If the system is in open loop, the MIL lamp flashes quickly. When the system enters closed loop, the MIL lamp flashes at half the speed of the open loop flashes.

Nissan Flash Code Testing

In some Nissan electronic concentrated engine control systems (ECCS), the PCM has two light emitting diodes (LEDs), which flash a fault code if a defect occurs in the system. One of these LEDs is red, and the second LED is green. The technician observes the flashing pattern of the two LEDs to determine the DTC. If there are no DTCs in the ECCS, the LEDs flash a system pass code. The flash code procedure varies depending on the year and model of the vehicle, and the procedure in the manufacturer's service manual must be followed. Later model Nissan engine computers have a five-mode diagnostic procedure. Be sure the engine is at normal operating temperature and complete the Preliminary Diagnostic Procedure explained previously in this chapter. Turn the diagnosis mode selector in the PCM to obtain the diagnostic modes (Figure 12-52).

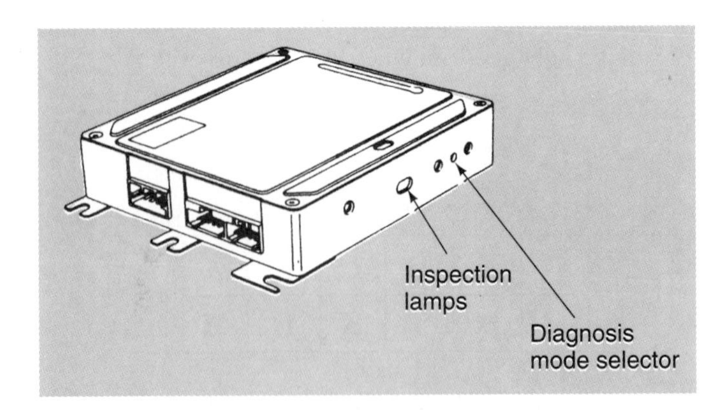

Figure 12-52 Diagnosis mode selector on PCM (Courtesy of Nissan Motor Corporation)

These diagnostic modes are available on some Nissan products:

Mode 1 — This mode checks the oxygen sensor signal. With the system in closed loop and the engine idling, the green light should flash on each time the oxygen sensor detects a lean condition. This light goes out when the oxygen sensor detects a rich condition. After 5-10 seconds, the PCM "clamps" on the ideal air-fuel ratio and pulse width, and the green light may be on or off. This PCM clamping of the pulse width only occurs at idle speed.

Mode 2 — In this mode, the green light comes on each time the oxygen sensor detects a lean mixture, and the red light comes on when the PCM receives this signal and makes the necessary correction in pulse width.

Mode 3 — This mode provides DTCs representing various defects in the system.

Mode 4 — Switch inputs to the PCM are tested in this mode. Mode 4 cancels codes available in mode 3.

Mode 5 — This mode increases the diagnostic sensitivity of the PCM for diagnosing intermittent faults while the vehicle is driven on the road.

After the defect is corrected, turn off the ignition switch, rotate the diagnosis mode selector counterclockwise, and install the PCM securely in the original position.

Scan Tester Diagnosis

Scan Tester Precautions

Several makes of scan testers are available to read the fault codes and perform other diagnostic functions. The exact tester buttons and test procedures vary on these testers, but many of the same basic diagnostic functions are completed regardless of the tester make. When test procedures are performed with a scan tester, these precautions must be observed:

1. Always follow the directions in the manual supplied by the scan tester manufacturer.
2. Do not connect or disconnect any connectors or components with the ignition switch on. This includes the scan tester power wires and the connection from the tester to the vehicle diagnostic connector.
3. Never short across or ground any terminals in the electronic system except those recommended by the vehicle manufacturer.
4. If the computer terminals must be removed, disconnect the scan tester diagnostic connector first.
5. Observe the service precautions listed previously in this chapter.

In the SAE J1930 terminology, the term scan tester (ST) replaces all previous terms of scan testers.

Scan Tester Features

Scan testers vary depending on the manufacturer, but many of these testers have the following features:

1. Display window — displays data and messages to the technician. Messages are displayed from left to right. Most scan testers display at least four readings on the display at the same time.
2. Memory cartridge — plugs into the scan tester. These memory cartridges are designed for specific vehicles and electronic systems. For example, a different cartridge may be required for the transmission computer and the engine computer. Most scan tester manufacturers supply memory cartridges for domestic and imported vehicles.
3. Power cord — connected from the scan tester to the battery terminals or cigarette lighter socket.
4. Adaptor cord — plugs into the scan tester, and connects to the DLC on the vehicle (Figure 12-53). A special adaptor cord is supplied with the tester for the diagnostic connector on each make of vehicle.

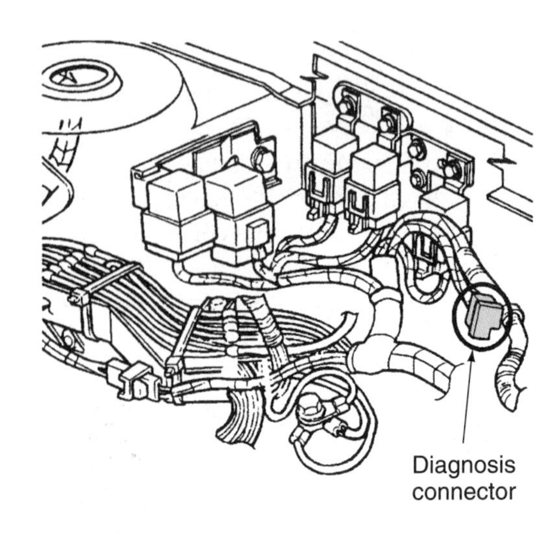

Figure 12-53 The scan tester adaptor cord has various ends to connect to the DLC on the vehicle. (Courtesy of Chrysler Corporation)

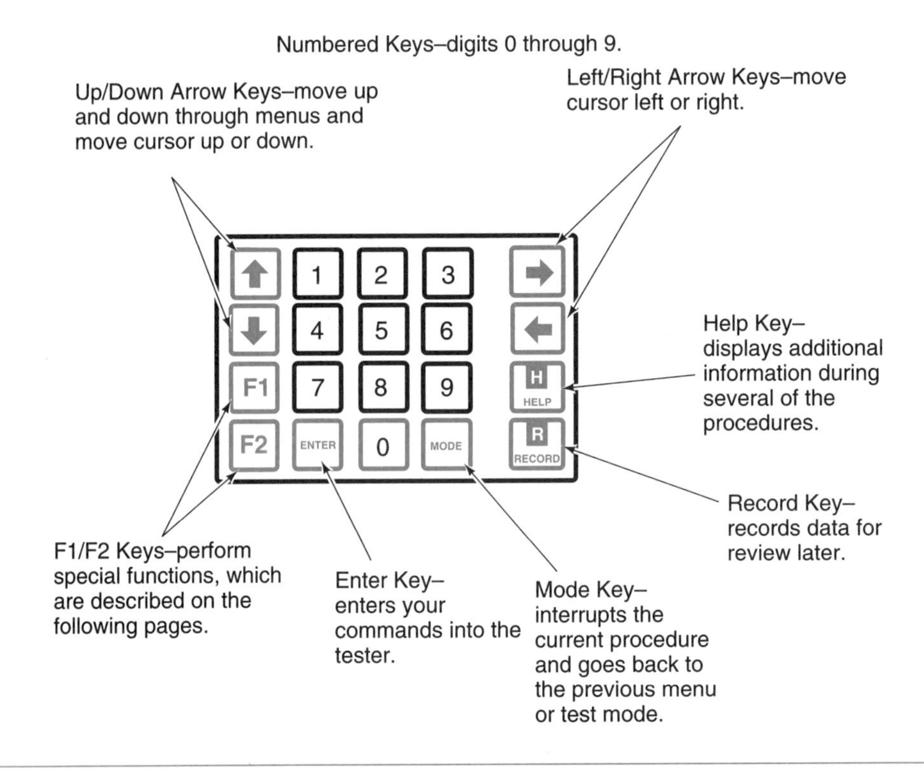

Figure 12-54 Scan tester features (Courtesy of OTC Division, SPX Corporation)

5. Serial interface — optional devices such as a printer, terminal, or personal computer may be connected to this terminal.
6. Keypad — allows the technician to enter data and reply to tester messages.

A typical scan tester keypad contains these buttons (Figure 12-54):

1. Numbered keys — digits 0 through 9.
2. UP/DOWN arrow keys — allow the technician to move back and forward through test modes and menus.
3. ENTER key — enters information into the tester.

4. MODE key — allows the technician to interrupt the current procedure and go back to the previous modes.
5. F1 and F2 keys — allow the technician to perform special functions described in the scan tester manufacturer's manuals.

Scan Tester Initial Entries

Scan tester operation varies depending on the make of the tester, but a typical example of initial entries follows:

Special Tools

Scan tester, modules, and connectors

1. Be sure the engine is at normal operating temperature and the ignition switch is off. With the correct module in the scan tester (Figure 12-55), connect the power cord to the vehicle battery.
2. Enter the vehicle year.—The tester displays enter 199X, and the technician presses the correct single digit and the ENTER key.
3. Enter VIN code.—This is usually a two-digit code based on the model year and engine type. These codes are listed in the scan tester operator's manual. The technician enters the appropriate two-digit code and presses the ENTER key.
4. Connect the scan tester adaptor cord to the diagnostic connector on the vehicle being tested.

Scan Tester Initial Selections

When the technician has programmed the scan tester by performing the initial entries, some entry options appear on the screen. These entry options vary depending on the scan tester and the vehicle being tested. Following is a typical list of initial entry options:

1. Engine
2. Antilock brake system (ABS)
3. Suspension
4. Transmission
5. Data line
6. Deluxe CDR
7. Test mode

The technician presses the number beside the desired selection to proceed with the test procedure. In the first four selections, the tester is asking the technician to select the computer system to be tested. If data line is selected, the scan tester provides a voltage reading from each input sensor in the system. If the technician selects Chrysler digital readout (CDR) on a Chrysler product, some of the next group of selections appear on the screen. When the technician selects the test mode number on a General Motors product, each test mode provides voltage or status readings from specific sensors or components.

Figure 12-55 Installing proper module in the scan tester for the vehicle being tested (Courtesy of OTC Division, SPX Corporation)

Scan Tester Test Selections

When the technician makes a selection from the initial test selection, the scan tester moves on to the actual test selections. These selections vary depending on the scan tester and the vehicle being tested. The following list includes many of the possible test selections and a brief explanation:

1. Faults codes — displays fault codes on the scan tester display.
2. Switch tests — allows the technician to operate switch inputs such as the brake switch to the PCM. Each switch input should change the reading on the tester.
3. ATM tests — forces the PCM to cycle all the solenoids and relays in the system for 5 minutes or until the ignition switch is turned off.
4. Sensor tests — provides a voltage reading from each sensor.
5. Automatic idle speed (AIS) motor — forces the PCM to operate the AIS motor when the up and down arrows are pressed. The engine speed should increase 100 rpm each time the up arrow is touched. RPM is limited to 1,500 or 2,000 rpm.
6. Solenoid state tests or output state tests — displays the on or off status of each solenoid in the system.
7. Emission maintenance reminder (EMR) tests — allows the technician to reset the EMR module. The EMR light reminds the driver when emission maintenance is required.
8. Wiggle test — allows the technician to wiggle solenoid and relay connections. An audible beep is heard from the scan tester if a loose connection is present.
9. Key on, engine off (KOEO) test — allows the technician to perform this test with the scan tester on Ford products.
10. Computed timing check — forces the PCM to move the spark advance 20° ahead of the initial timing setting on Ford products.
11. Key on, engine running (KOER) test — allows the technician to perform this test with the scan tester on Ford products.
12. Clear memory or erase codes — quickly erases fault codes in the PCM memory.
13. Code library — reviews fault codes.
14. Basic test — allows the technician to perform a faster test procedure without prompts.
15. Cruise control test — allows the technician to test the cruise control switch inputs, if the cruise module is in the PCM.

Snapshot Testing

Many scan testers have snapshot capabilities on some vehicles, which allow the technician to operate the vehicle under the exact condition when a certain problem occurs and freeze the sensor voltage readings into the tester memory. The vehicle may be driven back to the shop, and the technician can play back the recorded sensor readings. During the play back, the technician watches closely for a momentary change in any sensor reading, which indicates a defective sensor or wire. This action is similar to taking a series of sensor reading "snapshots," and then reviewing the pictures later. The snapshot test procedure may be performed on most vehicles with a data line from the computer to the DLC.

Mass Air Flow Sensor Testing, General Motors

While diagnosing a General Motors vehicle, one test mode displays grams per second from the MAF sensor. This mode provides an accurate test of the MAF sensor. The grams per second reading should be 4 to 7 with the engine idling. This reading should gradually increase as the engine speed increases. When the engine speed is constant, the grams-per-second reading should remain constant. If the grams-per-second reading is erratic at a constant engine speed or if this reading varies when the sensor is tapped lightly, the sensor is defective. A MAF sensor fault code may not be present with an erratic grams-per-second reading, but the erratic reading indicates a defective sensor.

Block Learn and Integrator Diagnosis, General Motors

While using a scan tester on General Motors vehicles, one test mode displays block learn and integrator. A scale of 0 to 255 is used for both of these displays and a mid-range reading of 128 is preferred. The oxygen (O_2) sensor signal is sent to the integrator chip and then to the pulse-width calculation chip in the PCM, and the block learn chip is connected parallel to the integrator chip. If the O_2 sensor voltage changes once, the integrator chip and the pulse width calculation chip change the injector pulse width. If the O_2 sensor provides four continually high or low voltage signals, the block learn chip makes a further injector pulse width change. When the integrator, or block learn, numbers are considerably above 128, the PCM is continually attempting to increase fuel; therefore, the O_2 sensor voltage signal must be continually low, or lean. If the integrator, or block learn, numbers are considerably below 128, the PCM is continually decreasing fuel, which indicates that the O_2 sensor voltage must be always high, or rich.

Cylinder Output Test, Ford

A cylinder output test may be performed on electronic engine control IV (EEC IV) SFI engines at the end of the KOER test. The cylinder output test is available on many SFI Ford products regardless of whether the KOER test was completed with a scan tester or with the flash code method.

When the KOER test is completed, momentarily push the throttle wide open to start the cylinder output test. After this throttle action, the PCM may require up to 2 minutes to enter the cylinder output test. In the cylinder output test, the PCM stops grounding each injector for about 20 seconds, and this action causes each cylinder to misfire.

While the cylinder is misfiring, the PCM looks at the rpm that the engine slows down. If the engine does not slow down, there is a problem in the injector, ignition system, engine compression, or a vacuum leak. If the engine does not slow down as much on one cylinder, a fault code is set in the PCM memory. For example, code 50 indicates a problem in number 5 cylinder. The correct DTC list must be used for each model year.

Ford Breakout Box Testing

On Chrysler and General Motors products, a data line is connected from the computer to the DLC, and the sensor voltage signals are transmitted on this data line and read on the scan tester. If these sensor voltage signals are normal on the scan tester, the technician knows that these signals are reaching the computer.

In 1989, Ford introduced data links on some Lincoln Continental models. Since that time, Ford has gradually installed data links on their engine computers. If a Ford vehicle has data links, there two extra pairs of wires in the DLC. Ford has introduced a new generation star (NGS) tester that has the capability to read data on some Ford engine computers. This technology is now available on other scan testers.

Since most Ford products previous to 1990 do not have a data line from the computer to the DLC, the sensor voltage signals cannot be displayed on the scan tester. Therefore, some other method must be used to prove that the sensor voltage signals are received by the computer. A breakout box is available from Ford Motor Company and some other suppliers (Figure 12-56).

Two large wiring connectors on the breakout box allow the box to be connected in series with the PCM wiring. Once the breakout box is connected, each PCM terminal is connected to a corresponding numbered breakout box terminal. These terminals match the terminals on Ford PCM wiring diagrams. For example, on many Ford EEC IV systems, PCM terminals 37 and 57 are 12-V supply terminals from the power relay to the PCM. Therefore, with the ignition switch on and the power relay closed, a digital voltmeter may be connected from breakout box terminals 37 and 57 to ground, and 12 V should be available at these terminals. The wiring diagram for the model year and system being tested must be used to identify the breakout box terminals.

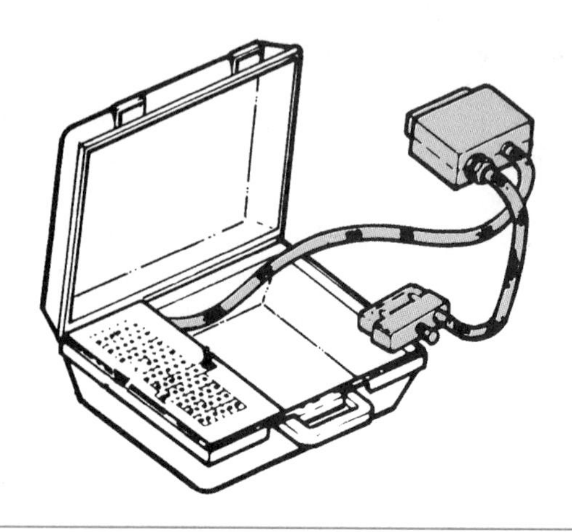

Figure 12-56 A breakout box may be connected in series with the PCM terminals on Ford vehicles to allow meter test connections. (Reprinted with the permission of Ford Motor Company)

PCM Service, General Motors

Service Bulletins. General Motors has a number of programmable read only memory (PROM) changes for their PCMs to correct various performance problems such as engine detonation and stalling. General Motors service bulletins provide the information regarding PROM changes. Several automotive test equipment suppliers and publishers also provide General Motors PROM identification (ID) and service bulletin books. This PROM ID and service bulletin information is absolutely necessary when diagnosing fuel injection problems on General Motors vehicles. Technicians must not have the idea they should immediately change a PROM to correct a problem. Most of the service bulletins relating to PROM changes provide a diagnostic procedure that directs the technician to test and eliminate all other causes of the problem before changing the PROM.

PROM Removal and Replacement. If a service bulletin or a fault code indicates that a programmable read only memory (PROM), calibration package (CALPAK), or memory calibrator (MEM-CAL) chip replacement is necessary, these chips are serviced separately from the PCM. A fault code representing one of these chips may indicate that the chip is defective or that the chip is improperly installed. The PROM and CALPAK chips are used in 160-baud PCMs, and the MEM-CAL chip is found in P4 PCMs. The MEM-CAL chip is a combined PROM, CALPAK, and electronic spark control (ESC) module. If PCM replacement is required, these chips are not supplied with the replacement PCM. Therefore, the chip or chips from the old PCM have to be installed in the new PCM. Always disconnect the negative battery cable before any chip replacement is attempted. Follow these steps for PROM replacement:

1. Remove the PROM access cover from the PCM (Figure 12-57).
2. Attach the PROM removal tool to the PROM and use a rocking action on alternate ends of the PROM carrier to remove the PROM (Figure 12-58).
3. Align the notch on the PROM carrier with the notch in the PROM socket in the PCM. Press on the PROM carrier only and push the PROM and carrier into the socket (Figure 12-59). Never remove the PROM from the carrier.

MEM-CAL Removal and Replacement.

1. Remove the MEM-CAL access cover on the ECM (Figure 12-60).
2. Use two fingers to push the retaining clips back away from the MEM-CAL, then grasp the MEM-CAL at both ends and lift it upward out of the socket.
3. Align the small notches on the MEM-CAL with the notches in the MEM-CAL socket in the PCM.
4. Press on the ends of the MEM-CAL and push it into the socket until the retaining clips push into the ends of the MEM-CAL (Figure 12-61).

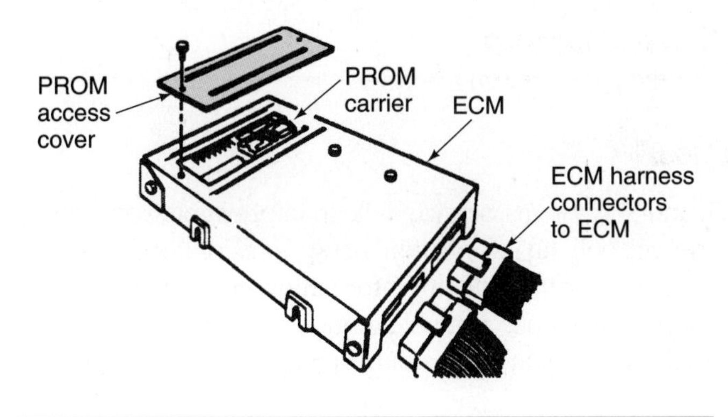

Figure 12-57 Removing PROM access cover (Courtesy of Chevrolet Motor Division, General Motors Corporation)

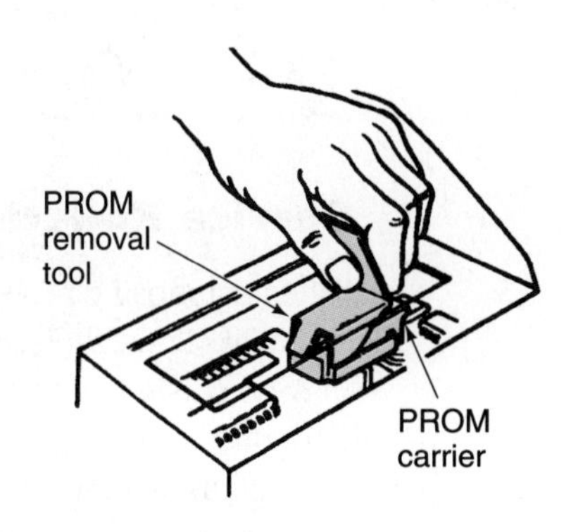

Figure 12-58 PROM removal tool (Courtesy of Chevrolet Motor Division, General Motors Corporation)

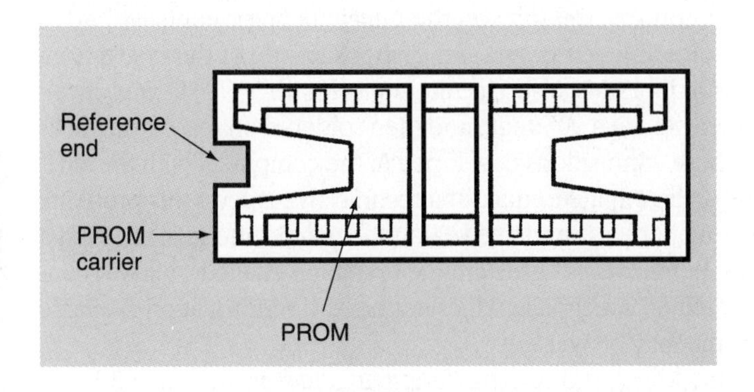

Figure 12-59 PROM installation (Courtesy of Chevrolet Motor Division, General Motors Corporation)

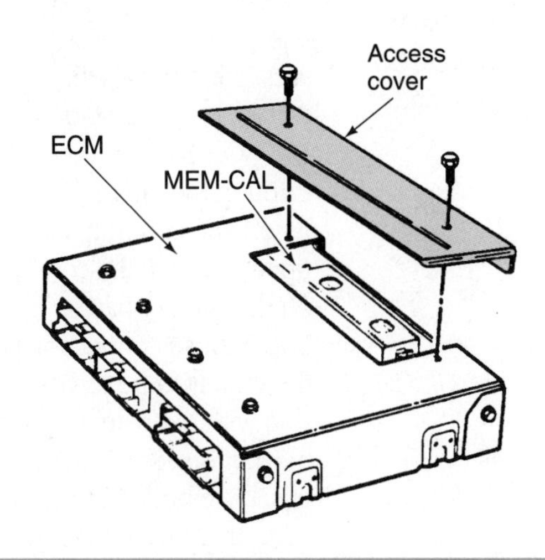

Figure 12-60 Removing MEM-CAL access cover (Courtesy of Oldsmobile Division, General Motors Corporation)

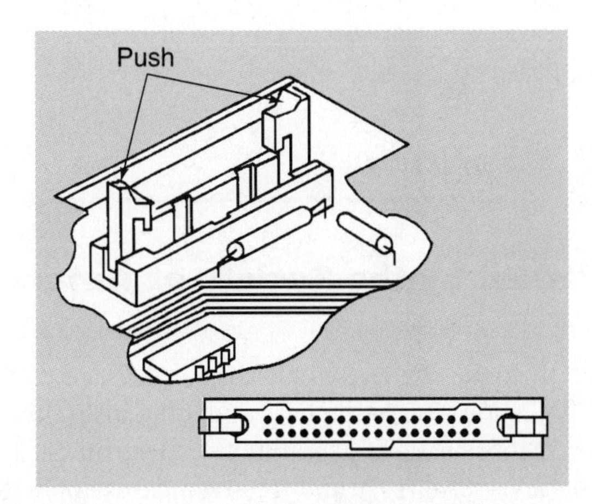

Figure 12-61 MEM-CAL installation (Courtesy of Oldsmobile Division, General Motors Corporation)

TBI, MFI, and SFI Diagnosis

Service Bulletin Information

When diagnosing problems in TBI, MFI, and SFI systems, service bulletin information is absolutely essential. If a technician does not have service bulletin information, many hours of diagnostic time may be wasted. Of course, we cannot include service bulletin information on all domestic and imported vehicles in this publication. This information is available from different suppliers on CD. We will discuss the solutions to three problems found in service bulletin information to emphasize the importance of this information.

A quad driver is a group of transistors in a computer that switches certain components on and off.

Many General Motors engines are equipped with Multec injectors. Some of these injectors experienced shorting problems in the windings, especially if the fuel contained some alcohol. If the injectors become shorted, they draw excessive current. General Motors P4 PCMs have a sense line connected to the quad driver that operates the injectors. When this sense line experiences excessive current flow from the shorted injectors, the quad driver shuts off and stops operating the injectors. This action protects the quad driver, but also causes the engine to stall. After a few minutes, the engine will usually restart. If a technician does not have this information available in a service bulletin, he or she may waste a great deal of time locating the problem.

On 160-baud General Motors computers, the pins on the internal components extended through the circuit board tracks, and soldering was done on the opposite side of the board from where the components were located. On P4 PCMs, a surface mount technology (SMT) was developed in which the component pins were bent at a 90° angle and then soldered on top of the tracks on the circuit boards. In some cases, loose connections developed in the computers with the SMT. These loose connections usually caused the engine to quit. If a technician suspects this problem, the PCM may be removed with the wiring harness connected. Start the engine and give the PCM a slap with the palm of your hand. If the engine stalls or the engine operation changes, a loose connection is present on the circuit board. When a technician does not have this information available in service bulletins, much diagnostic time may be wasted.

In 1991 Chrysler experienced some low-speed surging during engine warm-up on 3.3-L and 3.8-L engines. On these engines, the port fuel injectors sprayed against a hump in the intake port. As a result, fuel puddled behind this hump, especially while the engine was cold. When the engine temperature increased, this fuel evaporated and caused a rich air-fuel ratio and engine surging. Chrysler corrected this problem by introducing angled injectors with the orifices positioned at an angle so the fuel sprayed over the hump in the intake. When angled injectors are installed, the wiring connector must be positioned vertically. Angled injectors have beige exterior bodies. Technicians must have service bulletin information regarding problems like this.

OBD II EEC V Diagnosis

Standards Established by the Society of Automotive Engineers (SAE)

The SAE J1962 standards apply to the connectors such as the DLC and diagnostic repair tools. The J1962 standards also specify the location of the DLC under the instrument panel. The SAE J1930 standards apply to terms and abbreviations in automotive electronics. These standards provide universal terminology in this area. SAE J1979 and J2190 standards apply to test modes in automotive electronic systems. SAE J2012 standards define specific requirements for trouble codes.

Diagnostic Trouble Code (DTC) Interpretation

The SAE J2012 standards specify that all DTCs will have a five-digit alphanumeric numbering and lettering system. The following prefixes indicate the general area to which the DTC belongs:

1. P — power train
2. B — body
3. C — chassis

The first number in the DTC indicates who is responsible for the DTC definition.

1. 0 — SAE
2. 1 — manufacturer

The third digit in the DTC indicates the subgroup to which the DTC belongs. The possible subgroups are:

0 — Total system
1 — Fuel-air control
2 — Fuel-air control
3 — Ignition system misfire
4 — Auxiliary emission controls
5 — Idle speed control
6 — PCM and I/O
7 — Transmission
8 — Non-EEC power train

The fourth and fifth digits indicate the specific area where the trouble exists. Code P1711 has this interpretation:

P — Power train DTC
1 — Manufacturer-defined code
7 — Transmission subgroup
11 — Transmission oil temperature (TOT) sensor and related circuit

Scan Tester Diagnosis

The scan tester must have the appropriate connector to fit the DLC on an OBD II system, and the tester requires the proper software for the vehicle being tested. The vehicle make, model year, and engine size must be selected in the scan tester. Always follow the instructions in the manuals supplied by the scan tester manufacturer. Many similar tests, such as the KOEO and KOER tests, are performed on OBD II EEC V systems as on previous systems.

Diagnosis of Specific Problems

SERVICE TIP: Never replace a computer until the ground wires and voltage supply wires to the computer are checked and proven to be in satisfactory condition. High resistance in computer ground wires may cause unusual problems.

No-Start

1. Low compression
2. Improper valve timing
3. Defective ignition
4. Defective fuel system, fuel pump, filter, injectors

Hard Starting

1. Low compression
2. Lean air-fuel ratio, vacuum leak, injectors
3. Rich air-fuel ratio, injectors, cold start injector, input sensors
4. Leaking pressure regulator, fuel pump check valve, injectors
5. Defective ignition system

Rough Idle

1. Low compression
2. EGR valve (stuck open)
3. Vacuum leak
4. Dirty injectors
5. Dirty throttle body
6. Defective cold start injector

High Idle Speed

1. Coolant temperature sensor
2. Air charge temperature sensor
3. Thermostat stuck open
4. Low coolant level
5. P/N switch
6. Low battery and charging system voltage
7. Low voltage to computer (resistance in battery 12 V or ignition on 12 V wires)
8. Vacuum leak
9. Stickingor defective idle air control motor
10. Improper TPS adjustment or faulty TPS

Low Idle Speed

1. Coolant temperature sensor
2. Air charge temperature sensor
3. Sticking or defective idle air control motor
4. Improper TPS adjustment or faulty TPS
5. P/N switch

Rich Air-Fuel Mixture, Low Fuel Economy, Excessive Catalytic Converter Odor

1. Low compression
2. Defective ignition
3. High fuel pump pressure
4. Running in limp-in mode (defective sensor) (MAP sensor)
5. Coolant temperature sensor
6. Low coolant level
7. Air charge temperature sensor
8. Insufficient spark advance
9. Air pump air always upstream to exhaust ports with engine hot

Lean Air-Fuel Mixture

1. Low fuel pump pressure, pump, filter, regulator
2. Vacuum leak, especially on MAF applications (PCV valve)
3. Dirty injectors

Surging at Idle

1. Vacuum leak
2. Defective MAP sensor

3. Defective MAF sensor (also surges on acceleration)
4. Dirty injectors

Detonation

1. Lean air-fuel mixture
2. Excessive spark advance
3. Defective knock sensor or ESC module
4. Spark plug heat range too hot
5. Plug wires routed incorrectly
6. PROM change required, GM
7. Remove octane adjust connector — this action reduces spark advance on some Ford products

Engine Stalling

1. Defective injectors
2. Defective cold start injector
3. Defective idle air control motor
4. Improper idle speed
5. Improper minimum idle speed adjustment
6. Improper TPS adjustment or faulty TPS
7. Carbon and gum deposits in throttle body

Engine Surging After Torque Converter Clutch Lockup

1. Spark plugs
2. Spark plug wires
3. Distributor cap and rotor (distributor-type ignition)
4. Ignition coil
5. Fuel injectors
6. Vacuum leaks
7. EGR valve
8. MAF or MAP sensor
9. Worn camshaft lobes
10. Oxygen sensor
11. Low fuel pump pressure
12. Worn engine mounts
13. Front drive axle joints
14. TPS sensor
15. Low cylinder compression
16. Contaminated fuel

Engine Dieseling

1. Leaking injectors
2. Leaking cold start injector

Cylinder Misfiring

1. Low compression
2. Defective ignition system, spark plugs, plug wires, coil
3. Defective injectors
4. Vacuum leak

Engine Power Loss

1. Low compression
2. Improper EGR valve operation
3. Ignition defects

4. Reduced spark advance
5. Computer operating in limp-in mode
6. Low fuel pump pressure
7. Injectors
8. Restricted exhaust

Hesitation on Acceleration

Classroom Manual
Chapter 12, page 312

1. Lean air-fuel ratio, low fuel pump pressure, injectors, filter, vacuum leak
2. Improper TPS adjustment or faulty TPS
3. Reduced spark advance
4. Computer operating in limp-in mode

CUSTOMER CARE: Diagnosing is an extremely important part of a technician's job on today's high-tech vehicles. Always take time to diagnose a customer's vehicle accurately. Fast, inaccurate diagnosis of automotive problems leads to unnecessary, expensive repairs and unhappy customers who may take their business to another shop. Accurate diagnosis may take more time, but in the long term, it will improve customer relations and bring customers back to the shop.

Guidelines for Servicing TBI, MFI, and SFI systems

1. Prior to any diagnosis of TBI, MFI, or SFI systems, a preliminary diagnostic procedure must be performed, which includes checking such items as vacuum leaks and emission devices.
2. If there is no fuel pump pressure, always check the inertia switch, fuse, or fuse link first.
3. Fuel pressure should be relieved prior to disconnecting fuel system components.
4. Many vehicles have a test connector that allows the technician to operate the fuel pump continually while testing the fuel pump.
5. Low fuel pump pressure may be caused by a restricted line or filter or a defective fuel pump.
6. Low fuel pump pressure causes a lean air-fuel ratio.
7. High fuel pump pressure is caused by a sticking pressure regulator or a restricted return fuel line.
8. High fuel pump pressure results in a rich air-fuel ratio.
9. Port injectors may be tested with a balance test or a flow test.
10. Injectors should be tested for leakage and ohms resistance in the windings.
11. When injectors are cleaned with a pressurized container connected to the fuel rail, the fuel pump must be disabled and the fuel return line must be plugged.
12. A minimum idle speed adjustment is possible on some TBI, MFI, or SFI systems.
13. Throttle body components such as the IAC motor, TPS, and O-rings must not come in contact with throttle body cleaner.
14. On many vehicles, two terminals in the DLC must be connected to obtain flash codes from the check engine light.
15. On Chrysler products, the flash codes are obtained by cycling the ignition switch three times in a five-second interval.
16. A scan tester may be connected to the DLC on many vehicles to obtain fault codes and perform many other diagnostic functions.
17. A breakout box may be connected in series with the PCM terminals on Ford products to obtain voltage readings from the input sensors.
18. In General Motors PCMs, the PROM, CAL-PAK, and MEM-CAL chips may be replaced.

CASE STUDY

A customer phoned to say that he was having his SFI Cadillac towed to the shop because the engine had stopped and would not restart. Before the technician started working on the car, he routinely checked the oil and coolant. The engine oil dipstick indicated the crankcase was severely overfilled with oil, and the oil had a strong odor of gasoline. The technician checked the ignition system with a test spark plug and found the system to be firing normally. Of course, the technician thought the no-start problem must be caused by the fuel system, and the most likely problems would be the fuel filter or fuel pump.

The technician removed the air cleaner hose from the throttle body, and removed the air cleaner element to perform a routine check of the air cleaner element and throttle body. The throttle body showed evidence of gasoline lying at the lower edge of the throttle bore. The technician asked a co-worker to crank the engine while he looked in the throttle body. While cranking the engine, gasoline was flowing into the throttle body below the throttle. The technician thought this situation was impossible. An SFI system cannot inject fuel into the throttle body!

The technician began thinking about how fuel could be getting into the throttle body on this SFI system, and he reasoned the fuel had to be coming through one of the vacuum hoses. Next, he thought about which vacuum hose could be a source of this fuel, and he remembered the pressure regulator vacuum hose is connected to the intake. The pressure regulator vacuum hose was removed from the throttle body and placed in a container. When the engine was cranked, fuel squirted out of the vacuum hose, indicating the regulator diaphragm had a hole in it.

The technician installed a new pressure regulator and changed the engine oil and filter. After this service, the engine started and ran normally.

Terms to Know

Throttle body injection (TBI)
Multiport fuel injection (MFI)
Sequential fuel injection (SFI)
Power train control module (PCM)
Data link connector (DLC)
Schrader valve
Idle air control (IAC) motor
Fuel cut rpm
Diagnostic trouble code (DTC) diagnosis
Malfunction indicator light (MIL)
Field service mode
Scan tester
Snapshot testing
Block learn
Integrator
Breakout box
Memory calibrator (MEM-CAL)
Programmable read only memory (PROM)
Calibrator package (CAL-PAK)
Quad driver

ASE Style Review Questions

1. While discussing fuel pump pressure diagnosis:
Technician A says higher-than-specified fuel pump pressure may be caused by a sticking pressure regulator.
Technician B says the water in the fuel tank may prevent the fuel pump from pumping fuel.
Who is correct?
A. A only
B. B only
C. Both A and B
D. Neither A nor B

2. While discussing injector testing:
Technician A says a defective injector may cause cylinder misfiring at idle speed.
Technician B says restricted injector tips may result in acceleration stumbles.
Who is correct?
A. A only
B. B only
C. Both A and B
D. Neither A nor B

3. While discussing flash code diagnosis:
Technician A says the ignition switch must be turned off for 10 seconds between test sequences on a Ford EEC IV system.
Technician B says after one test sequence is completed on a Ford EEC IV system, another test may be started immediately.
Who is correct?
A. A only **C.** Both A and B
B. B only **D.** Neither A nor B

4. While discussing flash code diagnosis:
Technician A says the fault codes in a Chrysler PCM are erased in 10 seconds if the quick-disconnect is disconnected at the battery positive terminal.
Technician B says the fault codes in a PCM are erased in 30 minutes if the quick-disconnect is disconnected at the battery positive cable.
Who is correct?
A. A only **C.** Both A and B
B. B only **D.** Neither A nor B

5. While discussing flash code diagnosis:
Technician A says in a General Motors MFI system, the check engine light flashes each fault code four times.
Technician B says in a General Motors MFI system, terminals A and D must be connected in the DLC to obtain the fault codes.
Who is correct?
A. A only **C.** Both A and B
B. B only **D.** Neither A nor B

6. While discussing the effects of disconnecting battery cables:
Technician A says in most later model SFI systems, the battery cables may be disconnected without any adverse effects on the vehicle electronic system.
Technician B says on these systems, disconnecting the battery cables erases the adaptive memory in the computer.
Who is correct?
A. A only **C.** Both A and B
B. B only **D.** Neither A nor B

7. While discussing scan tester diagnosis of TBI, MFI, and SFI systems:
Technician A says the scan tester will erase fault codes quickly on many systems.
Technician B says many scan testers will store sensor readings during a road test and then play back the results in a snapshot test mode.
Who is correct?
A. A only **C.** Both A and B
B. B only **D.** Neither A nor B

8. While discussing block learn and integrator when the integrator number is 180 and the block learn number is 185:
Technician A says these numbers indicate a normal condition.
Technician B says these numbers indicate the PCM is trying to increase fuel delivery; therefore, the oxygen (O_2) sensor signal must be continually lean.
Who is correct?
A. A only **C.** Both A and B
B. B only **D.** Neither A nor B

9. While discussing a high idle speed problem:
Technician A says higher-than-normal idle speed may be caused by low electrical system voltage.
Technician B says higher-than-normal idle speed may be caused by a defective coolant temperature sensor.
Who is correct?
A. A only **C.** Both A and B
B. B only **D.** Neither A nor B

10. While discussing the causes of a rich air-fuel ratio:
Technician A says a rich air-fuel ratio may be caused by low fuel pump pressure.
Technician B says a rich air-fuel ratio may be caused by a defective coolant temperature sensor.
Who is correct?
A. A only **C.** Both A and B
B. B only **D.** Neither A nor B

Table 12-1 ASE TASK

Diagnose hot or cold no starting, hard starting incorrect idle speed, poor idle, flooding, hesitation, surging, engine misfire, power loss, stalling, poor mileage, and dieseling problems on vehicles with injection-type fuel systems; determine needed repairs.

Problem Area	Symptoms	Possible Causes	Classroom Manual	Shop Manual
ENGINE PERFORMANCE	Starting problems	Engine compression, ignition, or fuel system	277	289
	Improper idle speed, rough idle, stalling	Engine compression, ignition, or fuel system	287	289
	Flooding, hesitation, misfiring, power loss, dieseling	Engine compression, ignition, or fuel system	285	289

Table 12-2 ASE TASK

Inspect, test, and repair or replace fuel pressure regulation system and components of a fuel injection system.

Problem Area	Symptoms	Possible Causes	Classroom Manual	Shop Manual
ENGINE PERFORMANCE	Rich air-fuel ratio, low fuel economy	1. Defective pressure regulator, high pressure	284	264
		2. Restricted return fuel line	284	264
	Lean air-fuel ratio, hesitation	1. Defective pressure regulator, low pressure	284	264
		2. Defective fuel pump, restricted fuel filter	284	264

Table 12-3 ASE TASK

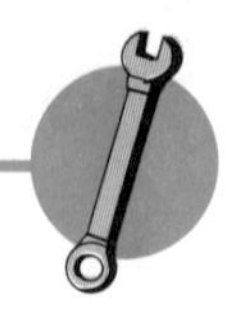

Inspect, test, adjust, and repair or replace fuel injection cold enrichment systems.

Problem Area	Symptoms	Possible Causes	Classroom Manual	Shop Manual
ENGINE	Hard starting	Defective cold start injector or thermo-time switch	292	266
	Rough idle	Defective cold start injector or thermo-time switch	292	266

Table 12-4 ASE TASK

Inspect, test, and replace fuel injection deceleration, fuel reduction, or shut off system and components.

Problem Area	Symptoms	Possible Causes	Classroom Manual	Shop Manual
ENGINE PERFORMANCE	Stalling, high emission levels	Defective fuel cut system	292	274

Table 12-5 ASE TASK

Remove, clean, and replace fuel injection throttle body and adjust related linkages.

Problem Area	Symptoms	Possible Causes	Classroom Manual	Shop Manual
ENGINE PERFORMANCE	Stalling, erratic idle	Carbon and gum deposits in throttle body	282	269

Table 12-6 ASE TASK

Inspect, test, clean, and replace fuel injectors.

Problem Area	Symptoms	Possible Causes	Classroom Manual	Shop Manual
ENGINE PERFORMANCE	Stalling, erratic idle, hesitation, hard starting loss of power	Restricted injector orifices	291	259

Table 12-7 ASE TASK

Inspect, clean, or replace throttle body mounting plates, fuel injection air induction system, intake manifold, and gaskets.

Problem Area	Symptoms	Possible Causes	Classroom Manual	Shop Manual
ENGINE PERFORMANCE	Stalling, erratic idle	1. Carbon and gum deposits in throttle body	282	269
		2. Vacuum leak at throttle mounting plate gasket	282	269

Table 12-8 ASE TASK

Inspect, test, clean, adjust, and replace components of fuel injection closed-loop fuel control systems.

Problem Area	Symptoms	Possible Causes	Classroom Manual	Shop Manual
ENGINE PERFORMANCE ECONOMY	Hard starting, low fuel economy, performance defects	Engine compression, ignition, or fuel system	277	253

Table 12-9 ASE TASK

Remove, clean, inspect/test, and repair or replace vacuum and electrical components and connections of fuel injection systems.

Problem Area	Symptoms	Possible Causes	Classroom Manual	Shop Manual
ENGINE PERFORMANCE ECONOMY	Hard starting, low fuel economy, performance defects	Engine compression, ignition, or fuel system	277	268

Table 12-10 ASE TASK

Perform analytic/diagnostic procedures on vehicles with on-board or self-diagnostic type computer systems; determine needed repairs.

Problem Area	Symptoms	Possible Causes	Classroom Manual	Shop Manual
ENGINE PERFORMANCE ECONOMY	Low fuel economy, performance problems, MIL light on	Engine compression, computer system faults ignition system defects	302	275

CHAPTER 13

Idle Speed Control Systems Service and Diagnosis

Upon completion and review of this chapter, you should be able to:

- ❑ Diagnose vacuum-operated decel valves.
- ❑ Perform an A/C idle speed check on a vacuum throttle kicker.
- ❑ Perform an idle speed check and adjustment on an electric throttle kicker solenoid.
- ❑ Perform an antidieseling adjustment on a vehicle with an electric throttle kicker solenoid.
- ❑ Diagnose causes on improper idle speed on vehicles with an idle air control motor.
- ❑ Diagnose idle contact switches and related circuits.
- ❑ Diagnose idle air control motors and idle air control by-pass air motors with a scan tester.
- ❑ Diagnose idle air control by-pass air motors and valves with a jumper wire connected at the data link connector.
- ❑ Remove, replace, and clean idle air control by-pass air motors and related throttle body passages.
- ❑ Diagnose idle air control by-pass air valves.
- ❑ Diagnose fast idle thermo valves.
- ❑ Diagnose starting air valves.

Vacuum-Operated Decel Valve Diagnosis

Basic Tools

Basic technician's tool set

Service manual

Jumper wires

Throttle body cleaner

Machinist's rule

Before diagnosing the vacuum-operated decel valve, always inspect all the hoses connected to the valve for cracks, leaks, and loose connections. Disconnect the clean air hose from the valve and accelerate the engine to 2,500 rpm (Figure 13-1). Release the throttle suddenly, and listen for air intake at the clean air inlet on the valve. As the engine decelerates, there should be an audible rush of air through this clean air inlet for a few seconds. If this audible rush of air is present, the valve is operating normally. Be sure the clean air hose is not restricted before reinstalling the hose.

If an audible rush of air flow into the air inlet is not present on deceleration, connect a vacuum gauge to the vacuum signal hose on the valve. With the engine idling, there should be full

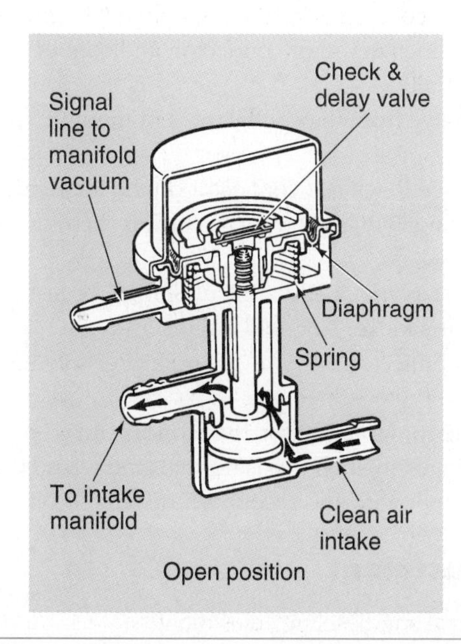

Figure 13-1 Vacuum-operated decel valve (Courtesy of Pontiac Motor Division, General Motors Corporation)

Special Tools

Vacuum gauge

manifold vacuum at the valve. If the manifold vacuum is normal at the valve signal hose, replace the valve. When the vacuum is lower than specified, check the vacuum signal hose for leaks. If this hose is satisfactory, check for intake manifold vacuum leaks and other causes of low manifold vacuum, such as late ignition timing and low engine compression.

Classroom Manual
Chapter 13, page 325

Service and Diagnosis of Combination Throttle Kicker and Idle Stop Solenoid

A/C Idle Speed Check

The idle speed adjustment procedure varies with each vehicle, engine, or model year. Always follow the vehicle manufacturer's recommended procedure in the service manual. The following A/C idle rpm check applies to a carburetor with a combined vacuum and electric throttle kicker. The vacuum throttle kicker is activated when the driver selects the A/C mode, and the electric kicker solenoid is energized when the ignition switch is turned on. Prior to any idle speed check, the engine must be at normal operating temperature. Follow these steps for the A/C idle speed check:

An electric throttle kicker solenoid may be called an idle stop solenoid.

1. Select the A/C mode, and set the temperature control to the coldest position.
2. Each time the A/C compressor cycles off and on, the kicker solenoid should be energized and the vacuum kicker plunger should move in and out. If the plunger reacts properly, the system is satisfactory. There is no adjustment on the vacuum kicker stem. When the vacuum kicker plunger does not react properly, check all the vacuum hoses and the solenoid, and test the vacuum kicker diaphragm for leaks.

Engine Idle Speed Check

The ignition timing should be checked and adjusted as necessary, prior to the idle speed check. The engine idle speed adjustment procedure varies depending on the vehicle make and model year. Always follow the recommended procedure in the vehicle manufacturer's service manual. Follow these steps for a typical idle rpm check:

Special Tools

Tachometer

1. With the transaxle in neutral and the parking brake applied, turn all the accessories and lights off. Be sure the engine is at normal operating temperature, and connect a tachometer from the coil negative primary terminal to ground.
2. Disconnect the cooling fan motor connector, and connect 12 V to the motor terminal so the fan runs continually.
3. Remove the PCV valve from the crankcase vent module, and allow this valve to draw in underhood air.
4. Disconnect the O_2 feedback test connector on the left fender shield.
5. Disconnect the wiring connector from the kicker vacuum solenoid on the left fender shield.
6. If the idle rpm is not as specified on the underhood emission label, adjust the idle rpm with the screw on the kicker solenoid (Figure 13-2).
7. Reconnect the O_2 connector, PCV valve, and kicker solenoid connector on the left fender shield.
8. Increase the engine rpm to 2,500 for 15 seconds, and then allow the engine to idle. If the idle speed changes slightly, this is normal, and a readjustment is not required.
9. Disconnect the jumper wire and reconnect the fan motor connector.

Antidieseling Adjustment

Follow these steps for a typical antidieseling adjustment:

1. Be sure the engine is at normal operating temperature and place the transaxle in neutral with the parking brake applied. Turn off all the lights and accessories.

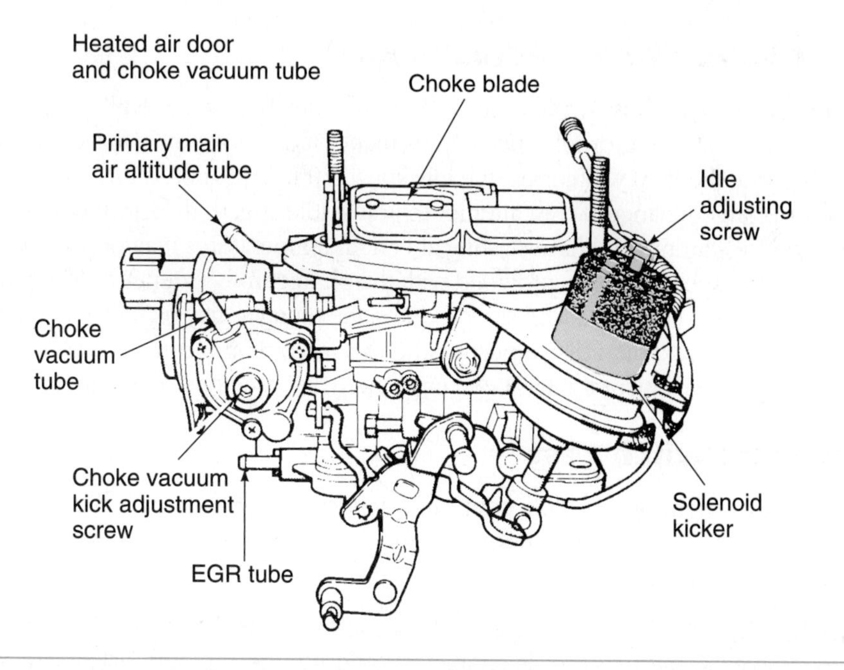

Figure 13-2 Idle speed screw on throttle kicker solenoid (Courtesy of Chrysler Corporation)

2. Remove the red wire from the six-way carburetor connector on the carburetor side of the connector, and disconnect the O_2 feedback test connector on the left fender shield.
3. Adjust the throttle stop screw to 700 rpm.
4. Reconnect the red wire in the six-way carburetor connector and the O_2 feedback test connector.

Idle Air Control Motor Service and Diagnosis

Improper Idle Speed General Diagnosis

Since the PCM operates the IAC or IAC BPA motor in response to the input signals, idle speed is controlled automatically. If the idle speed is not correct, the problem is in one of these areas:

1. An intake manifold vacuum leak
2. A defective input sensor or switch
3. The IAC or IAC BPA motor
4. Connecting wires from the IAC or IAC BPA motor to the PCM
5. The PCM

If an intake manifold vacuum leak occurs, the PCM senses the increase in manifold pressure caused by the vacuum leak. Under this condition, the PCM supplies more fuel and the idle speed increases. When the idle speed is higher than specified, always check for intake manifold vacuum leaks before proceeding with any further diagnosis.

SERVICE TIP: Never replace a PCM until you have diagnosed all other possible causes of the problem and then checked the power supply terminals and ground terminals on the PCM. Some PCMs have been replaced needlessly because of a ground or power supply problem.

If the IAC or IAC BPA motor is defective or seized, the idle speed is fixed at all temperatures. All other possible causes of the problem must be eliminated and the PCM must be diagnosed as the problem before PCM replacement.

Classroom Manual
Chapter 13, page 326

The Society of Automotive Engineers (SAE) J1930 terminology is an attempt to standardize terminology in automotive electronics.

In the SAE J1930 terminology, the term idle speed control (ISC) motor is replaced with the term idle air control (IAC) motor. This term applies to IAC motors that control idle speed by moving the throttle linkage.

In the SAE J1930 terminology, the term idle air control by-pass air (IAC BPA) refers to the type of IAC motor that by-passes air around the throttle to control idle speed.

The term IAC BPA motor refers to an idle speed control motor with a tapered pintle that moves inward and outward to control by-pass air and idle speed.

Idle Air Control Motor Adjustment

Some IAC motors, such as those on Chrysler TBIs, have a hex bolt on the end of the motor plunger. However, this hex bolt is not for idle speed adjustment. If this hex bolt is turned, it will not affect idle speed, because the PCM will correct the idle speed. If the hex bolt is turned and the length of the IAC motor plunger changed in an attempt to adjust idle speed, the throttle may not be in the proper position for starting, and hard starting at certain temperatures may occur. On some applications, an idle rpm specification is provided with the plunger fully extended. The plunger can be adjusted under this condition. On Chrysler products, the plunger may be fully extended by turning the ignition switch off and then disconnecting the IAC motor connector. The IAC motor hex bolt should only require adjustment if it has been improperly adjusted or if a new motor is installed.

Idle Contact Switch Test

An idle contact switch may be called a nose switch.

An idle contact switch may be called a throttle position switch.

The procedure for diagnosing the idle contact switch varies depending on the vehicle make and model year. Always follow the procedure recommended in the vehicle manufacturer's service manual. Following is a typical idle contact switch test:

WARNING: Never connect a 12-V source across the idle contact switch terminals on the ISC motor. If these contacts are closed, the contacts will be ruined.

CAUTION: Connecting a pair of jumper wires from the terminals of a 12-V battery to the IAC motor idle contact switch terminals may result in very high current flow and jumper wire heating, which could result in burns to your hands.

Special Tools

Digital multimeter

1. Backprobe terminal B on the IAC motor, and turn on the ignition switch. Connect a digital voltmeter from terminal B to ground, and hold the throttle approximately half open (Figure 13-3). The voltage supplied from the PCM to the idle contact switch should be 4.5 V to 8 V, depending on the system. Always refer to the vehicle manufacturer's specifications.
2. If the voltage at terminal B on the IAC motor is not within specifications, turn off the ignition switch and disconnect the IAC motor connector and the PCM connector. Connect the ohmmeter leads from terminal B in the IAC motor connector to terminal 2D8

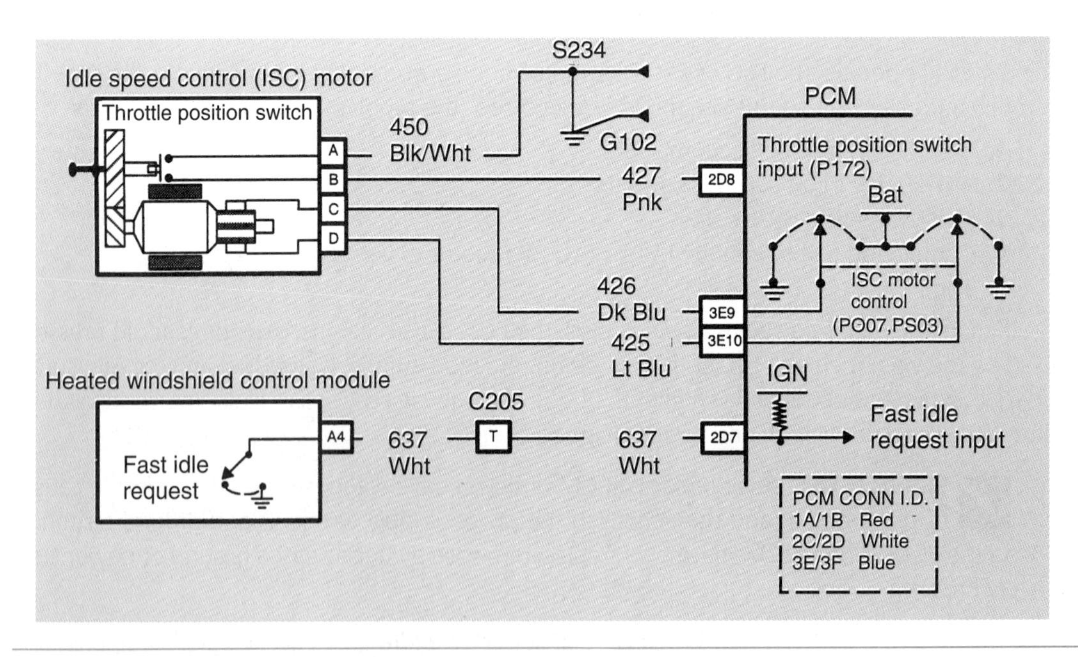

Figure 13-3 Idle air control motor circuit (Courtesy of Cadillac Motor Car Division, General Motors Corporation)

in the PCM connector. The ohm reading should be less than 0.5 Ω. If the reading is more than this value, repair the resistance problem or open circuit in the wire from the PCM to the idle contact switch.

3. Connect the ohmmeter leads from terminal B on the IAC motor connector to ground. The ohmmeter reading should be infinite. If the reading is not infinite, repair the ground in the wire from the PCM to the idle contact switch.
4. If the voltage at terminal B on the IAC motor is not within specifications, and the ohmmeter readings in steps 2 and 3 are satisfactory, reconnect the PCM and IAC motor connectors. Backprobe terminal 2D8 at the PCM, and turn on the ignition switch. Connect a digital voltmeter from terminal 2D8 to ground, and observe the voltage. If the voltage is not within specifications, check all the power supply and ground wires to the PCM. When the power supply and ground wires are satisfactory, replace the PCM.
5. If the voltage in step 1 is satisfactory, return the throttle to the idle position, and observe the voltmeter. The reading should be less than 1 V with the throttle in the idle position and the idle contact switch closed. If the voltage reading is within specifications, the idle contact switch is satisfactory.
6. When the voltage in step 5 is not within specifications, connect the voltmeter from terminal A on the IAC motor to ground. If the voltage at this point is above 0.2 V, repair the resistance problem or open circuit in the ground wire connected to terminal A. When the voltage at terminal A is 0.2 V or less and the voltage in step 5 is above specifications, replace the IAC motor.

Scan Tester Diagnosis, Idle Air Control and Idle Air Control By-pass Motors

If the idle speed is not within specifications, the input sensors and switches should be checked carefully with the scan tester. For example, if the throttle position sensor (TPS) voltage is lower than specified at idle speed, the PCM interprets this condition as the throttle being closed too much. Under this condition, the PCM opens the IAC or IAC BPA motor to increase idle speed.

Special Tools

Scan tester

The term IAC BPA valve refers to an idle speed control valve that the computer pulses open and closed to control by-pass air.

If the engine coolant temperature sensor resistance is higher than normal, it sends a higher-than-normal voltage signal to the PCM. The PCM thinks the coolant is colder than it actually is, and under this condition, the PCM operates the IAC or IAC BPA motor to increase idle speed. Many input sensor defects cause other problems in engine operation besides improper idle rpm.

Defective input switches result in improper idle rpm. For example, if the A/C switch is always closed, the PCM thinks the A/C is on continually. This action results in the PCM operating the IAC or IAC BPA motor to provide a higher idle rpm. On many vehicles, the scan tester indicates the status of the input switches as closed or open, or high or low. Most input switches provide a high-voltage signal to the PCM when they are open and a low voltage signal if they are closed.

On some vehicles, a fault code is set in the PCM memory if the IAC or IAC BPA motor or connecting wires is defective. On other systems, a fault code is set in the PCM memory if the idle rpm is out of range. On Chrysler products, the actuation test mode (ATM), or actuate outputs mode, may be entered with the ignition switch on. The IAC or IAC BPA motor may be selected in the actuate outputs mode, and the PCM is forced to extend and retract the IAC or IAC BPA motor plunger every 2.8 seconds. When this plunger extends and retracts properly, the motor, connecting wires, and PCM are in normal condition. If the plunger does not extend and retract, further diagnosis is necessary to locate the cause of the problem.

WARNING: When performing a set engine rpm mode test on an IAC motor, always be sure the transmission selector is in the park position and the parking brake is applied.

On some IAC or IAC BPA motors, a set engine rpm mode may be entered on the scan tester. In this mode, each time a specified scan tester button is touched, the rpm should increase 100 rpm to a maximum of 2,000 rpm. Another specified scan tester button may be touched to decrease the speed in 100 rpm steps. On some scan testers, the up and down arrows are used to increase and decrease the engine rpm during this test. If the IAC or IAC BPA motor responds properly during this diagnosis, the PCM, motor, and connecting wires are in satisfactory condition, and further diagnosis of the inputs is required.

SERVICE TIP: If the IAC or IAC BPA motor counts are zero on the scan tester, the circuit is likely open between the PCM and the motor. Wiggle the wires on the IAC or IAC BPA motor and observe the scan tester reading. If the count reading changes while wiggling the wires, you have found the problem.

Classroom Manual
Chapter 13, page 329

On some systems, the scan tester reads the IAC or IAC BPA motor counts, and the count range is provided in the scan tester instruction manual. Some of the input switches, such as the A/C, may be operated and the scan tester counts should change. If the scan tester counts change when the A/C is turned on and off, the motor, connecting wires, and PCM, are operational. When the scan tester counts do not change under this condition, further diagnosis is required.

Photo Sequence 10 shows a typical procedure for performing a scan tester diagnosis of an idle air control motor.

Idle Air Control By-pass Air Motor Service and Diagnosis

Idle Air Control By-pass Air Motor and Valve Diagnosis with Jumper Wire at the Data Link Connector

On some systems, such as Toyota, a jumper wire may be connected to terminals E1 and TE1 in the data link connector (DLC) to diagnose the IAC BPA valve with the engine at normal operating temperature (Figure 13-4). When the engine is started with this jumper wire connection, the engine speed should increase to 1,000 rpm to 1,300 rpm for 5 seconds, and then return to idle speed. If the IAC BPA valve does not respond as specified, further diagnosis of the IAC BPA valve, wires, and PCM is required.

On General Motors vehicles, the IAC BPA motor extends fully when terminals A and B are connected in the DLC, and the ignition switch is turned on. With the IAC BPA motor removed from the throttle body, this jumper connection may be completed while observing the IAC BPA motor. If the motor does not extend, further diagnosis of the motor, connecting wires, and PCM is required.

Idle Air Control By-pass Air Motor Removal and Cleaning

Carbon deposits in the IAC BPA motor air passage in the throttle body or on the IAC BPA motor pintle result in erratic idle operation and engine stalling. Remove the motor from the throttle body,

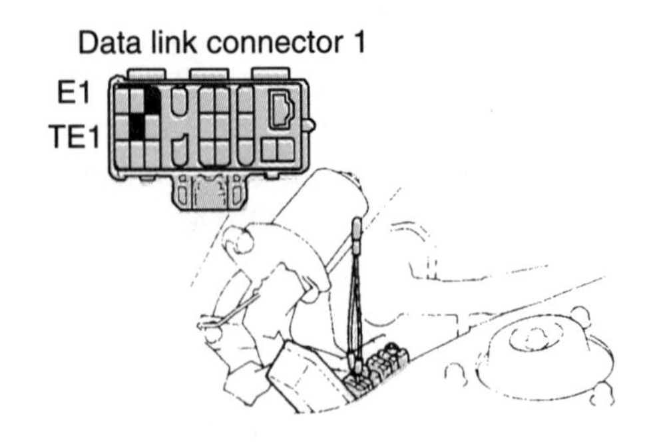

Figure 13-4 Data link connector connection to check IAC BPA valve (Courtesy of Toyota Motor Corporation)

Photo Sequence 10
Typical Procedure for Performing a Scan Tester Diagnosis of an Idle Air Control Motor

P10-1 Be sure the ignition switch is off.

P10-2 Connect the scan tester leads to the battery terminals with the correct polarity.

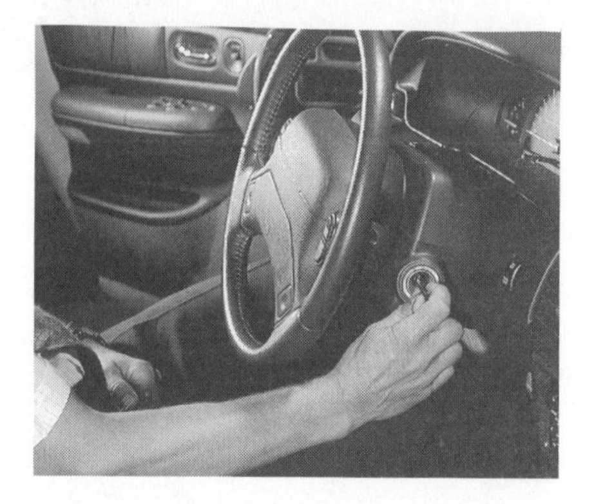

P10-3 Turn the key on, but don't start the engine.

P10-4 Program the scan tester as required for the vehicle being tested.

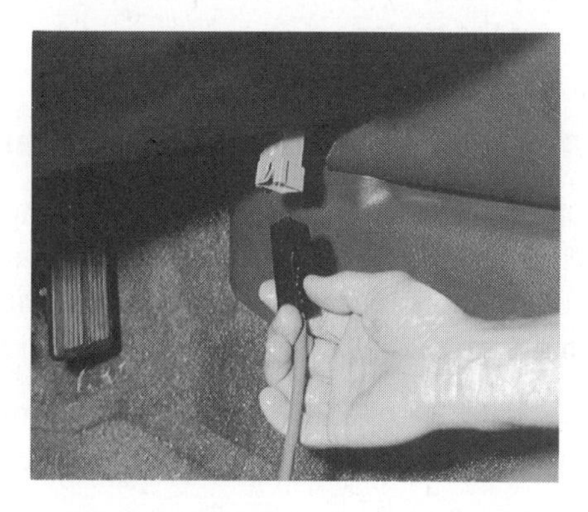

P10-5 Use the proper adaptor to connect the scan tester lead to the data link connector (DLC).

P10-6 Select idle air control (IAC) motor test on the scan tester.

P10-7 Press the appropriate scan tester button to increase engine rpm, and observe the rpm on the scan tester. The scan tester will automatically perform a non-running actuation test of the IAC hardware and software.

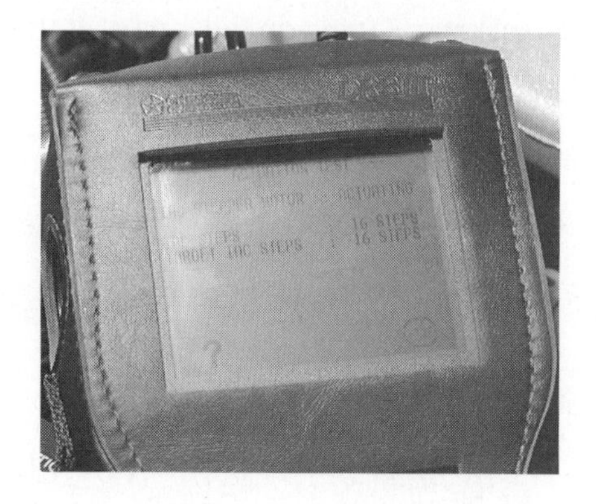

P10-8 At the low range, the IAC motor has a target range of 16 steps. The scan tester steps the motor through and monitors those steps.

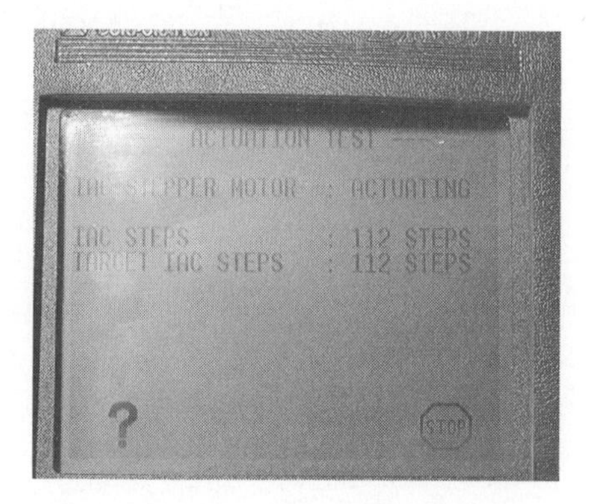

P10-9 At the high range, the scan tester steps the motor through and monitors 112 steps.

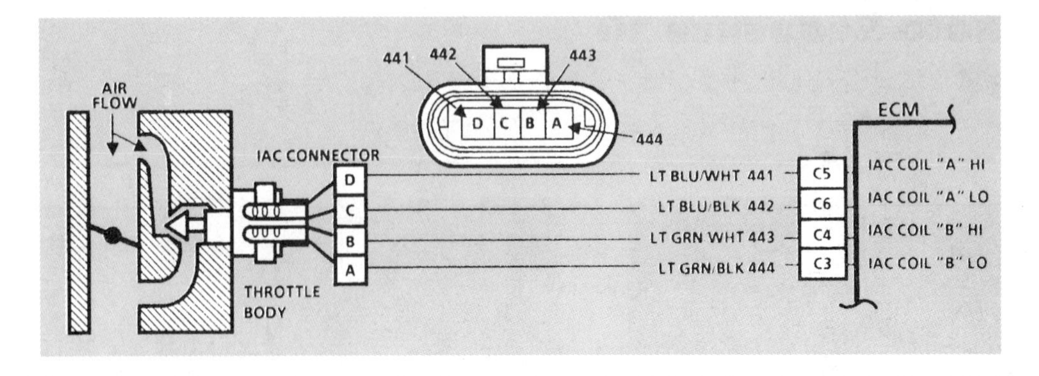

Figure 13-5 IAC BPA motor terminals (Courtesy of Chevrolet Motor Division, General Motors Corporation)

and inspect the throttle body air passage for carbon deposits. If heavy carbon deposits are present, remove the complete throttle body for cleaning. Clean the throttle body IAC BPA air passage, motor sealing surface, and pintle seat with throttle body cleaner. Clean the motor pintle with throttle body cleaner.

WARNING: IAC BPA motor damage may result if throttle body cleaner is allowed to enter the motor.

IAC BPA Motor Diagnosis and Installation

Connect a pair of ohmmeter leads across terminals D and C on the motor connector, and observe the ohmmeter reading (Figure 13-5). Repeat this ohmmeter test on terminals A and B on the motor. In each test, the ohmmeter reading should equal the manufacturer's specifications, usually 40 to 80 ohms. If the ohmmeter reading is not within specifications, replace the IAC BPA motor.

WARNING: IAC BPA motor damage may occur if the motor is installed with the pintle extended more than the specified distance.

If a new IAC BPA motor is installed, be sure the part number, pintle shape, and diameter are the same as those on the original motor. Measure the distance from the end of the pintle to the shoulder of the motor casting (Figure 13-6). If this distance exceeds 1.125 in (28 mm), use hand pressure to push the pintle inward until this specified distance is obtained.

Install a new gasket or O-ring on the motor. If the motor is sealed with an O-ring, lubricate this ring with transmission fluid, and install the motor. If the motor is threaded into the throttle

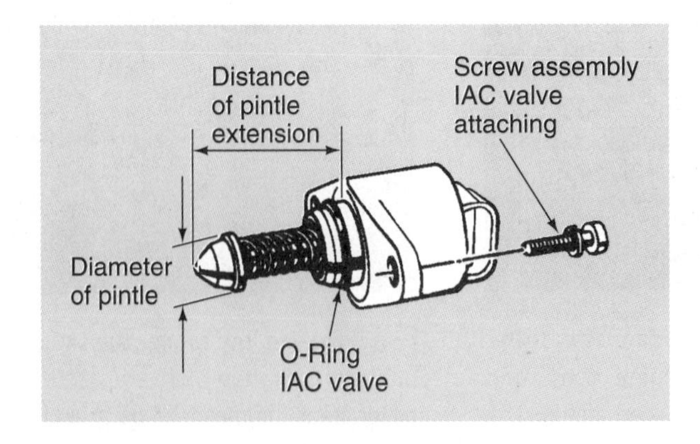

Figure 13-6 Measuring the distance from the pintle tip to the shoulder on the IAC BPA motor (Courtesy of Chevrolet Motor Division, General Motors Corporation)

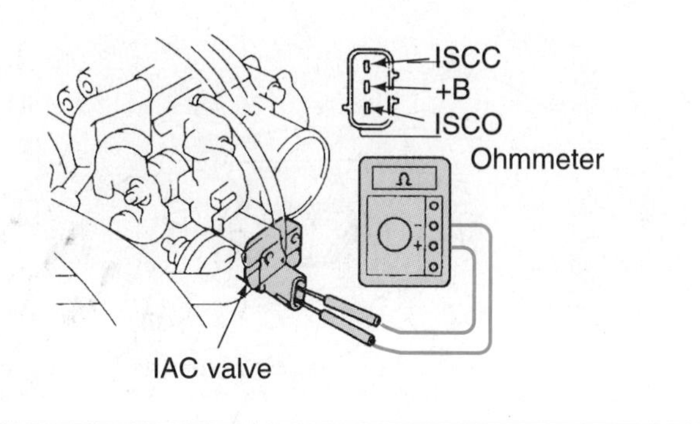

Figure 13-7 Ohmmeter connections to test the IAC BPA valve windings (Courtesy of Toyota Motor Corporation)

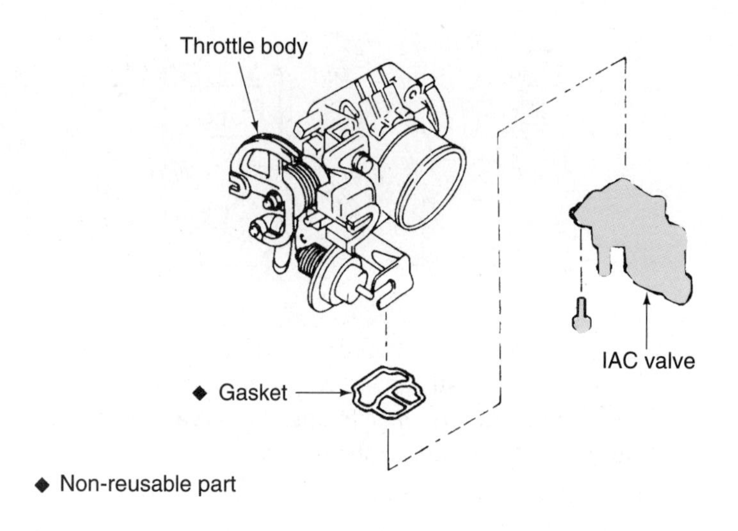

Figure 13-8 IAC BPA valve removed from throttle body (Courtesy of Toyota Motor Corporation)

body, tighten the motor to the specified torque. When the motor is bolted to the throttle body, tighten the mounting bolts to the specified torque.

Idle Air Control By-pass Air Valve Service and Diagnosis

Disconnect the IAC BPA valve connector and connect a pair of ohmmeter leads to terminals B+ and ISCC on the valve. Repeat the test between terminals B+ and ISCO on the valve (Figure 13-7). In both these tests on the valve windings, the ohmmeter should read 19.3 Ω to 22.3 Ω. If the resistance of the valve windings is not within specifications, replace the valve.

Disconnect all throttle body linkages and vacuum hoses. Remove the throttle body, and remove the IAC BPA valve mounting bolts. Remove the valve and gasket (Figure 13-8). The cleaning procedure for the IAC BPA motor may be followed on the IAC BPA valve.

Connect a jumper wire from the battery positive terminal to the valve B+ terminal. Connect another jumper wire from the battery negative terminal to the valve ISCC terminal. Be careful not to short the jumper wires together. With this connection, the valve must be closed (Figure 13-9).

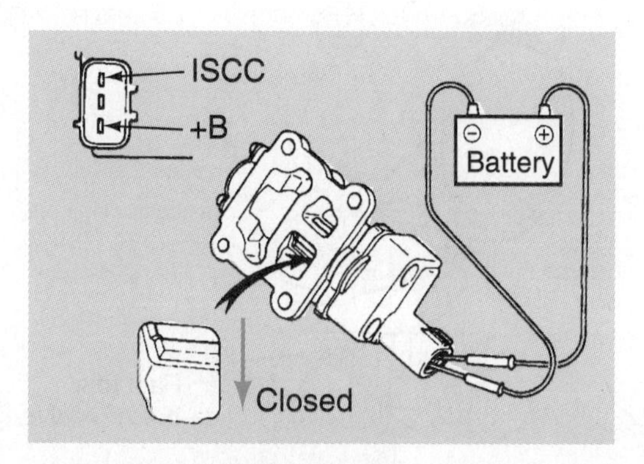

Figure 13-9 Test connections from the battery terminals to the IAC BPA valve to close the valve (Courtesy of Toyota Motor Corporation)

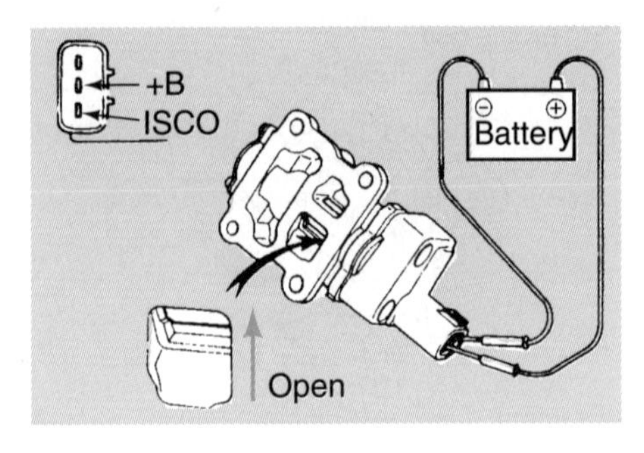

Figure 13-10 Test connections from the battery terminals to the IAC BPA valve to open the valve (Courtesy of Toyota Motor Corporation)

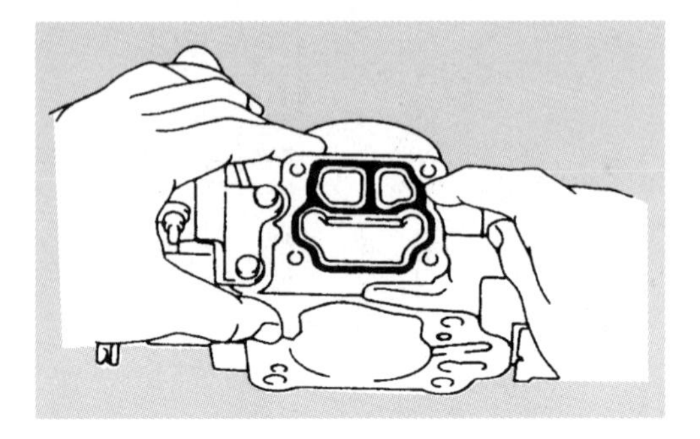

Figure 13-11 Installing a new IAC valve-to-throttle body gasket (Courtesy of Toyota Motor Corporation)

Leave the jumper wire connected from the battery positive terminal to the valve B+ terminal, and connect the jumper wire from the battery negative terminal to the ISCO terminal. Under this condition, the valve must be open (Figure 13-10). If the valve does not open and close properly, replace the valve.

Clean the throttle body and valve mounting surfaces, and install a new gasket between the valve and the throttle body (Figure 13-11). Install the valve, and tighten the mounting bolts to the specified torque. Be sure the throttle body and intake mounting surfaces are clean, and install a new gasket between the throttle body and the intake. Tighten the throttle body mounting bolts to the specified torque. Reconnect the IAC BPA valve wiring connector and all throttle body linkages and hoses.

Classroom Manual
Chapter 13, page 330

Diagnosis of Fast Idle Thermo Valve

The fast idle thermo valve is factory adjusted and should not be disassembled. Remove the air duct from the throttle body, and be sure the engine temperature is below 86°F (30°C). Start the engine and place your finger over the lower port in the throttle body (Figure 13-12). Under this condition, there should be air flow through this lower port and the fast idle thermo valve. If there is no air flow through the lower port, replace the fast idle thermo valve.

As the engine temperature increases, the air flow through the lower throttle body port should decrease. When the engine approaches normal operating temperature, the air flow should stop

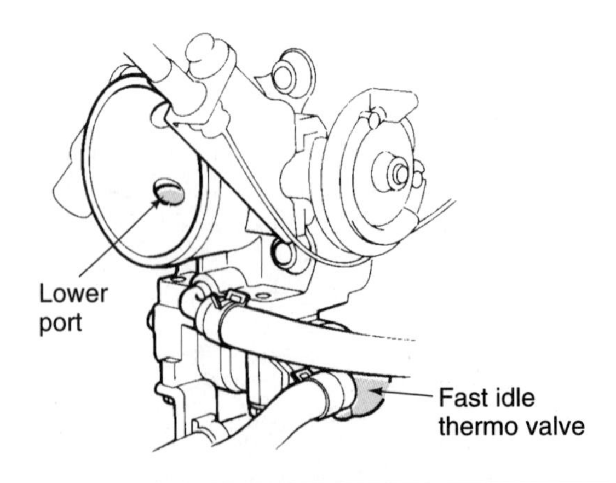

Figure 13-12 Lower port in throttle body for fast idle thermo valve air intake (Courtesy of American Honda Motor Co., Inc.)

flowing through the lower port and the fast idle thermo valve. If there is air flow through the lower throttle body port with the engine at normal operating temperature, check the cooling system for proper operation and temperature. The fast idle thermo valve is heated by engine coolant. If the cooling system operation is normal, replace the fast idle thermo valve.

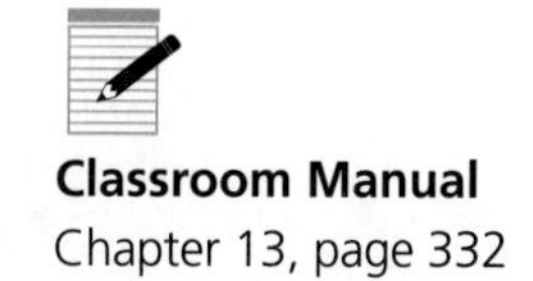

Classroom Manual
Chapter 13, page 332

Diagnosis of Starting Air Valve

If the starting air valve is open with the engine running, the idle rpm may be higher than specified. Disconnect the vacuum signal hose from the starting air valve, and connect a vacuum gauge to this hose (Figure 13-13). With the engine idling, this vacuum should be above 16 in hg. If the vacuum is lower than specified, check for leaks or restrictions in the signal hose. When the hose is satisfactory, check for late ignition timing or engine conditions such as low compression, that result in low vacuum.

If the signal vacuum is satisfactory, check the hoses from the starting air valve to the intake manifold and the air cleaner for restrictions and leaks. When the hoses are satisfactory, remove the hose from the starting air valve to the air cleaner. With the engine idling, there should be no air flow through this hose. When air flow is present, replace the starting air valve.

With the engine cranking, there should be air flow through the hose from the starting air valve to the air cleaner. If no air flow is present, replace the starting air valve.

Classroom Manual
Chapter 13, page 333

CUSTOMER CARE: Always concentrate on quality workmanship and customer satisfaction. Most customers do not mind paying for vehicle repairs if the work is done properly, and their vehicle problem is corrected. A follow-up phone call to determine customer satisfaction a few days after repairing a car indicates that you are interested in the customer's car, and that you consider quality work and satisfied customers a priority.

Guidelines for Servicing and Diagnosing Idle Speed Control Systems

1. Vacuum-operated decel valves may be diagnosed by the air flow through the valve as the engine is decelerated.
2. If a vacuum-operated decel valve is open all the time, idle rpm is higher than specified.
3. When a vacuum-operated decel valve does not open on deceleration, exhaust emissions are higher than normal on deceleration.
4. The vacuum-operated throttle kicker may be diagnosed by checking the idle rpm with the A/C on.

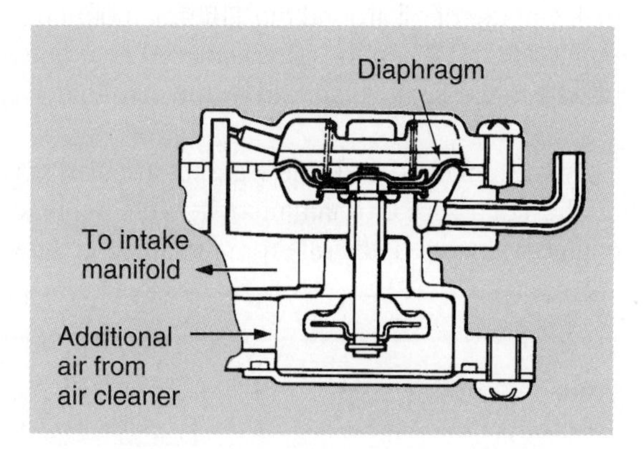

Figure 13-13 Starting air valve (Courtesy of American Honda Motor Co., Inc.)

5. During the antidieseling adjustment, the throttle stop screw is turned to obtain the proper throttle shutdown position.
6. On an engine with an IAC or IAC BPA motor, improper idle speed may be caused by a vacuum leak, defective input sensor or switch, defective IAC or IAC BPA motor, defective wires between the PCM and motor, or the PCM.
7. If an IAC motor has a hex bolt on the end of the motor plunger, this bolt must not be used to adjust idle speed.
8. The idle contact switch operated by the IAC motor plunger may be checked by measuring the voltage at each of the switch terminals with the ignition switch on.
9. A scan tester may be used to check input sensors and switches that affect IAC or IAC BPA motor operation.
10. On some vehicles, the scan tester may be used to force the PCM to operate the IAC or IAC BPA motor while observing the engine rpm.
11. On some vehicles, a jumper wire may be connected to specific terminals on the data link connector (DLC) to check IAC BPA valve response.
12. Erratic idle speed and engine stalling may be caused by carbon deposits on the IAC motor pintle and seat or in the IAC motor air passages in the throttle body.
13. The windings in an IAC BPA motor or valve may be checked with an ohmmeter.
14. Some IAC BPA valves may be checked by supplying 12 V to the valve windings and checking valve opening and closing.
15. A fast idle thermo valve may be diagnosed by checking the air flow through the valve with the engine cold and checking to be sure there is no air flow through the valve with the engine at normal operating temperature.
16. A starting air valve may be diagnosed by checking air flow through the valve while cranking the engine and checking to be sure there is no air flow through the valve with the engine running.

CASE STUDY

A customer complained about high idle speed on a Chevrolet one-half ton truck with TBI and a 5.7-L engine. Further questioning of the customer indicated that the idle rpm was sometimes normal when the throttle was returned to the idle position from higher speed. However, in a very short time, the idle rpm would gradually increase.

The technician visually checked all vacuum hose and wiring connections on the engine without finding any problems. The technician connected the scan tester to the DLC, and checked all the input sensors and switches. All the scan tester readings were normal.

The technician checked the torque on the intake manifold and throttle body retaining bolts and found these bolts were tightened to the specified torque. The technician used an oil can to squirt a small amount of oil around the TBI unit mounting surface on the intake manifold with the engine idling. When the oil was placed near one of the rear retaining bolts, the oil was pulled into the intake manifold by the manifold vacuum, indicating a vacuum leak.

The TBI assembly was removed and a new gasket installed between this assembly and the intake manifold. The TBI assembly retaining bolts were tightened to the specified torque. When the engine was started, the idle rpm remained at the vehicle manufacturer's specified idle rpm.

Terms to Know

Vacuum-operated decel valve
Vacuum throttle kicker
Electric throttle kicker
Idle stop solenoid
Idle speed control motor (ISC)
Idle contact switch
Nose switch
Throttle position switch
Idle air control (IAC) motor
Idle air control by-pass air (IAC BPA) motor
Idle air control by-pass air (IAC BPA) valve
Fast idle thermo valve
Starting air valve

ASE Style Review Questions

1. While discussing vacuum-operated decel valves:
Technician A says if the vacuum-operated decel valve is open all the time, engine idle speed is higher than normal.
Technician B says if the vacuum-operated decel valve is closed all the time, engine idle speed is lower than normal.
Who is correct?
A. A only **C.** Both A and B
B. B only **D.** Neither A nor B

2. While discussing the throttle stop screw adjustment:
Technician A says if the throttle stop screw adjustment provides a higher rpm than specified, the engine may detonate on acceleration.
Technician B says if the throttle stop screw adjustment provides a higher rpm than specified, the engine may diesel, or after-run, when the ignition switch is turned off.
Who is correct?
A. A only **C.** Both A and B
B. B only **D.** Neither A nor B

3. While discussing the causes of higher-than-specified idle rpm:
Technician A says an intake manifold vacuum leak may cause higher than specified idle rpm.
Technician B says if the throttle position sensor voltage signal is higher than specified, the idle rpm may be higher than specified.
Who is correct?
A. A only **C.** Both A and B
B. B only **D.** Neither A nor B

4. While discussing IAC motors:
Technician A says the hex nut on the IAC motor plunger may be rotated to adjust idle rpm.
Technician B says rotating this hex nut does not change idle rpm.
Who is correct?
A. A only **C.** Both A and B
B. B only **D.** Neither A nor B

5. While discussing idle contact switch diagnosis:
Technician A says when the idle contact switch is closed, the voltage at both switch terminals is low.
Technician B says when the idle contact switch is open, the voltage is high at the switch terminal connected to the PCM.
Who is correct?
A. A only **C.** Both A and B
B. B only **D.** Neither A nor B

6. While discussing scan tester diagnosis of IAC and IAC BPA motors:
Technician A says on some vehicles, the scan tester may be used to signal the PCM and increase the engine rpm in 100 rpm increments.
Technician B says on some vehicles the scan tester reads the IAC or IAC BPA motor counts, indicating the amount of motor opening.
Who is correct?
A. A only **C.** Both A and B
B. B only **D.** Neither A nor B

7. While discussing IAC BPA valve diagnosis:
Technician A says on some vehicles, a jumper wire may be connected to specific DLC terminals to check the IAC BPA valve operation.
Technician B says if the scan tester indicates zero IAC BPA valve counts, there may be an open circuit between the PCM and the IAC valve.
Who is correct?
A. A only **C.** Both A and B
B. B only **D.** Neither A nor B

8. While discussing IAC BPA motor removal, service, and replacement:
Technician A says throttle body cleaner may be used to clean the IAC BPA motor internal components.
Technician B says on some vehicles, IAC BPA motor damage occurs if the pintle is extended more than specified during installation.
Who is correct?
A. A only **C.** Both A and B
B. B only **D.** Neither A nor B

9. While discussing fast idle thermo valve operation:
Technician A says the fast idle thermo valve may be disassembled and serviced.
Technician B says there should be air flow through the fast idle thermo valve when the engine is at normal operating temperature.
Who is correct?
A. A only **C.** Both A and B
B. B only **D.** Neither A nor B

10. While discussing starting air valve diagnosis:
Technician A says there should be air flow through the valve with the engine running.
Technician B says if a vacuum leak is present in the vacuum signal hose, the valve will not close properly.
Who is correct?
A. A only **C.** Both A and B
B. B only **D.** Neither A nor B

Table 13-1 ASE TASK

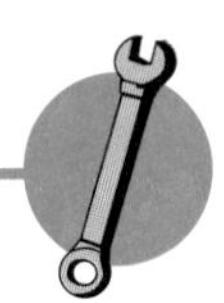

Test the operation of idle speed control systems.

Problem Area	Symptoms	Possible Causes	Classroom Manual	Shop Manual
ENGINE PERFORMANCE	Fixed idle speed, stalling	1. Loose wiring connections	326	302
		2. Defective wires, PCM-to-idle speed control device	328	302
		3. Defective idle speed device	329	302
		4. Defective PCM	330	302

Table 13-2 ASE TASK

Inspect, test, adjust, and replace sensors, solenoids, vacuum valves, motors, switches, wiring, and hoses of idle speed control systems.

Problem Area	Symptoms	Possible Causes	Classroom Manual	Shop Manual
ENGINE PERFORMANCE	Faster-than-specified idle rpm	1. Vacuum leak	326	301
		2. Defective input sensors, switches	328	303
		3. Defective wires, PCM-to-idle speed control device	328	303
		4. Defective idle speed control device	329	303
		5. Defective PCM	330	303
	Slower-than-specified idle rpm	1. Defective input sensors, switches	328	303
		2. Defective wires, PCM-to-idle speed control device	328	307
		3. Defective idle speed control device	329	307
		4. Defective PCM	330	307

Table 13-3 ASE TASK

Test the operation of deceleration control systems.

Problem Area	Symptoms	Possible Causes	Classroom Manual	Shop Manual
ENGINE PERFORMANCE	Faster-than-specified idle rpm	Defective deceleration control device	325	299
	High emissions during deceleration	Defective deceleration control device	326	300

Table 13-4 ASE TASK

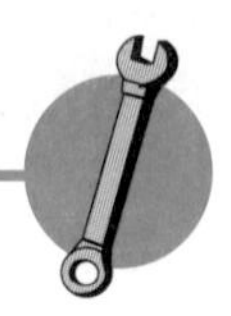

Inspect, test, adjust, and replace electrical components, circuits, vacuum components, and hoses of deceleration control systems.

Problem Area	Symptoms	Possible Causes	Classroom Manual	Shop Manual
ENGINE PERFORMANCE	Faster-than-specified idle speed	Defective deceleration control device	325	299
	High emissions during deceleration	Defective deceleration control device	326	300

Exhaust Gas Recirculation, Secondary Air Injection, and Evaporative Emission Control Systems, Diagnosis and Service

Upon completion and review of this chapter, you should be able to:

- ❑ Diagnose engine performance problems caused by improper EGR operation.
- ❑ Use a scan tester to diagnose an EGR system.
- ❑ Diagnose and service port EGR valves.
- ❑ Diagnose and service negative backpressure EGR valves.
- ❑ Diagnose and service positive backpressure EGR valves.
- ❑ Diagnose and service digital EGR valves.
- ❑ Diagnose and service linear EGR valves.
- ❑ Diagnose EGR vacuum regulator (EVR) solenoids.
- ❑ Diagnose exhaust gas temperature sensors.
- ❑ Diagnose EGR pressure transducers (EPT).
- ❑ Diagnose and service pulsed secondary air injection systems.
- ❑ Diagnose and service secondary air injection systems.
- ❑ Use a scan tester to diagnose secondary air injection systems.
- ❑ Diagnose engine performance problems caused by secondary air injection system defects.
- ❑ Diagnose and service evaporative (EVAP) systems.
- ❑ Use a scan tester to diagnose EVAP systems.
- ❑ Diagnose EVAP system thermal vacuum valves.

Diagnosis of Exhaust Gas Recirculation (EGR) Valves

Basic Tools

Basic technician's tool set

Service manual

Wire brush

Lengths of vacuum hose

Heat-resistant water container

Thermometer

General EGR System Diagnosis

If the EGR valve remains open at idle and low engine speed, the idle operation is rough and surging occurs at low speed. When this problem is present, the engine may hesitate on low-speed acceleration or stall after deceleration or after a cold start. If the EGR valve does not open, engine detonation occurs. When a defect occurs in the EGR system, a diagnostic trouble (DTC) is usually set in the PCM memory. The actual defects required to set a DTC may vary depending on the vehicle make and model year.

When diagnosing any EGR system, the first step is to check all the vacuum hoses and electrical connections in the system. In many EGR systems, the PCM uses inputs from the ECT, TPS, and MAP sensors to operate the EGR valve. Improper EGR operation may be caused by a defect in one of these sensors. A scan tester may be connected to the data link connector (DLC) to check for an EGR DTC or a DTC from another sensor which may affect EGR operation. The cause of any DTCs should be corrected before any further EGR diagnosis.

Diagnosis of Port EGR Valve

The EGR system diagnostic procedure varies depending on the vehicle make and model year. Always follow the diagnostic procedure in the vehicle manufacturer's service manual.

Special Tools

Vacuum hand pump

With the engine at normal operating temperature and operating at idle speed, disconnect the vacuum hose from the EGR valve. Supply 18 in Hg of vacuum to the valve with a vacuum hand pump and observe the EGR diaphragm movement. In some applications, a mirror may be held under the EGR valve to see the diaphragm movement. When the vacuum is applied, the EGR valve should open and idle operation should become very rough. If the valve diaphragm does not hold the vacuum, replace the valve. When the valve does not open, remove the valve and check for carbon in the passages under the valve. Clean the passages as required.

CAUTION: If the engine has been operating recently, the EGR valve may be very hot. Wear protective gloves when diagnosing or servicing this valve.

WARNING: Do not wash an EGR valve in any type of solvent. This action will damage the valve diaphragm.

WARNING: Sandblasting an EGR valve may damage valve components and plug orifices.

Carbon may be cleaned from the lower end of the EGR valve with a wire brush, but do not immerse the valve in solvent, and do not sandblast the valve.

Diagnosis of Negative Backpressure EGR Valve

With the engine at normal operating temperature and the ignition switch off, disconnect the vacuum hose from the EGR valve, and connect a hand vacuum pump to the vacuum fitting on the valve. Supply 18 in Hg of vacuum to the EGR valve and observe the valve operation and the vacuum gauge. The EGR valve should open and hold the vacuum for 20 seconds. When the valve does not operate properly, replace the valve.

With 18 in Hg supplied to the EGR valve from the hand pump, start the engine. The vacuum should drop to zero, and the valve should close. If the valve does not operate properly, replace the valve.

Diagnosis of Positive Backpressure EGR Valve

With the engine at normal operating temperature and running at idle speed, disconnect the vacuum hose from the EGR valve. Connect a hand vacuum pump to the EGR valve vacuum fitting, and operate the hand pump to supply vacuum to the valve. The vacuum should be bled off, and the EGR valve diaphragm and stem should not move. If the EGR valve does not operate properly, replace the valve.

Disconnect the EGR vacuum supply hose from the TBI unit and connect a long hose from this port directly to the EGR valve. Accelerate the engine to 2,000 rpm and observe the EGR valve. The valve should open at this engine speed. Allow the engine to return to idle speed. The EGR valve should close. If the EGR valve does not open properly, remove the valve and check for a plugged or restricted exhaust passage under the valve. When these passages are not restricted, replace the valve.

Photo Sequence 11 shows a typical procedure for diagnosing a positive backpressure exhaust gas recirculation valve.

Digital EGR Valve Diagnosis

Special Tools

Scan tester

The digital EGR valve may be diagnosed with a scan tester. With the engine at normal operating temperature and the ignition switch off, connect the scan tester to the DLC. Start the engine and allow the engine to operate at idle speed. Select EGR control on the scan tester, and then energize EGR solenoid #1 with the scan tester. When this action is taken, the engine rpm should decrease slightly. The engine rpm should drop slightly as each EGR solenoid is energized with the scan tester. When the

Photo Sequence 11

Typical Procedure for Diagnosing a Positive Backpressure Exhaust Gas Recirculation Valve

P11-1 With the engine at normal operating temperature, disconnect the vacuum hose from the EGR valve.

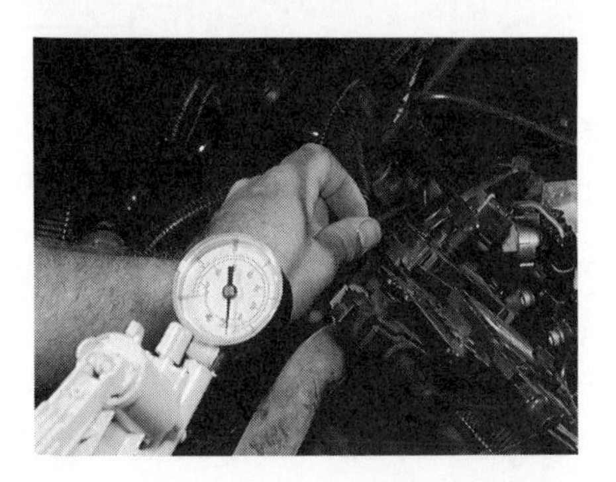

P11-2 Connect the vacuum hose from a hand pump to the EGR valve vacuum fitting.

P11-3 With the engine idling, operate the hand pump to supply vacuum to the EGR valve. The vacuum should be bled off to zero on the hand pump gauge, and the EGR valve should not open. No change in engine operation should occur.

P11-4 Shut off the engine.

P11-5 Disconnect the EGR vacuum supply hose from the throttle body.

P11-6 Connect a long vacuum hose from the EGR vacuum port on the throttle body directly to the EGR valve vacuum fitting.

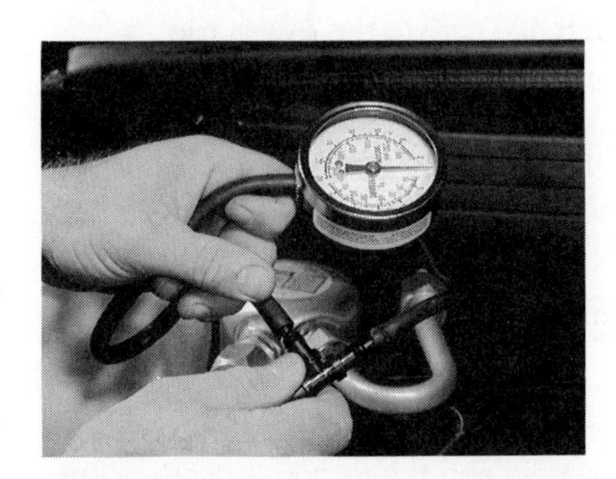

P11-7 Use a T fitting to connect a vacuum gauge in the direct vacuum line to the EGR valve.

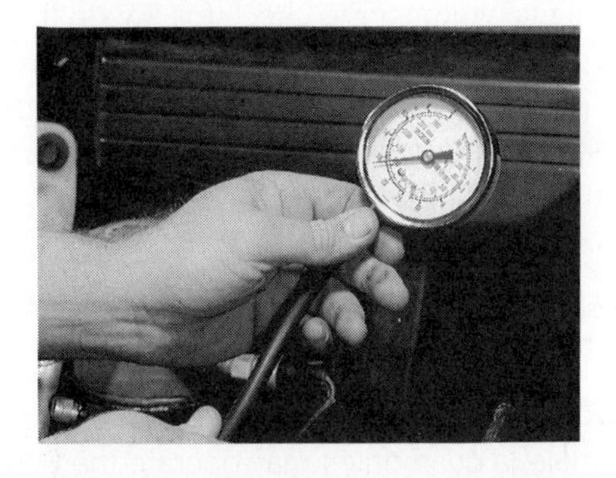

P11-8 Start the engine, increase the rpm to 2,000, and observe the vacuum gauge. Vacuum should now be supplied to the EGR valve, and this valve should be open.

P11-9 Shut off the engine, disconnect the long vacuum hose and vacuum gauge, and reconnect the original EGR hoses.

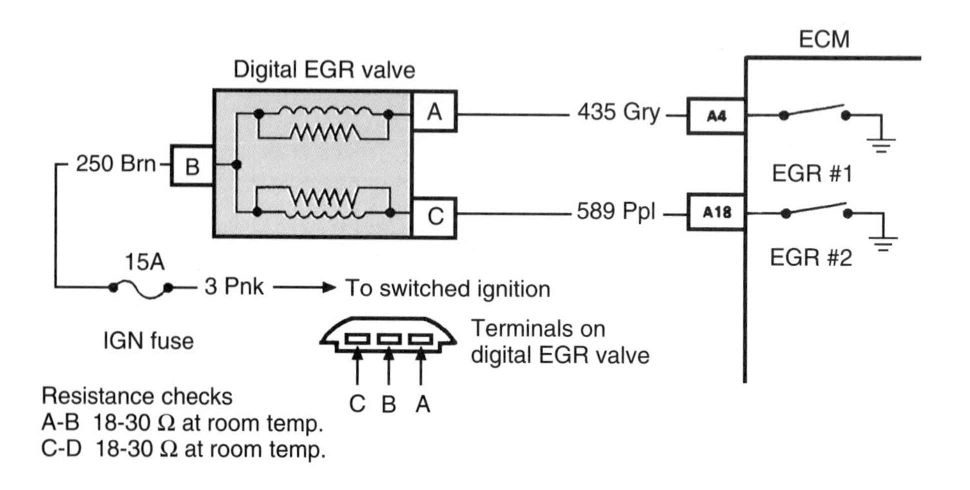

Figure 14-1 Digital EGR valve wiring diagram (Courtesy of Oldsmobile Division, General Motors Corporation)

EGR valve does not operate properly, check these items before replacing the EGR valve:

1. Check for 12 V at the power supply wire on the EGR valve (Figure 14-1).
2. Check the wires between the EGR valve and the PCM.
3. Remove the EGR valve, and check for plugged passages under the valve.

Linear EGR Valve Diagnosis

The linear EGR valve diagnostic procedure varies depending on the vehicle make and model year. Always follow the recommended procedure in the vehicle manufacturer's service manual. The scan tester may be used to diagnose a linear EGR valve. The engine should be at normal operating temperature prior to EGR valve diagnosis. Since the linear EGR valve has an EVP sensor, the actual pintle position may be checked on the scan tester. The pintle position should not exceed 3% at idle speed. The scan tester may be operated to command a specific pintle position, such as 75%, and this commanded position should be achieved within 2 seconds. With the engine idling, select various pintle positions and check the actual pintle position. The pintle position should always be within 10% of the commanded position. When the linear EGR valve does not operate properly:

Special Tools

Digital multimeter

1. Check the fuse in the 12-V supply wire to the EGR valve.
2. Check for open circuits, grounds, and shorts in the wires connected from the EGR valve winding to the PCM.
3. Use a digital voltmeter to check for 5 V on the reference wire to the EVP sensor.
4. Check for excessive resistance in the EVP sensor ground wire.
5. Leave the wiring harness connected to the valve, and remove the valve. Connect a digital voltmeter from the pintle position wire at the EGR valve to ground, and manually push the pintle upward (Figure 14-2). The voltmeter reading should change from approximately 1 V to 4.5 V.

If the EGR valve did not operate properly on the scan tester and tests 1 through 5 are satisfactory, replace the valve.

SERVICE TIP: The same quad driver in a PCM may operate several outputs. For example, a quad driver may operate the EVR solenoid and the torque converter clutch solenoid. On General Motors computers, the quad drivers sense high current flow. If a solenoid winding is shorted and the quad driver senses high current flow, the quad driver shuts off all the outputs it controls, rather than being damaged by the high current flow. When the PCM fails to operate an output or outputs, always check the resistance of the solenoid wind-

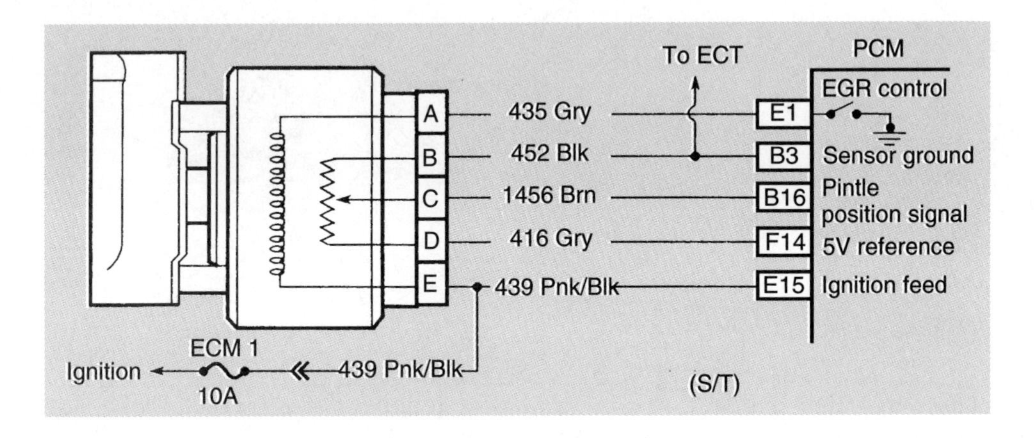

Figure 14-2 Linear EGR valve wiring diagram (Courtesy of Chevrolet Motor Division, General Motors Corporation)

ings in the outputs before replacing the PCM. A lower-than-specified resistance in a solenoid winding indicates a shorted condition, and this problem may explain why the PCM quad driver stops operating the outputs.

EGR Vacuum Regulator (EVR) Tests

Connect a pair of ohmmeter leads to the EVR terminals to check the winding for open circuits and shorts (Figure 14-3). An infinite ohmmeter reading indicates an open circuit, whereas a lower-than-specified reading means the winding is shorted.

Connect the ohmmeter leads from one of the EVR solenoid terminals to ground on the solenoid case (Figure 14-4). A low ohmmeter reading indicates a grounded winding, and an infinite reading indicates the winding is not grounded.

A scan tester may be used to diagnose the EVR solenoid operation. In the appropriate mode, the scan tester displays the EVR solenoid status as on or off. With the engine idling, the EVR solenoid should remain off. Road test the vehicle and drive the vehicle until the conditions required to open the EVR solenoid are present. When these conditions are present, the scan tester should indicate that the EVR solenoid is on.

SERVICE TIP: In some EGR systems, the PCM energizes the EVR solenoid at idle and low speeds. Under this condition, the solenoid shuts off vacuum to the EGR valve. When the proper input signals are available, the PCM deenergizes the EVR solenoid and allows vacuum to the EGR valve.

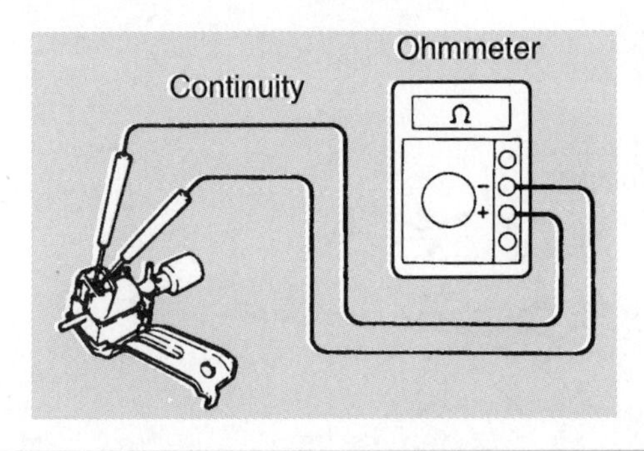

Figure 14-3 Ohmmeter connected to test the EVR solenoid winding for an open or shorted circuit (Courtesy of Toyota Motor Corporation)

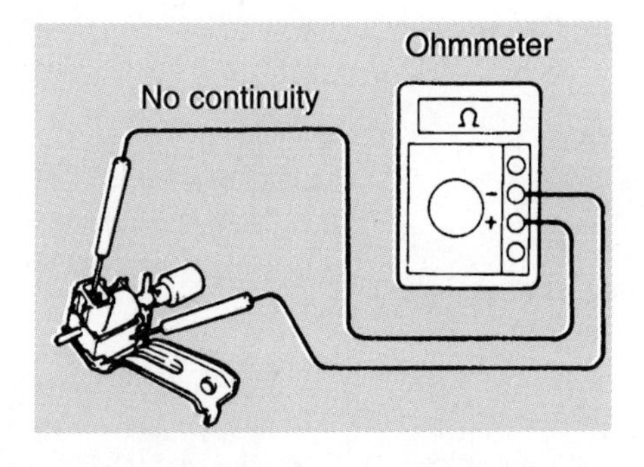

Figure 14-4 Ohmmeter connected to test the EVR solenoid winding for a grounded condition (Courtesy of Toyota Motor Corporation)

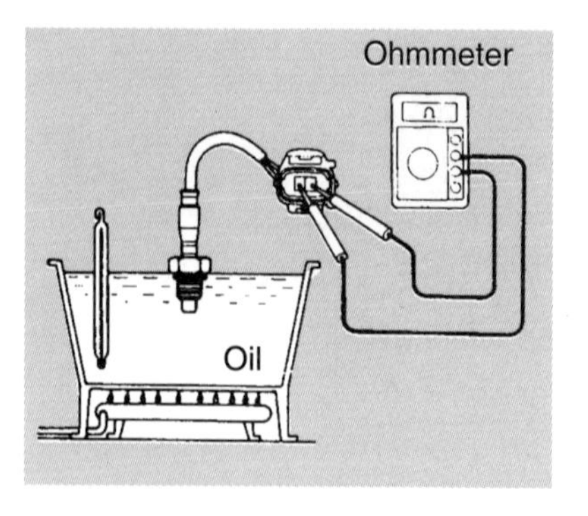

Figure 14-5 Testing exhaust gas temperature sensor (Courtesy of Toyota Motor Corporation)

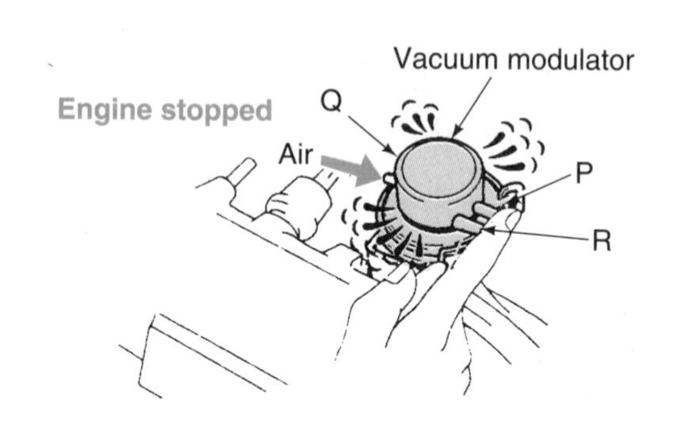

Figure 14-6 Testing EPT at idle speed (Courtesy of Toyota Motor Corporation)

Exhaust Gas Temperature Sensor Diagnosis

Remove the exhaust gas temperature sensor, and place the sensor in a container of water. Place a thermometer in the water and heat the container (Figure 14-5). Connect the ohmmeter leads to the exhaust gas temperature sensor terminals. The exhaust gas temperature sensor should have the specified resistance at various temperatures.

EGR Pressure Transducer (EPT) Diagnosis

CAUTION: If the engine has been running, EGR components, including the EPT, and especially the exhaust pressure supply pipe to the EPT, are hot. Wear protective gloves during diagnosis.

The EPT diagnosis varies depending on the vehicle make and model year. Always follow the instructions in the vehicle manufacturer's service manual. The following procedure is for a 1993 Toyota Camry:

1. Disconnect the vacuum hoses from ports P, Q, and R on the EPT.
2. With the ignition switch off, block ports P and R with your finger.
3. Connect a length of vacuum hose to port Q, and blow air into this port (Figure 14-6). Air should pass freely through the air filter on the side of the EPT.
4. Operate the engine at 2,500 rpm, and repeat step 3. There should be strong resistance to air flow (Figure 14-7). If the EPT does not operate properly, check for restriction in the exhaust pressure tube to the EPT. If there are no restrictions in this tube, replace the EPT.

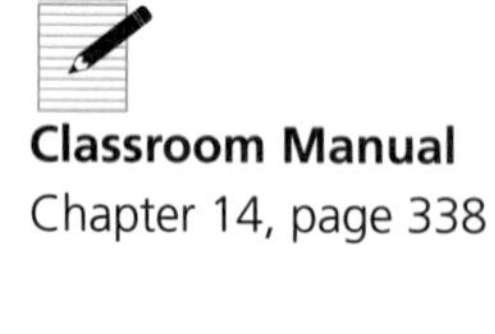

Classroom Manual
Chapter 14, page 338

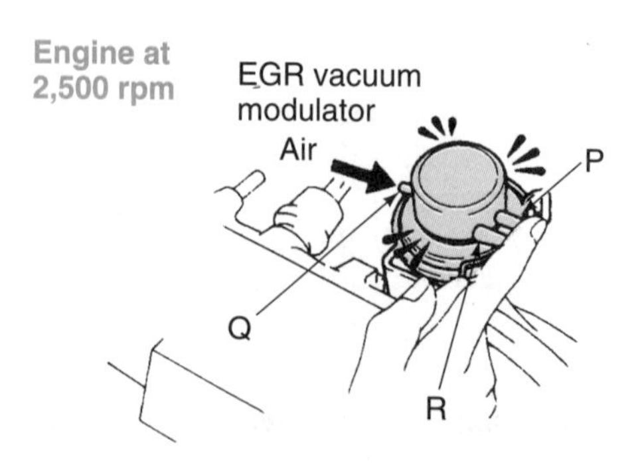

Figure 14-7 Testing EPT at 2,500 rpm (Courtesy of Toyota Motor Corporation)

Pulsed Secondary Air Injection System Diagnosis

SERVICE TIP: If the metal container or the clean air hose from this container show evidence of burning, some of the one-way check valves are allowing exhaust into this container and clean air hose.

Check all the hoses and pipes in the system for looseness and rusted or burned conditions. Remove the clean air hose from the air cleaner, and start the engine. With the engine idling, there should be steady audible pulses at the end of the hose. If these pulses are erratic, check for cylinder misfiring. When the cylinders are not misfiring, check for sticking one-way check valves or restricted exhaust inlet air tubes in the exhaust manifold.

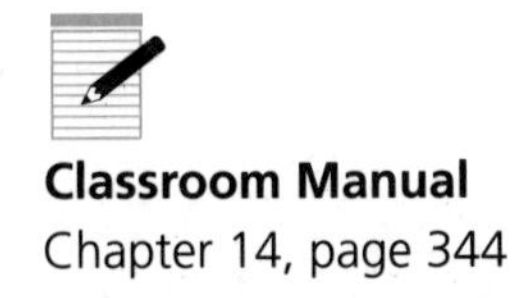

Classroom Manual
Chapter 14, page 344

Secondary Air Injection System Service and Diagnosis

General AIR System Diagnosis

The first step in diagnosing a secondary air injection system is to check all vacuum hoses and electrical connections in the system. Many AIR system pumps have a centrifugal filter behind the pulley. Air flows through this filter into the pump, and the filter keeps dirt out of the pump. The pulley and filter are bolted to the pump shaft, and these two components are serviced separately (Figure 14-8). If the pulley or filter is bent, worn, or damaged, it should be replaced. The pump assembly is usually not serviced.

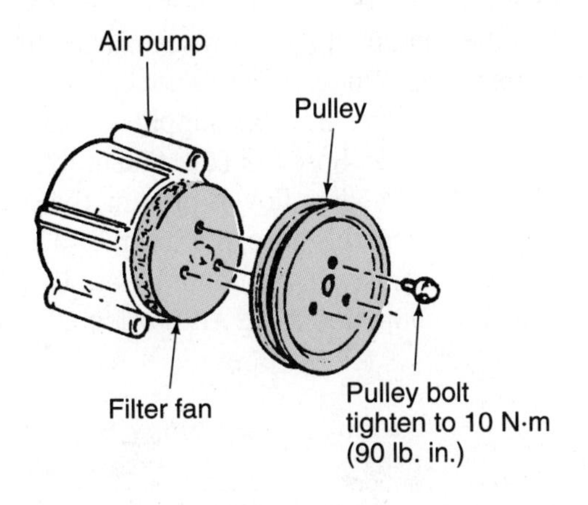

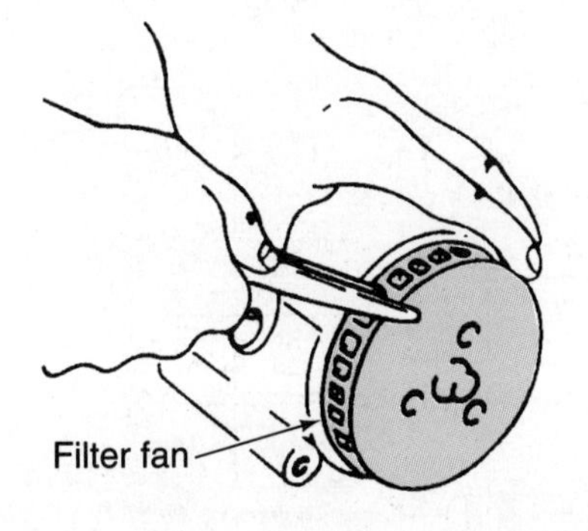

Figure 14-8 AIR pump pulley and centrifugal filter (Courtesy of Chevrolet Motor Division, General Motors Corporation)

The AIR pump belt must have the specified tension. A loose belt or a defective AIR system may result in high emission levels and/or excessive fuel consumption.

In some AIR systems, pressure relief valves are mounted in the AIRB and AIRD valves. Other AIR systems have a pressure relief valve in the pump. If the pressure relief valve is stuck open, air flow from the pump is continually exhausted through this valve, which causes high tail pipe emissions.

If the hoses in the AIR system show evidence of burning, the one-way check valves are leaking, which allows exhaust to enter the system. Leaking air manifolds and pipes result in exhaust leaks and excessive noise.

Some AIR systems will set DTCs in the PCM if there is a fault in the AIRB or AIRD solenoids and related wiring. In some AIR systems, DTCs are set in the PCM memory if the air flow from the pump is continually upstream or downstream. Always use a scan tester to check for any DTCs related to the AIR system, and correct the causes of these codes before proceeding with further system diagnosis.

Upstream air refers to air flow from the AIR pump to the exhaust ports.

Downstream air refers to air flow from the AIR pump to the catalytic converters.

If the AIR system does not pump air into the exhaust ports during engine warm-up, HC emissions are high during this mode, and the O_2 sensor, or sensors, takes longer to reach normal operating temperature. Under this condition, the PCM remains in open loop longer. Since the air-fuel ratio is richer in open loop, fuel economy is reduced.

When the AIR system pumps air into the exhaust ports with the engine at normal operating temperature, the additional air in the exhaust stream causes lean signals from the O_2 sensor, or sensors. The PCM responds to these lean signals by providing a rich air-fuel ratio from the injectors. This action increases fuel consumption.

AIR System Component Diagnosis

AIRB Solenoid and Valve Diagnosis. When the engine is started, listen for air being exhausted from the AIRB valve for a short period of time. If this air is not exhausted, remove the vacuum hose from the AIRB and start the engine. If air is now exhausted from the AIRB valve, check the AIRB solenoid and connecting wires. When air is still not exhausted from the AIRB valve, check the air supply from the pump to the valve. If the air supply is available, replace the AIRB valve.

During engine warm-up, remove the hose from the AIRD valve to the exhaust ports and check for air flow from this hose (Figure 14-9). If air flow is present, the system is operating normally in this mode. When air is not flowing from this hose, remove the vacuum hose from the AIRD valve and connect a vacuum gauge to this hose. If vacuum is above 12 in Hg, replace the AIRD valve. When the vacuum is zero, check vacuum hoses, the AIRD solenoid, and connecting wires.

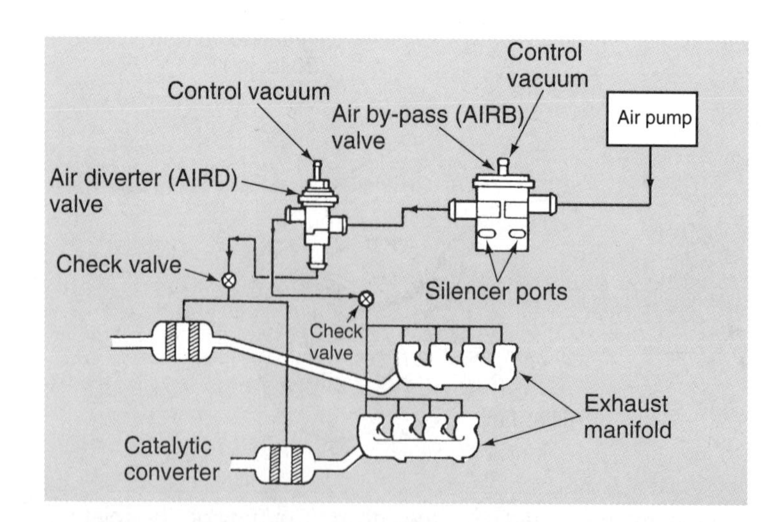

Figure 14-9 AIR system (Reprinted with the permission of Ford Motor Company)

SERVICE TIP: With the engine at normal operating temperature, the AIR system sometimes goes back into the upstream mode with the engine idling. It may be necessary to increase the engine speed to maintain the downstream mode.

With the engine at normal operating temperature, disconnect the air hose between the AIRD valve and the catalytic converters and check for air flow from this hose. When air flow is present, system operation in the downstream mode is normal. If there is no air flow from this hose, disconnect the vacuum hose from the AIRD valve and connect a vacuum gauge to the hose. When the vacuum gauge indicates zero vacuum, replace the AIRD valve. If some vacuum is indicated on the gauge, check the hose, the AIRD solenoid, and connecting wires.

Evaporative (EVAP) System Diagnosis and Service

General EVAP System Diagnosis

CAUTION: Gasoline vapors are extremely explosive. Do not smoke or allow any other sources of ignition near EVAP system components. A gasoline vapor explosion may result in personal injury and/or property damage.

CAUTION: If gasoline odor occurs in, or around, a vehicle, check the EVAP system for cracked or disconnected hoses, and check the fuel system for leaks. Gasoline leaks or escaping vapors may result in an explosion causing personal injury and/or property damage. The cause of fuel leaks or fuel vapor leaks should be repaired immediately.

EVAP system diagnosis varies depending on the vehicle make and model year. Always follow the service and diagnostic procedure in the vehicle manufacturer's service manual. If the EVAP system is purging vapors from the charcoal canister when the engine is idling or operating at very low speed, rough engine operation will occur, especially at higher atmospheric temperatures. Cracked hoses or a canister saturated with gasoline may allow gasoline vapors to escape to the atmosphere, resulting in gasoline odor in, and around, the vehicle.

All the hoses in the EVAP system should be checked for leaks, restrictions, and loose connections. The electrical connections in the EVAP system should be checked for looseness, corroded terminals, and worn insulation. When a defect occurs in the canister purge solenoid and related circuit, a DTC is usually set in the PCM memory. If a DTC related to the EVAP system is set in the PCM memory, always correct the cause of this code before further EVAP system diagnosis.

A scan tester may be used to diagnose the EVAP system. In the appropriate scan tester mode, the tester indicates whether the purge solenoid is on or off. Connect the scan tester to the DLC and start the engine. With the engine idling, the purge solenoid should be off. Leave the scan tester connected and road test the vehicle. Be sure all the conditions required to energize the purge solenoid are present, and observe this solenoid status on the scan tester. The tester should indicate the purge solenoid is on when all the conditions are present for canister purge operation. If the purge solenoid is not on under the necessary conditions, check the power supply wire to the solenoid, solenoid winding, and the wire from the solenoid to the PCM.

EVAP System Component Diagnosis

The canister purge solenoid winding may be checked with an ohmmeter in the same way as the EVR solenoid. With the tank pressure control valve removed, try to blow air through the valve with your mouth from the tank side of the valve (Figure 14-10). Some restriction to air flow should be felt until the air pressure opens the valve. Connect a vacuum hand pump to the vacuum fitting on the valve and apply 10 in Hg to the valve. Now try to blow air through the valve from the tank

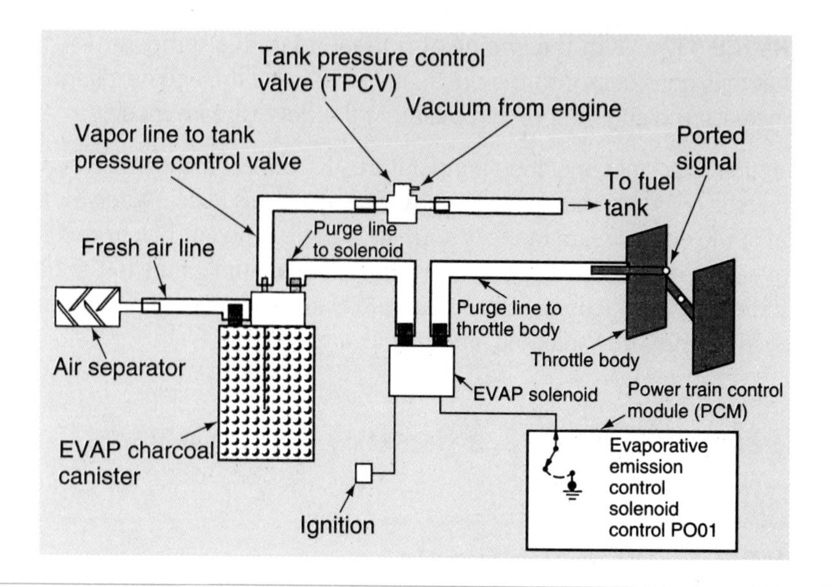

Figure 14-10 EVAP system (Courtesy of Cadillac Motor Car Division, General Motors Corporation)

side. Under this condition, there should be no restriction to air flow. If the tank pressure control valve does not operate properly, replace the valve.

If the fuel tank has a pressure and vacuum valve in the filler cap, check these valves for dirt contamination and damage. The cap may be washed in clean solvent. When the valves are sticking or damaged, replace the cap.

When the charcoal filter has a replaceable filter, check the filter for dirt contamination. Replace the filter as required.

EVAP System Thermal Vacuum Valve (TVV) Diagnosis

Follow these steps for TVV diagnosis:

1. Drain the coolant from the radiator into a suitable container.
2. Remove the TVV from the water outlet.
3. Place the TVV and a thermometer in a container filled with water. The water temperature must be below 95°F (35°C).
4. Connect a length of vacuum hose to the upper TVV port and try to blow air through the TVV with your mouth. The TVV should be closed and no air should flow through the TVV (Figure 14-11). If the TVV allows air flow, replacement is necessary.
5. Heat the water and observe the thermometer reading. When the water temperature is above 129°F (54°C), you should be able to blow air through the TVV. Replace the TVV if it does not open at the specified temperature.
6. Install thread sealant to two or three threads on the TVV, and install the TVV. Be sure this component is tightened to the specified torque.
7. Refill the radiator with coolant.

CUSTOMER CARE: Customers should be advised that maintaining their vehicles according to the vehicle manufacturer's recommended maintenance schedule is one of the best ways to keep their vehicles meeting emission standards. Always advise the customer when you find emission devices that are not operating properly. When customers are advised that their vehicles are causing excessive emissions and air pollution, they are usually

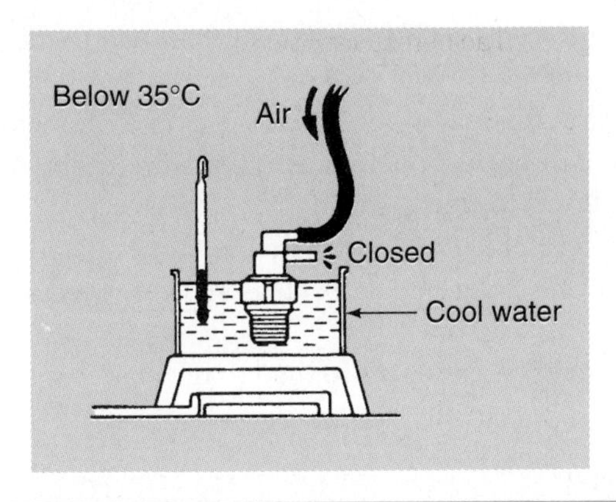

Figure 14-11 Testing TVV in EVAP system (Courtesy of Toyota Motor Corporation)

willing to have the necessary repairs completed. Familiarize yourself with state vehicle inspection and maintenance programs so you can advise customers regarding the necessary maintenance so their vehicles meet the standards of these programs.

Guidelines for Servicing and Diagnosing EGR, AIR, and EVAP Systems

1. Rough engine idle and surging at low speed occur if the EGR valve is open under these conditions.
2. If the EGR valve is open at idle and low speeds, the engine may hesitate on acceleration, stall after deceleration, or stall after a cold start.
3. The PCM uses information from the ECT, TPS, and MAP to operate the EGR valve.
4. Some EGR system faults will set a code in the PCM memory.
5. When 18 in Hg of vacuum are supplied to a port EGR valve with the engine idling, the valve should open, and the engine idle operation should become very rough.
6. When 18 in Hg of vacuum are supplied to a negative backpressure EGR valve with the ignition switch off, the valve should open and hold the vacuum for 20 seconds. When the engine starts, the vacuum should drop to zero and the valve should close.
7. With the engine idling, if vacuum from a hand pump is supplied to a positive backpressure EGR valve, the vacuum should be bled off.
8. With manifold vacuum supplied to a positive backpressure EGR valve, the valve should open when the engine is accelerated to 2,000 rpm.
9. A scan tester may be used to command the PCM to operate each solenoid in a digital EGR valve, and the change in engine operation should be noticeable as each solenoid in the EGR valve is opened.
10. A scan tester may be used to command the PCM to provide a specific linear EGR valve opening, and the resulting EGR valve opening may be observed on the scan tester.
11. An EGR vacuum regulator (EVR) solenoid winding may be checked for open circuits, shorts, and grounds with an ohmmeter.
12. A scan tester will indicate whether the EVR solenoid is on or off during a road test.
13. An ohmmeter may be used to check an exhaust gas temperature sensor as it is heated to various temperatures in a container of water.
14. The vent port in an EGR pressure transducer (EPT) should be closed at 2,500 rpm.
15. Steady audible pulses should be heard at the clean air hose in a pulsed secondary air injection system when this hose is removed from the air cleaner.

16. Some secondary air injection (AIR) systems exhaust air to the atmosphere for a brief interval when the engine is started.
17. During engine warm-up, many AIR systems deliver air flow to the exhaust ports.
18. When the engine is at normal operating temperature, many AIR systems deliver air flow to the catalytic converters.
19. In many EVAP systems, the PCM energizes the purge solenoid and provides canister purging when specific input signals, such as vehicle speed above 20 mph (32 km/h), are present.
20. In some EVAP systems, vacuum is supplied to the charcoal canister by a thermal vacuum valve (TVV).

CASE STUDY

A customer complained about a hesitation on acceleration on a Chevrolet truck with a 5.7-L engine. It was not necessary to road test the vehicle to experience the symptoms. Each time the engine was accelerated with the transmission in park, there was a very noticeable hesitation.

The technician removed the air cleaner and observed the injector spray pattern, which appeared normal. A fuel pressure test indicated the specified fuel pressure. The technician connected the scan tester to the DLC and checked the PCM for DTCs. There were no DTCs in the PCM memory.

The technician checked all the sensor readings with the scan tester, and all readings were within specifications. Next, the technician visually checked the operation of the EGR valve while accelerating the engine. The EGR valve remained closed at idle speed, but each time the engine was accelerated, the EGR valve moved to the wide-open position and remained there. With the hose removed from the EGR valve, the engine accelerated normally.

The technician checked the letters on top of the EGR valve and found it was a negative backpressure valve. A vacuum hand pump was connected to the EGR valve, and 18 in Hg of vacuum were supplied to the valve with the ignition switch off. The valve opened and held the vacuum for 20 seconds. With 18 in Hg of vacuum supplied to the EGR valve, the engine was started. The vacuum dropped slightly, but the valve remained open, indicating the exhaust pressure was not keeping the bleed valve open and the passages in the valve stem or under the valve were restricted.

The EGR valve was removed, and since there was no carbon in the passages under the valve, the valve was replaced. When the replacement EGR valve was installed, the engine accelerated normally.

Terms to Know

Exhaust gas recirculation (EGR) valve
Diagnostic trouble code (DTC)
Port EGR valve
Negative backpressure EGR valve
Positive backpressure EGR valve
Digital EGR valve
Linear EGR valve
EGR vacuum regulator (EVR) solenoid
Exhaust gas temperature sensor
EGR pressure transducer (EPT)
Pulsed secondary air injection system
Secondary air injection (AIR) system
AIR by-pass (AIRB) solenoid
AIR diverter (AIRD) solenoid
Upstream air
Downstream air
Evaporative (EVAP) system
Canister purge solenoid
Thermal vacuum valve (TVV)

ASE Style Review Questions

1. While discussing EGR system diagnosis:
   Technician A says if the EGR valve is open at idle and low speeds, the engine may surge during low-speed operation.
   Technician B says if the EGR valve is open at idle and low speeds, the engine may stall on deceleration.
   Who is correct?
   A. A only
   B. B only
   C. Both A and B
   D. Neither A nor B

2. While discussing EGR valve diagnosis:
   Technician A says if the EGR valve does not open, the engine may hesitate on acceleration.
   Technician B says if the EGR valve does not open, the engine may detonate on acceleration.
   Who is correct?
   A. A only
   B. B only
   C. Both A and B
   D. Neither A nor B

3. While discussing EGR valve diagnosis:
   Technician A says a defective throttle position sensor (TPS) may affect the EGR valve operation.
   Technician B says a defective engine coolant temperature (ECT) sensor may affect the EGR valve operation.
   Who is correct?
   A. A only
   B. B only
   C. Both A and B
   D. Neither A nor B

4. When discussing the diagnosis of a positive backpressure EGR valve:
   Technician A says with the engine running at idle speed, if a hand pump is used to supply vacuum to the EGR valve, the valve should open at 12 in Hg of vacuum.
   Technician B says with the engine running at idle speed, any vacuum supplied to the EGR valve should be bled off, and the valve should not open.
   Who is correct?
   A. A only
   B. B only
   C. Both A and B
   D. Neither A nor B

5. While discussing digital EGR valve diagnosis:
   Technician A says a scan tester may be used to command the PCM to open each solenoid in the EGR valve.
   Technician B says the EGR valve should open when 18 in Hg of vacuum are supplied to the valve at idle speed.
   Who is correct?
   A. A only
   B. B only
   C. Both A and B
   D. Neither A nor B

6. While discussing EGR vacuum regulator (EVR) diagnosis:
   Technician A says a scan tester will indicate whether the EVR is on or off.
   Technician B says the EVR winding is shorted if it has less resistance than specified.
   Who is correct?
   A. A only
   B. B only
   C. Both A and B
   D. Neither A nor B

7. While discussing exhaust gas temperature sensor diagnosis:
   Technician A says the resistance of this sensor should increase as the sensor temperature increases.
   Technician B says the resistance of this sensor should increase as the sensor temperature decreases.
   Who is correct?
   A. A only
   B. B only
   C. Both A and B
   D. Neither A nor B

8. While discussing EGR pressure transducer (EPT) diagnosis:
   Technician A says the vent in the EPT should be open at high engine rpm.
   Technician B says exhaust pressure is supplied to the top of the diaphragm in the EPT.
   Who is correct?
   A. A only
   B. B only
   C. Both A and B
   D. Neither A nor B

9. While discussing secondary air injection systems:
Technician A says the air flow from the pump should be directed to the catalytic converter during engine warm-up.
Technician B says the air flow from the pump should be directed to the exhaust ports during engine warm-up.
Who is correct?
A. A only
B. B only
C. Both A and B
D. Neither A nor B

10. While discussing secondary air injection systems:
Technician A says if the air flow is directed to the exhaust ports all the time, the air-fuel ratio is rich.
Technician B says if the air flow is by-passed to the atmosphere all the time, the PCM will remain in open loop too long.
Who is correct?
A. A only
B. B only
C. Both A and B
D. Neither A nor B

Table 14-1 ASE TASK

Test the operation of exhaust gas recirculation (EGR) systems.

Problem Area	Symptoms	Possible Causes	Classroom Manual	Shop Manual
ENGINE PERFORMANCE	Rough idle, surging at low speed	EGR valve remaining open at idle, low speed	338	315
	Stalling on deceleration or after a cold start	EGR valve remaining open at idle, low speed	339	315
	Detonation on acceleration	EGR valve remaining closed all the time	339	315

Table 14-2 ASE TASK

Inspect, test, repair, and replace valve, valve manifold, and exhaust passages of exhaust gas recirculation (EGR) systems.

Problem Area	Symptoms	Possible Causes	Classroom Manual	Shop Manual
ENGINE PERFORMANCE	Rough idle, surging at low speed	EGR valve stuck open at idle, low speed	338	316
	Stalling on deceleration or after a cold start	EGR valve stuck open at idle, low speed	339	316
	Detonation on acceleration	EGR valve stuck closed, passage filled with carbon	339	316

Table 14-3 ASE TASK

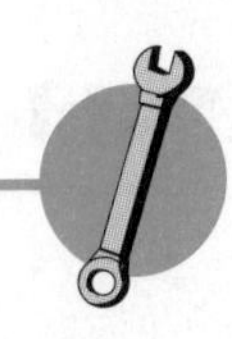

Inspect, test, repair, and replace vacuum/pressure controls, filters, and hoses of exhaust gas recirculation (EGR) systems.

Problem Area	Symptoms	Possible Causes	Classroom Manual	Shop Manual
ENGINE PERFORMANCE	Rough idle, surging at low speed	EVR solenoid vent plugged, holding EGR valve open at idle	342	319
	Stalling on deceleration or after a cold start	EVR solenoid vent plugged, holding EGR valve open at idle	342	319
	Detonation on acceleration	EVR solenoid faulty, leaking hose to EVR	343	319

Table 14-4 ASE TASK

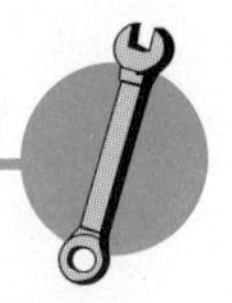

Inspect, test, repair, and replace electrical/electronic sensors, controls, and wiring of exhaust gas recirculation (EGR) systems.

Problem Area	Symptoms	Possible Causes	Classroom Manual	Shop Manual
ENGINE PERFORMANCE	Detonation on acceleration	Defective wires EVR solenoid to PCM	342	319
	Rough idle, surging at low speed	Defective input sensor causing improper EGR valve opening warm-up	343	315

Table 14-5 ASE TASK

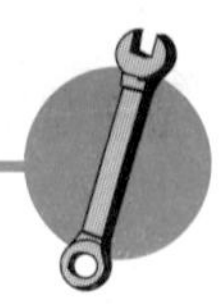

Test the operation of pump-type air injection systems.

Problem Area	Symptoms	Possible Causes	Classroom Manual	Shop Manual
EMISSION LEVELS	High emission levels	Inoperative AIR pump	345	322
FUEL CONSUMPTION	Excessive fuel consumption	1. AIR pump air flow continually directed upstream	346	322
		2. AIR pump air flow not directed to exhaust ports during warm-up	346	322

Table 14-6 ASE TASK

Inspect, test, service, and replace pump, pressure relief valve, filter, pulley, and belt of pump-type air injection systems.

Problem Area	Symptoms	Possible Causes	Classroom Manual	Shop Manual
EMISSION LEVELS	High emission levels	1. Defective AIR pump	345	321
		2. Loose AIR pump belt	346	322
		3. Plugged or damaged AIR pump filter	346	322
		4. AIR system relief valve stuck open	346	322

Table 14-7 ASE TASK

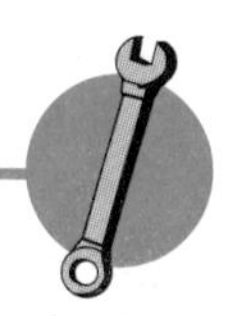

Inspect, test, and replace vacuum-operated air control valves and vacuum hoses of pump-type air injection systems.

Problem Area	Symptoms	Possible Causes	Classroom Manual	Shop Manual
EMISSION LEVELS	High emission levels	AIR by-pass (AIRB) valve continually by-passing air to atmosphere	345	322
FUEL ECONOMY	High fuel consumption	Pump air flow directed to exhaust ports with hot engine	346	322

Table 14-8 ASE TASK

Inspect and replace electrically/electronically operated components and circuits of pump-type air injection systems.

Problem Area	Symptoms	Possible Causes	Classroom Manual	Shop Manual
EMISSION LEVELS	High emission levels	AIR by-pass (AIRB) valve continually by-passing air to atmosphere	346	322
FUEL ECONOMY	High fuel consumption	Pump air flow directed to exhaust ports with hot engine	346	322

Table 14-9 ASE TASK

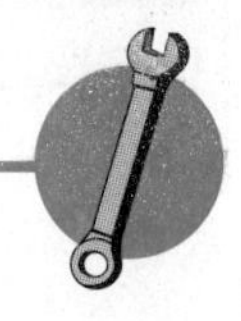

Inspect, service, and replace hoses, check valves, air manifolds, and nozzles of pump-type air injection systems.

Problem Area	Symptoms	Possible Causes	Classroom Manual	Shop Manual
NOISE	Exhaust leak noise, worse on acceleration	Leak in air manifolds, hoses	346	322
SYSTEM DAMAGE	Burned hoses	Leaking one-way valves	345	322

Table 14-10 ASE TASK

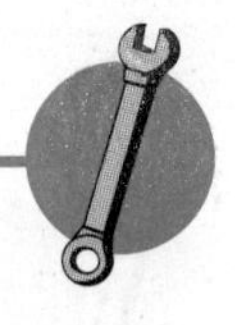

Test the operation of exhaust pulse-type air injection systems.

Problem Area	Symptoms	Possible Causes	Classroom Manual	Shop Manual
EMISSION LEVELS	High emission levels at low speed	Inoperative pulse-type air injection system	344	321
NOISE	Exhause leak noise worse on acceleration	Leak in air manifolds, hoses	345	321
SYSTEM DAMAGE	Burned hoses	Leaking one-way valves	345	321

Table 14-11 ASE TASK

Inspect, test, and replace pulse air valve(s), filters, silencers, and hoses of exhaust pulse-type air injection systems.

Problem Area	Symptoms	Possible Causes	Classroom Manual	Shop Manual
NOISE	Exhaust leak noise worse on acceleration	Leak in air manifolds, hoses	345	321
SYSTEM DAMAGE	Burned hoses	Leaking one-way valves	345	321

Table 14-12 ASE TASK

Test the operation of fuel vapor control systems.

Problem Area	Symptoms	Possible Causes	Classroom Manual	Shop Manual
ENGINE PERFORMANCE	Rough idle and low-speed operation	Canister purging at idle and low speed	347	323
ODOR	Gasoline smell in, or around, vehicle	Defective system, leaking hoses	347	323

Table 14-13 ASE TASK

Inspect and replace fuel tank cap, liquid/vapor separator, liquid check valve, lines, hoses of fuel vapor control systems.

Problem Area	Symptoms	Possible Causes	Classroom Manual	Shop Manual
ODOR	Gasoline smell in, or around, vehicle	Defective system, leaking hoses	347	324

Table 14-14 ASE TASK

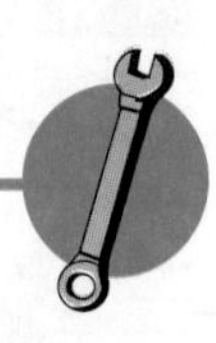

Inspect, service, and replace canister, filter, and purge lines of fuel vapor control systems.

Problem Area	Symptoms	Possible Causes	Classroom Manual	Shop Manual
ENGINE PERFORMANCE	Rough idle and low-speed operation	Canister purging at idle and low speed	348	323
ODOR	Gasoline smell in, or around, vehicle	Defective system, leaking hoses	347	323

Table 14-15 ASE TASK

Inspect, test, and replace thermal, vacuum, and electrical controls of fuel vapor control systems.

Problem Area	Symptoms	Possible Causes	Classroom Manual	Shop Manual
ENGINE PERFORMANCE	Rough idle and low-speed operation	Canister purging at idle and low speed	348	323
ODOR	Gasoline smell in, or around, vehicle	Defective system, leaking hoses	349	323

Positive Crankcase Ventilation, Spark Timing Control, and Intake Manifold Heat Control Systems, Service and Diagnosis

Upon completion and review of this chapter, you should be able to:

- ❏ Diagnose complete PCV systems.
- ❏ Diagnose PCV valves.
- ❏ Diagnose thermal vacuum valves (TVVs).
- ❏ Diagnose vacuum delay valves.
- ❏ Diagnose knock sensors, knock sensor modules, and the related wiring.
- ❏ Diagnose mixture heaters, relays, and temperature switches in mechanically controlled mixture heater systems.
- ❏ Diagnose mixture heaters, relays, and related wiring in computer-controlled mixture heater systems.
- ❏ Diagnose heat riser valves, diaphragms, solenoids, related wiring, and hoses in computer-controlled heat riser valve systems.

Basic Tools

Basic technician's tool set

Service manual

Various lengths and sizes of hoses

Heat-resistant water container and thermometer

12-V test light

Positive Crankcase Ventilation (PCV) System Service and Diagnosis

PCV System Diagnosis and Service

If the PCV valve is stuck in the open position, excessive air flow through the valve causes a lean air-fuel ratio and possible rough idle operation or engine stalling. When the PCV valve or hose is restricted, excessive crankcase pressure forces blow-by gases through the clean air hose and filter into the air cleaner. Worn rings or cylinders cause excessive blow-by gases and increased crankcase pressure, which forces blow-by gases through the clean air hose and filter into the air cleaner. A restricted PCV valve or hose may result in the accumulation of moisture and sludge in the engine and engine oil.

Leaks at engine gaskets, such as rocker arm cover or crankcase gaskets, will result in oil leaks and the escape of blow-by gases to the atmosphere. However, the PCV system also draws unfiltered air through these leaks into the engine. This action could result in wear of engine components, especially when the vehicle is operating in dusty conditions.

When diagnosing a PCV system, the first step is to check all the engine gaskets for signs of oil leaks (Figure 15-1). Be sure the oil filler cap fits and seals properly. Check the clean air hose and the PCV hose for cracks, deterioration, loose connections, and restrictions.

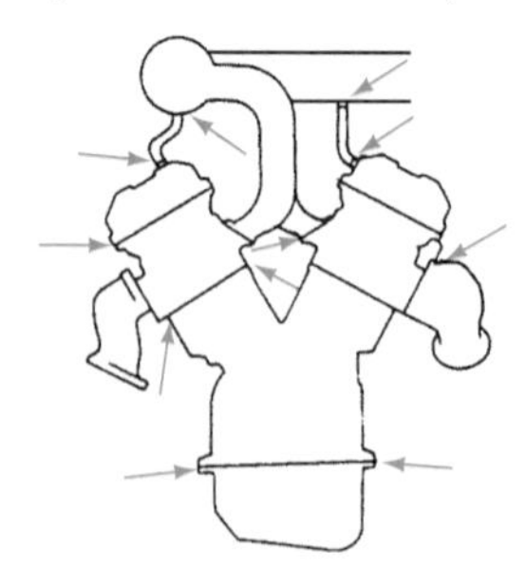

Figure 15-1 Checking engine gaskets, PCV hose, and clean air hose (Courtesy of Toyota Motor Corporation)

Check the PCV clean air filter for contamination and replace this filter if necessary (Figure 15-2). If there is evidence of oil in the air cleaner, check the PCV valve and hose for restriction. When the PCV valve and hose are in satisfactory condition, perform a cylinder compression test to check for worn cylinders and piston rings.

PCV Valve Diagnosis and Service

CAUTION: Do not attempt to suck through a PCV valve with your mouth. Sludge and other deposits inside the valve are harmful to the human body.

Vehicle manufacturers recommend different PCV valve checking procedures. Always follow the procedure in the vehicle manufacturer's service manual. Some vehicle manufacturer's recommend removing the PCV valve from the rocker arm cover and the hose. Connect a length of hose to the inlet side of the PCV valve, and blow air through the valve with your mouth while holding your finger near the valve outlet (Figure 15-3). Air should pass freely through the valve. If air does not pass freely through the valve, replace the valve.

Connect a length of hose to the outlet side of the PCV valve and try to blow back through the valve (Figure 15-4). It should be difficult to blow air through the PCV valve in this direction. When air passes easily through the valve, replace the valve.

Some vehicle manufacturers recommend removing the PCV valve from the rocker arm cover and placing your finger over the valve with the engine idling. When there is no vacuum at the PCV valve, the valve, hose, or manifold inlet is restricted. Replace the restricted component, or components.

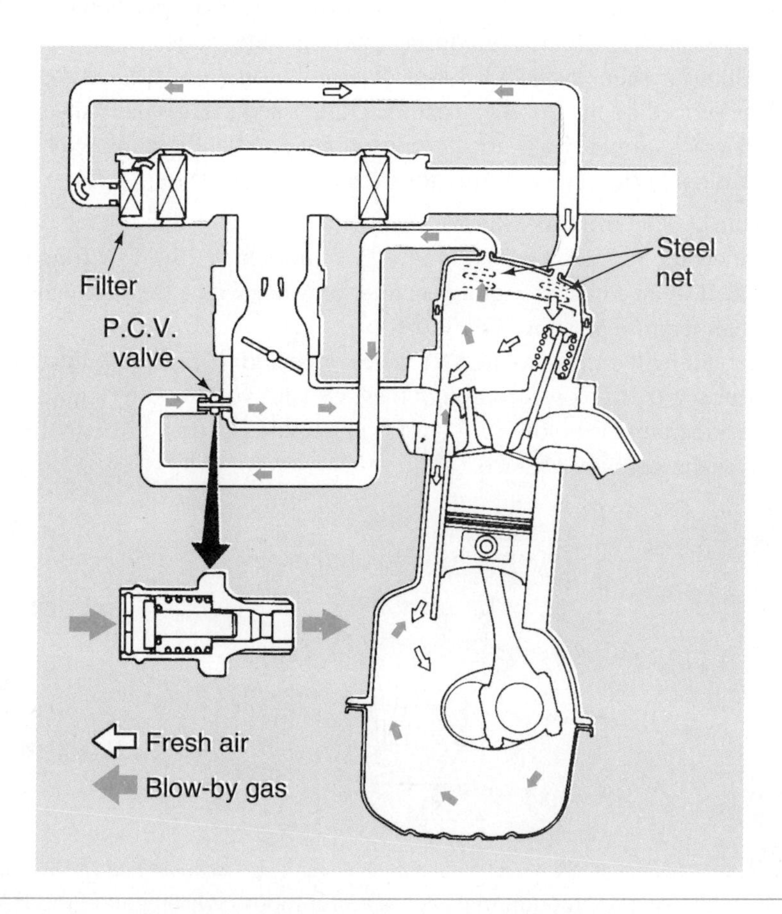

Figure 15-2 PCV air filter and complete system (Courtesy of Nissan Motor Co., Ltd.)

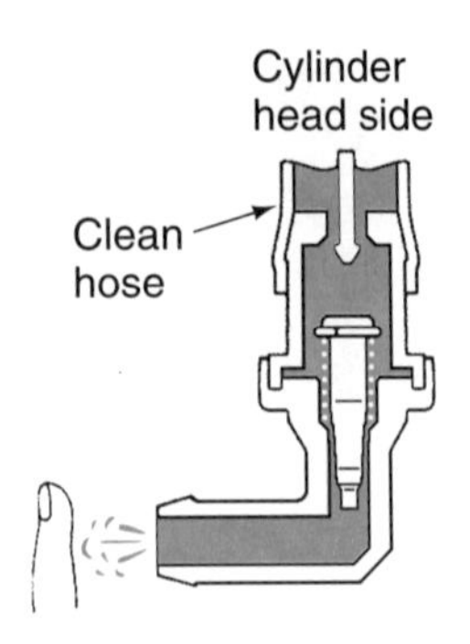

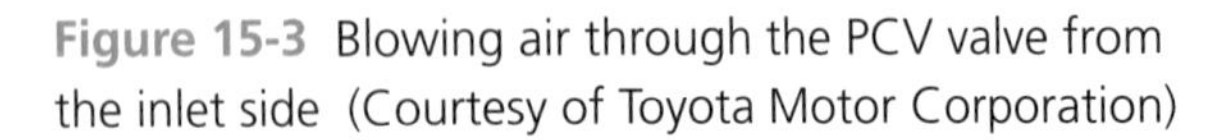
Figure 15-3 Blowing air through the PCV valve from the inlet side (Courtesy of Toyota Motor Corporation)

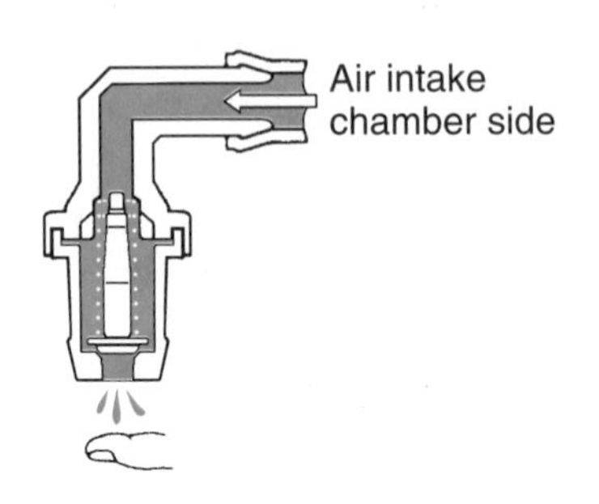

Figure 15-4 Blowing air through the PCV valve from the outlet side (Courtesy of Toyota Motor Corporation)

Classroom Manual
Chapter 15, page 354

Remove the PCV valve from the rocker arm cover and the hose. Shake the valve beside your ear and listen for the tapered valve rattling inside the valve housing. If no rattle is heard, PCV valve replacement is required.

Spark Control System Diagnosis

Thermal Vacuum Valve (TVV) Diagnosis

If a TVV connected in the distributor vacuum advance hose continually vents the vacuum advance, the vacuum advance is inoperative, which results in reduced fuel economy and performance. When the TVV does not vent the distributor vacuum advance at the specified temperatures during engine warm-up, emission levels are high during this operating mode.

The procedure for checking a TVV varies depending on the vehicle make and model year. Always follow the procedure in the vehicle manufacturer's service manual. Always check the TVV vacuum hoses for cracks, kinks, leaks, loose connections, and deterioration. Follow these steps to check a TVV that is connected in the distributor vacuum advance hose:

1. Drain the coolant from the radiator.
2. Disconnect the vacuum hoses from the TVV and remove the TVV from the engine.
3. Place the TVV in a heat-resistant container of water with a thermometer. Be sure the water temperature is below 50°F (10°C).
4. Connect a length of hose to the TVV port connected to the distributor vacuum advance, and try to blow air through the TVV with your mouth (Figure 15-5). If the water temperature is below 50°F (10°C), air should not flow through the port connected to the vacuum advance.

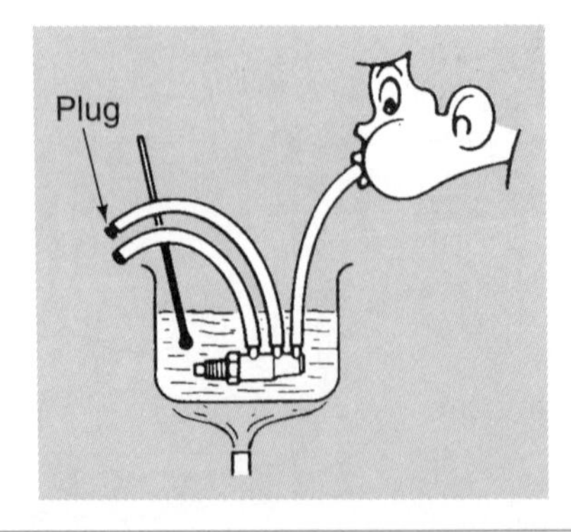

Figure 15-5 Blowing air through the TVV with your mouth while heating the TVV in a container of water (Courtesy of Nissan Motor Co., Ltd.)

5. Heat the water container and continue trying to blow through the TVV outlet connected to the vacuum advance. When the water temperature is between 50°F to 122°F (10°C to 50°C), the port in the TVV should be open, and you should be able to blow through the hose. If the water temperature is above 122°F (50°C), the TVV should be closed, and you should not be able to blow through the hose. When the TVV does not open and close at the specified temperatures, replace the TVV.

Vacuum Delay Valve Diagnosis

If a vacuum delay valve is plugged, the distributor advance is inoperative. Under this condition, fuel economy and engine performance are reduced. When a vacuum delay valve does not delay the vacuum advance after the throttle is opened, emission levels may be higher than normal during acceleration.

A vacuum delay valve may be referred to as a spark delay valve.

Always check the vacuum delay valve hoses for cracks, leaks, kinks, loose connections, and deterioration. Check the vacuum delay valve for proper installation. These valves are usually marked carburetor and distributor.

Special Tools

Vacuum hand pump

Vacuum gauge

Remove the vacuum delay valve from the vacuum hose, and connect a vacuum hand pump to the intake manifold side of the valve (Figure 15-6). Connect a vacuum gauge to the vacuum advance side of the valve. Supply 18 in Hg of vacuum with the hand pump. Use a second hand on a watch to observe the length of time for this vacuum to appear on the vacuum gauge. Compare the vacuum build-up time recorded with the watch to the vehicle manufacturer's specifications. If the vacuum delay valve does not provide the proper delay time, replace the valve.

Release the vacuum in the hand pump, and observe the vacuum gauge. When this vacuum is released, the vacuum delay valve should immediately release the vacuum on the gauge. If the vacuum on the gauge is not released immediately, replace the vacuum delay valve.

Diagnosis of Knock Sensor and Knock Sensor Module

WARNING: Operating an engine with a detonation problem for a sufficient number of miles may result in piston, ring, and cylinder wall damage.

If the knock sensor system does not provide an engine detonation signal to the PCM, the engine detonates, especially on acceleration. When the knock sensor system provides excessive spark retard, fuel economy and engine performance are reduced.

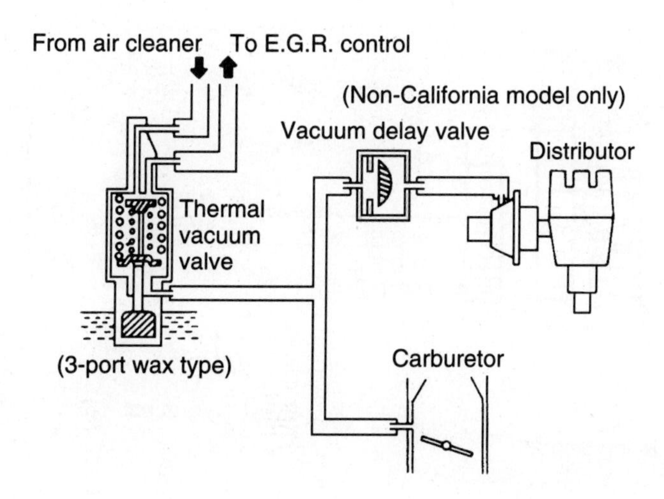

Figure 15-6 Vacuum delay valve connected in vacuum advance hose (Courtesy of Nissan Motor Co., Ltd.)

The first step in diagnosing the knock sensor and knock sensor module is to check all the wires and connections in the system for loose connections, corroded terminals, and damage. With the ignition switch on, be sure 12 V are supplied through the fuse to terminal B on the knock sensor module. Repair or replace the wires, terminals, and fuse, as required.

The diagnostic procedure for the knock sensor system varies depending on the vehicle make and model year. Always follow the procedure in the vehicle manufacturer's service manual.

Special Tools

Scan tester

Connect a scan tester to the DLC and check for DTCs related to the knock sensor system. If DTCs are present, diagnose the cause of these codes. When no DTCs related to the knock sensor system are present, follow these steps to diagnose the system:

1. Connect the scan tester to the DLC, and be sure the engine is at normal operating temperature.

SERVICE TIP: If the knock sensor torque is more than specified, the sensor may become too sensitive and provide an excessively high voltage signal, resulting in more spark retard than required. When the knock sensor torque is less than specified, the knock sensor signal is lower than normal, resulting in engine detonation.

2. Operate the engine at 1,500 rpm and observe the knock sensor signal on the scan tester. If a knock sensor signal is present, disconnect the wire from the knock sensor and repeat the test at the same engine speed. If the knock sensor signal is no longer present, the engine has an internal knock or the knock sensor is defective. When the knock sensor signal is still present on the scan tester, check the wire from the knock sensor to the knock sensor module for picking up false signals from an adjacent wire. Reroute the knock sensor wire as necessary.
3. If the knock sensor signal is not indicated on the scan tester in step 2, tap on the engine block near the knock sensor with a small hammer. When the knock sensor signal is now present, the knock sensor system is satisfactory.
4. When a knock sensor signal is not present in step 3, turn off the ignition switch and disconnect the knock sensor module wiring connector. Connect a 12-V test light from 12 V to terminal D in this wiring connector (Figure 15-7). If the light is off, repair the wire connected from this terminal to ground. When the light is on, proceed to step 5.
5. Reconnect the knock sensor module wiring connector, and disconnect the knock sensor wire. Operate the engine at idle speed and momentarily connect a 12-V test light from 12 V to the knock sensor wire. If a knock sensor signal is now generated on the

The knock sensor module may be referred to as an electronic spark control (ESC) module.

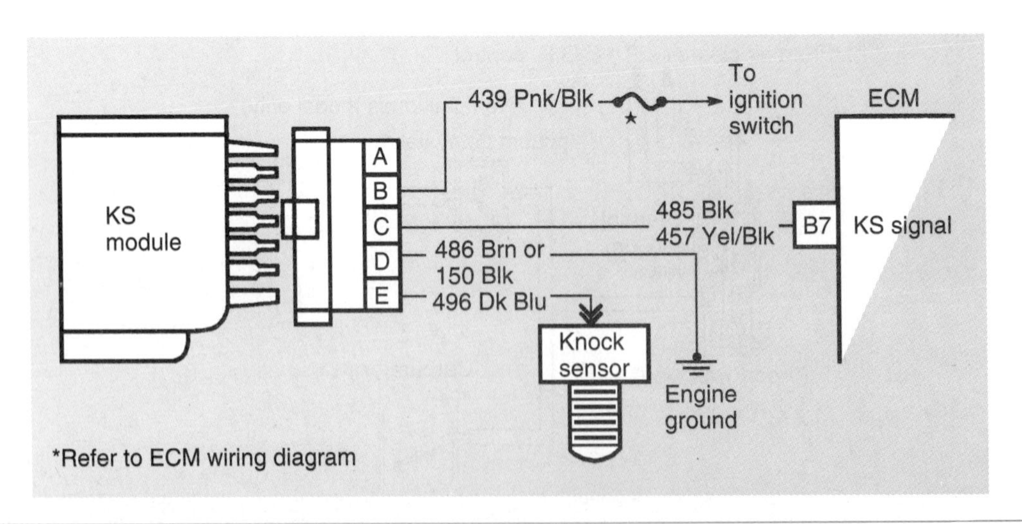

Figure 15-7 Knock sensor, knock sensor module, and wiring connections (Courtesy of Chevrolet Motor Division, General Motors Corporation)

scan tester, there is a faulty connection at the knock sensor or the knock sensor is defective. When a knock sensor signal is not generated, check for faulty wires from the knock sensor to the module or from the module to the PCM. Check the wiring connections at the module. If the wires and connections are satisfactory, the knock sensor module is likely defective.

Classroom Manual
Chapter 15, page 358

Photo Sequence 12 shows a typical procedure for diagnosing knock sensors and knock sensor modules.

Diagnosis of Intake Manifold Heater Systems

Diagnosis of Mechanically Controlled Intake Manifold Heater System

Initial Diagnosis. If the mixture heater system is inoperative, the engine may hesitate on acceleration during engine warm-up, and fuel consumption may be higher than normal. The diagnostic procedure for a mechanically operated intake manifold heater system varies depending on the vehicle. Always check all the wires and wiring terminals in the system for loose connections, corroded terminals, or damaged wires. With the ignition switch on, use a digital voltmeter to check for 12 V at the battery side of the mixture heater relay contacts and the battery side of this relay winding. If 12 V are not present at either of these terminals, check the fuses and fuse link in these circuits (Figure 15-8).

Mixture Heater Relay Diagnosis. Disconnect the mixture heater relay, and connect a pair of ohmmeter leads across terminals 1 and 2 on the relay (Figure 15-9). If the resistance is not within the manufacturer's specifications, replace the relay. When the resistance between terminals 1 and 2 is normal, supply 12 V to terminal 1 and ground terminal 2. After these connections are completed, connect a pair of ohmmeter leads to terminals 3 and 4. When the ohmmeter indicates zero ohms, the relay is satisfactory. If the ohmmeter does not indicate zero ohms, replace the relay.

Special Tools

Multimeter

Temperature Switch Diagnosis. Drain the coolant from the radiator and remove the temperature switch. Place the temperature switch and a thermometer in a heat-resistant container filled

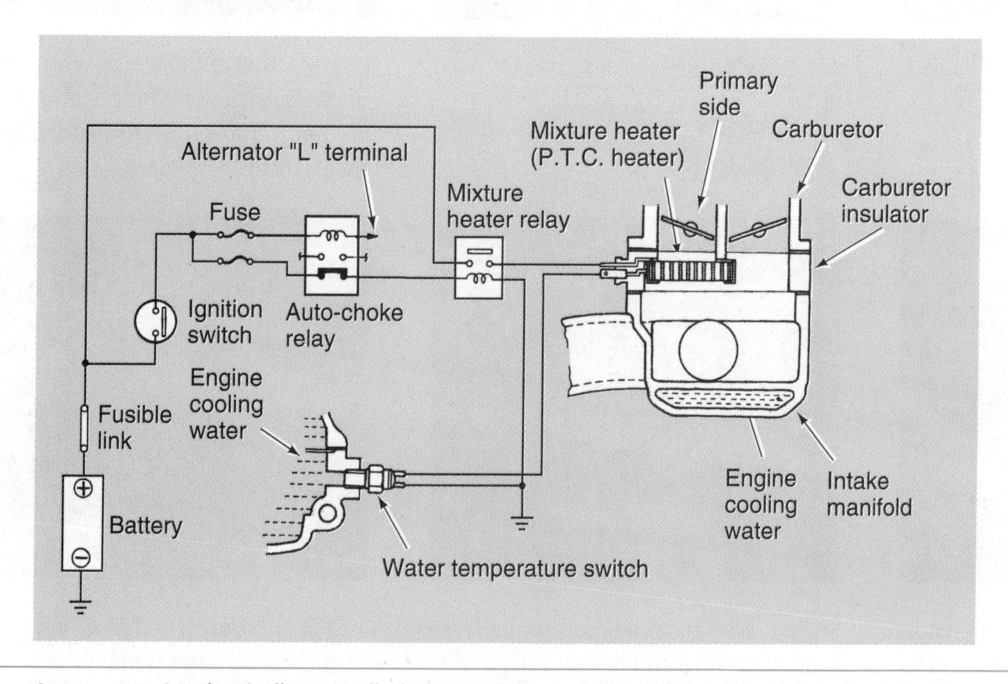

Figure 15-8 Mechanically controlled mixture heater system (Courtesy of Nissan Motor Co., Ltd.)

Photo Sequence 12

Typical Procedure for Diagnosing Knock Sensors and Knock Sensor Modules

P12-1 Be sure the engine is at normal operating temperature and the ignition switch is off.

P12-2 Connect the scan tester leads to the battery terminals with the correct polarity.

P12-3 Program the scan tester for the vehicle being tested.

P12-4 Use the proper adapter to connect the scan tester lead to the data link connector (DLC).

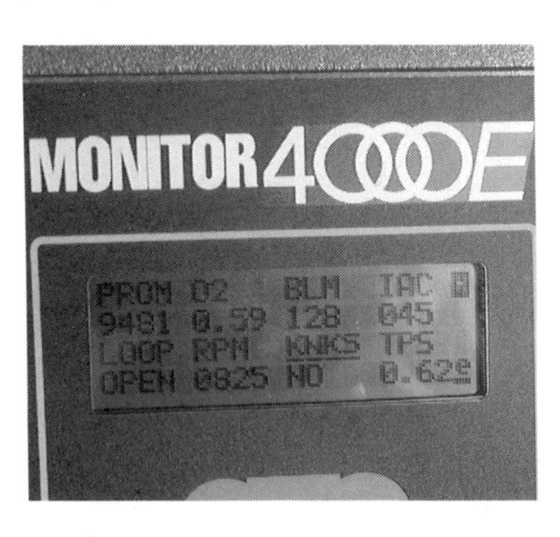

P12-5 Select knock sensor on the scan tester.

P12-6 Observe the knock sensor signal on the scan tester with the engine running at 1,500 rpm. The knock sensor should indicate no signal.

P12-7 Tap on the right exhaust manifold above the sensor with a small hammer and observe the scan tester reading.

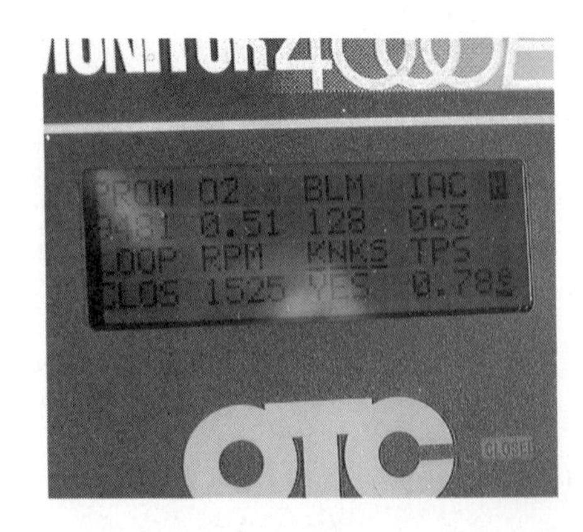

P12-8 The knock sensor should now indicate a signal on the scan tester.

P12-9 Shut off the engine and disconnect the scan tester leads.

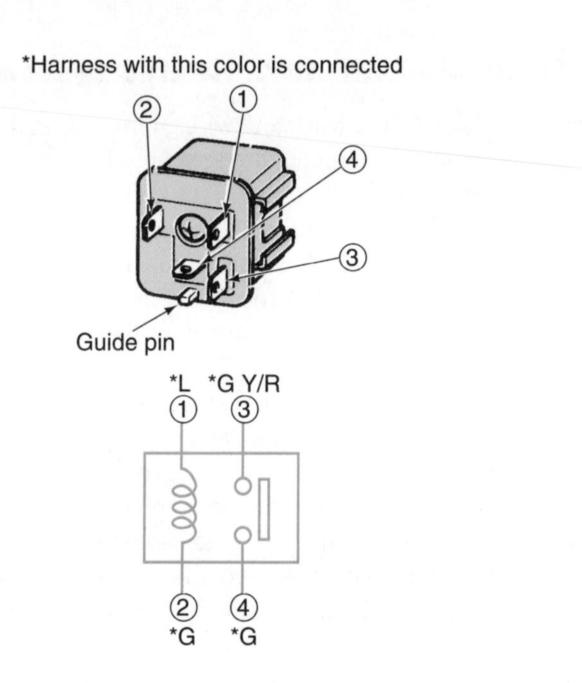

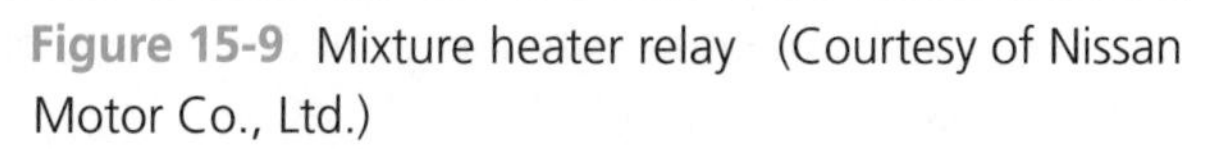

Figure 15-9 Mixture heater relay (Courtesy of Nissan Motor Co., Ltd.)

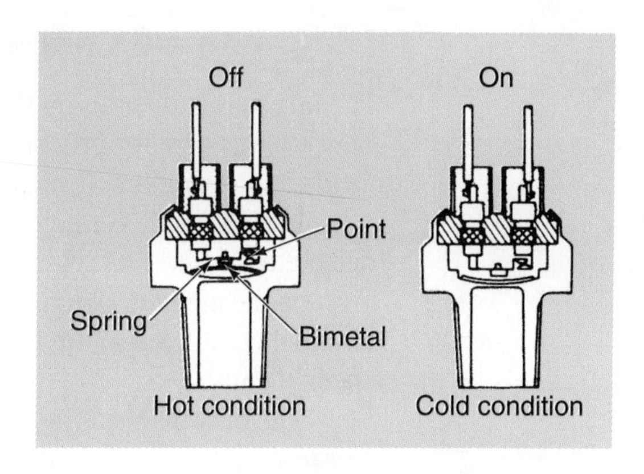

Figure 15-10 Temperature switch operation (Courtesy of Nissan Motor Co., Ltd.)

with water. Be sure the water temperature is below 99°F (37°C). Connect an ohmmeter to the temperature switch terminals and heat the water while observing the thermometer. When the water temperature is below 99°F (37°C), the temperature switch contacts should be closed and the ohmmeter should read zero ohms (Figure 15-10). Above this temperature, the ohmmeter reading should be infinite.

Mixture Heater Diagnosis. Disconnect the mixture heater wires and connect the ohmmeter leads to the mixture heater terminals (Figure 15-11). Be sure the ohmmeter switch is on the X1 scale. If the mixture heater does not have the specified resistance, replace the heater assembly. Connect the ohmmeter leads from one of the mixture heater terminals to ground. An infinite ohmmeter reading indicates the heater is not grounded. When the ohmmeter reading is less than infinite, replace the heater.

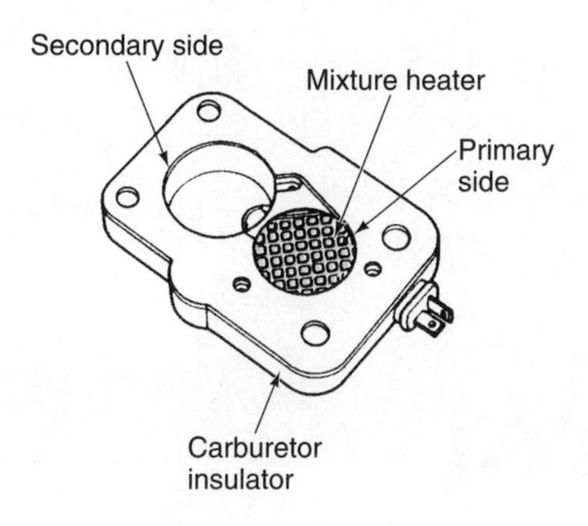

Figure 15-11 Mixture heater terminals (Courtesy of Nissan Motor Co., Ltd.)

Diagnosis of Computer-Controlled Mixture Heater System

The first step in diagnosing this system is to check all the wires and wiring terminals in the system. Check the system for DTCs, such as an engine coolant temperature sensor code, that could affect the system operation. If such a DTC is present, diagnose the cause of the code.

Use a digital voltmeter to check for 12 V at the mixture heater relay contact terminal connected to the fuse link (Figure 15-12). If 12 V are not available at this terminal, check the fuse link. Turn on the ignition switch, and use a digital voltmeter to check for 12 V at the mixture heater relay terminal connected to the fuse and ignition switch. When 12 V are not available at this terminal, check the fuse.

The mixture heater relay and the mixture heater may be tested in the same way as these components on the mechanically operated mixture heater system explained previously.

The PCM should ground the mixture heater relay winding and close the relay contacts when the coolant temperature is below 122°F (50°C). At this coolant temperature, the PCM grounds the mixture heater relay winding while cranking the engine and when the engine is running. When the coolant temperature is above 122°F (50°C), the PCM does not provide a ground for the mixture heater relay winding, and the relay contacts open. Under this condition, current is no longer supplied to the mixture heater.

SERVICE TIP: Before replacing any PCM, always be sure the ground wires and power supply wires to the PCM are satisfactory.

When the mixture heater and the relay are satisfactory, be sure the coolant temperature is below 122°F (50°C), and use a digital voltmeter to check for 12 V at the mixture heater terminal connected to the relay contacts with the engine idling. If 12 V are not present at this terminal, shut off the ignition switch, and disconnect the PCM connector. With the ignition switch on, use a digital voltmeter to check for 12 V at terminal 16 in the PCM connector. If 12 V are not available at this terminal, repair the open circuit in the wire from the relay winding to the PCM. When 12 V are available at PCM terminal 16, use a jumper wire to connect this terminal to the battery ground. If 12 V are now available at the mixture heater terminal connected to the relay contacts, the PCM is likely defective.

Classroom Manual
Chapter 15, page 360

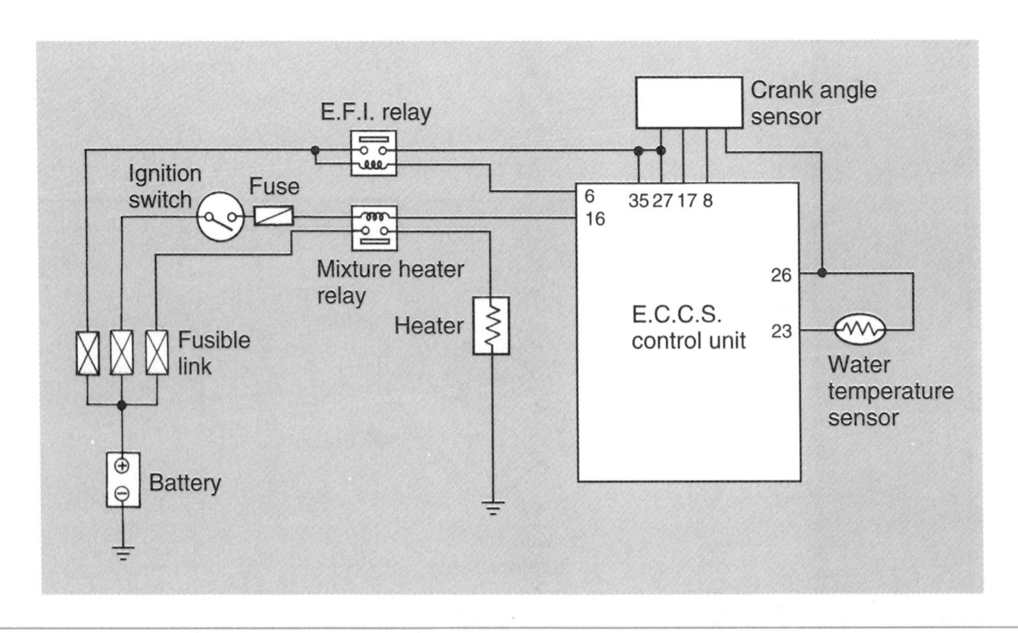

Figure 15-12 Computer-controlled mixture heater system (Courtesy of Nissan Motor Co.,Ltd.)

Computer-Controlled Heat Riser Valve Diagnosis

Vacuum Diaphragm and Heat Riser Diagnosis

CAUTION: If the engine has been running, heat riser valves and operating rods are extremely hot. Wear protective gloves to avoid burns when servicing these components.

If the heat riser valve remains wide open at all engine temperatures, the engine may hesitate on acceleration during engine warm-up, and an increase in fuel consumption and emission levels may be experienced. When the heat riser valve remains closed at all engine temperatures, the engine has a loss of power on acceleration, and detonation may occur.

On General Motors products, a vacuum-operated heat riser system is called an early fuel evaporation (EFE) system.

Disconnect the vacuum hose from the heat riser vacuum diaphragm, and connect a vacuum hand pump to the vacuum fitting on this diaphragm (Figure 15-13). Supply 18 in Hg of vacuum to the heat riser diaphragm with the hand pump. If the diaphragm does not hold the vacuum, replace the diaphragm. If the heat riser valve does not move to the closed position, disconnect the diaphragm rod from the diaphragm and pull on the rod. If the heat riser valve is seized, squirt some penetrating oil on the ends of the valve shaft. When this action does not loosen the valve, replace the heat riser valve assembly.

Heat Riser Solenoid Diagnosis

Always check the vacuum hoses connected to the heat riser solenoid for cracks, leaks, kinks, loose connections, and deterioration. Check the wires and wiring terminals connected to this solenoid. Connect a scan tester to the DLC and check for DTCs, such as an engine coolant temperature sensor code, which could affect the operation of this system.

Disconnect the heat riser solenoid wiring connector, and connect an ohmmeter to the solenoid terminals. A lower-than-specified reading indicates a shorted winding, whereas an infinite reading indicates an open winding. Connect the ohmmeter leads from one of the solenoid terminals to the solenoid case. An infinite reading indicates the winding is not grounded, and a low reading indicates a grounded winding.

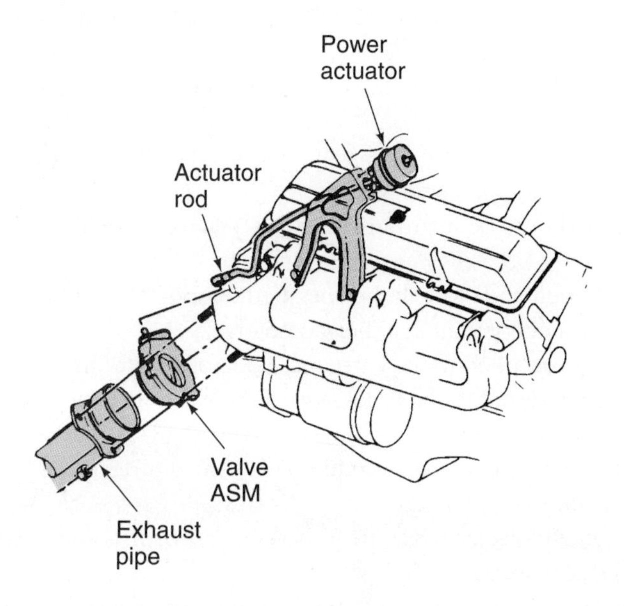

Figure 15-13 Vacuum diaphragm and heat riser valve (Courtesy of Pontiac Motor Division, General Motors Corporation)

With the ignition switch on, use a digital voltmeter to check for 12 V at the solenoid input terminal. If 12 V are not available at this terminal, check the fuse connected in this circuit. When the fuse is satisfactory, check for an open wire between the ignition switch and the solenoid.

Remove the vacuum hoses from the heat riser solenoid, and connect a vacuum gauge to the vacuum hose connected to the intake manifold. With the engine idling, the vacuum should be above 16 in Hg. When the vacuum is lower than 16 in Hg, check the hose from the solenoid to the intake manifold for restrictions and leaks.

Be sure the coolant temperature is below 75°F (24°C), and connect a vacuum gauge to the vacuum fitting on the heat riser solenoid connected to the vacuum diaphragm. With the engine idling, the vacuum gauge should indicate over 16 in Hg. If this vacuum is not present, use a jumper wire to ground the solenoid terminal connected to the PCM. When the specified vacuum is now available on the gauge, use an ohmmeter to check the wire from the solenoid to the PCM for an open circuit or a ground. If this wire is satisfactory, the PCM is probably faulty.

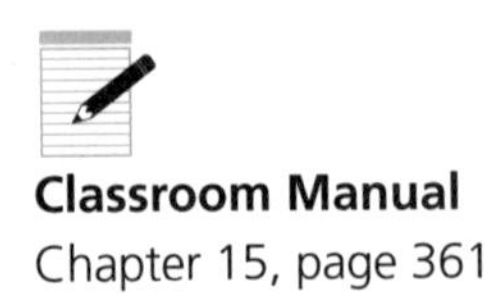

Classroom Manual
Chapter 15, page 361

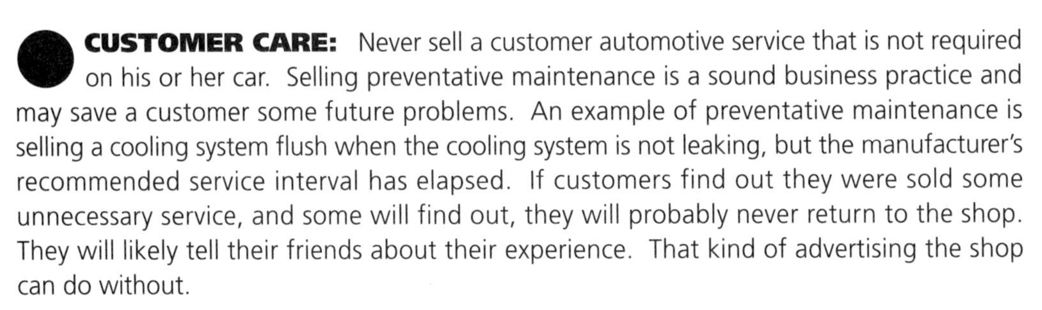

CUSTOMER CARE: Never sell a customer automotive service that is not required on his or her car. Selling preventative maintenance is a sound business practice and may save a customer some future problems. An example of preventative maintenance is selling a cooling system flush when the cooling system is not leaking, but the manufacturer's recommended service interval has elapsed. If customers find out they were sold some unnecessary service, and some will find out, they will probably never return to the shop. They will likely tell their friends about their experience. That kind of advertising the shop can do without.

Guidelines for Diagnosing and Servicing Positive Crankcase Ventilation, Spark Timing Control, and Intake Manifold Heat Control Systems

1. Proper operation of the PCV system depends on adequate engine gasket sealing.
2. If oil is present in the air cleaner, the PCV system may be restricted or the cylinders may have excessive blow-by.
3. Air should move freely from the inlet side of the PCV valve to the outlet side of the valve.
4. With the engine idling and the PCV valve removed from the rocker arm cover, vacuum should be available at the valve inlet.
5. A rattling noise should be heard inside the PCV valve when the valve is moved beside your ear.
6. A TVV should open and close at the specified temperatures when the TVV is heated in a container of water containing a thermometer.
7. When vacuum is supplied to the carburetor side of a vacuum delay valve and a vacuum gauge is connected to the other side of the valve, the vacuum delay valve should delay the vacuum for the specified time.
8. When the vacuum supplied to a vacuum delay valve is decreased, the valve should release the vacuum immediately.
9. When the engine block is tapped near the knock sensor, a knock sensor signal should appear on the scan tester.
10. If a 12-V test light is connected from the battery positive terminal to the disconnected knock sensor wire with the engine running, a knock sensor signal should appear on the scan tester.

11. If the mixture heater system is inoperative, the engine may hesitate on acceleration during warm-up, and fuel consumption may be increased.
12. When the heat riser valve remains open at all engine temperatures, the engine may hesitate on acceleration during engine warm-up, and an increase in fuel consumption and emission levels may be experienced.
13. If the heat riser valve remains closed at all engine temperatures, the engine may detonate and have a loss of power.

CASE STUDY

A customer complained about severe engine detonation on a GMC 1,500 series truck with a 5.7-L engine. The technician road tested the truck and found a very noticeable detonation even during moderate acceleration. Apart from the detonation problem, the engine performance was normal.

The technician checked the basic timing and found it was set to specifications. When a scan tester was connected to the DLC, the technician did not find any DTCs in the PCM memory.

The technician checked the knock sensor, knock sensor module, and related circuit. When the engine block was tapped near the knock sensor with the engine running at 1,500 rpm, a knock sensor signal appeared on the scan tester indicating the knock sensor, knock sensor module, and connecting wires were sending a knock sensor signal to the PCM.

The technician checked the programmable read only memory (PROM) number with the scan tester and compared this number to the latest PROM number in the PROM identification (ID) manual. The PROM ID manual indicated an updated PROM was available for the computer in this vehicle. The technician checked the bulletin numbers provided by the scan tester for this vehicle, and discovered one of these bulletins recommended the PROM change to correct detonation. The new PROM was installed, and a road test indicated the detonation problem was eliminated.

Terms to Know

Positive crankcase ventilation (PCV) valve
Thermal vacuum valve (TVV)
Vacuum delay valve
Knock sensor
Knock sensor module
Mixture heater
Temperature switch

ASE Style Review Questions

1. While discussing PCV system diagnosis:
Technician A says an accumulation of oil in the air cleaner indicates the PCV valve is stuck open.
Technician B says an accumulation of oil in the air cleaner may indicate the piston rings and cylinders are worn.
Who is correct?
A. A only
B. B only
C. Both A and B
D. Neither A nor B

2. While discussing PCV system diagnosis:
Technician A says a defective PCV valve may cause rough idle operation.
Technician B says satisfactory PCV system operation depends on a properly sealed engine.
Who is correct?
A. A only
B. B only
C. Both A and B
D. Neither A nor B

3. While discussing PCV valve diagnosis:
Technician A says you should be able to blow air freely from the inlet side to the outlet side of the PCV valve.
Technician B says you should be able to blow air freely from the outlet side to the inlet side of the PCV valve.
Who is correct?
A. A only **C.** Both A and B
B. B only **D.** Neither A nor B

4. While discussing the diagnosis of a thermal vacuum valve (TVV) connected to the distributor advance hose:
Technician A says if the TVV vents the vacuum supplied to the vacuum advance all the time, the engine may hesitate on low-speed acceleration.
Technician B says the TVV should not vent the vacuum supplied to the vacuum advance when the coolant temperature is below 50°F (10°C).
Who is correct?
A. A only **C.** Both A and B
B. B only **D.** Neither A nor B

5. While discussing vacuum delay valve diagnosis:
Technician A says when vacuum is supplied to the carburetor side of the vacuum delay valve with a hand pump, and a vacuum gauge is connected to the distributor side of this valve, the vacuum should appear instantly on the gauge.
Technician B says when the vacuum supplied to the carburetor side of the vacuum delay valve is decreased, this valve should delay the decrease in vacuum on the distributor side of the valve.
Who is correct?
A. A only **C.** Both A and B
B. B only **D.** Neither A nor B

6. While discussing knock sensor and knock sensor module diagnosis:
Technician A says with the engine running at 1,500 rpm, if the engine block is tapped near the knock sensor, a knock sensor signal should appear on the scan tester.
Technician B says with the engine running at 1,500 rpm, if a 12-V test light is connected from a 12-V source to the disconnected knock sensor wire, a knock sensor signal should appear on the scan tester.
Who is correct?
A. A only **C.** Both A and B
B. B only **D.** Neither A nor B

7. While discussing knock sensor and knock sensor module diagnosis:
Technician A says if the knock sensor torque is more than specified, the spark advance provided by the PCM may be reduced.
Technician B says when the PCM memory contains DTCs related to the knock sensor system, these codes should be diagnosed before further system and component diagnosis.
Who is correct?
A. A only **C.** Both A and B
B. B only **D.** Neither A nor B

8. While discussing mechanically controlled mixture heater system diagnosis:
Technician A says if 12 V are not available at the input terminal on the mixture heater relay contacts, the ignition switch may be defective.
Technician B says if 12 V are not available at the input terminal on the mixture heater relay contacts, there may be an open wire between the ignition switch and the relay contacts.
Who is correct?
A. A only **C.** Both A and B
B. B only **D.** Neither A nor B

9. While discussing computer-controlled mixture heater system diagnosis:
Technician A says if the coolant temperature is below 122°F (50°C), the PCM should provide a ground for the mixture heater relay winding.
Technician B says if the coolant temperature is above 150°F (65°C), the PCM should provide a ground for the mixture heater relay winding.
Who is correct?
A. A only **C.** Both A and B
B. B only **D.** Neither A nor B

10. While discussing computer-controlled heat riser system diagnosis:
Technician A says when intake manifold vacuum is supplied to heat riser vacuum diaphragm, the heat riser valve should be closed.
Technician B says the PCM operates the heat riser solenoid to supply vacuum to the heat riser diaphragm when the engine is at normal operating temperature.
Who is correct?
A. A only **C.** Both A and B
B. B only **D.** Neither A nor B

Table 15-1 ASE TASK

Test the operation of positive crankcase ventilation (PCV) systems.

Problem Area	Symptoms	Possible Causes	Classroom Manual	Shop Manual
ENGINE PERFORMANCE	Rough idle operation, stalling	PCV valve sticking open	354	334
AIR CLEANER CONTAMINATION	Oil in air cleaner	1. PCV valve or hose restricted	354	334
		2. Worn piston rings and cylinders	354	334
ENGINE CONTAMINATION	Sludge in engine and engine oil	PCV valve or hose restricted	356	334

Table 15-2 ASE TASK

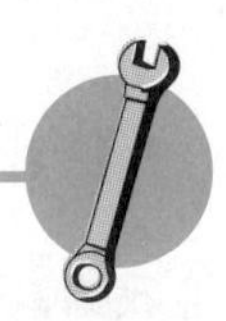

Inspect, service, and replace positive crankcase ventilation (PCV) filter/breather cap, valve, tubes, orifices, and hoses.

Problem Area	Symptoms	Possible Causes	Classroom Manual	Shop Manual
ENGINE PERFORMANCE	Rough idle operation, stalling	PCV valve sticking open	356	335
AIR CLEANER CONTAMINATION	Oil in air cleaner	1. PCV valve or hose restricted	356	335
		2. Worn piston rings and cylinders	356	335
ENGINE CONTAMINATION	Sludge in engine and engine oil	PCV or hose restricted	356	335

Table 15-3 ASE TASK

Test the operation of spark timing systems.

Problem Area	Symptoms	Possible Causes	Classroom Manual	Shop Manual
ENGINE PERFORMANCE	Hesitation on acceleration	Inoperative spark control system, reduced spark advance	357	336
	Detonation	Defective spark control system, excessive spark advance	357	337
EMISSION LEVELS	High emission levels	Defective spark control system, reduced spark advance	357	337

Table 15-4 ASE TASK

Inspect, test, repair, and replace electrical/electronic components and circuits of spark timing control systems.

Problem Area	Symptoms	Possible Causes	Classroom Manual	Shop Manual
ENGINE PERFORMANCE	Hesitation on acceleration	Inoperative spark control system, reduced spark advance	358	338
	Detonation	Defective spark control system, excessive spark advance	358	338
EMISSION LEVELS	High emission levels	Defective spark control system, reduced spark advance	358	338

Table 15-5 ASE TASK

Inspect, test, repair, and replace thermal, mechanical, or vacuum components and hoses of spark timing control systems.

Problem Area	Symptoms	Possible Causes	Classroom Manual	Shop Manual
ENGINE PERFORMANCE	Hesitation on acceleration	Inoperative spark control system, reduced spark advance	357	336
	Detonation	Defective spark control system, excessive spark advance	358	337
EMISSION LEVELS	High emission levels	Defective spark control system, reduced spark advance	357	337

Table 15-6 ASE TASK

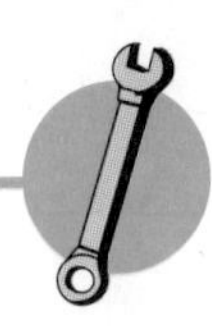

Test the operation of electrical/vacuum/coolant-type intake manifold temperature control systems.

Problem Area	Symptoms	Possible Causes	Classroom Manual	Shop Manual
ENGINE PERFORMANCE	Hesitation on acceleration during warm-up	Inoperative mixture heater system	359	339

Table 15-7 ASE TASK

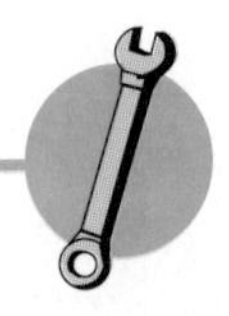

Inspect, test, repair, or replace components of electrical/vacuum/coolant-type intake manifold temperature control systems.

Problem Area	Symptoms	Possible Causes	Classroom Manual	Shop Manual
ENGINE PERFORMANCE	Hesitation on acceleration during warm-up	Inoperative mixture heater system	360	339

Table 15-8 ASE TASK

Inspect, service, and replace manifold temperature control (heat riser) systems.

Problem Area	Symptoms	Possible Causes	Classroom Manual	Shop Manual
ENGINE PERFORMANCE	Hesitation on acceleration during warm-up	Heat riser valve always open	361	343
	Loss of engine power, detonation	Heat riser valve remains closed at all times	301	343

Engine Diagnosis With Infrared Analyzer, Engine Analyzer, and Oscilloscope

Upon completion and review of this chapter, you should be able to:

- ❑ Operate a four-gas infrared analyzer to check emission levels.
- ❑ Diagnose high levels of hydrocarbons (HC), carbon monoxide (CO), oxygen (O_2), and carbon dioxide (CO_2).
- ❑ Use the infrared analyzer for state inspection, maintenance (I/M) testing.
- ❑ Test catalytic converter efficiency with an infrared analyzer.
- ❑ Test the secondary air injection system efficiency with an infrared analyzer.
- ❑ Test the positive crankcase ventilation (PCV) system with an infrared analyzer.
- ❑ Diagnose cylinder misfiring and vacuum leaks with the infrared analyzer.
- ❑ Diagnose head gasket and combustion chamber leaks with the infrared analyzer.
- ❑ Diagnose evaporative system leaks with the infrared analyzer.
- ❑ Use an engine analyzer to test cylinder performance, ignition system condition, battery starting and charging system performance, emission levels, and engine computer systems.
- ❑ Diagnose ignition system condition with oscilloscope patterns.

Infrared Emissions Testing

Basic Tools

Basic technician's tool set

Service manual

Insulated pliers

CAUTION: The infrared analyzer probe and sample hose may be very hot after the vehicle has been running. Wear protective gloves when handling these components.

WARNING: Always clean the water trap and change the filters regularly to provide accurate analyzer readings and prevent analyzer damage.

WARNING: Operate the analyzer for 15 minutes after vehicle testing is completed to dry moisture out of the tester and provide longer equipment life.

WARNING: Do not use an infrared analyzer to sample exhaust on a diesel engine. This action will damage the analyzer.

WARNING: Remove the infrared analyzer probe from the tail pipe when using combustion chamber cleaners. If the probe is not removed under this condition, the analyzer will be damaged.

WARNING: Store the analyzer probe off the floor to avoid contamination and damage. Do not drop the analyzer probe.

SERVICE TIP: Leaks in the exhaust system cause inaccurate infrared analyzer readings. Always inspect the exhaust system for leaks prior to infrared exhaust analysis.

Infrared Analyzer Calibration

Special Tools

Infrared analyzer

Infrared emission analyzers have a warm-up and calibration period, which is usually about 15 minutes. Modern infrared emission analyzers perform this calibration automatically. Always be sure the analyzer is calibrated properly so it provides accurate readings. Some older infrared analyzers had to be calibrated manually with calibration controls on the analyzer.

On most infrared analyzers, a warning light is illuminated if the exhaust flow through the analyzer is restricted because of a plugged filter, probe, or hose. This warning light must be off before proceeding with the infrared exhaust analysis.

Most vehicle manufacturers and test equipment manufacturers recommend disconnecting the AIR pump or pulsed secondary air injection system during an infrared exhaust gas analysis. Always follow the recommended procedure in the vehicle manufacturer's service manual or the equipment operator's manual.

Common Causes of Excessive Emissions

Excessive Hydrocarbon (HC) Emissions. Higher-than-normal HC emissions may be caused by one or more of these conditions:

1. Ignition system misfiring
2. Improper ignition timing
3. Excessively lean or rich air-fuel ratio
4. Low cylinder compression
5. Leaking head gasket
6. Defective valves, guides, or lifters
7. Defective rings, pistons, or cylinders

Excessive Carbon Monoxide (CO) Emissions. Higher-than-normal CO emissions may be caused by one of these items:

1. Rich air-fuel ratio
2. Dirty air filter
3. Faulty injectors
4. Higher-than-specified fuel pressure
5. Defective input sensor

Excessive HC and CO Emissions. When HC and CO emissions are higher than normal, check these items:

1. Plugged PCV system
2. Heat riser valve stuck open
3. AIR pump inoperative or disconnected
4. Engine oil diluted with gasoline

Lower-Than-Normal Carbon Dioxide (CO_2) Emissions. Lower-than-normal CO_2 levels may result from one of these problems:

1. Exhaust gas sample dilution because of leaking exhaust system
2. Rich air-fuel ratio

Lower-Than-Normal Oxygen (O_2) Readings and Higher-Than-Normal CO Readings. When O_2 readings are lower than normal and CO readings are higher than normal, check these items:

1. Rich air-fuel ratio
2. Defective injectors, pintles not seating properly, dripping fuel
3. Higher-than-specified fuel pressure
4. Defective input sensor
5. Restricted PCV system
6. Carbon canister purging at idle and low speeds

Higher-Than-Normal Oxygen (O_2) Readings, and Lower-Than-Normal CO Readings. When O_2 readings are higher than normal and CO readings are lower than normal, check these items:

1. Lean air-fuel ratio
2. Vacuum leak

3. Lower-than-specified fuel pressure
4. Defective injectors
5. Defective input sensor
6. AIR pump or pulsed secondary air injection system connected during infrared exhaust gas analysis

To pinpoint the exact cause of the improper infrared analyzer readings, perform a detailed test of individual systems and components.

State Inspection, Maintenance (I/M) Testing

CAUTION: When performing any test with the engine running, be sure the parking brake is applied with the automatic transmission in park or a manual transmission in neutral. Ignoring these precautions may cause personal injury and/or property damage.

CAUTION: If a vehicle is equipped with a vacuum parking brake release, disconnect and plug the vacuum hose to the release diaphragm before applying the parking brake prior to engine running tests. Personal injury and/or property damage may occur if this precaution is not followed.

The infrared analyzer may be used to check emission levels for state I/M testing. Infrared analyzers must meet specific standards, such as California BAR-80 or ETI-80, to be used in state I/M testing. Some analyzers may require a software change or the addition of a printer to comply with these programs.

Catalytic Converter Efficiency

With the engine at normal operating temperature, operate the engine at idle speed and observe the O_2 and CO readings. When the O_2 is above 0.5%, the catalytic converter is receiving enough oxygen to function properly. If the CO reading is above 0.5%, the catalytic converter is not oxidizing CO properly.

Secondary Air Injection (AIR) System Efficiency

Operate the engine at normal temperature and idle speed. Observe the O_2 readings and then disconnect the AIR system. Record the O_2 level, and compare the reading to the level when the AIR system was operational. Most AIR or PAIR systems should increase the O_2 level by 2% to 5% when they are operational.

Positive Crankcase Ventilation (PCV) Test

With the engine at normal operating temperature and running at idle speed, observe the O_2 and CO readings. Remove the PCV valve from the rocker arm cover, and allow this valve to pull in fresh air. The O_2 reading should increase, and the CO reading should decrease. If the O_2 reading does not change, the PCV system is restricted. When the O_2 reading increases more than 1% or the CO reading decreases more than 1%, there are excessive blow-by gases or the oil is contaminated with fuel.

Cylinder Misfiring and Vacuum Leaks

Many engine analyzers have an infrared emission analyzer combined with an oscilloscope and other test equipment. The engine analyzer can stop each spark plug from firing momentarily and record the rpm drop. If a cylinder is not contributing to engine power, there is very little rpm drop when the spark plug stops firing. Some of these analyzers also record the HC change when each spark plug stops firing. When a cylinder is misfiring and not contributing to engine power, the HC emissions from that cylinder are high. Therefore, when the spark plug stops firing, there is not much change in HC emissions. A vacuum leak results in higher-than-normal O_2 levels.

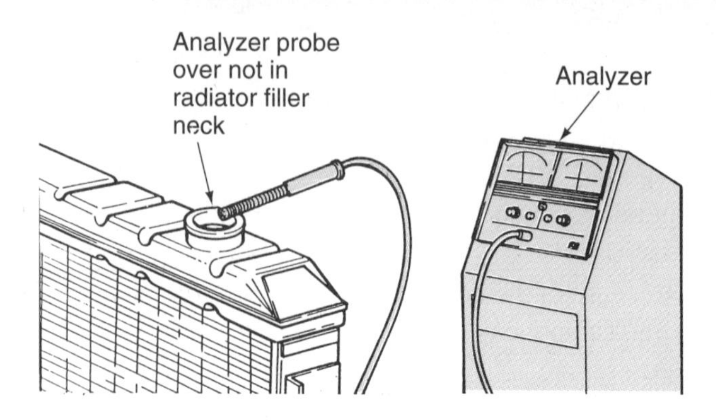

Figure 16-1 Using the infrared emission analyzer to check for head gasket and combustion chamber leaks (Courtesy of Chrysler Corporation)

Head Gasket and Combustion Chamber Leaks

Remove the radiator cap when the engine is cold. Place the infrared analyzer probe above the radiator filler neck and run the engine. Do not immerse the probe in the coolant. Head gasket or combustion chamber leaks into the cooling system result in an HC reading (Figure 16-1).

Evaporative (EVAP) System Leaks

With the engine at normal operating temperature, place the analyzer probe near any component in the EVAP system where a gasoline vapor leak is suspected. When the HC reading increases, the probe is near the gasoline vapor leak.

Infrared Emission Analyzer with Diagnostic Assistance

Classroom Manual
Chapter 16, page 367

Some infrared emission analyzers provide diagnostic assistance for the technician in the tester software program. If one or more of the emission levels is high, the technician may press the appropriate tester button and request assistance in diagnosing the excessive emission level (Figure 16-2).

Diagnosis With Engine Analyzer

Initial Selections and Main Test Areas

Special Tools
Engine analyzer

On some engine analyzers, a service test screen that lists all the tests that may be performed appears (Figure 16-3). The technician may select any individual test. For example, the technician

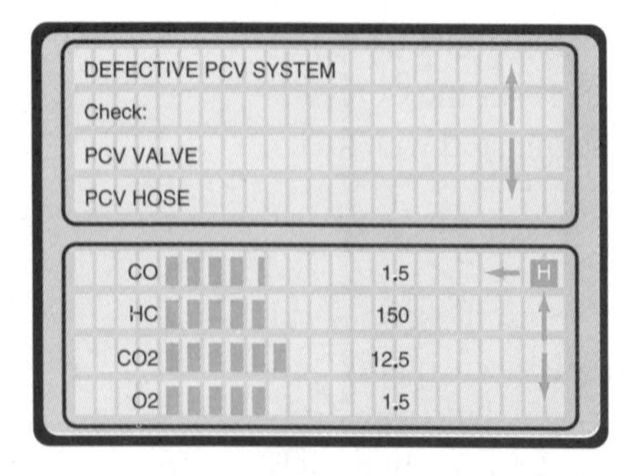

Figure 16-2 Infrared emission analyzer with diagnostic assistance in the software program (Courtesy of OTC Division, SPX Corporation)

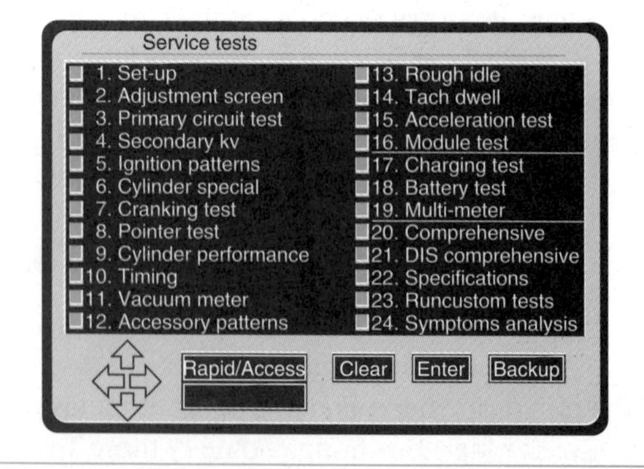

Figure 16-3 The service test screen displays all the engine analyzer tests. (Courtesy of Bear Automotive Service Equipment Company)

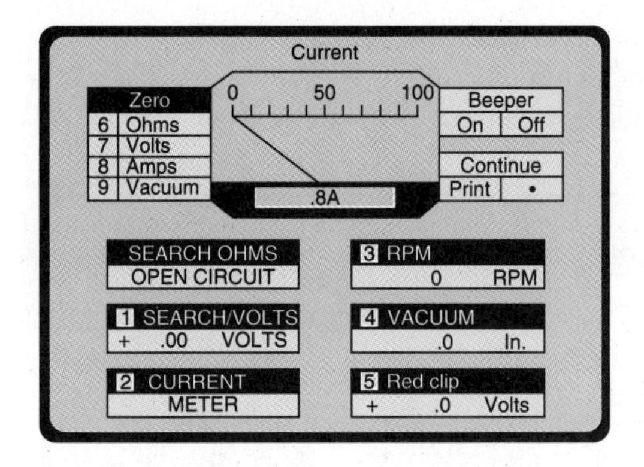

Figure 16-4 Multimeter screen (Courtesy of Bear Automotive Service Equipment Company)

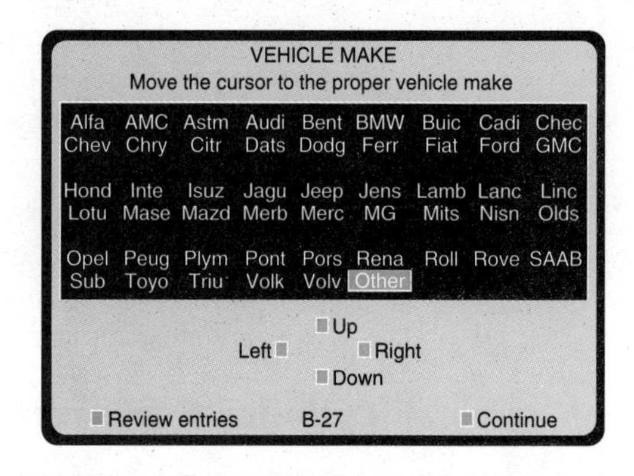

Figure 16-5 Vehicle manufacturer selection (Courtesy of Allen Test Products Division)

may want to use the multimeter to measure the ohms resistance in the coil windings. The technician simply moves the cursor, or highlighter, to multimeter and makes this selection to display the multimeter readings on the screen (Figure 16-4). Now the multimeter leads may be connected to the desired component or circuit to test volts, amperes, or ohms.

If the technician selects COMPREHENSIVE TESTS on the service test screen, the analyzer automatically performs a complete test sequence. When some action is required from the technician, this action is requested on the screen.

When the technician selects DIS COMPREHENSIVE on the service test screen, the analyzer automatically performs a test sequence on the EI system.

If the technician selects RUN CUSTOM TESTS, the analyzer allows the technician to design his or her own test sequence.

The technician must inform the engine analyzer regarding the type of vehicle being tested. When the technician selects SPECIFICATIONS on the service test menu, information is entered regarding the vehicle being tested. Some analyzers will display the specifications for the vehicle being tested and allow the technician to change specifications. In some cases, the latest service bulletins may include a specification change.

When some analyzers are turned on, the technician is requested to select the vehicle manufacturer with the cursor on the screen (Figure 16-5). After the vehicle make selection, the technician is requested to enter more specific vehicle information such as vehicle make, model year, and engine size.

Some analyzers provide a visual inspection screen where the technician uses the cursor to enter PASS or FAIL for various components and systems on the vehicle (Figure 16-6). The results of this visual inspection are printed out in the diagnostic report after the diagnosis is completed.

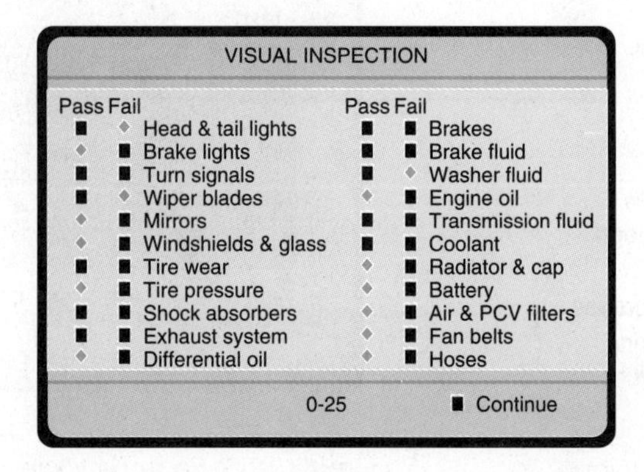

Figure 16-6 Visual inspection screen (Courtesy of Allen Test Products Division)

Engine analyzer test capabilities vary depending on the equipment manufacturer. Always follow the test procedures in the equipment manufacturer's manual.

Many engine analyzers perform diagnostic tests in these main areas:

1. Cylinder performance
2. Ignition performance tests
3. Battery, starting, and charging systems
4. Emission level analysis
5. Engine computer system

Cylinder Performance Test

During the cylinder performance tests, the analyzer momentarily stops the ignition system from firing each cylinder. During this brief time, the rpm drop is recorded. When a cylinder is not contributing to engine power because it has low compression or some other problem, there is very little rpm drop when the spark plug in that cylinder stops firing.

During the cylinder performance test, some analyzers record the actual rpm drop, while others indicate the percentage of rpm drop. Many analyzers also record the amount of hydrocarbon (HC) change in parts per million (ppm) when each cylinder stops firing. If a cylinder is misfiring prior to the cylinder performance test, the cylinder has high HC emissions. Therefore, when this cylinder stops firing during the cylinder performance test, there is not much change in HC emissions. A cylinder with low compression will not have much rpm drop or HC change during the cylinder performance test (Figure 16-7).

Ignition Performance Tests

On many engine analyzers, ignition performance tests include primary circuit tests, secondary kilovolt (kV) tests, acceleration test, scope patterns, and cylinder miss recall. Primary circuit tests include coil input voltage, coil primary resistance, dwell, curb idle speed, and idle vacuum. Secondary kV tests include the average kV, maximum kV, and minimum kV for each cylinder (Figure 16-8). The kV is the voltage required to start firing the spark plug. A high-resistance problem in a spark plug or spark plug wire causes a higher firing kV, whereas a fouled spark plug, or a cylinder with low compression, results in a lower firing kV.

Some secondary kV tests include a snap kV test in which the analyzer directs the technician to accelerate the engine suddenly. When this action is taken, the firing kV should increase evenly on each cylinder. Some engine analyzers also display circuit gap for each cylinder (Figure 16-9). The circuit gap is the voltage to fire all the gaps in the secondary circuit, such as the rotor gap, but the spark plug gap is excluded.

6%	7%	6.5%	3%	6%	6%	6.5%	7%	Low Speed
1825	1725	1750	950	1825	1750	1800	1800	HC Increase
1	5	4	2	6	3	7	8	Firing Order

Figure 16-7 Cylinder performance test results (Courtesy of Allen Test Products Division)

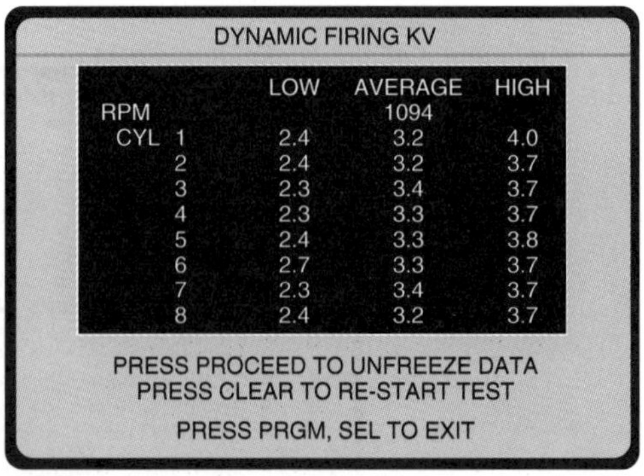

DYNAMIC FIRING KV

	LOW	AVERAGE	HIGH
RPM		1094	
CYL 1	2.4	3.2	4.0
2	2.4	3.2	3.7
3	2.3	3.4	3.7
4	2.3	3.3	3.7
5	2.4	3.3	3.8
6	2.7	3.3	3.7
7	2.3	3.4	3.7
8	2.4	3.2	3.7

PRESS PROCEED TO UNFREEZE DATA
PRESS CLEAR TO RE-START TEST
PRESS PRGM, SEL TO EXIT

Figure 16-8 Secondary kilovolt (kV) readings (Courtesy of Sun Electric Corporation)

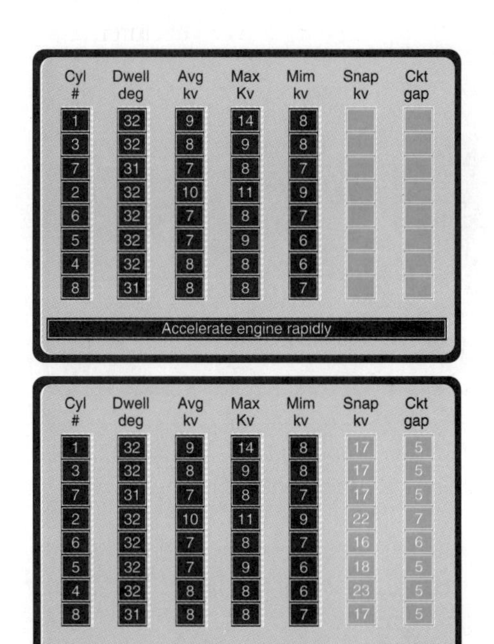

Figure 16-9 Secondary kilovolt (kV) readings, including snap kV and circuit gap values (Courtesy of Bear Automotive Service Equipment Company)

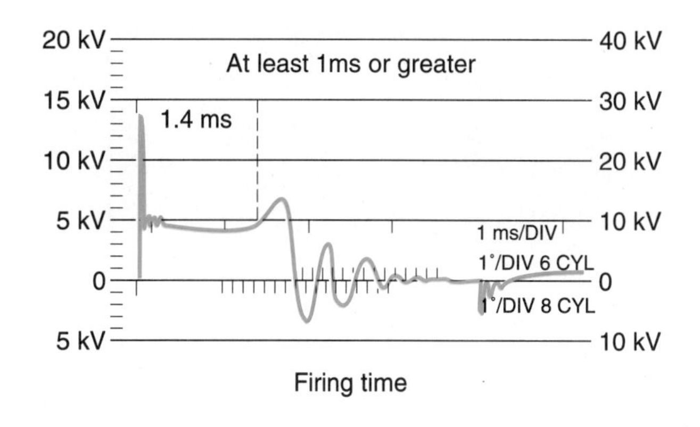

Figure 16-10 Millisecond burn time scale on the scope screen (Courtesy of Allen Test Products Division)

In the SAE J1930 terminology, the term distributorless ignition system (DIS) is replaced by the term electronic ignition (EI).

A cylinder performance test may be called a cylinder balance test or power check.

Burn time refers to the length of time in milliseconds that a spark plug continues firing.

Some analyzers display the burn time for each cylinder with the secondary kV tests. The burn time is the length of the spark line in milliseconds (ms). When using other analyzers, the burn time is read on the scope screen, preferably using a raster pattern. While reading the burn time, the scope screen has 5 divisions of 1 ms each. The average burn time should be 1 ms to 1.5 ms (Figure 16-10).

The secondary kV display from an EI system includes average kV for each cylinder on the compression stroke and average kV for each matching cylinder that fires at the same time on the exhaust stroke. The burn time is also included on the secondary kV display from an EI system (Figure 16-11).

Many engine analyzers also provide primary and secondary scope patterns (Figure 16-12). Some analyzers are capable of freezing the scope patterns into the analyzer memory and recalling

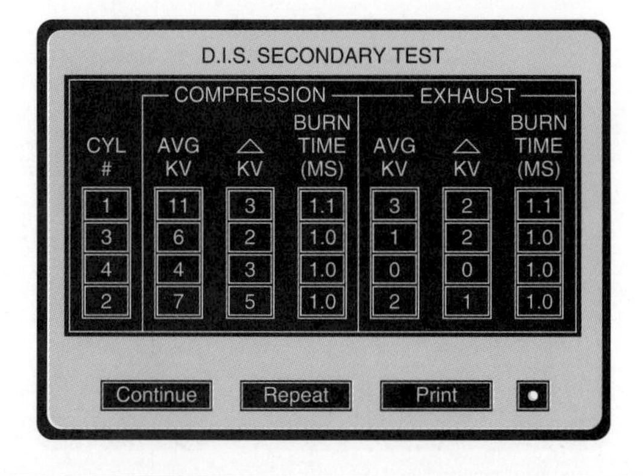

Figure 16-11 Secondary kV display on an EI system (Courtesy of Bear Automotive Service Equipment Company)

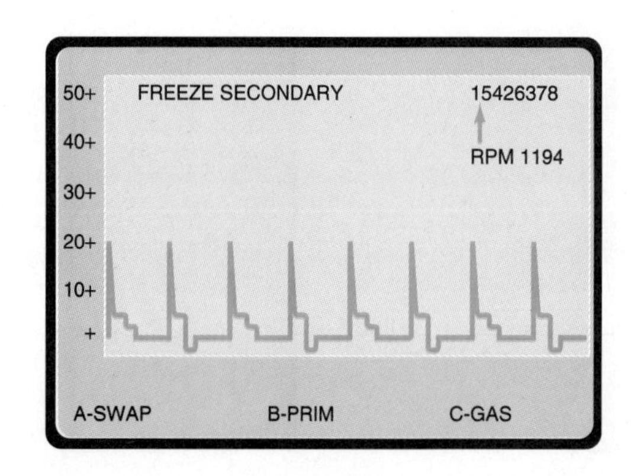

Figure 16-12 Secondary scope pattern display (Courtesy of Sun Electric Corporation)

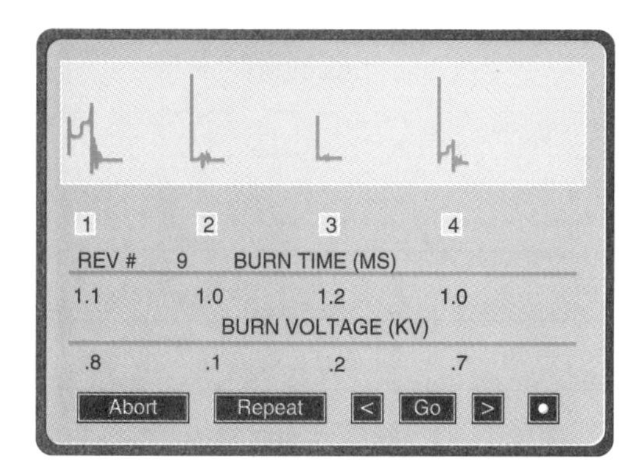

Figure 16-13 Recalling individual cylinder firings to easily identify intermittent cylinder misfires (Courtesy of Bear Automotive Service Equipment Company)

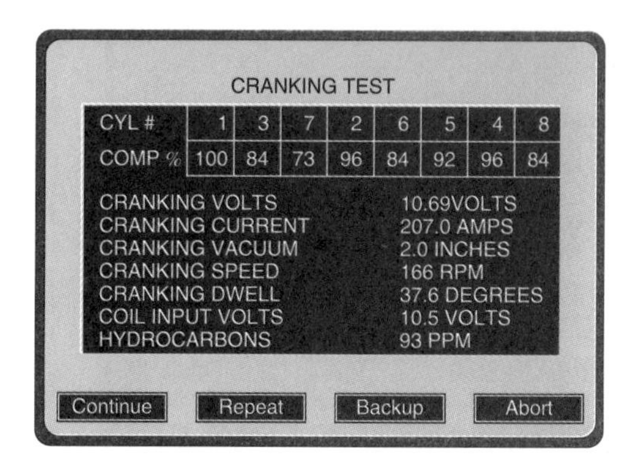

Figure 16-14 Cranking tests (Courtesy of Bear Automotive Service Equipment Company)

this information on request. During the recall procedure, individual cylinder firings are played back to easily identify intermittent cylinder misfirings (Figure 16-13).

Battery, Starting, and Charging System Tests

During a battery test, the engine analyzer places a load on the battery and many analyzers record the battery open circuit voltage, load voltage, and available cold cranking amperes.

While performing the starter, or cranking, test, some analyzers record cranking volts, cranking current, cranking vacuum, cranking speed, cranking dwell, coil input volts, and HC.

On some analyzers, the cranking amperes are displayed for each cylinder. When a cylinder has low compression, the starter draw in amperes will be reduced for that cylinder. Other analyzers assign a value of 100% to the cylinder with the highest cranking amperes, and then provide a percentage reading for each of the other cylinders in relation to the cylinder with the 100% reading (Figure 16-14).

Charging circuit tests include regulator voltage, alternator current, and alternator waveforms (Figure 16-15).

Emission Level Analysis

The emission level analysis involves diagnosis using the four-gas infrared analyzer as described previously in this chapter. The four-gas analysis screen on some analyzers includes the HC, CO, O_2, CO_2, engine temperature, exhaust temperature, vacuum, and engine rpm (Figure 16-16).

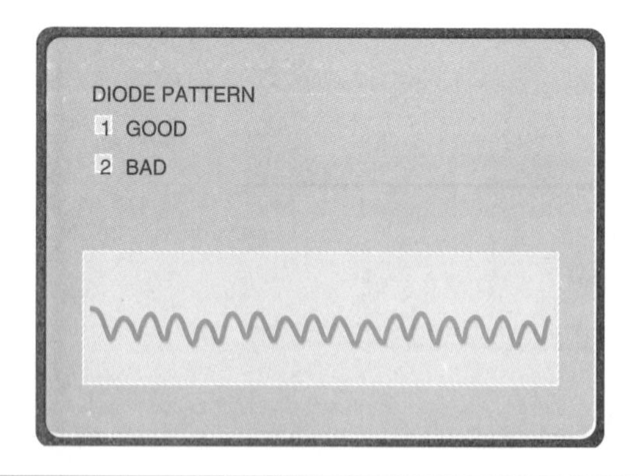

Figure 16-15 Alternator waveform (Courtesy of Bear Automotive Service Equipment Company)

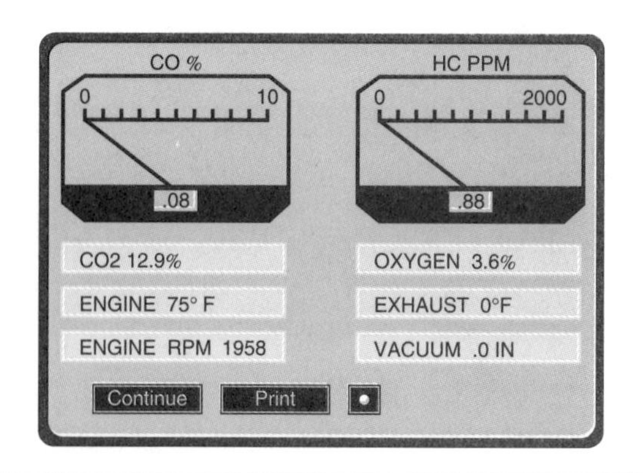

Figure 16-16 Four-gas infrared analyzer display (Courtesy of Bear Automotive Service Equipment Company)

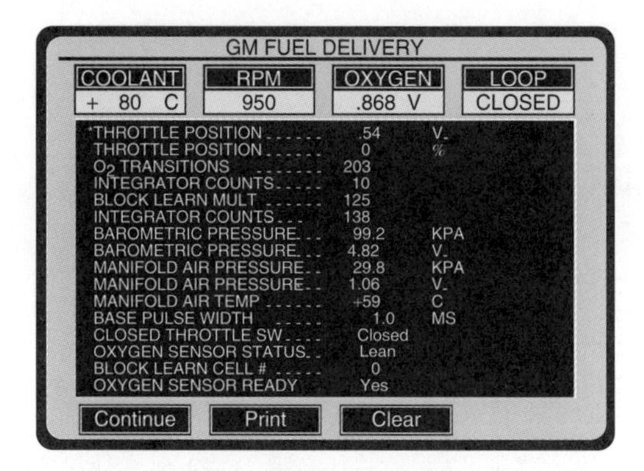

Figure 16-17 Engine analyzer screen for General Motors engine computer system tests (Courtesy of Bear Automotive Service Equipment Company)

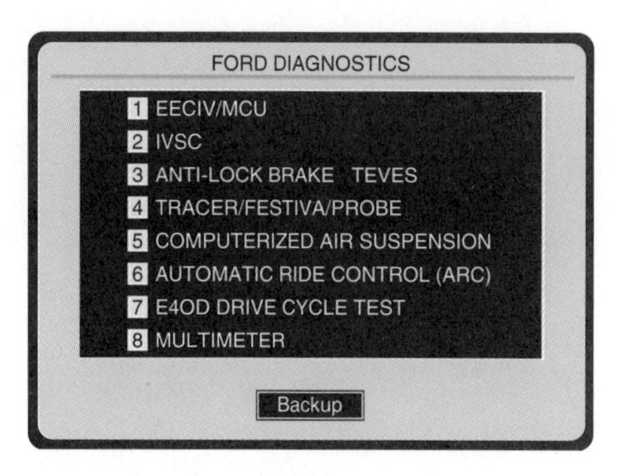

Figure 16-18 Engine analyzer screen for Ford engine computer system tests (Courtesy of Bear Automotive Service Equipment Company)

Engine Computer System Diagnosis

Many engine analyzers will diagnose engine computer systems in much the same way as a scan tester. The analyzer performs tests on General Motors (Figure 16-17), Ford (Figure 16-18), and Chrysler (Figure 16-19) engine computer systems.

Classroom Manual
Chapter 16, page 371

Scope Pattern Diagnosis

Maximum Secondary Coil Voltage

WARNING: Removing a spark plug wire with the engine running on an electronic ignition (EI) system may cause a leakage defect in the secondary insulation in the coil or spark plug wires.

SERVICE TIP: An engine analyzer checks maximum secondary coil voltage automatically without removing a spark plug wire.

Manufacturers of catalytic converter equipped vehicles usually do not recommend removing a spark plug wire with the engine running. If this action is recommended, it will be for a very short time period. Always follow the instructions in the vehicle manufacturer's service manual.

Some manufacturers recommend removing a spark plug wire with a pair of insulated pliers and

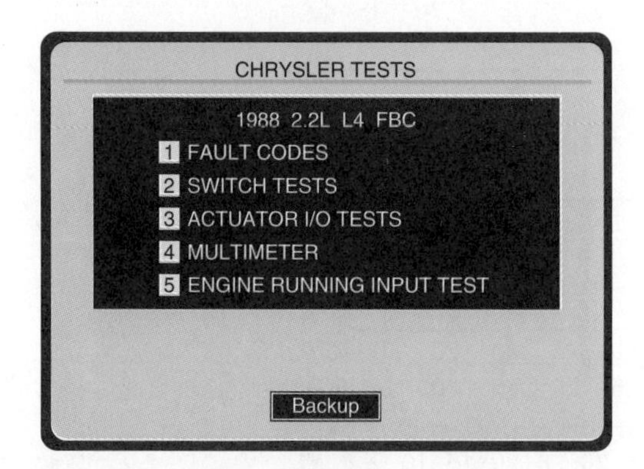

Figure 16-19 Engine analyzer screen for Chrysler engine computer system tests (Courtesy of Bear Automotive Service Equipment Company)

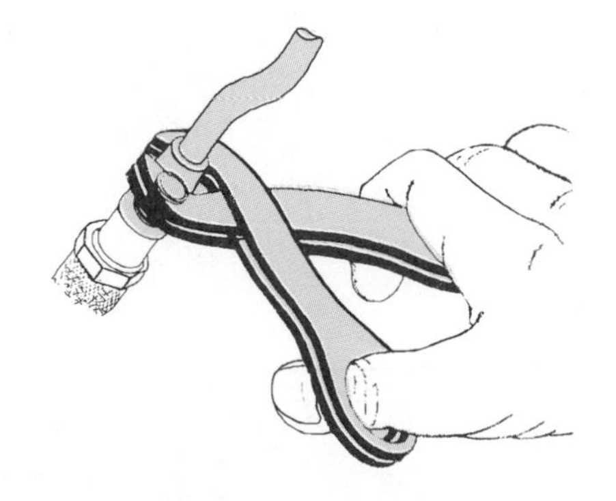

Figure 16-20 Removing a spark plug wire with a pair of insulated pliers (Courtesy of Sun Electric Corporation)

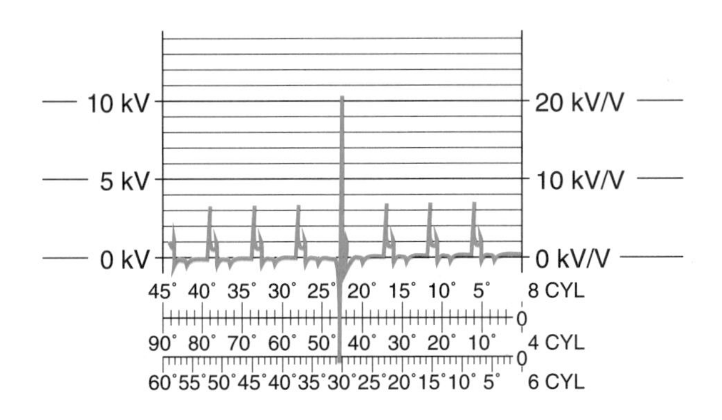

Figure 16-21 Maximum secondary coil voltage (Courtesy of Sun Electric Corporation)

observing the maximum coil voltage on the screen (Figure 16-20). When this action is taken, all the energy in the coil should be produced in the form of voltage, and the maximum voltage displayed on the scope screen should exceed the maximum specified secondary coil voltage (Figure 16-21).

Lower-than-specified maximum secondary coil voltage may be caused by these defects:

1. Defective coil
2. Secondary insulation leakage
3. Low primary input voltage
4. High primary resistance
5. Improper dwell

Secondary Insulation Leakage

The secondary insulation leakage test may be used on some DI systems. Since vehicle manufacturers do not recommend removing spark plug wires with the engine running on EI systems, this test is not recommended on these systems.

When a spark plug wire is removed and the maximum secondary coil voltage appears on the screen, the downward sweep below the zero line on the screen should equal one-half the height of the maximum voltage line. When the maximum secondary coil voltage is low and the downward sweep is reduced, there is insulation leakage in the secondary circuit (Figure 16-22).

Secondary insulation leakage may be caused by leakage in the coil secondary insulation, secondary coil wire, distributor cap and rotor, or spark plug wires.

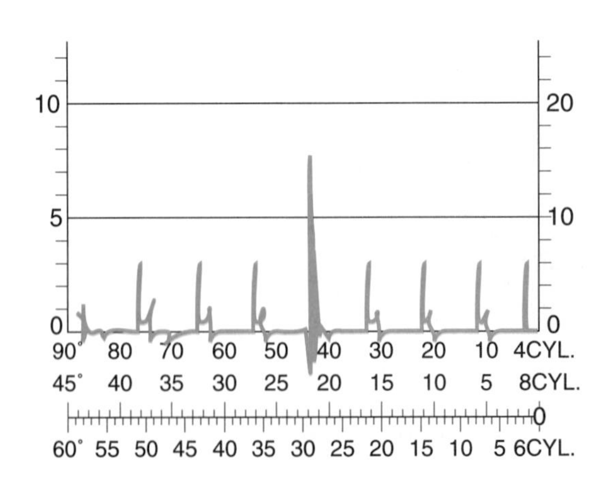

Figure 16-22 Insulation leakage in the secondary circuit reduces downward sweep while checking maximum secondary coil voltage. (Courtesy of Sun Electric Corporation)

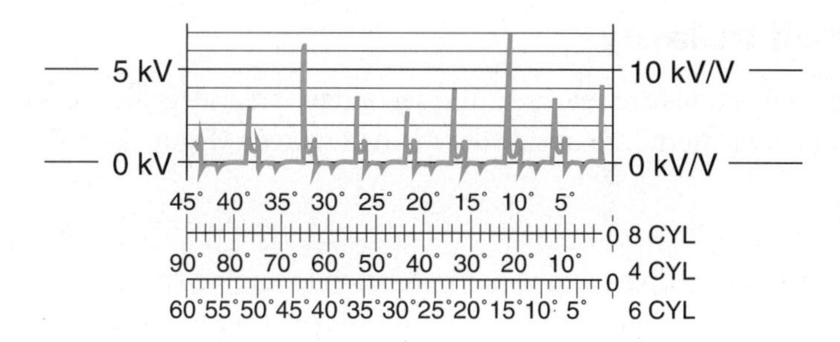

Figure 16-23 High firing lines caused by high secondary resistance (Courtesy of Sun Electric Corporation)

Secondary Resistance

SERVICE TIP: High resistance in the secondary coil wire in a DI system causes reduced oscillations after the spark line on a secondary scope pattern.

High resistance in the secondary ignition circuit results in high firing lines (Figure 16-23) and high spark lines (Figure 16-24). This high resistance may be caused by excessive resistance in the spark plug wires, spark plugs, or distributor cap and rotor.

Low resistance in the secondary ignition circuit causes low firing lines and long, low spark lines (Figure 16-25). This type of scope pattern may be caused by a fouled spark plug, low cylinder compression, or a grounded spark plug wire.

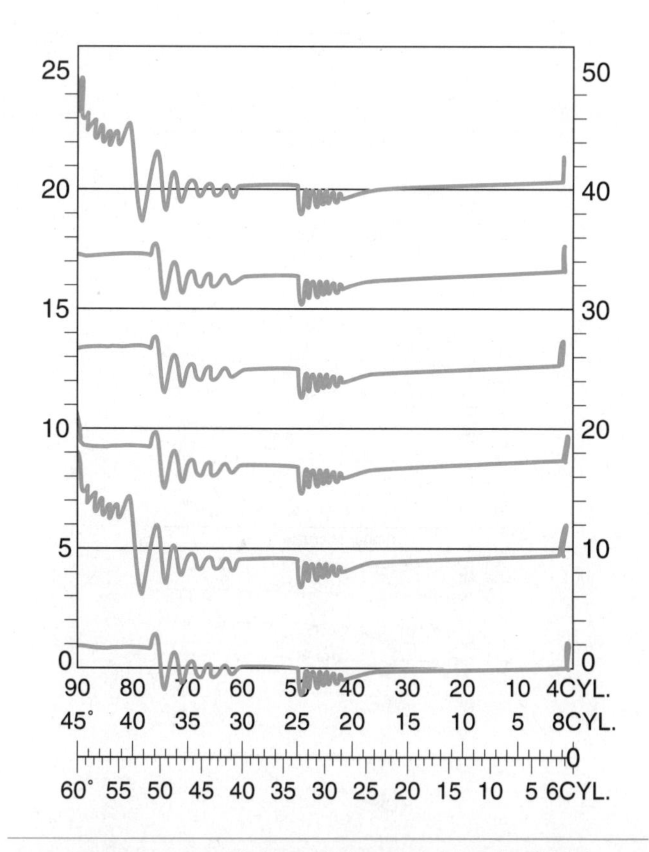

Figure 16-24 High spark lines resulting from high secondary resistance (Courtesy of Sun Electric Corporation)

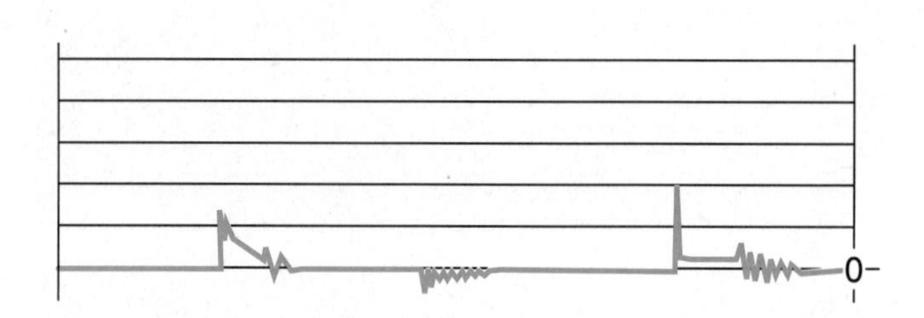

Figure 16-25 Low firing line and long, low spark line caused by low secondary resistance (Courtesy of Sun Electric Corporation)

Secondary Coil Polarity

If the primary coil connections are reversed, the secondary polarity is also reversed. Under this condition, the secondary patterns are upside down on the screen (Figure 16-26).

Dwell

Some ignition systems have a dwell that varies in relation to engine speed. In these systems, the dwell may be 15° at idle speed and 30° to 35° at 2,500 rpm (Figure 16-27). The dwell may be checked using the primary or secondary pattern. Since dwell is a function of the module on DI and EI systems, there is rarely a problem in this area. On point-type ignition systems, the dwell is adjusted by opening or closing the point gap. Excessive variation in the dwell with the engine running at low speed may indicate worn distributor components on a point-type ignition system.

Primary Coil Voltage

Classroom Manual
Chapter 16, page 372

If the primary coil voltage on the scope screen is higher than normal, there may be low primary resistance or excessive charging system voltage. When the primary coil voltage on the scope screen is lower than normal, there may be excessive resistance in the primary circuit or low primary input voltage.

Photo Sequence 13 shows a typical procedure for diagnosing the engine, ignition, electrical, and fuel systems with an engine analyzer.

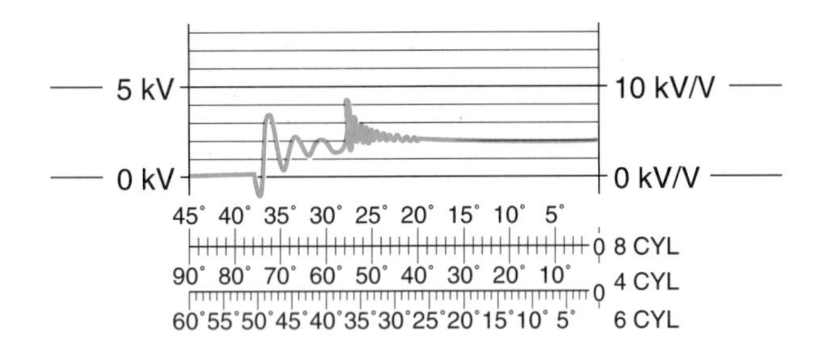

Figure 16-26 Reversed primary coil connections cause reversed secondary polarity and upside-down secondary scope patterns. (Courtesy of Sun Electric Corporation)

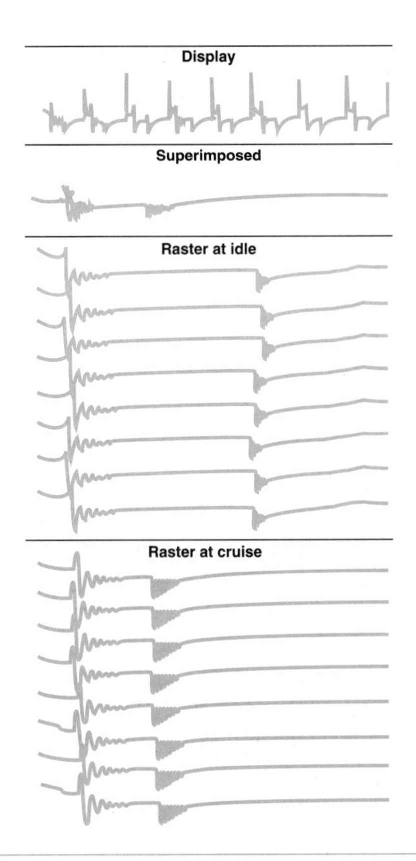

Figure 16-27 On some ignition systems, the dwell increases in relation to engine speed. (Courtesy of Sun Electric Corporation)

Photo Sequence 13

Typical Procedure for Diagnosing Engine, Ignition, Electrical, and Fuel Systems With an Engine Analyzer

P13-1 Connect the analyzer leads and hoses to the vehicle.

P13-2 With the engine at normal operating temperature, enter the necessary information in the analyzer regarding the vehicle being tested.

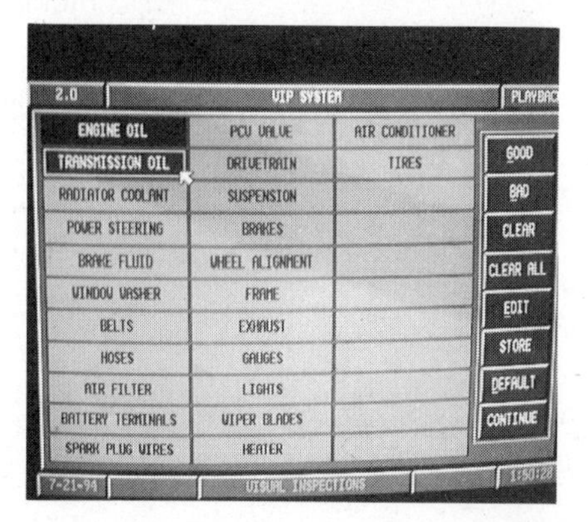

P13-3 Perform a visual inspection and enter the results on the analyzer screen.

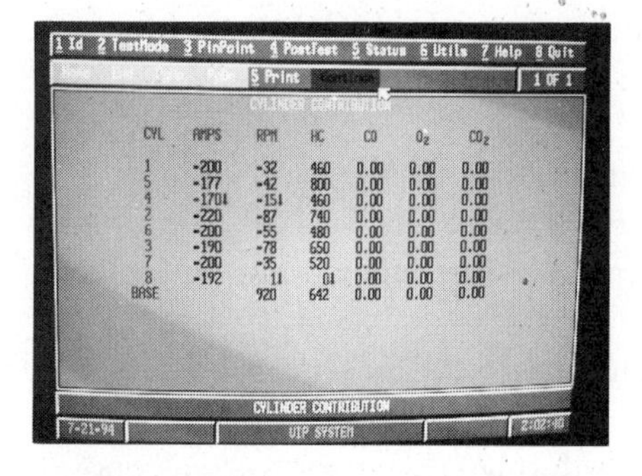

P13-4 Perform a cylinder performance test and observe the results on the analyzer screen.

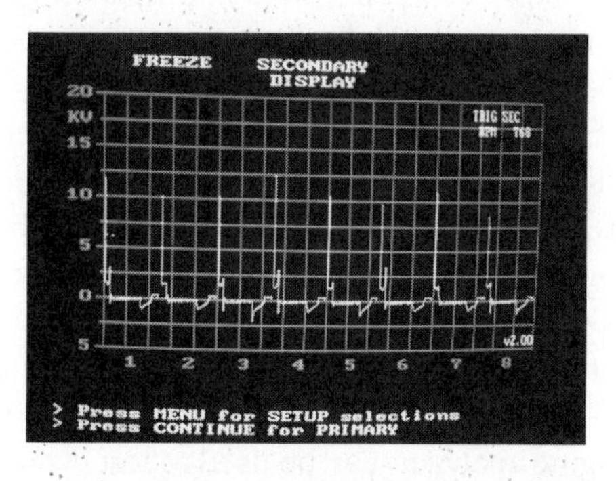

P13-5 Perform secondary kV tests and observe the results on the screen.

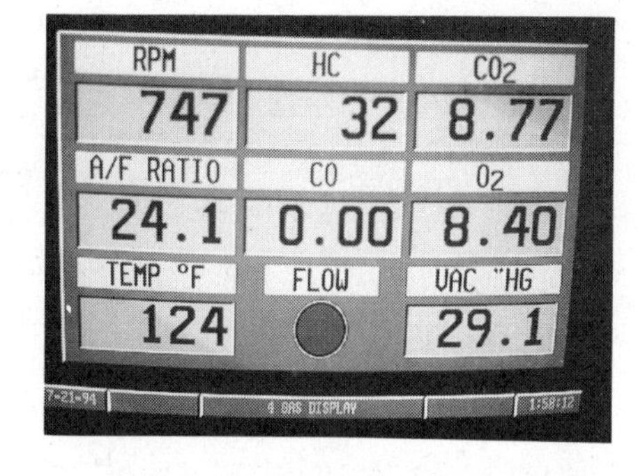

P13-6 Perform an emission level analysis and observe the results on the screen.

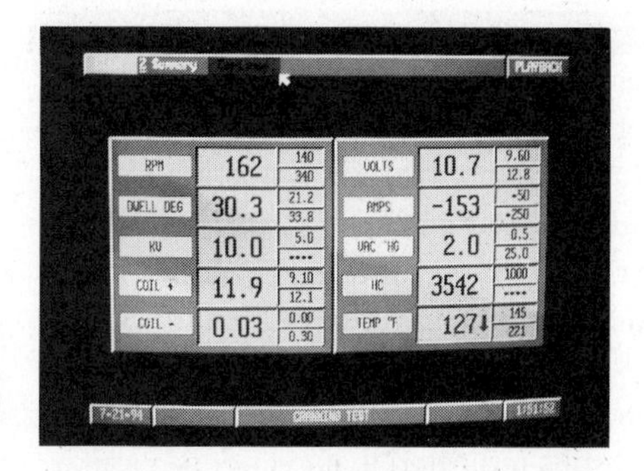

P13-7 Perform cranking tests and observe the results on the screen.

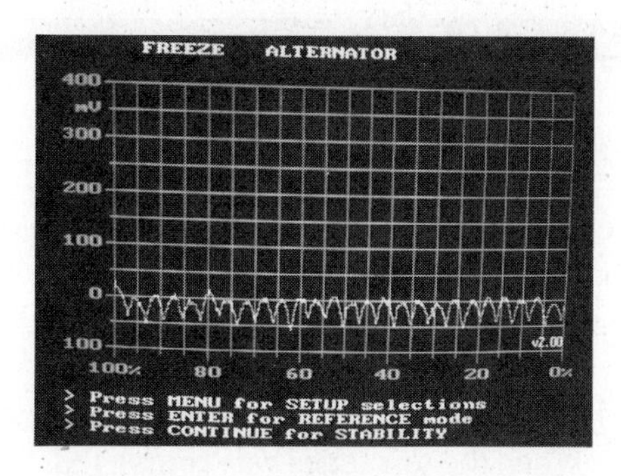

P13-8 Display the alternator waveform on the screen and check the alternator condition from the waveform display.

P13-9 Shut off the engine and disconnect the analyzer leads and hoses.

CUSTOMER CARE: Always be willing to spend a few minutes explaining problems, including safety concerns, regarding the customer's vehicle. When customers understand why certain repairs are necessary, they feel better about spending the money. For example, if you explain that a leaking fuel vapor line on an evaporative system may result in an explosion, the customer appreciates this warning and is willing to spend the money for the necessary repairs.

Guidelines for Engine Diagnosis With Infrared Analyzer, Engine Analyzer, and Scope Patterns

1. Many infrared analyzers have a warm-up period and an automatic calibration process.
2. High HC emissions may be caused by cylinder misfiring because of an ignition defect, lean or rich air-fuel ratio, or low compression.
3. High CO emissions may result from a rich air-fuel ratio.
4. Excessive HC and CO emissions may be caused by a plugged PCV system, heat riser valve stuck open, AIR pump inoperative, or engine oil diluted with gasoline.
5. Lower-than-normal CO_2 levels may be caused by exhaust system leaks or a rich air-fuel ratio.
6. Lower-than-normal O_2 readings and higher-than-normal CO readings indicate a rich air-fuel ratio, restricted PCV system, or canister purging at idle and low speeds.
7. Higher-than-normal O_2 readings and lower-than-normal CO readings indicate a lean air-fuel ratio or an AIR system in operation during the infrared analyzer tests.
8. Some four-gas infrared analyzers meet the requirements for use in state inspection, maintenance (I/M) programs.
9. An infrared analyzer may be used to test catalytic converter efficiency, secondary air injection system operation, PCV system operation, cylinder misfiring and vacuum leaks, EVAP system leaks, and combustion chamber leaks.
10. During a comprehensive test procedure, an engine analyzer automatically performs a series of tests.
11. Custom tests in an engine analyzer allow the technician to design a specific test procedure.
12. An engine analyzer may be used to test cylinder performance, ignition system condition, battery starting and charging system performance, fuel and emission system performance, and engine computer system condition.
13. Scope patterns may be used to test the secondary maximum coil voltage, leakage, resistance, coil polarity, and dwell.
14. Scope patterns may be used to test the primary voltage, dwell, and dwell variation.

CASE STUDY

A customer brought a Dodge truck into the shop with a cylinder-misfiring complaint. When the technician listened to the engine, he concluded that a road test was not necessary since the engine was misfiring especially on acceleration. The technician visually checked the ignition wiring, including spark plug wires, without finding any problems. A check of the spark plug wire installation indicated that the spark plug wires were installed properly in relation to the engine firing order.

The technician connected the engine analyzer to the engine and ran a comprehensive test series. When the test series was completed, the results indicated that cylinders 3 and 6 had low output. There was very little rpm drop on these two cylinders during the cylinder performance test. The secondary kV tests indicated these two cylinders had low normal kV and low maximum kV.

Since 3 and 6 follow each other in the firing order, the technician thought the most likely cause of this problem was a leakage problem in the secondary ignition circuit. The technician had already visually inspected the spark plug wires. When the distributor cap was removed to inspect the cap and rotor, the technician found the cap had a crack between the terminals connected to the #3 and #6 spark plug wires. The rotor and other distributor components appeared to be in normal condition.

After replacing the distributor cap, the technician repeated the cylinder performance and secondary kV tests. All cylinders performed normally in the tests, and the misfiring condition was eliminated.

Terms to Know

Hydrocarbons (HC)
Carbon monoxide (CO)
Carbon dioxide (CO_2)
Oxygen (O_2)
Inspection, maintenance (I/M) testing
Comprehensive tests
Custom tests
Parts per million (ppm)
Burn time
Kilovolts (kV)

ASE Style Review Questions

1. While discussing the use of an infrared analyzer:
Technician A says an infrared analyzer should not be used on a diesel engine.
Technician B says leaks in the exhaust system affect the infrared analyzer readings.
Who is correct?
A. A only **C.** Both A and B
B. B only **D.** Neither A nor B

2. While discussing infrared analyzer readings:
Technician A says high HC readings may be caused by ignition system misfiring.
Technician B says high CO readings may be caused by a lean air-fuel ratio.
Who is correct?
A. A only **C.** Both A and B
B. B only **D.** Neither A nor B

3. While discussing infrared analyzer readings:
Technician A says low CO_2 readings may be caused by a rich air-fuel ratio.
Technician B says low CO_2 readings may be caused by vacuum leak.
Who is correct?
A. A only **C.** Both A and B
B. B only **D.** Neither A nor B

4. While discussing infrared analyzer readings:
Technician A says high O_2 readings indicate a rich air-fuel ratio.
Technician B says high O_2 readings indicate a lean air-fuel ratio.
Who is correct?
A. A only **C.** Both A and B
B. B only **D.** Neither A nor B

5. While discussing secondary air injection (AIR) system operation and infrared analyzer testing:
Technician A says the O_2 emissions should be 2% to 5% higher with the AIR pump disconnected compared to these emissions with the AIR pump operational.
Technician B says the O_2 emissions should be 1% higher with the AIR pump disconnected compared to these emissions with the AIR pump operational.
Who is correct?
A. A only **C.** Both A and B
B. B only **D.** Neither A nor B

6. While discussing the use of the infrared analyzer:
Technician A says the infrared analyzer may be used to check for combustion chamber leaks.
Technician B says the infrared analyzer may be used to check for leaks in the EVAP system.
Who is correct?
A. A only **C.** Both A and B
B. B only **D.** Neither A nor B

7. While discussing cylinder performance tests with an engine analyzer:
Technician A says a cylinder with low compression causes a greater rpm drop during the cylinder performance test than a cylinder with normal compression.
Technician B says a cylinder with low compression causes a greater increase in HC emissions during the cylinder performance test than a cylinder with normal compression.
Who is correct?
A. A only **C.** Both A and B
B. B only **D.** Neither A nor B

8. While discussing ignition secondary kilovolt (kV) readings:
Technician A says a fouled spark plug causes a lower-than-normal spark plug firing kV.
Technician B says a cylinder with low compression causes a lower-than-normal spark plug firing kV.
Who is correct?
A. A only **C.** Both A and B
B. B only **D.** Neither A nor B

9. While discussing scope patterns:
Technician A says low maximum secondary coil voltage may be caused by low primary voltage input.
Technician B says a leakage problem in the secondary ignition circuit causes low maximum secondary coil voltage.
Who is correct?
A. A only **C.** Both A and B
B. B only **D.** Neither A nor B

10. While discussing scope patterns:
Technician A says a grounded spark plug wire causes a low spark plug firing kV and a low spark line.
Technician B says excessive resistance in a spark plug wire causes low spark plug firing kV.
Who is correct?
A. A only **C.** Both A and B
B. B only **D.** Neither A nor B

Table 16-1 ASE TASK

Inspect and adjust exhaust gas analyzer, inspect vehicle's exhaust system; obtain exhaust gas readings.

Problem Area	Symptoms	Possible Causes	Classroom Manual	Shop Manual
EMISSION LEVELS	High HC emissions	Cylinder misfiring, rich air-fuel ratio, ignition misfiring	368	352
	High CO emissions	Rich air-fuel ratio, defective fuel system or emission systems	369	352
	High O_2 emissions	Lean air-fuel ratio, defective fuel system, or emission systems	370	352
	Low CO_2 emissions	Rich air-fuel ratio, defective fuel system or emission systems	370	352

Table 16-2 ASE TASK

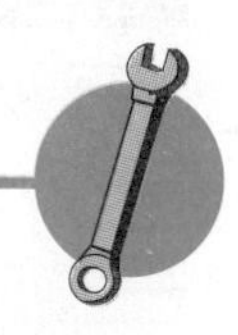

Diagnose engine mechanical, ignition, fuel, and emission control problems with an exhaust gas analyzer; determine needed repairs.

Problem Area	Symptoms	Possible Causes	Classroom Manual	Shop Manual
ENGINE PERFORMANCE	Cylinder misfiring	Cylinder misfiring, rich air-fuel ratio, ignition misfiring	367	352
FUEL CONSUMPTION	Reduced fuel economy	Rich air-fuel ratio, defective fuel system or emission systems	368	352
ENGINE PERFORMANCE	Rough idle, acceleration hesitation	Lean air-fuel ratio, defective fuel system or emission systems	370	352
FUEL CONSUMPTION	Reduced fuel economy	Rich air-fuel ratio, defective fuel system or emission systems	368	352

Table 16-3 ASE Task

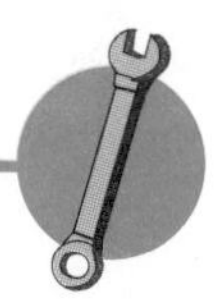

Diagnose engine mechanical, electrical, and fuel problems with an oscilloscope and/or engine analyzer; determine needed repairs.

Problem Area	Symptoms	Possible Causes	Classroom Manual	Shop Manual
ENGINE PERFORMANCE	Cylinder misfiring	1. Defective ignition components	373	354
		2. Low compression, defective rings, valves	371	356
		3. Vacuum leak	371	358
	Rough idle	1. Vacuum leak, lean air-fuel ratio	371	358
		2. Defective emission systems	371	358
	Misfiring on acceleration	Defective ignition components	373	359
	Failure to start, slow cranking	Defective starter or related circuit	371	358
	Overcharged, or undercharged battery	Defective charging system	371	359

Turbocharger and Supercharger Diagnosis and Service

Upon completion and review of this chapter, you should be able to:

- ❑ Perform a turbocharger inspection.
- ❑ Diagnose intake and exhaust leaks that affect turbocharger operation.
- ❑ Test turbocharger boost pressure.
- ❑ Remove and replace turbochargers.
- ❑ Measure axial shaft movement in a turbocharger.
- ❑ Measure wastegate stroke.
- ❑ Inspect and replace turbocharger components.
- ❑ Prelubricate turbocharger bearings.

Turbocharger Diagnosis

Basic Tools

Basic technician's tool set

Service manual

Straightedge

Basic Turbocharger Inspection and Diagnosis

CAUTION: If the engine has been running, turbochargers and related components are extremely hot. Use caution and wear protective gloves to avoid burns when servicing these components.

The first step in turbocharger diagnosis is to check all linkages and hoses connected to the turbocharger. Inspect the wastegate diaphragm linkage for looseness and binding, and check the hose from the wastegate diaphragm to the intake manifold for cracks, kinks, and restrictions. Check the coolant hoses and oil line connected to the turbocharger for leaks.

Excessive blue smoke in the exhaust may indicated worn turbocharger seals. The technician must remember that worn valve guide seals or piston rings also cause oil consumption and blue smoke in the exhaust.

Check all turbocharger mounting bolts for looseness. A rattling noise may be caused by loose turbocharger mounting bolts. Some whirring noise is normal when the turbocharger shaft is spinning at high speed. Excessive internal turbocharger noise may be caused by too much end play on the shaft, which allows the blades to strike the housings.

Check for exhaust leaks in the turbine housing and related pipe connections. If exhaust gas is escaping before it reaches the turbine wheel, turbocharger effectiveness is reduced. Check for intake system leaks. If there is a leak in the intake system before the compressor housing, dirt may enter the turbocharger and damage the compressor or turbine wheel blades. When a leak is present in the intake system between the compressor wheel housing and the cylinders, turbocharger pressure is reduced.

CAUTION: When using a propane cylinder to check for intake system leaks, maintain the cylinder in an upright position and keep the cylinder away from rotating components, hot components, or sources of ignition. Failure to observe this precaution may result in personal injury and/or property damage.

CAUTION: When using a propane cylinder to check for intake system leaks, do not smoke and do not place the end of the hose near any source of ignition. Failure to observe this precaution may result in an explosion, causing personal injury and/or property damage.

A propane cylinder with a metering valve and hose may be used to check the intake system for leaks. With the engine idling, open the metering valve a small amount and position the end of the hose near the locations of any suspected leaks in the intake manifold. When a leak is present, the engine speed increases (Figure 17-1). Close the metering valve when the test is completed.

When turbochargers have computer-controlled boost pressure, a diagnostic trouble code (DTC) is stored in the PCM memory if a fault is present in the boost control solenoid or solenoid-to-PCM wiring.

Special Tools

Propane cylinder with metering valve and hose

Testing Boost Pressure

SERVICE TIP: If the engine has low cylinder compression, there is reduced air flow through the cylinders, which results in lower turbocharger shaft speed and boost pressure.

SERVICE TIP: Excessive boost pressure causes engine detonation and possible engine damage.

Connect a pressure gauge to the intake manifold to check the boost pressure. The pressure gauge hose should be long enough so the gauge may be positioned in the passenger compartment. One of the front windows may be left down enough to allow the gauge hose to extend into the passenger compartment. Road test the vehicle at the speed specified by the vehicle manufacturer and observe the boost pressure. Some vehicle manufacturers recommend accelerating from a stop to 60 mph (96 kph) at wide-open throttle while observing the boost pressure.

Higher-than-specified boost pressure may be caused by a defective wastegate system. Low boost pressure may be caused by the wastegate system or turbocharger defects such as damaged wheel blades or worn bearings. An engine with low cylinder compression will usually have low boost pressure.

Special Tools

Pressure gauge

Boost pressure is the amount of pressure supplied from the turbocharger to the intake manifold. This pressure is measured in pounds per square inch (psi) or kilopascals (kPa).

Turbocharger Service

Turbocharger Removal

The turbocharger removal procedure varies depending on the engine. For example, on some cars, such as a Nissan 300 ZX, the manufacturer recommends the engine be removed to gain access to the turbocharger. On other applications, the turbocharger may be removed with the engine in the

Figure 17-1 Checking for intake system leaks with a propane cylinder and metering valve (Courtesy of Chrysler Corporation)

vehicle. Always follow the turbocharger removal procedure in the vehicle manufacturer's service manual. Following is a typical turbocharger removal procedure:

1. Disconnect the negative battery cable and drain the cooling system.
2. Disconnect the exhaust pipe from the turbocharger.
3. Remove the support bracket between the turbocharger and the engine block.
4. Remove the bolts from the oil drain back housing on the turbocharger.
5. Disconnect the turbocharger coolant inlet tube nut at the block outlet and remove the tube-support bracket (Figure 17-2).
6. Remove the air cleaner element, air cleaner box, bracket, and related components.
7. Disconnect the accelerator linkage, throttle body electrical connector, and vacuum hoses.
8. Loosen the throttle body-to-turbocharger inlet hose clamps, and remove the three throttle body-to-intake manifold attaching screws. Remove the throttle body.
9. Loosen the lower turbocharger discharge hose clamp on the compressor wheel housing.
10. Remove the fuel rail-to-intake manifold screws and the fuel line bracket screw. Remove the two fuel rail bracket-to-heat shield retaining clips and pull the fuel rail and injectors upward out of the way. Tie the fuel rail in this position with a piece of wire.
11. Disconnect the oil supply line from the turbocharger housing.
12. Remove the intake manifold heat shield.

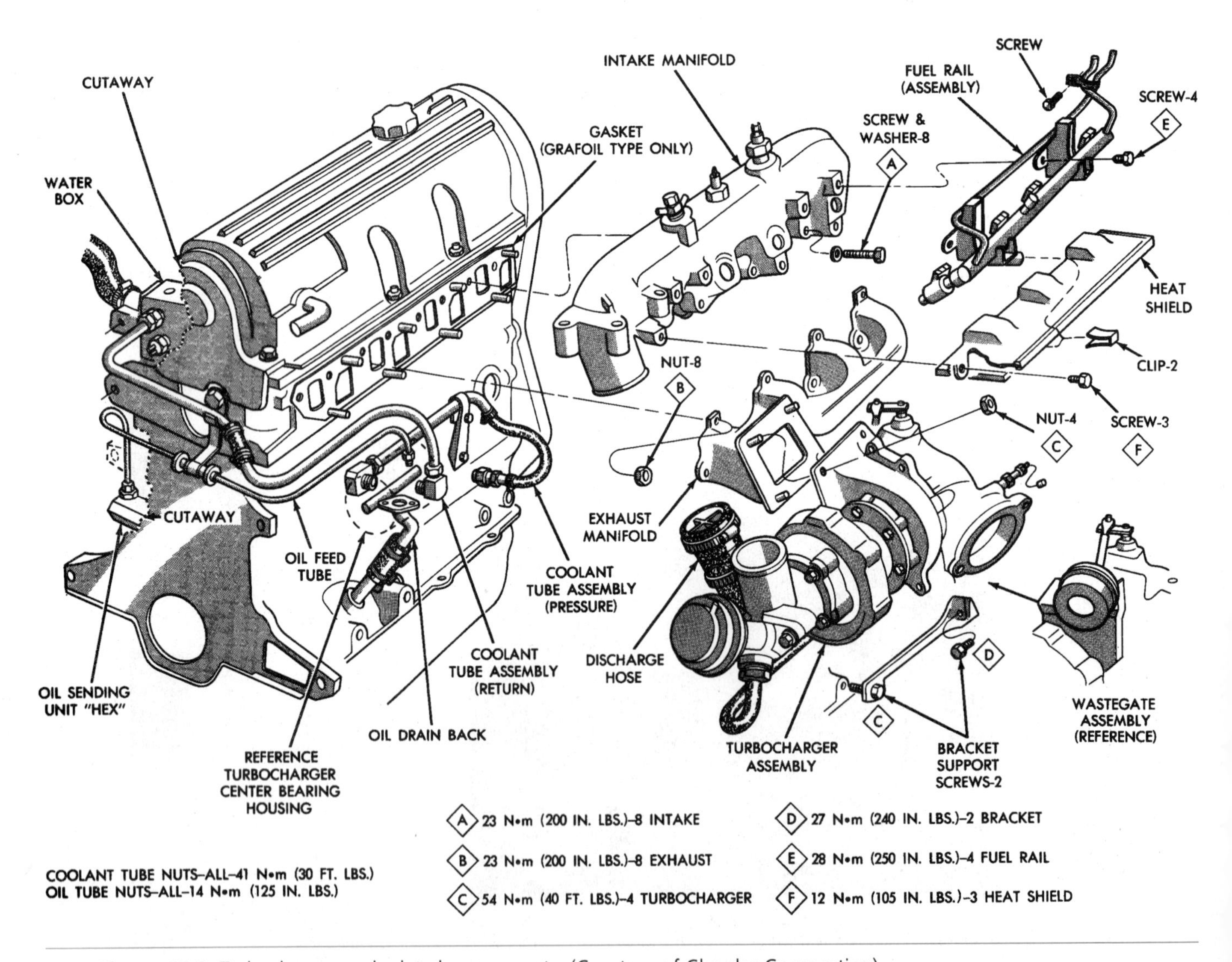

Figure 17-2 Turbocharger and related components (Courtesy of Chrysler Corporation)

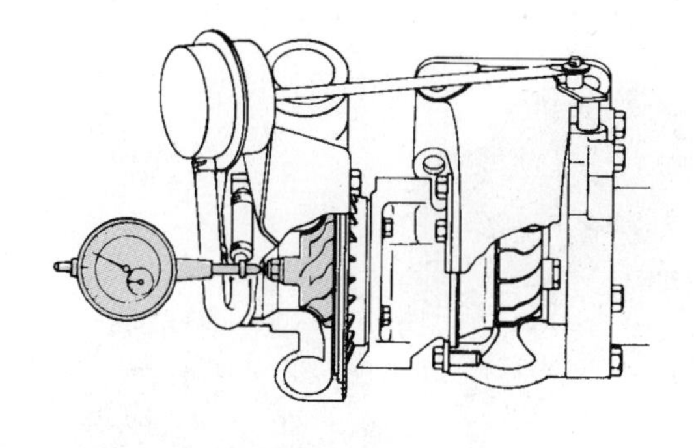

Figure 17-3 Measuring turbocharger shaft axial movement (Courtesy of Nissan Motor Co., Ltd.)

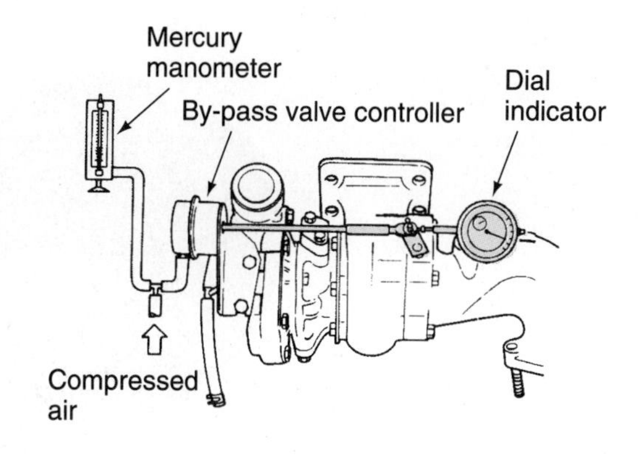

Figure 17-4 Measuring wastegate rod movement (Courtesy of Nissan Motor Co., Ltd.)

13. Disconnect the coolant return line from the turbocharger and the water box. Remove the line-support bracket from the cylinder head and remove the line.
14. Remove the four nuts retaining the turbocharger to the exhaust manifold, and remove the turbocharger from the manifold studs. Move the turbocharger downward towards the passenger side of the vehicle, and then lift the unit up and out of the engine compartment.

Measuring Turbocharger Shaft Axial Movement

Special Tools

Dial indicator

After the turbocharger is removed from the engine, remove the turbine outlet elbow from the turbine housing. Position a dial indicator against the shaft, and move the shaft inward and outward while observing the dial indicator reading (Figure 17-3). The axial shaft movement should not exceed the vehicle manufacturer's specifications. On some turbochargers, the maximum axial shaft movement is 0.003 in (0.91 mm). If the axial shaft movement exceeds the manufacturer's specifications, the turbocharger must be repaired or replaced. Some manufacturers recommend complete turbocharger replacement, whereas other manufacturers recommend replacing the center housing assembly as a unit. Some turbocharger manufacturers recommend replacement of individual components. Always follow the service procedures in the vehicle manufacturer's service manual.

Measuring Wastegate Stroke

Special Tools

Hand pressure pump

If the wastegate stroke is reduced, wastegate valve opening is decreased and boost pressure is increased. Connect a hand pressure pump and a pressure gauge to the wastegate diaphragm. Position a dial indicator against the outer end of the wastegate diaphragm rod (Figure 17-4). Supply the specified pressure to the wastegate diaphragm and observe the dial indicator movement. If the wastegate rod movement is less than specified, disconnect the rod from the wastegate valve linkage, and check the linkage for binding. If this linkage moves freely, replace the wastegate diaphragm. When the wastegate valve linkage is binding, turbocharger repair or replacement is required.

Turbocharger Component Inspection

If the vehicle manufacturer recommends turbocharger disassembly, inspect the wheels and shaft after the end housings are removed. Lack of lubricant or lubrication with contaminated oil results in bearing failure, which leads to wheel rub on the end housings. A contaminated cooling system may provide reduced turbocharger bearing cooling and premature bearing failure. Bearing failure will likely lead to seal damage. Inspect the shaft and bearings for a burned condition (Figure 17-5). If the shaft and bearings are burned, replace the complete center housing assembly or individual parts as recommended by the manufacturer.

If the shaft and bearings are in satisfactory condition, but the blades are damaged, check the air intake system for leaks or a faulty air cleaner element. When the blades or shaft and bearings

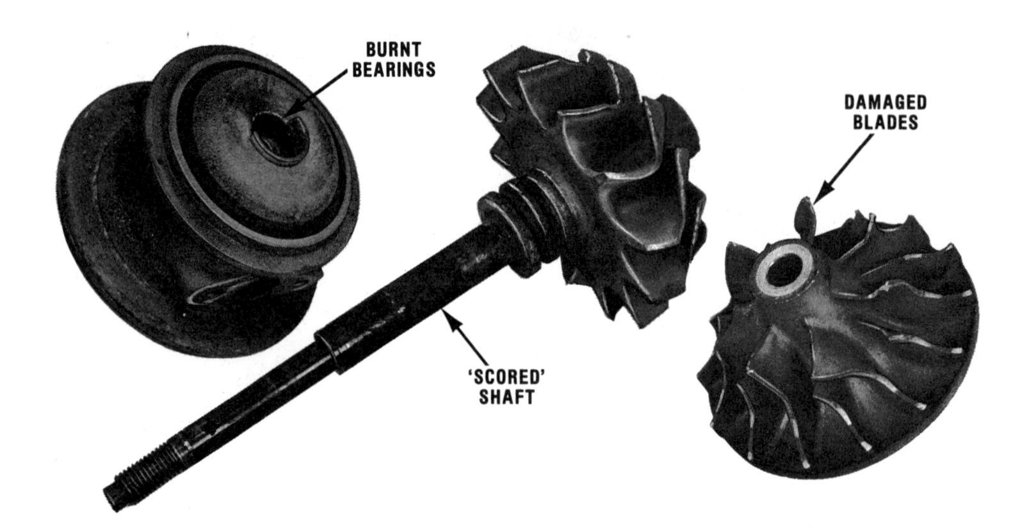

Figure 17-5 Inspecting turbocharger shaft, bearings, and wheels (Courtesy of Chrysler Corporation)

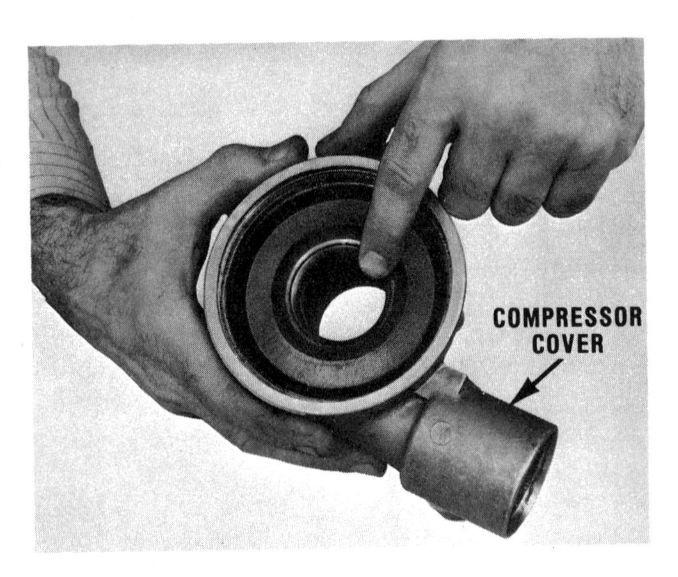

Figure 17-6 Inspecting turbocharger end housings (Courtesy of Chrysler Corporation)

must be replaced, always check the end housings for damage (Figure 17-6). When these housings are marked or scored, replacement is necessary. Since turbocharger components are subjected to extreme heat, use a straightedge to check all mating surfaces for a warped condition. Replace warped components as necessary.

Turbocharger Installation and Prelubrication

Turbocharger bearing prelubrication refers to lubricating these bearings before starting the engine after the turbocharger has been replaced or serviced.

WARNING: Failure to prelubricate turbocharger bearings may result in premature bearing failure.

Prior to reinstalling the turbocharger, be sure the engine oil and filter are in satisfactory condition. Change the oil and filter as required, and be sure the proper oil level is indicated on the dipstick. Check the coolant for contamination. Flush the cooling system if contamination is present. Reverse the turbocharger removal procedure explained previously in this chapter to install the turbocharger. Replace all gaskets and be sure all fasteners are tightened to the specified torque.

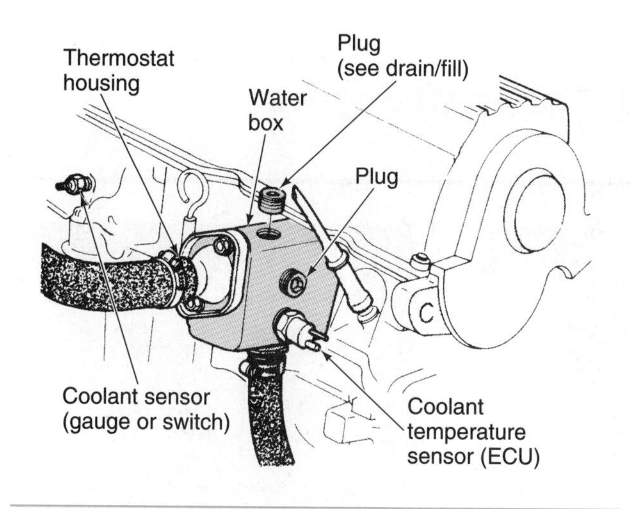

Figure 17-7 Removing plug in the top of the water box while filling cooling system (Courtesy of

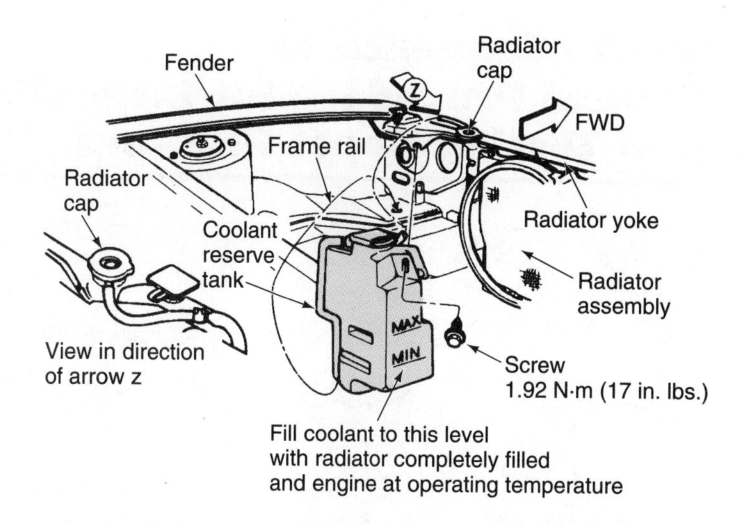

Figure 17-8 Coolant level marks on the reserve tank

Follow the vehicle manufacturer's recommended procedure for filling the cooling system. On some Chrysler engines, this involves removing a plug on top of the water box and pouring coolant into the radiator filler neck until coolant runs out the hole in the top of the water box (Figure 17-7). Install the plug in the top of the water box and tighten the plug to the specified torque. Continue filling the cooling system to the maximum level mark on the reserve tank (Figure 17-8).

The turbocharger bearings must be prelubricated before starting the engine to prevent bearing damage. Some vehicle manufacturers recommend removing the turbocharger oil supply pipe and pouring a half pint of the specified engine oil into the turbocharger to prelubricate the bearings (Figure 17-9).

Other vehicle manufacturers recommend disabling the ignition system by disconnecting the positive primary coil wire and cranking the engine for 10 to 15 seconds to allow the engine lubrication system to lubricate the turbocharger bearings. Always follow the turbocharger prelubrication instructions in the vehicle manufacturer's service manual.

Classroom Manual
Chapter 17, page 379

Photo Sequence 14 shows a typical procedure for inspecting turbochargers and testing boost pressure.

Figure 17-9 Pouring engine oil into the turbocharger oil inlet to prelubricate the turbocharger bearings (Courtesy of Chrysler Corporation)

Photo Sequence 14
Typical Procedure for Inspecting Turbochargers and Testing Boost Pressure

P14-1 Check all turbocharger linkages of looseness, and check all turbocharger hoses for leaks, cracks, kinks, and restrictions.

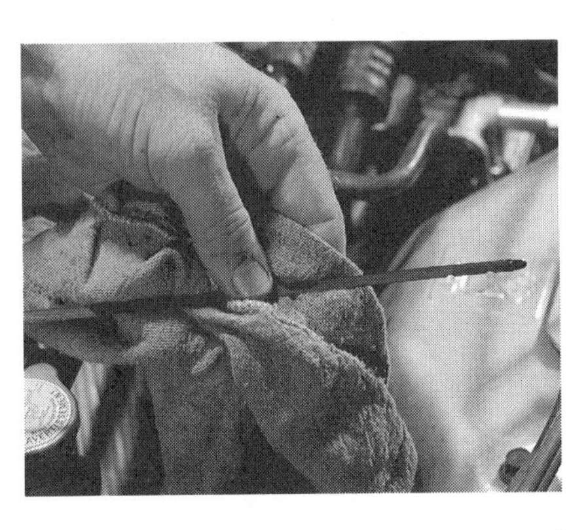

P14-2 Check the level and condition of the engine oil on the dipstick.

P14-3 Check the exhaust for evidence of blue smoke when the engine is accelerated.

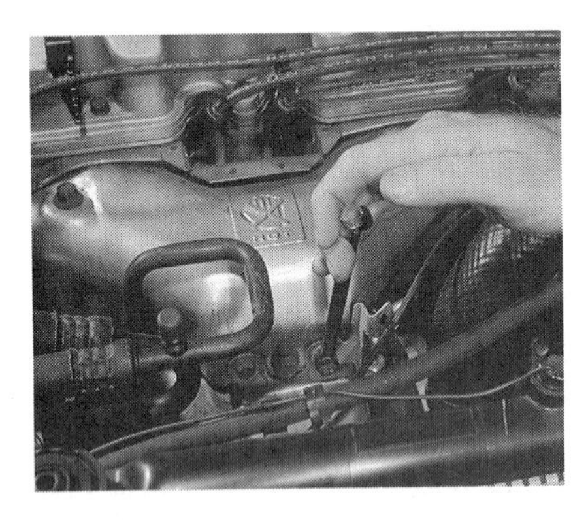

P14-4 Check all turbocharger mounting bolts for looseness.

P14-5 Check for exhaust leaks between the engine and the turbocharger, and use a stethoscope to listen for excessive turbocharger noise.

P14-6 Use a propane cylinder, metering valve, and hose to check for intake manifold vacuum leaks.

P14-7 Connect a pressure gauge to the intake manifold and locate the gauge in the passenger compartment where it can be easily seen by the driver.

P14-8 Road test the car and accelerate from 0 to 60 mph at wide-open throttle while observing the pressure gauge.

P14-9 Disconnect the pressure gauge from the intake manifold and remove the gauge from the passenger compartment.

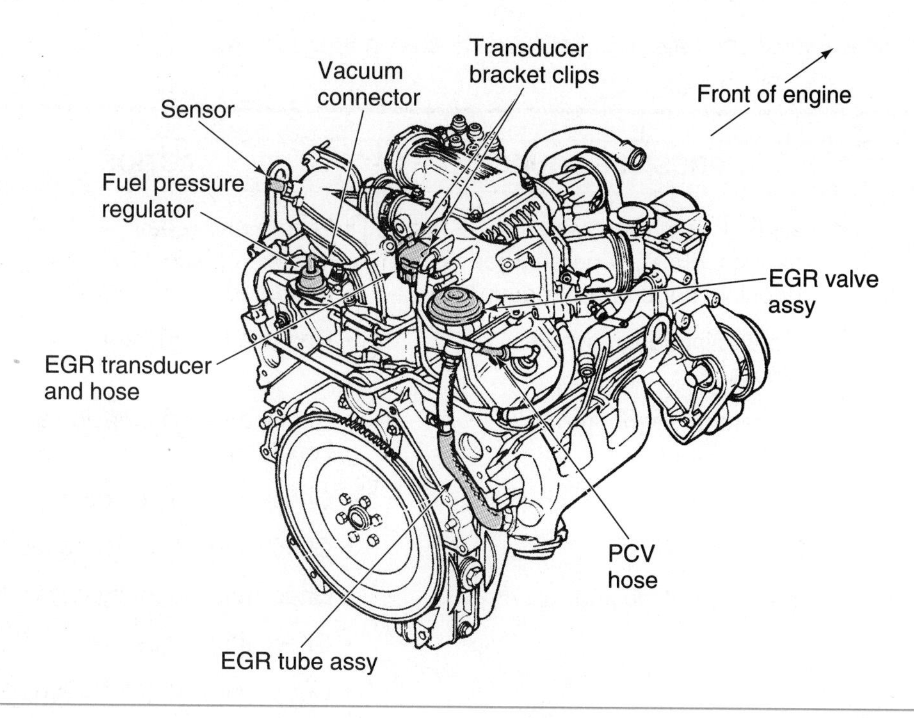

Figure 17-10 Component removal prior to supercharger removal (Reprinted with the permission of Ford Motor Company)

Supercharger Diagnosis and Service

Supercharger Lubrication

The fluid in the front supercharger housing lubricates the rotor drive gears. This fluid does not require changing for the life of the vehicle. However, this fluid level should be checked at 30,000-mi (48,000-km) intervals. To check the fluid level in the front supercharger housing, remove the Allen head plug in the top of this housing. The fluid should be level with the bottom of the threads in the plug opening. If the fluid level is low, add the required amount of synthetic fluid that meets the vehicle manufacturer's specifications.

Supercharger Diagnosis

The supercharger on Ford cars is serviced only as an assembly. Therefore, the technician must diagnose supercharger problems and replace the supercharger if necessary. Supercharger problems include low boost, high boost, reduced vehicle response and/or fuel economy, noise, and oil leaks (Table 17-1).

Supercharger Removal

Follow these steps for supercharger removal:

1. Disconnect the negative battery cable.
2. Remove the air inlet tube from the throttle body.
3. Remove the cowl vent screens.
4. Drain the coolant from the radiator.
5. Disconnect the spark plug wires from the spark plugs in the right cylinder head and position them out of the way.
6. Remove the electrical connections from the air charge temperature sensor, throttle position sensor, and idle air control valve.

Table 17-1 **Supercharger Diagnosis (Reprinted with the permission of Ford Motor Company)**

CONDITION	POSSIBLE SOURCE	ACTION
• Low Boost	• Air leak at intercooler, flanges, ducts, supercharger housing, supercharger outlet.	• Locate and repair leak or replace damaged component.
	• Contamination in system, blockage.	• Remove obstruction.
	• Supercharger not turning.	• Check drive belt tension and condition. • Check coupling for damage. • Check for pulley slipping on shaft.
	• By-pass not closing.	• Check function of by-pass actuator. • Check stop adjustment. • Check vacuum hose condition and installation.
	• Insufficient flow from supercharger.	• Check supercharger for incorrect clearances—wear from contamination. • Check for correct pulley diameter.
• High Boost	• Too much flow.	• Check for exhaust restrictions or damage. • Check catalyst for damage. • Check for correct pulley diameter.
• Vehicle Response too "Touchy" and/or Poor Fuel Economy	• By-pass not opening.	• Check function of by-pass actuator. • Check for stuck or restricted valve. • Check vacuum hose condition and installation • Check actuator diaphragm for damage or leak.
• Supercharger Noisy	• Mechanical damage to supercharger.	• Replace supercharger.
• Noise in Air Handling Systems	• Air leaks.	• Check all flanges for proper fit and position. • Check for proper isolation of components.
• Oil on Outside of Supercharger	• Leaking seals.	• Replace supercharger.
	• Loose fill plug.	• Tighten fill plug.
	• Input shaft damaged at seal.	• Replace supercharger.

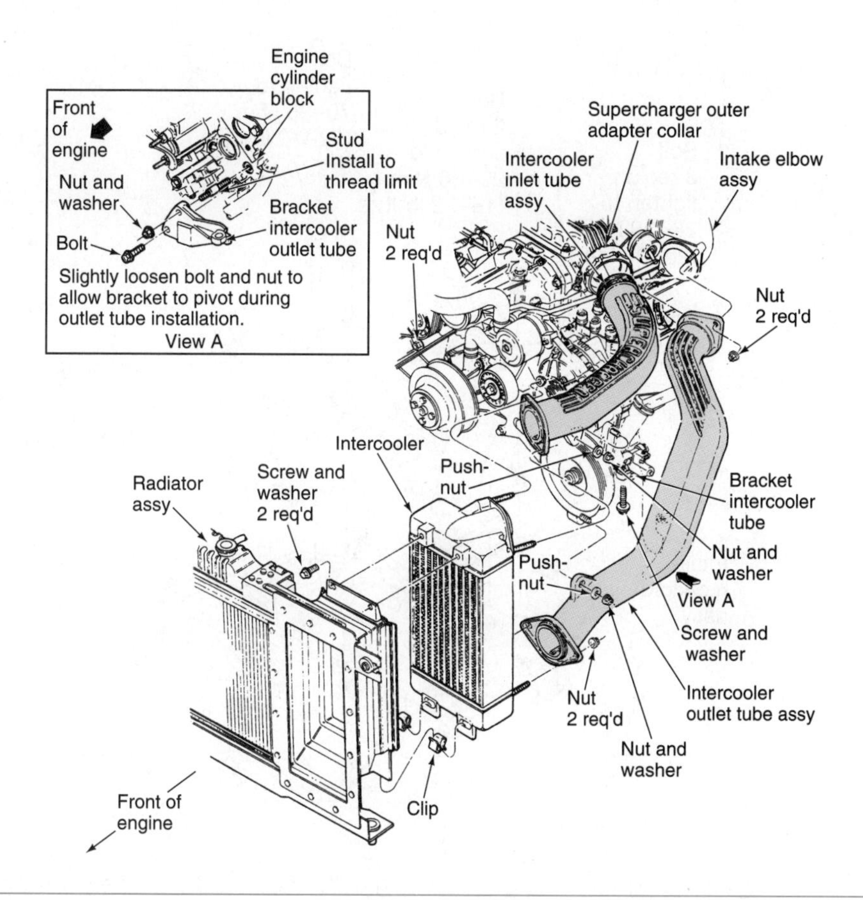

Figure 17-11 Removing intercooler inlet and outlet tubes (Reprinted with the permission of Ford Motor Company)

7. Disconnect the vacuum hoses from the supercharger air inlet plenum.
8. Remove the EGR transducer from the bracket and disconnect the vacuum hose to this component (Figure 17-10).
9. Disconnect the PCV tube from the supercharger air inlet plenum.
10. Disconnect the throttle linkage and remove the throttle linkage bracket. Position this linkage and bracket out of the way. Remove the cruise control linkage if equipped.
11. Remove the two EGR valve attaching bolts and place this valve out of the way.
12. Remove the coolant hoses from the throttle body.
13. Remove the supercharger drive belt.
14. Remove the inlet and outlet tubes from the intercooler (Figure 17-11).
15. Remove three intercooler adapter attaching bolts from the intake manifold.
16. Remove three supercharger mounting bolts (Figure 17-12).
17. Remove the supercharger and air intake plenum as an assembly.

This procedure should be followed for supercharger installation:

1. Clean all gasket surfaces and inspect these surfaces for scratches and metal burrs. Remove metal burrs as required.
2. Place a new gasket on the intake manifold surface that mates with the supercharger intercooler adapter.
3. Install the supercharger, throttle body, and air intake plenum as an assembly.
4. Install three supercharger mounting bolts and tighten these bolts to the specified torque.
5. Install three bolts that retain the intercooler adapter to the intake manifold, and tighten these bolts to the specified torque.

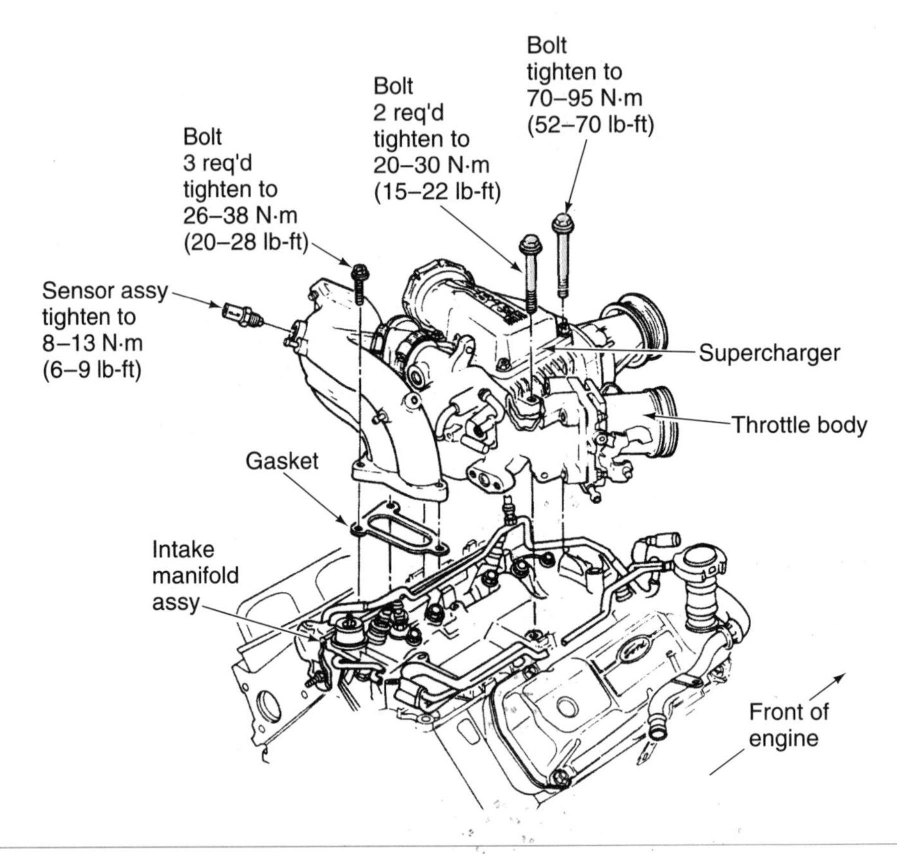

Figure 17-12 Removing supercharger mounting bolts (Reprinted with the permission of Ford Motor Company)

6. Install the intercooler inlet and outlet tubes and tighten all mounting bolts to the specified torque.
7. Install the coolant hoses on the throttle body and tighten these hose clamps.
8. Install a new EGR valve gasket and install the EGR valve. Tighten the EGR valve mounting bolts to the specified torque.
9. Install the throttle linkage and bracket, and tighten the bracket bolts to the specified torque.
10. Connect all the vacuum hoses to the original locations on the air intake plenum, EGR valve, and EGR transducer.
11. Install the spark plug wires on the spark plugs in the right cylinder head, and connect the electrical connectors to the throttle position sensor, air charge temperature sensor, and idle air control valve.
12. Install the cowl covers.
13. Install and tighten the air inlet tube on the throttle body.
14. Install the supercharger drive belt.
15. Refill the cooling system to the proper level.
16. Connect the battery ground cable.
17. Start the engine and check for air leaks in the supercharger system.

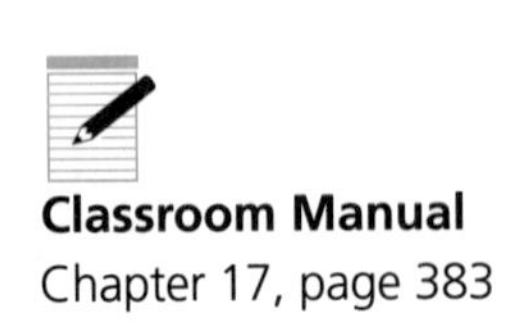

Classroom Manual
Chapter 17, page 383

CUSTOMER CARE: When talking to customers, always remember the two *P*s: pleasant and polite. There may be many days when we do not feel like being pleasant and polite. Perhaps we have several problem vehicles in the shop with symptoms that are difficult to diagnose and correct. Some service work may be behind schedule, and customers may be irate because their vehicles are not ready on time. However, we should always remain pleasant and polite with customers. Our attitude does much to make the customer feel better and to realize their business is appreciated. A customer may not feel very happy about an expensive repair bill, but a pleasant attitude on our part may help to improve their feelings. When the two *P*s are remembered by service personnel, customer relations are enhanced, and the customer feels like returning to the shop. Conversely, if service personnel have a grouchy, indifferent attitude, the customer may be turned off and take his or her business to another shop.

Guidelines for Turbocharger Diagnosis and Service

1. A basic turbocharger inspection involves inspecting all turbocharger linkages, hoses, and lines, and checking for intake system and exhaust leaks.
2. Leaks in the air intake system may allow dirt particles to enter the turbocharger and damage the blades.
3. Exhaust leaks between the cylinders and the turbocharger decrease turbocharger efficiency.
4. Blue smoke in the exhaust may be caused by worn turbocharger seals, worn valve stem seals and guides, or worn piston rings.
5. Low cylinder compression reduces air flow through the engine and decreases turbocharger shaft speed and boost pressure.
6. Higher-than-specified boost pressure may be caused by a defective wastegate system.
7. Lower-than-specified boost pressure may be caused by a faulty wastegate system, worn turbocharger bearings or blades, or low cylinder compression.
8. When the axial movement on the turbocharger shaft is more than specified, the blades may strike the end housings.
9. Premature turbocharger bearing failure may be caused by lack of lubrication, contaminated oil, or a contaminated cooling system.
10. After the turbocharger has been replaced, it must be prelubricated before starting the engine.

CASE STUDY

A customer complained about lack of power on a Dodge Spirit with a 2.5-L turbocharged engine. The customer said the engine had much slower acceleration compared to what it had had previously.

The technician performed a basic turbocharger inspection and found that all linkages, hoses, and lines were in normal condition. A check for intake and exhaust leaks did not reveal any problems, and the turbocharger did not have an excessive noise level. The technician checked the turbocharger boost pressure during a road test and discovered it was 3 psi below the specified boost pressure. There was some evidence of blue smoke in the exhaust during the road test.

The technician decided to check the engine compression before any further turbocharger service. The compression test revealed that all four cylinders had much lower-than-specified compression. The technician placed a small amount of oil in each cylinder and repeated the compression test. With the oil in the cylinders, the compression pressure improved considerably on each cylinder, indicating worn piston rings and cylinders. The customer was advised that the low engine compression was the cause of the reduced power, and the engine required rebuilding.

Terms to Know

Boost pressure

Wastegate stroke

Prelubrication

ASE Style Review Questions

1. While discussing turbocharger inspection and diagnosis:
   Technician A says excessive turbocharger shaft end play may cause a rattling noise from the turbocharger.
   Technician B says an intake system air leak upstream from the compressor wheel may allow dirt particles to enter the turbocharger.
   Who is correct?
   A. A only **C.** Both A and B
   B. B only **D.** Neither A nor B

2. While discussing basic turbocharger inspection and diagnosis:
   Technician A says blue smoke in the exhaust may indicate worn turbocharger seals.
   Technician B says an exhaust leak between the cylinders and the turbocharger has no effect on turbocharger operation.
   Who is correct?
   A. A only **C.** Both A and B
   B. B only **D.** Neither A nor B

3. While discussing turbocharger boost pressure:
   Technician A says low cylinder compression does not affect turbocharger operation.
   Technician B says a wastegate sticking in the closed position decreases boost pressure.
   Who is correct?
   A. A only **C.** Both A and B
   B. B only **D.** Neither A nor B

4. While discussing turbocharger service:
   Technician A says if the shaft axial movement is 0.008 in, the turbocharger must be repaired or replaced.
   Technician B says excessive shaft axial movement may be caused by worn bearings.
   Who is correct?
   A. A only **C.** Both A and B
   B. B only **D.** Neither A nor B

5. While discussing wastegate stroke measurement:
   Technician A says reduced wastegate stroke may cause excessive boost pressure.
   Technician B says reduced wastegate stroke may cause reduced boost pressure.
   Who is correct?
   A. A only **C.** Both A and B
   B. B only **D.** Neither A nor B

6. While discussing turbocharger service:
   Technician A says turbocharger bearing failure may be caused by a contaminated cooling system.
   Technician B says turbocharger bearing failure may be caused by contaminated engine oil.
   Who is correct?
   A. A only **C.** Both A and B
   B. B only **D.** Neither A nor B

7. While discussing turbocharger blade damage:
   Technician A says blade damage may be caused by an exhaust leak between the turbocharger and the catalytic converter.
   Technician B says blade damage may be caused by excessive shaft axial movement.
   Who is correct?
   A. A only **C.** Both A and B
   B. B only **D.** Neither A nor B

8. While discussing turbocharger service:
   Technician A says blade damage may be caused by a leaking gasket between the intake manifold and the cylinder head.
   Technician B says blade damage may be caused by a leaking gasket between the turbocharger and the exhaust pipe connected to the catalytic converter.
   Who is correct?
   A. A only **C.** Both A and B
   B. B only **D.** Neither A nor B

9. While discussing turbocharger service:
Technician A says if the end housings are scored, they must be replaced.
Technician B says a straightedge should be used to check all turbocharger mating surfaces for a warped condition.
Who is correct?
A. A only
B. B only
C. Both A and B
D. Neither A nor B

10. While discussing turbocharger replacement:
Technician A says after the turbocharger is installed on some engines, the bearings are prelubricated by disabling the ignition and cranking the engine for 10 to 15 seconds.
Technician B says after the turbocharger is installed on the engine, bearing prelubrication is not required.
Who is correct?
A. A only
B. B only
C. Both A and B
D. Neither A nor B

Table 17-2 ASE TASK

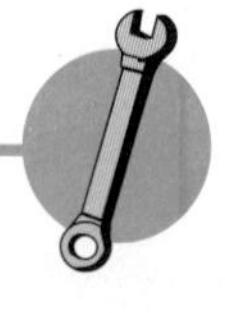

Test the operation of turbocharger/supercharger systems; determine needed repairs.

Problem Area	Symptoms	Possible Causes	Classroom Manual	Shop Manual
ENGINE PERFORMANCE	Lack of power	Low boost pressure	379	368
OIL CONSUMPTION	Blue smoke in exhaust	Worn turbocharger seals	382	368
NOISE	Rattling noise in turbocharger	1. Worn turbocharger bearings	382	371
		2. Excessive axial shaft movement	382	370

Table 17-3 ASE TASK

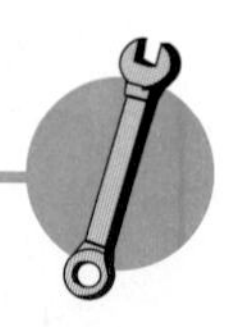

Remove, clean, inspect, and repair or replace turbocharger/supercharger system components.

Problem Area	Symptoms	Possible Causes	Classroom Manual	Shop Manual
TURBOCHARGER DAMAGE	Worn bearings	Contaminated oil or cooling system	382	371
	Damaged blades	1. Leaking air intake system	382	371
		2. Excessive axial shaft movement	382	370
	Scored end housings	Excessive axial shaft movement	382	370

Table 17-4 ASE TASK

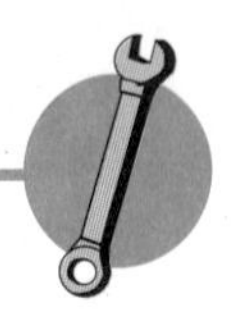

Identify the causes of turbocharger/supercharger failure; determine needed repairs.

Problem Area	Symptoms	Possible Causes	Classroom Manual	Shop Manual
TURBOCHARGER DAMAGE	Worn bearings	Contaminated oil or cooling system	382	379
	Damaged blades	1. Leaking air intake system	382	371
		2. Excessive axial shaft movement	382	370
	Scored end housings	1. Excessive axial shaft movement	382	370

APPENDIX A

Engine Performance Special Tool Suppliers

Sun Electric Corporation
One Sun Parkway
Crystal Lake, IL 60014

Allen Testproducts Division
Kalamazoo, MI 49007

Bear Automotive Inc.
A Company of SPX Corporation
12121 West Ferrick Place
Milwaukee, WI 53222

OTC A Division of SPX Corporation
655 Eisenhower Dr.
Owatonna, MN 55060

Fluke Corporation
P.O. Box 9090
Everett, WA 98206

SK Hand Tool Corporation
3535 W. 47th St.
Chicago, IL 60632-2998

Mac Tools, Inc.
South Fayette St.
Washington Court House, OH 43160

Snap-on Tools Corporation
Kenosha, WI 53141-1410

Ferret Instruments
1310 Higgins Dr.
Cheboygan, MI 49721

Kleer-Flo Company
15151 Technology Dr.
Eden Prairie, MN 55344

APPENDIX B

Metric Conversions

	to convert these	to these,	multiply by:
TEMPERATURE	Centigrade Degrees	Fahrenheit Degrees	1.8 then + 32
	Fahrenheit Degrees	Centigrade Degrees	0.556 after − 32
LENGTH	Millimeters	Inches	0.03937
	Inches	Millimeters	25.4
	Meters	Feet	3.28084
	Feet	Meters	0.3048
	Kilometers	Miles	0.62137
	Miles	Kilometers	1.60935
AREA	Square Centimeters	Square Inches	0.155
	Square Inches	Square Centimeters	6.45159
VOLUME	Cubic Centimeters	Cubic Inches	0.06103
	Cubic Inches	Cubic Centimeters	16.38703
	Cubic Centimeters	Liters	0.001
	Liters	Cubic Centimeters	1000
	Liters	Cubic Inches	61.025
	Cubic Inches	Liters	0.01639
	Liters	Quarts	1.05672
	Quarts	Liters	0.94633
	Liters	Pints	2.11344
	Pints	Liters	0.47317
	Liters	Ounces	33.81497
	Ounces	Liters	0.02957
WEIGHT	Grams	Ounces	0.03527
	Ounces	Grams	28.34953
	Kilograms	Pounds	2.20462
	Pounds	Kilograms	0.45359
WORK	Centimeter Kilograms	Inch-Pounds	0.8676
	Inch-Pounds	Centimeter-Kilograms	1.15262
	Meter Kilograms	Foot-Pounds	7.23301
	Foot-Pounds	Newton-Meters	1.3558
PRESSURE	Kilograms/Square Centimeter	Pounds/Square Inch	14.22334
	Pounds/Square Inch	Kilograms/Square Centimeter	0.07031
	Bar	Pounds/Square Inch	14.504
	Pounds/Square Inch	Bar	0.06895

GLOSSARY

Note: Terms are highlighted in color, followed by Spanish translation in bold.

Accelerator pump adjustment An adjustment that sets the accelerator pump height and stroke.
Ajuste de la bomba del acelerador Ajuste que fija la altura y la carrera de la bomba del acelerador.

Actuation test mode A scan tester mode used to cycle the relays and actuators in a computer system.
Modo de prueba de activación Instrumento de pruebas de exploración utilizado para ciclar los relés y los accionadores en una computadora.

Adjusting pads, mechanical lifters Metal discs that are available in various thicknesses, and positioned in the end of the mechanical lifter to adjust valve clearance.
Cojines de ajuste, elevadores mecánicos Discos metálicos disponibles en diferentes espesores que se colocan en el extremo del elevador mecánico para ajustar el espacio libre de la válvula.

Advance-type timing light A timing light that is capable of checking the degrees of spark advance.
Luz de ensayo de regulación del encendido tipo avance Luz de ensayo de regulación del encendido capaz de verificar la cantidad del avance de la chispa.

AIR by-pass (AIRB) solenoid A computer-controlled solenoid that directs air to the atmosphere or to the AIR diverter solenoid.
Solenoide de paso AIR Solenoide controlado por computadora que conduce el aire hacia la atmósfera o hacia el solenoide derivador AIR.

Air charge temperature (ACT) sensor A sensor that sends a signal to the computer in relation to intake air temperature.
Sensor de la temperatura de la carga de aire Sensor que le envía una señal a la computadora referente a la temperatura del aire aspirado.

AIR diverter (AIRD) solenoid A computer-controlled solenoid in the secondary air injection system that directs air upstream or downstream.
Solenoide derivador AIRD Solenoide controlado por computadora en el sistema de inyección secundaria de aire que conduce el aire hacia arriba o hacia abajo.

Air-operated vacuum pump A vacuum pump operated by air pressure that may be used to pump liquids such as gasoline.
Bomba de vacío accionada hidráulicamente Bomba de vacío accionada por la presión del aire que puede utilizarse para bombear líquidos, como por ejemplo la gasolina.

Analog meter A meter with a movable pointer and a meter scale.
Medidor analógico Medidor provisto de un indicador móvil y una escala métrica.

Antidieseling adjustment An adjustment that prevents dieseling when the ignition switch is turned off.
Ajuste antiautoencendido Ajuste que evita el autoencendido cuando el botón conmutador de encendido está apagado.

ASE blue seal of excellence A seal displayed by an automotive repair facility that employs ASE certified technicians.
Sello azul de excelencia de la ASE Logotipo exhibido en talleres de reparación de automóviles donde se emplean mecánicos certificados por la ASE.

ASE technician certification Certification of automotive technicians in various classifications by the National Institute for Automotive Service Excellence (ASE).
Certificación de mecánico de la ASE Certificación de mecánico de automóviles en áreas diferentes de especialización otorgada por el Instituto Nacional para la Excelencia en la Reparación de Automóviles (ASE).

Automatic shutdown (ASD) relay A computer-operated relay that supplies voltage to the fuel pump, coil primary, and other components on Chrysler fuel injected engines.
Relé de parada automática Relé accionado por computadora que les suministra tensión a la bomba del combustible, al bobinado primario, y a otros componentes en motores de inyección de combustible fabricados por la Chrysler.

Barometric (Baro) pressure sensor A sensor that sends a signal to the computer in relation to barometric pressure.
Sensor de la presión barométrica Sensor que le envía una señal a la computadora referente a la presión barométrica.

Belt tension gauge A gauge designed to measure belt tension.
Calibrador de tensión de la correa de transmisión Calibrador diseñado para medir la tensión de una correa de transmisión.

Bimetal sensor A temperature-operated sensor used to control vacuum.
Sensor bimetal Sensor accionado por temperatura utilizado para controlar el vacío.

Block learn A chip responsible for fuel control in a General Motors PCM.
Control de combustible en bloque Pastilla responsable de controlar el combustible en un módulo del control del tren transmisor de potencia de la General Motors.

Blowgun A device attached to the end of an air hose to control and direct air flow while cleaning components.
Soplete Dispositivo fijado en el extremo de una manguera de aire para controlar y conducir el flujo de aire mientras se lleva a cabo la limpieza de los componentes.

Boost pressure The amount of intake manifold pressure created by a turbocharger or supercharger.
Presión de sobrealimentación Cantidad de presión en el colector de aspriación producida por un turbocompresor o un compresor.

Breakout box A terminal box that is designed to be connected in series at Ford PCM terminals to provide access to these terminals for test purposes.
Caja de desenroscadura Caja de borne diseñada para conectarse en serie a los bornes del módulo del control del tren transmisor de potencia de la Ford, con el objetivo de facilitar el acceso a dichos bornes para propósitos de prueba.

Burn time The length of the spark line while the spark plug is firing measured in milliseconds.
Duración del encendido Espacio de tiempo que la línea de chispas de la bujía permanece encendida, medido en milisegundos.

Calibrator package (CAL-PAK) A removable chip in some computers that usually contains a fuel backup program.
Paquete del calibrador Pastilla desmontable en algunas computadoras; normalmente contiene un programa de reserva para el combustible.

Canister purge solenoid A computer-operated solenoid connected in the evaporative emission control system.

Solenoide de purga de bote Solenoide accionado por computadora conectado en el sistema de control de emisiones de evaporación.

Canister-type pressurized injector cleaning container A container filled with unleaded gasoline and injector cleaner and pressurized during the manufacturing process or by the shop air supply.

Recipiente de limpieza del inyector presionizado tipo bote Recipiente lleno de gasolina sin plomo y limpiador de inyectores, presionizado durante el proceso de fabricación o mediante el suministro de aire en el taller mecánico.

Carbon dioxide (CO_2) A gas formed as a by-product of the combustion process.

Bióxido de carbono (CO_2) Gas que es un producto derivado del proceso de combustión.

Carbon monoxide A gas formed as a by-product of the combustion process in the engine cylinders. This gas is very dangerous or deadly to the human body in high concentrations.

Monóxido de carbono Gas que es un producto derivado del proceso de combustión en los cilindros del motor. Este gas es muy peligroso y en altas concentraciones podría ocasionar la muerte.

Catalytic converter vibrator tool A tool used to remove the pellets from catalytic converters.

Herramienta vibradora del convertidor catalítico Herramienta utilizada para remover los granos gordos de los convertidores catalíticos.

Choke control lever adjustment An adjustment that provides proper choke spring tension.

Ajuste de la palanca de control del estrangulador Ajuste que proporciona la tensión adecuada del resorte del estrangulador.

Choke diaphragm connector rod adjustment An adjustment that assures proper cushioning of the secondary air valves by the vacuum break diaphragm.

Ajuste de la biela con el diafragma del estrangulador Ajuste que asegura el acojinamiento adecuado de las válvulas de aire secundarias mediante el diafragma del interruptor de vacío.

Choke unloader adjustment An adjustment that assures the proper choke position when the throttles are held wide open and the choke spring is cold.

Ajuste del abridor del estrangulador Ajuste que asegura la posición adecuada del estrangulador cuando se mantienen las mariposas abiertas de par en par y el resorte del estrangulador está frío.

Choke vacuum kick adjustment An adjustment that assures the proper choke opening when the vacuum kick diaphragm is bottomed after a cold engine is started.

Ajuste del retardador de vacío del estrangulador Ajuste que asegura la apertura adecuada del estrangulador cuando el diafragma del retardador de vacío se sumerje luego del arranque de un motor frío.

Choke valve angle gauge A gauge with a degree scale and a level bubble for performing carburetor adjustments.

Calibrador del ángulo de la válvula de estrangulación Calibrador provisto de una escala medida en grados y un tubo de burbuja para nivelación con el que se realizan ajustes al carburador.

Closed loop A computer operating mode in which the computer uses the oxygen sensor signal to help control the air-fuel ratio.

Bucle cerrado Modo de funcionamiento de una computadora en el que se utiliza la señal del sensor de oxígeno para ayudar a controlar la relación de aire y combustible.

Comprehensive tests A complete series of battery, starting, charging, ignition, and fuel system tests performed by an engine analyzer.

Pruebas comprensivas Serie completa de pruebas realizadas en los sistemas de la batería, del arranque, de la carga, del encendido, y del combustible con un analizador de motores.

Compression gauge A gauge used to test engine compression.

Manómetro de compresión Calibrador utilizado para revisar la compresión de un motor.

Computed timing test A computer system test mode on Ford products that checks spark advance supplied by the computer.

Prueba de regulación del avance calculado Modo de prueba en una computadora de productos fabricados por la Ford que verifica el avance de la chispa suministrado por la computadora.

Computer-controlled carburetor performance test A test that determines the general condition of a computer-controlled carburetor system.

Prueba de rendimiento del carburador controlado por computadora Prueba que determina la condición general de un sistema de carburador controlado por computadora.

Concealment plug A plug installed over the idle mixture screw to prevent carburetor tampering by inexperienced service personnel.

Tapón obturador Tapón instalado sobre el tornillo de mezcla de la marcha lenta para evitar que mecánicos inexpertos alteren el carburador.

Continuous self-test A computer system test mode on Ford products that provides a method of checking defective wiring connections.

Prueba automática continua Modo de prueba en una computadora de productos fabricados por la Ford que proporciona un método de verificar conexiones defectuosas del alambrado.

Coolant hydrometer A tester designed to measure coolant specific gravity and determine the amount of antifreeze in the coolant.

Hidrómetro de refrigerante Instrumento de prueba diseñado para medir la gravedad específica del refrigerante y determinar la cantidad de anticongelante en el refrigerante.

Cooling system pressure tester A tester used to test cooling system leaks and radiator pressure caps.

Instrumento de prueba de la presión del sistema de enfriamiento Instrumento de prueba utilizado para revisar fugas en el sistema de enfriamiento y en las tapas de presión del radiador.

Custom tests A series of tests programmed by the technician and performed by an engine analyzer.

Pruebas de diseño específico Serie de pruebas programadas por el mecánico y realizadas por un analizador de motores.

Cylinder leakage tester A tester designed to measure the amount of air leaking from the combustion chamber past the piston rings or valves.

Instrumento de prueba de la fuga del cilindro Instrumento de prueba diseñado para medir la cantidad de aire que se escapa desde la cámara de combustión y que sobrepasa los anillos de pistón o las válvulas.

Data link connector (DLC) A computer system connector to which the computer supplies data for diagnostic purposes.

Conector de enlace de datos Conector de computadora al que ésta suministra datos para propósitos diagnósticos.

Diagnostic trouble code (DTC) A code retained in a computer memory representing a fault in a specific area of the computer system.

Códigos indicadores de fallas para propósitos diagnósticos Código almacenado en la memoria de una computadora que representa una falla en un área específica de la computadora.

Diesel particulates Small carbon particles in diesel exhaust.
Partículas de diesel Pequeñas partículas de carbón presentes en el escape de un motor diesel.

Digital EGR valve An EGR valve that contains a computer-operated solenoid or solenoids.
Válvula EGR digital Una válvula EGR que contiene un solenoide o solenoides accionados por computadora.

Digital meter A meter with a digital display.
Medidor digital Medidor con lectura digital.

Distributor ignition (DI) system SAE J1930 terminology for any ignition system with a distributor.
Sistema de encendido con distribuidor Término utilizado por la SAE J1930 para referirse a cualquier sistema de encendido que tenga un distribuidor.

Downstream air Air injected into the catalytic converter.
Aire conducido hacia abajo Aire inyectado dentro del convertidor catalítico.

EGR pressure transducer (EPT) A vacuum switching device operated by exhaust pressure that opens and closes the vacuum passage to the EGR valve.
Transconductor de presión EGR Dispositivo de conmutación de vacío accionado por la presión del escape que abre y cierra el paso del vacío a la válvula EGR.

EGR vacuum regulator (EVR) solenoid A solenoid that is cycled by the computer to provide a specific vacuum to the EGR valve.
Solenoide regulador de vacío EGR Solenoide ciclado por la computadora para proporcionarle un vacío específico a la válvula EGR.

Electric throttle kicker An electric solenoid that controls throttle opening at idle speed and prevents dieseling.
Nivelador eléctrico de la mariposa Solenoide eléctrico que controla la apertura de la mariposa a una velocidad de marcha lenta y evita el autoencendido.

Electronic fuel injection (EFI) A generic term applied to various types of fuel injection systems.
Inyección electrónica de combustible Término general aplicado a varios sistemas de inyección de combustible.

Electronic ignition (EI) system SAE J1930 terminology for any ignition system without a distributor.
Sistema de encendido electrónico Término utilizado por la SAE J1930 para referirse a cualquier sistema de encendido que no tenga distribuidor.

Engine analyzer A tester designed to test engine systems such as battery, starter, charging, ignition, and fuel, plus engine condition.
Analizador de motores Instrumento de prueba diseñado para revisar sistemas de motores, como por ejemplo los de la batería, del arranque, de la carga, del encendido, y del combustible, además de la condición del motor.

Engine coolant temperature (ECT) sensor A sensor that sends a voltage signal to the computer in relation to coolant temperature.
Sensor de la temperatura del refrigerante del motor Sensor que le envía una señal de tensión a la computadora referente a la temperatura del refrigerante.

Engine lift A hydraulically operated piece of equipment used to lift the engine from the chassis.
Elevador de motores Equipo accionado hidráulicamente que se utiliza para levantar el motor del chasis.

Evaporative (EVAP) system A system that collects fuel vapors from the fuel tank and directs them into the intake manifold rather than allowing them to escape to the atmosphere.
Sistema de evaporación Sistema que acumula los vapores del combustible que escapan del tanque del combustible y los conduce hacia el colector de aspiración en vez de permitir que los mismos se escapen hacia la atmósfera.

Exhaust gas analyzer A tester that measures carbon monoxide, carbon dioxide, hydrocarbons, and oxygen in the engine exhaust.
Analizador del gas del escape Instrumento de prueba que mide el monóxido de carbono, el bióxido de carbono, los hidrocarburos, y el oxígeno en el escape del motor.

Exhaust gas recirculation (EGR) valve A valve that circulates a specific amount of exhaust gas into the intake manifold to reduce NOx emissions.
Válvula de recirculación del gas del escape Válvula que hace circular una cantidad específica del gas del escape hacia el colector de aspiración para disminuir emisiones de óxidos de nitrógeno.

Exhaust gas recirculation valve position (EVP) sensor A sensor that sends a voltage signal to the computer in relation to EGR valve position.
Sensor de la posición de la válvula de recirculación del gas del escape Sensor que le envía una señal de tensión a la computadora referente a la posición de la válvula EGR.

Exhaust gas temperature sensor A sensor that sends a voltage signal to the computer in relation to exhaust temperature.
Sensor de la temperatura del gas del escape Sensor que le envía una señal de tensión a la computadora referente a la temperatura del escape.

Fast idle cam position adjustment An adjustment that positions the fast idle cam properly in relation to the choke valve position.
Ajuste de la posición de árbol de levas de marcha lenta rápida Ajuste que coloca el árbol de levas de marcha lenta rápida adecuadamente según la posición de la válvula de estrangulación.

Fast idle thermo valve A valve operated by a thermo-wax element that allows more air into the intake manifold to increase idle speed when the engine is cold.
Termoválvula de la marcha lenta rápida Válvula accionada por un elemento de termocera que permite que una mayor cantidad de aire entre en el colector de aspiración para aumentar la velocidad de la marcha lenta cuando el motor está frío.

Feeler gauge Metal strips with a specific thickness for measuring clearances between components.
Calibrador de espesores Láminas metálicas de un espesor específico para medir espacios libres entre componentes.

Field service mode A computer diagnostic mode that indicates whether the computer is in open or closed loop on General Motors PCMs.
Modo de servicio de campo Modo diagnóstico de una computadora que indica si la misma se encuentra en bucle abierto o cerrado en módulos del control del tren transmisor de potencia de la General Motors.

Flash code diagnosis Reading computer system diagnostic trouble codes (DTCs) from the flashes of the malfunction indicator light (MIL).
Diagnosis con código de destello La lectura de códigos indicativos de fallas para propósitos diagnósticos de una computadora mediante los destellos de la luz indicadora de funcionamiento defectuoso.

Float drop adjustment An adjustment that provides the proper maximum downward float position.
Ajuste del descenso del flotador Ajuste que proporciona la posición descendente máxima adecuada del flotador.

Float level adjustment A float adjustment that assures the proper level of fuel in the float bowl.
Ajuste del nivel del flotador Ajuste del flotador que asegura el nivel adecuado de combustible en el depósito del flotador.

Floor jack A hydraulically operated device mounted on casters and used to raise one end or corner of the chassis.
Gato de pie Dispositivo accionado hidráulicamente montado en rolletes y utilizado para levantar un extremo o una esquina del chasis.

Four-gas emissions analyzer An analyzer designed to test carbon monoxide, carbon dioxide, hydrocarbons, and oxygen in the exhaust.
Analizador de cuatro tipos de emisiones Analizador diseñado para revisar el monóxido de carbono, el bióxido de carbono, los hidrocarburos, y el oxígeno en el escape.

Fuel cut rpm The rpm range in which the computer stops operating the injectors during deceleration.
Detención de combustible según las rpm Margen de revoluciones al que la computadora detiene el funcionamiento de los inyectores durante la desaceleración.

Fuel pressure test port A threaded port on the fuel rail to which a pressure gauge may be connected to test fuel pressure.
Lumbrera de prueba de la presión del combustible Lumbrera fileteada que se encuentra en el carril del combustible a la que puede conectársele un calibrador de presión para revisar la presión del combustible.

Fuel pump volume The amount of fuel the pump delivers in a specific time period.
Volumen de la bomba del combustible Cantidad de combustible que la bomba envía dentro de un espacio de tiempo específico.

Fuel tank purging Removing fuel vapors and foreign material from the fuel tank.
Purga del tanque del combustible La remoción de vapores de combustible y de material extraño del tanque del combustible.

Graphite oil An oil with a graphite base that may be used for special lubricating requirements such as door locks.
Aceite de grafito Aceite con una base de grafito que puede utilizarse para necesidades de lubrificación especiales, como por ejemplo en cerraduras de puertas.

Hall effect pickup A pickup containing a Hall element and a permanent magnet with a rotating blade between these components.
Captación de efecto Hall Captación que contiene un elemento Hall y un imán permanente entre los cuales está colocada una aleta giratoria.

Hand-held digital pyrometer A tester for measuring component temperature.
Pirómetro digital de mano Instrumento de prueba para medir la temperatura de un componente.

Hand press A hand-operated device for pressing precision-fit components.
Prensa de mano Dispositivo accionado manualmente para prensar componentes con un fuerte ajuste de precisión.

Heated resistor-type MAF sensor A MAF sensor that uses a heated resistor to sense air intake volume and temperature, and sends a voltage signal to the computer in relation to the total volume of intake air.
Sensor MAF tipo resistor térmico Sensor MAF que utiliza un resistor térmico para advertir el volumen y la temperatura del aire aspirado, y que le envía una señal de tensión a la computadora referente al volumen total de aire aspirado.

Hot wire-type MAF sensor A MAF sensor that uses a heated wire to sense air intake volume and temperature, and sends a voltage signal to the computer in relation to the total volume of intake air.
Sensor MAF tipo térmico Sensor MAF que utiliza un alambre térmico para advertir el volumen y la temperatura del aire aspirado, y que le envía una señal de tensión a la computadora referente al volumen total de aire aspirado.

Hydraulic press A hydraulically operated device for pressing precision-fit components.
Prensa hidráulica Dispositivo accionado hidráulicamente que se utiliza para prensar componentes con un fuerte ajuste de precisión.

Hydraulic valve lifters Round, cylindrical, metal components used to open the valves. These components are operated by oil pressure to maintain zero clearance between the valve stem and rocker arm.
Desmontaválvulas hidráulicas Componentes metálicos, cilíndricos y redondos utilizados para abrir las válvulas. Estos componentes se accionan mediante la presión del aceite para mantener cero espacio libre entre el vástago de válvula y el balancín.

Hydrocarbons (HC) Left over fuel from the combustion process.
Hidrocarburos El combustible restante después del proceso de combustión.

Hydrometer A tester designed to measure the specific gravity of a liquid.
Hidrómetro Instrumento de prueba diseñado para medir la gravedad específica de un líquido.

Idle air control by-pass air (IAC BPA) motor An IAC motor that controls idle speed by regulating the amount of air by-passing the throttle.
Motor para el control de la marcha lenta con el paso de aire Un motor IAC que controla la velocidad de la marcha lenta regulando la cantidad de aire que se desvía de la mariposa.

Idle air control by-pass air (IAC BPA) valve A valve operated by the IAC BPA motor that regulates the air by-passing the throttle to control idle speed.
Válvula para el control de la marcha lenta con el paso de aire Válvula accionada por el motor IAC BPA que regula el aire que se desvía de la mariposa para controlar la velocidad de la marcha lenta.

Idle air control (IAC) motor A computer-controlled motor that controls idle speed under all conditions.
Motor para el control de la marcha lenta con aire Motor controlado por computadora que controla la velocidad de la marcha lenta bajo cualquier condición del funcionamiento del motor.

Idle contact switch A switch in the IAC motor stem that informs the computer when the throttle is in the idle position.
Conmutador de contacto de la marcha lenta Conmutador en el vástago del motor IAC que le advierte a la computadora cuándo la mariposa está en la posición de marcha lenta.

Idle stop solenoid An electric solenoid that maintains the throttle in the proper idle speed position and prevents dieseling when the ignition switch is turned off.
Solenoide de detención de la marcha lenta Solenoide eléctrico que mantiene la mariposa en la posición de velocidad de marcha lenta adecuada y evita el autoencendido al desconectarse el botón conmutador de encendido.

Ignition crossfiring Ignition firing between distributor cap terminals or spark plug wires.
Encendido por inducción Encendido entre los bornes de la tapa del distribuidor o los alambres de las bujías.

Ignition module tester An electronic tester designed to test ignition modules.

Instrumento de prueba del módulo del encendido Instrumento de prueba electrónico diseñado para revisar módulos del encendido.

Injector balance tester A tester designed to test port injectors.

Instrumento de prueba del equilibro del inyector Instrumento de prueba diseñado para revisar inyectores de lumbreras.

Inspection, maintenance (I/M) testing Emission inspection and maintenance programs that are usually administered by various states.

Pruebas de inspección y mantenimiento Programas de inspección y mantenimiento de emisiones que normalmente administran diferentes estados.

Integrator A chip responsible for fuel control in a General Motors PCM.

Integrador Pastilla responsable de controlar el combustible en un módulo del control del tren transmisor de potencia de la General Motors.

International System (SI) A system of weights and measures in which each unit may be divided by 10.

Sistema internacional Sistema de pesos y medidas en el que cada unidad puede dividirse entre 10.

Jack stand A metal stand used to support one corner of the chassis.

Soporte de gato Soporte de metal utilizado para apoyar una esquina del chasis.

Key on engine off (KOEO) test A computer system test mode on Ford products that displays diagnostic trouble codes (DTCs) with the key on and the engine stopped.

Prueba con la llave en la posición de encendido y el motor apagado Modo de prueba en una computadora de productos fabricados por la Ford que muestra códigos indicadores de fallas para propósitos diagnósticos cuando la llave está en la posición de encendido y el motor está apagado.

Key on engine running (KOER) test A computer system test mode on Ford products that displays diagnostic trouble codes (DTCs) with the engine running.

Prueba con la llave en la posición de encendido y el motor encendido Modo de prueba en una computadora de productos fabricados por la Ford que muestra códigos indicadores de fallas para propósitos diagnósticos cuando el motor está encendido.

Kilovolts (kV) Thousands of volts.

Kilovoltios (kV) Miles de voltios.

Knock sensor A sensor that sends a voltage signal to the computer in relation to engine detonation.

Sensor de golpeteo Sensor que le envía una señal de tensión a la computadora referente a la detonación del motor.

Knock sensor module An electronic module that changes the analog knock sensor signal to a digital signal and sends it to the PCM.

Módulo del sensor de golpeteo Módulo electrónico que convierte la señal analógica del sensor de golpeteo en una señal digital y se la envía al módulo del control del tren transmisor de potencia.

Lift A device used to raise a vehicle.

Elevador Dispositivo utilizado para levantar un vehículo.

Linear EGR valve An EGR valve containing an electric solenoid that is pulsed on and off by the computer to provide a precise EGR flow.

Válvula EGR lineal Válvula EGR con un solenoide eléctrico que la computadora enciende y apaga para proporcionar un flujo exacto de EGR.

Magnetic-base thermometer A thermometer that may be retained to metal components with a magnetic base.

Termómetro con base magnética Termómetro que puede sujetarse a componentes metálicos por medio de una base magnética.

Magnetic probe-type digital tachometer A digital tachometer that reads engine rpm and uses a magnetic probe pickup.

Tacómetro digital tipo sonda magnética Tacómetro digital que lee las rpm del motor y utiliza una captación de sonda magnética.

Magnetic probe-type digital timing meter A digital reading that displays crankshaft degrees and uses a magnetic-type pickup probe mounted in the magnetic timing probe receptacle.

Medidor de regulación digital tipo sonda magnética Lectura digital que muestra los grados del cigüeñal y utiliza una sonda de captación tipo magnético montada en el receptáculo de la sonda de regulación magnética.

Magnetic sensor A sensor that produces a voltage signal from a rotating element near a winding and a permanent magnet. This voltage signal is often used for ignition triggering.

Sensor magnético Sensor que produce una señal de tensión desde un elemento giratorio cerca de un devanado y de un imán permanente. Esta señal de tensión se utiliza con frecuencia para arrancar el motor.

Magnetic timing offset An adjustment to compensate for the position of the magnetic receptacle opening in relation to the TDC mark on the crankshaft pulley.

Desviación de regulación magnética Ajuste para compensar la posición de la abertura del receptáculo magnético de acuerdo a la marca TDC en la roldana del cigüeñal.

Magnetic timing probe receptacle An opening in which the magnetic timing probe is installed to check basic timing.

Receptáculo de la sonda de regulación magnética Abertura en la que está montada la sonda de regulación magnética para verificar la regulación básica.

Malfunction indicator light (MIL) A light in the instrument panel that is illuminated by the PCM if certain defects occur in the computer system.

Luz indicadora de funcionamiento defectuoso Luz en el panel de instrumentos que el módulo del control del tren transmisor de potencia ilumina si ocurren ciertas fallas en la computadora.

Manifold absolute pressure (MAP) sensor An input sensor that sends a signal to the computer in relation to intake manifold vacuum.

Sensor de la presión absoluta del colector Sensor de entrada que le envía una señal a la computadora referente al vacío del colector de aspiración.

Mechanical valve lifters, or solid tappets Round, cylindrical, metal components mounted between the camshaft lobes and the pushrods to open the valves.

Desmontaválvulas mecánicas, o alzaválvulas sólidas Componentes metálicos, cilíndricos y redondos montados entre los lóbulos del árbol de levas y las varillas de empuje para abrir las válvulas.

Memory calibrator (MEM-CAL) A removable chip in some computers that replaces the PROM and CAL-PAK chips.

Calibrador de memoria Pastilla desmontable en algunas computadoras que reemplaza las pastillas PROM y CAL-PAK.

Meter impedance The total internal electrical resistance in a meter.

Impedancia de un medidor La resistencia eléctrica interna total en un medidor.

Mixture heater An electric heater mounted between the carburetor and the intake manifold that heats the air-fuel mixture when the engine is cold.

Calentador de la mezcla Calentador eléctrico que se monta entre el carburador y el colector de aspriación para calentar la mezcla de aire y combustible cuando el motor está frío.

Muffler chisel A chisel that is designed for cutting muffler inlet and outlet pipes.
Cincel para silenciadores Cincel diseñado para cortar los tubos de entrada y salida del silenciador.

Multiport fuel injection (MFI) A fuel injection system in which the injectors are grounded in the computer in pairs or groups of three or four.
Inyección de combustible de paso múltiple Sistema de inyección de combustible en el que los inyectores se ponen a tierra en la computadora en pares o en grupos de tres o cuatro.

National Institute for Automotive Service Excellence (ASE) An organization responsible for certification of automotive technicians in the US.
Instituto Nacional para la Excelencia en la Reparación de Automóviles Organización que tiene a su cargo la certificación de mecánicos de automóviles en los Estados Unidos.

Negative backpressure EGR valve An EGR valve containing a vacuum bleed valve that is operated by negative pulses in the exhaust.
Válvula EGR de contrapresión negativa Válvula EGR que contiene una válvula de descarga de vacío accionada por impulsos negativos en el escape.

Neutral/drive switch (NDS) A switch that sends a signal to the computer in relation to gear selector position.
Conmutador de mando neutral Conmutador que le envía una señal a la computadora referente a la posición del selector de velocidades.

Nose switch A switch in the IAC motor stem that informs the computer when the throttle is in the idle position.
Conmutador de contacto de la marcha lenta Conmutador en el vástago del motor IAC que le advierte a la computadora cuándo la mariposa está en la posición de marcha lenta.

Oil pressure gauge A gauge used to test engine oil pressure.
Manómetro de la presión del aceite Calibrador utilizado para revisar la presión del aceite del motor.

Open loop A computer operating mode in which the computer controls the air-fuel ratio and ignores the oxygen sensor signal.
Bucle abierto Modo de funcionamiento de una computadora en el que se controla la relación de aire y combustible y se pasa por alto la señal del sensor de oxígeno.

Optical-type pickup A pickup that contains a photo diode and a light emitting diode with a slotted plate between these components.
Captación tipo óptico Captación que contiene un fotodiodo y un diodo emisor de luz entre los cuales está colocada una placa ranurada.

Oscilloscope A cathode ray tube (CRT) that displays voltage waveforms from the ignition system.
Osciloscopio Tubo de rayos catódicos que muestra formas de onda de tensión provenientes del sistema de encendido.

Output state test A computer system test mode on Ford products that turns the relays and actuators on and off.
Prueba del estado de producción Modo de prueba en una computadora de productos fabricados por la Ford que enciende y apaga los relés y los accionadores.

Oxygen (O_2) A gaseous element that is present in air.
Oxígeno (O_2) Elemento gaseoso presente en el aire.

Oxygen (O_2) sensor A sensor mounted in the exhaust system that sends a voltage signal to the computer in relation to the amount of oxygen in the exhaust stream.
Sensor de oxígeno (O_2) Sensor montado en el sistema de escape que le envía una señal de tensión a la computadora referente a la cantidad de oxígeno en el caudal del escape.

Park/neutral switch A switch connected in the starter solenoid circuit that prevents starter operation except in park or neutral.
Conmutador PARK/neutral Conmutador conectado en el circuito del solenoide del arranque que evita el funcionamiento del arranque si el selector de velocidades no se encuentra en las posiciones PARK o NEUTRAL.

Parts per million (ppm) The volume of a gas such as hydrocarbons in ppm in relation to one million parts of the total volume of exhaust gas.
Partes por millón (ppm) Volumen de un gas, como por ejemplo los hidrocarburos, en partes por millón de acuerdo a un millón de partes del volumen total del gas del escape.

Photoelectric tachometer A tachometer that contains an internal light source and a photoelectric cell. This meter senses rpm from reflective tape attached to a rotating component.
Tacómetro fotoeléctrico Tacómetro que contiene una fuente interna de luz y una célula fotoeléctrica. Este medidor advierte las rpm mediante una cinta reflectora adherida a un componente giratorio.

Pinging noise A shop term for engine detonation that sounds like a rattling noise in the engine cylinders.
Sonido agudo Término utilizado en el taller mecánico para referirse a la detonación del motor cuyo ruido se asemeja a un estrépito en los cilindros de un motor.

Pipe expander A tool designed to expand exhaust system pipes.
Expansor de tubo Herramienta diseñada para expandir los tubos del sistema de escape.

Polyurethane air cleaner cover A circular polyurethane ring mounted over the air cleaner element to improve cleaning capabilities.
Cubierta de poliuretano del filtro de aire Anillo circular de poliuretano montado sobre el elemento del filtro de aire para facilitar la limpieza.

Port EGR valve An EGR valve operated by ported vacuum from above the throttle.
Válvula EGR lumbrera Válvula EGR accionada por un vacío con lumbreras desde la parte superior de la mariposa.

Port fuel injection (PFI) A fuel injection system with an injector positioned in each intake port.
Inyección de combustible de lumbrera Sistema de inyección de combustible que tiene un inyector colocado en cada una de las lumbreras de aspiración.

Positive backpressure EGR valve An EGR valve with a vacuum bleed valve that is operated by positive pressure pulses in the exhaust.
Válvula EGR de contrapresión positiva Válvula EGR con una válvula de descarga de vacío accionada por impulsos de presión positiva en el escape.

Positive crankcase ventilation (PCV) valve A valve that delivers crankcase vapors into the intake manifold rather than allowing them to escape to the atmosphere.
Válvula de ventilación positiva del cárter Válvula que conduce los vapores del cárter hacia el colector de aspiración en vez de permitir que los mismos se escapen hacia la atmósfera.

Power balance tester A tester designed to stop each cylinder from firing for a brief time and record the rpm decrease.
Instrumento de prueba del equilibro de la potencia Instrumento de prueba diseñado para detener el encendido de cada cilindro por un breve espacio de tiempo y registrar el descenso de las rpm.

Power train control module (PCM) SAE J1930 terminology for an engine control computer.
Módulo del control del tren transmisor de potencia Término utilizado por la SAE J1930 para referirse a una computadora para el control del motor.

Prelubrication Lubrication of components such as turbocharger bearings prior to starting the engine.
Prelubrificación Lubrificación de componentes, como por ejemplo los cojinetes del turbocompresor, antes del arranque del motor.

Pressurized injector cleaning container A small, pressurized container filled with unleaded gasoline and injector cleaner for cleaning injectors with the engine running.
Recipiente presionizado para la limpieza del inyector Pequeño recipiente presionizado lleno de gasolina sin plomo y limpiador de inyectores para limpiar los inyectores cuando el motor está encendido.

Programmable read only memory (PROM) A computer chip containing some of the computer program. This chip is removable in some computers.
Memoria de solo lectura programable (PROM) Pastilla de memoria que contiene una parte del programa de la computadora. Esta pastilla es desmontable en algunas computadoras.

Propane-assisted idle mixture adjustment A method of adjusting idle mixture with fuel supplied from a small propane cylinder.
Ajuste de la mezcla de la marcha lenta asistido por propano Método de ajustar la mezcla de la marcha lenta utilizando el combustible suministrado desde un cilindro pequeño de propano.

Pulsed secondary air injection system A system that uses negative pressure pulses in the exhaust to move air into the exhaust system.
Sistema de inyección secundaria de aire por impulsos Sistema que utiliza impulsos de la presión negativa en el escape para conducir el aire hacia el sistema de escape.

Quad driver A group of transistors in a computer that controls specific outputs.
Excitador cuádruple Grupo de transistores en una computadora que controla salidas específicas.

Quick-disconnect fuel line fittings Fuel line fittings that may be disconnected without using a wrench.
Conexiones de la línea del combustible de desmontaje rápido Conexiones de la línea del combustible que se pueden desmontar sin la utilización de una llave de tuerca.

Radiator shroud A circular component positioned around the cooling fan to concentrate the air flow through the radiator.
Bóveda del radiador Componente circular que rodea el ventilador de enfriamiento para concentrar el flujo de aire a través del radiador.

Reference pickup A pickup assembly that is often used for ignition triggering.
Captación de referencia Conjunto de captación que se utiliza con frecuencia para el arranque del encendido.

Reference voltage A constant voltage supplied from the computer to some of the input sensors.
Tensión de referencia Tensión constante que le suministra la computadora a algunos de los sensores de entrada.

Revolutions per minute (rpm) drop The amount of rpm decrease when a cylinder stops firing for a brief time.
Descenso de las revoluciones por minuto (rpm) Cantidad que descienden las rpm cuando un cilindro detiene el encendido por un breve espacio de tiempo.

Ring ridge A ridge near the top of the cylinder created by wear in the ring travel area of the cylinder.
Reborde del anillo Reborde cerca de la parte superior del cilindro ocasionado por un desgaste en el área de carrera del anillo del cilindro.

Room temperature vulcanizing (RTV) sealant A type of sealant that may be used to replace gaskets, or to help to seal gaskets, in some applications.
Compuesto obturador vulcanizador a temperatura ambiente Tipo de compuesto obturador que puede utilizarse para reemplazar guarniciones, o para ayudar a sellarlas, en algunas aplicaciones.

Scan tester A tester designed to test automotive computer systems.
Instrumento de pruebas de exploración Instrumento de prueba diseñado para revisar computadoras de automóviles.

Schrader valve A threaded valve on the fuel rail to which a pressure gauge may be connected to test fuel pressure.
Válvula Schrader Válvula fileteada que se encuentra en el carril del combustible a la que puede conectársele un calibrador de presión para revisar la presión del combustible.

Secondary air injection (AIR) system A system that injects air into the exhaust system from a belt driven pump.
Sistema de inyección secundaria de aire Sistema que inyecta aire dentro del sistema de escape desde una bomba accionada por correa.

Secondary air valve alignment An adjustment that provides correct air valve alignment in the secondary bores.
Alineación de la válvula de aire secundaria Ajuste que proporciona una alineación adecuada de la válvula de aire en los calibres secundarios.

Secondary air valve opening adjustment An adjustment that provides the proper wide-open air valve position.
Ajuste de la apertura de la válvula de aire secundaria Ajuste que proporciona una posición abierta de par en par adecuada de la válvula de aire.

Secondary air valve spring adjustment An adjustment that provides the proper secondary air valve spring tension.
Ajuste del resorte de la válvula de aire secundaria Ajuste que proporciona la tensión adecuada del resorte de la válvula de aire secundaria.

Secondary lockout adjustment An adjustment that assures the secondary throttles are locked when the choke is partly or fully closed on a four-barrel carburetor.
Ajuste de fijación secundario Ajuste que asegura la fijación de las mariposas secundarias cuando el estrangulador está parcial o completamente cerrado en un carburador de cuatro cilindros.

Secondary throttle linkage adjustment An adjustment that assures the correct primary throttle opening when the secondary throttles begin to open.
Ajuste de la conexión de la mariposa secundaria Ajuste que asegura la apertura correcta de la mariposa primaria cuando las mariposas secundarias comienzan a abrirse.

Self-powered test light A test light powered by an internal battery.
Luz de prueba propulsada automáticamente Luz de prueba propulsada por una batería interna.

Self-test input wire A diagnostic wire located near the diagnostic link connector (DLC) on Ford vehicles.
Alambre de entrada de prueba automática Alambre diagnóstico ubicado cerca del conector de enlace diagnóstico en vehículos fabricados por la Ford.

Sequential fuel injection (SFI) A fuel injection system in which the injectors are individually grounded into the computer.
Inyección de combustible en ordenamiento Sistema de inyección de combustible en el que los inyectores se ponen individualmente a tierra en la computadora.

Shop layout The design of an automotive repair shop.
Arreglo del taller de reparación Diseño de un taller de reparación de automóviles.

Silicone grease A heat-dissipating grease placed on components such as ignition modules.
Grasa de silicón Grasa para disipar el calor utilizada en componentes, como por ejemplo módulos del encendido.

Slitting tool A special chisel designed for slitting exhaust system pipes.
Herramienta de hender Cincel especial diseñado para hendir los tubos del sistema de escape.

Snap shot testing The process of freezing computer data into the scan tester memory during a road test and reading this data later.
Prueba instantánea Proceso de capturar datos de la computadora en la memoria del instrumento de pruebas de exploración durante una prueba en carretera y leer dichos datos más tarde.

Specific gravity The weight of a liquid in relation to the weight of an equal volume of water.
Gravedad específica El peso de un líquido de acuerdo al peso de un volumen igual de agua.

Starting air valve A vacuum-operated valve that supplies more air into the intake manifold when starting the engine.
Válvula de aire para el arranque Válvula accionada por vacío que le suministra mayor cantidad de aire al colector de aspiración durante el arranque del motor.

Stethoscope A tool used to amplify sound and locate abnormal noises.
Estetoscopio Herramienta utilizada para amplificar el sonido y localizar ruidos anormales.

Sulfuric acid A corrosive acid mixed with water and used in automotive batteries.
Ácido sulfúrico Ácido sumamente corrosivo mezclado con agua y utilizado en las baterías de automóviles.

Switch test A computer system test mode that tests the switch input signals to the computer.
Prueba de conmutación Modo de prueba de una computadora que revisa las señales de entrada de conmutación hechas a la computadora.

Synchronizer (SYNC) pickup A pickup assembly that produces a voltage signal for ignition triggering or injector sequencing.
Captación sincronizadora Conjunto de captación que produce una señal de tensión para el arranque del encendido o para el ordenamiento del inyector.

Tach-dwellmeter A meter that reads engine rpm and ignition dwell.
Tacómetro y medidor de retraso Medidor que lee las rpm del motor y el retraso del encendido.

Tachometer (TACH) terminal The negative primary coil terminal.
Borne del tacómetro Borne negativo de la bobina primaria.

Temperature switch A mechanical switch operated by coolant or metal temperature.
Conmutador de temperatura Conmutador mecánico accionado por la temperatura del refrigerante o del metal.

Test spark plug A spark plug with the electrodes removed so it requires a much higher firing voltage for testing such components as the ignition coil.
Bujía de prueba Bujía que a consecuencia de habérsele removido los electrodos necesitará mayor tensión de encendido para revisar componentes, como por ejemplo la bobina del encendido.

Thermal vacuum switch (TVS) A vacuum switching device operated by heat applied to a thermo-wax element.
Conmutador de vacío térmico Dispositivo de conmutación de vacío accionado por el calor aplicado a un elemento de termocera.

Thermal vacuum valve (TVV) A valve that is opened and closed by a thermo-wax element mounted in the cooling system.
Válvula térmica de vacío Válvula de vacío que un elemento de termocera montado en el sistema de enfriamiento abre y cierra.

Thermostat tester A tester designed to measure thermostat opening temperature.
Instrumento de prueba del termostato Instrumento de prueba diseñado para medir la temperatura inicial del termostato.

Throttle body injection (TBI) A fuel injection system with the injector or injectors mounted above the throttle.
Inyección del cuerpo de la mariposa Sistema de inyección de combustible en el que el inyector, o los inyectores, están montados sobre la mariposa.

Throttle position sensor (TPS) A sensor mounted on the throttle shaft that sends a voltage signal to the computer in relation to throttle opening.
Sensor de la posición de la mariposa Sensor montado sobre el árbol de la mariposa que le envía una señal de tensión a la computadora referente a la apertura de la mariposa.

Throttle position switch A switch that informs the computer whether the throttle is in the idle position. This switch is usually part of the TPS.
Conmutador de la posición de la mariposa Conmutador que le advierte a la computadora si la mariposa se encuentra en la posición de marcha lenta. Este conmutador normalmente forma parte del sensor de la posición de la mariposa.

Timing connector A wiring connector that must be disconnected while checking basic ignition timing on fuel injected engines.
Conector de regulación Conector del alambrado que debe desconectarse mientras se verifica la regulación básica del encendido en motores de inyección de combustible.

Two-gas emissions analyzer An analyzer designed to measure hydrocarbons and carbon monoxide in the exhaust.
Analizador de dos tipos de emisiones Analizador diseñado para medir los hidrocarburos y el monóxido de carbono en el escape.

United States customary (USC) A system of weights and measures.
Sistema usual estadounidense (USC) Sistema de pesos y medidas.

Upstream air Air injected into the exhaust ports.
Aire conducido hacia arriba Aire inyectado dentro de las lumbreras del escape.

Vacuum delay valve A vacuum valve with a restrictive port that delays a vacuum increase through the valve.
Válvula de retardo de vacío Válvula de vacío con una lumbrera restrictiva que retarda el aumento de vacío a través de la válvula.

Vacuum-operated decel valve A valve that allows more air into the intake manifold during deceleration to improve emission levels.
Válvula de desaceleración accionada por vacío Válvula accionada por vacío que admite una mayor cantidad de aire en el colector de aspiración durante una desaceleración a fin de reducir los niveles de emisiones.

Vacuum pressure gauge A gauge designed to measure vacuum and pressure.
Manómetro de la presión del vacío Calibrador diseñado para medir el vacío y la presión.

Vacuum throttle kicker A vacuum diaphragm with a stem that provides more throttle opening under certain conditions such as deceleration or A/C on.

Nivelador de la mariposa de vacío Diafragma de vacío con un vástago que proporciona una apertura más extensa de la mariposa bajo ciertas condiciones, como por ejemplo una desaceleración o cuando el aire acondicionado está encendido.

Valve overlap The few degrees of crankshaft rotation when both valves are open and the piston is near TDC on the exhaust stroke.

Solape de la válvula Los pocos grados que gira el cigüeñal cuando ambas válvulas están abiertas y el pistón se encuentra cerca del punto muerto superior durante la carrera de escape.

Valve stem installed height The distance between the top of the valve retainer and the valve spring seat on the cylinder head.

Altura instalada del vástago de la válvula Distancia entre la parte superior del retenedor de la válvula y el asiento del resorte de la válvula en la culata del cilindro.

Vane-type MAF sensor A MAF sensor containing a pivoted vane that moves a pointer on a variable resistor. This resistor sends a voltage signal to the computer in relation to the total volume of intake air.

Sensor MAF tipo paleta Sensor MAF con una paleta articulada que mueve un indicador en un resistor variable. Este resistor le envía una señal de tensión a la computadora referente al volumen total de aire aspirado.

Vehicle speed sensor (VSS) A sensor that is usually mounted in the transmission and sends a voltage signal to the computer in relation to engine speed.

Sensor de la velocidad del vehículo Sensor que normalemente se monta en la transmisión y que le envía una señal de tensión a la computadora referente a la velocidad del motor.

Viscous-drive fan clutch A cooling fan drive clutch that drives the fan at higher speed when the temperature increases.

Embrague de mando viscoso del ventilador Embrague de mando del ventilador de enfriamiento que acciona el ventilador para que gire más rápido a temperaturas más altas.

Volt-amp tester A tester designed to test volts and amps in such circuits as battery, starter, and charging.

Instrumento de prueba de voltios y amperios Instrumento de prueba diseñado para revisar los voltios y los amperios en circuitos como por ejemplo, de la batería, del arranque y de la carga.

Wastegate stroke The amount of turbocharger wastegate diaphragm and rod movement.

Carrera de la compuerta de desagüe Cantidad de movimiento del diafragma y de la varilla de la compuerta de desagüe del turbocompresor.

Wet compression test A cylinder compression test completed with a small amount of oil in the cylinder.

Prueba húmeda de compresión Prueba de la compresión de un cilindro llevada a cabo con una pequeña cantidad de aceite en el cilindro.

INDEX

Note: Page numbers in **bold** print reference non-text material.